17/10/95
DONATED BY
B.O. HUGHES

# Avian Endocrinology

# Avian Endocrinology

Edited by
**P J Sharp**

A publication of the Society for Endocrinology

Journal of Endocrinology Limited, 17/18 The Courtyard, Woodlands,
Almondsbury, Bristol BS12 4NQ, UK

A publication of the Society for Endocrinology


*British Library Cataloguing in Publication Data*
A CIP catalogue record for this book is available from the British Library.

ISBN 1 898099 01 4



Cover: red grouse (*Lagopus lagopus scoticus*) reproduced with permission from Matthew Gloag and Son Ltd

Cover design by Argent
Printed in Great Britain by Burgess Science Press

## Preface

Avian endocrinology continues to expand in scope and diversity, particularly in those areas exploiting new techniques in molecular genetics. It contributes new understanding - of general significance, of basic molecular and cellular mechanisms involved in growth, reproduction and metabolism. Just as importantly, it provides novel insights into endocrine, environmental and behavioural interactions in whole animals. Of more specialized interest are analyses of endocrine mechanisms controlling poultry production traits. These are likely to have a major impact on the future development of the poultry industry when applied in conjunction with techniques to produce transgenic poultry and to map the chicken genome.

These aspects of avian endocrinology are covered in this book which contains expanded versions of selected papers presented at the Fifth International Symposium on Avian Endocrinology, held in Edinburgh, September 13-17th, 1992. Other papers presented at the Symposium on endocrine, neuroendocrine and behavioural processes are published in *Poultry Science Reviews*, volume 4, issue 4 and volume 5, issue 1 (Ed. R Dietert). *Avian Endocrinology* will be of interest to research scientists working on the molecular and cellular control of growth, reproduction and metabolism, neuroendocrine control mechanisms and environmental and behavioural interactions. It will also provide a valuable source of information for teachers and graduate students in animal physiology, cell and molecular biology, and poultry/avian science.

Financial support for the Fifth International Symposium on Avian Endocrinology and publication of this book is gratefully acknowledged from the following organisations: The Agricultural and Food Research Council, British Poultry Science Travel Fund, Eli Lilly, Journals of Reproduction and Fertility Ltd, Society for Endocrinology and The Royal Society (London). The Scottish Ornithological Union kindly granted permission to reproduce line drawings of birds.

Finally, I thank Mrs Sue Thorn of the Journal of Endocrinology Ltd for her encouragement and support during the preparation of this book and Mrs Jan Cherrie for her accurate typing and organisation of manuscripts on the word processor.

Peter J Sharp
Edinburgh, August 1993

# Contents

## Part 4 Endocrine control of metabolism and water balance



## Part 5 Endocrine control of bone and soft tissue growth



## Part 6 Endocrine control of ovarian and oviduct function

## Part 7 Techniques for transgenesis

# Part One
# Neuroendocrine Control Mechanisms

Avian Endocrinology
Ed. P J Sharp, pp 1-10
Journal of Endocrinology Ltd, Bristol (1993)

# The molecular cloning of the chicken pre-pro-gonadotrophin releasing hormone-I gene

I C Dunn, Y Chen, P J Sharp and H M Sang
Department of Reproduction and Development, AFRC Roslin Institute (Edinburgh), Roslin, Midlothian EH25 9PS, UK

## Introduction

The amino acid sequence of chicken gonadotrophin releasing hormone-I (cGnRH-I) was first described in the early 1980s (1, 2, 3) and the amino acid sequence of a second form of chicken GnRH (cGnRH-II) was reported in 1984 (4), however no information has been available on the genes which code for their pre-pro-peptides.

The cDNA and genomic sequences are known for human, rat, mouse and atlantic salmon GnRH (5-13), while cDNA sequences have been reported for GnRH from two other osteichthyes: a cichlid and masu salmon (14, 15). In this chapter, the features of the chicken cGnRH-I gene and the deduced structure of the cGnRH-I pre-pro-peptide (16) are contrasted with GnRH genes and their products from other species. No reports are currently available on the isolation of a gene encoding cGnRH-II despite the isolation of GnRH genes from species known to contain two forms of GnRH (13-16). It must be assumed that the cGnRH-II gene has substantially diverged from the other GnRH genes or is unrelated to them. The evolution of the GnRH peptides has been recently reviewed by Millar and King (17).

## Organization of the pre-pro-cGnRH-I gene

All the GnRH genes for which genomic sequences are available, including the cGnRH-I gene, contain 4 exons (Fig. 1). The first exon is always non-coding; the second exon codes for the signal peptide, GnRH decapeptide and its C-terminal cleavage site and the first amino acids (11 amino acids in birds and mammals, 10 in osteichthyes) of the cryptic peptide, GnRH associated peptide (GAP). The third exon codes for the middle portion of GAP and the fourth exon codes for the C-terminus of GAP. The remainder of exon IV is untranslated.

The cGnRH-I gene spans 3002 base pairs (bp) of genomic DNA from the transcription initiation site of exon I to the polyadenylation site at the end of exon IV (16). The mouse, rat and human GnRH genes span between 4130-4927 bp (8, 9, 12) and the atlantic salmon gene 1252 bp (13) of genomic DNA (Fig. 1). A very short untranslated region in exon IV makes the chicken GnRH gene mature mRNA transcript the shortest so far described, being 390 bp in length (16).

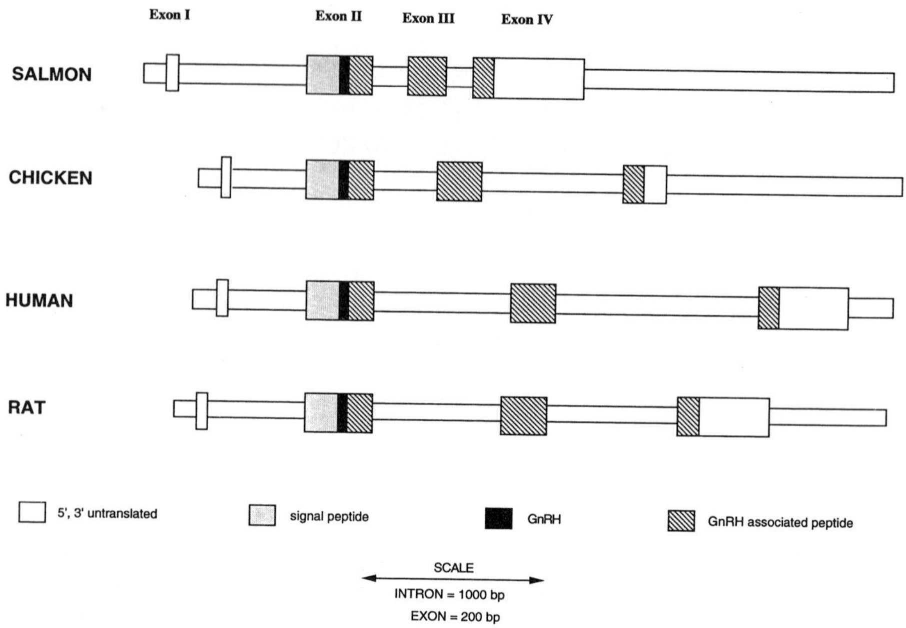


Fig. 1 The intron-exon organization of the GnRH genes of salmon, chicken, human and rat. The exon sequences are shown by the wider bars while the introns are shown as narrow bars. The functions of the exons are indicated by the shading shown in the key. Note that the scales for the intron and exons are different. (Modified from 13).

Comparison of the cDNA encoding cGnRH-I from the chicken with cDNAs from other species encoding mGnRH or sGnRH shows that exon II is the most highly conserved between species. Conservation between chicken exon II and human, rat and mouse exon II is 61-64% and between chicken exon II and cichlid and atlantic salmon exon II is 49-52%. Chicken exon III has a 47-53% identity with the human, rat and mouse exon IIIs and 30-31% identity with exon III in the cichlid and atlantic salmon. The coding region of exon IV is highly conserved between chicken and rat (69%), but not between chicken and salmon. There is 34-38% identity between the non-coding exon I in the chicken and exon I in human, rat and mouse and atlantic salmon.

Only one start site appears to be used for the initiation of transcription in the chicken GnRH-I gene (16) although, in rat and man, there are 1 major and 2 minor transcription initiation sites in the GnRH gene, with the principal one being at the most distal site (10, 11).

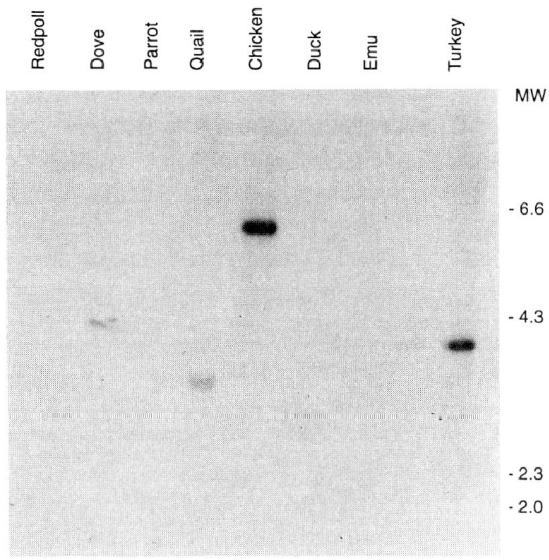


Fig. 2 Southern analysis of cGnRH-I in various avian species using a chicken GnRH cDNA probe. Genomic DNAs (10μg) isolated from red blood cells from emu (*Dromaius novaehollandia*), Port Lincoln parrot (*Barnardius zonarius semitorguatus*), quail (*Coturnix coturnix*), dove (*Streptopelia risoria*), redpoll (*Acanthis flammea*), turkey (*Meleagris gallopavo*) and chicken (*Gallus gallus domesticus*) were digested with EcoRI and separated on a 1% agarose gel. A 165 bp fragment of chicken cGnRH-I random-primed cDNA corresponding to exon II from the coding region and most of exon III of the chicken GnRH gene was used as a probe. Molecular sizes (MW) are indicated in kilobases.

## Comparisons with other avian species

Southern analyses of genomic DNAs isolated from emu (*Dromaius novaehollandia*), Port Lincoln parrot (*Barnardius zonarius semitorguatus*), quail (*Coturnix coturnix*), dove (*Streptopelia risoria*), redpoll (*Acanthis flammea*), turkey (*Meleagris gallopavo*) and chicken (*Gallus gallus domesticus*), digested with EcoRI and probed with a cGnRH-I cDNA comprising 165 bp from exons II and III starting at the decapeptide coding sequence, showed hybridisation to all species except emu, Port Lincoln parrot and redpoll (Fig. 2). The estimated sizes of the EcoRI fragments which hybridised to the probe were: turkey, 3.85 kb; quail, 3.4 kb; and dove, 4.2 kb (Fig. 2). Not surprisingly the species in which the hybridisation was strongest were the Galliformes (turkey and quail). The weaker hybridisation to the dove gene is consistent with this species being from a separate order, the Columbiformes. The failure of the cGnRH-I probe to hybridise to emu, parrot and redpoll DNA under the conditions used suggests that their GnRH-I genes are substantially different from the chicken GnRH-I gene. Since the probe corresponded to the region coding for the decapeptide, the cleavage site and 42 amino acids of GAP, it may be that the GAP sequence is not well conserved in the avian species distantly related to Galliformes.

```
               1                                                     50
CHICKEN        AGGAAAAAGG AACGCTGAAA ATCTAGTGGA ATCATTTCAA GAGATCGCAA
TURKEY         AGGAAAAAGG AATGTCGAAA ATCTGGTGGA ATCATTTCAA GAGATTGCAA
Consensus      AGGAAAAAGG AA.G..GAAA ATCT.GTGGA ATCATTTCAA GAGAT.GCAA

               51                                                   100
CHICKEN        ACGAAATGGA AAGTCTAGGA GAAGGGCAGA AGGCTGAATG CCCTGGCTCT
TURKEY         ACGAAATGGA AAATCTTGGG GAAGCGCAGA AGGCTGAATG CCCTGGATCA
Consensus      ACGAAATGGA AA.TCT.GG. GAAG.GCAGA AGGCTGAATG CCCTGG.TC.

               101                                                  150
CHICKEN        TACCAGCATC CCAGGCTGAG TGATCTGAAG GAAACCATGG CAAGCCTGAT
TURKEY         TACCAGCATC CCAGCCTGAG CAATCTGAAG GAAACCATGG CAAGTCTGAT
Consensus      TACCAGCATC CCAG.CTGAG ..ATCTGAAG GAAACCATGG CAAG.CTGAT

               151                                                  200
CHICKEN        CGAAGGAGAA GCCAGAAGAA AGGAGATTTA AATCCACAAC ACCATCAACA
TURKEY         CGGAGGAGAA GCCAGAAGAA AGAAGATTTA AATGCACAAA TCCAACAACA
Consensus      CG.AGGAGAA GCCAGAAGAA AG.AGATTTA AAT.CACAA. .CCA.CAACA

               201                               237
CHICKEN        AACCCAGAAT AAAATTTCAA AGCAATGTTG AAAACAGTT
TURKEY         GACTCAGAAT AAAATTTCAA AGCAATGTTG AAAACAG
Consensus      .AC.CAGAAT AAAATTTCAA AGCAATGTTG AAAACAG
```

Fig. 3 Comparison of the RTPCR-derived cDNA sequence for part of the turkey GnRH-I gene with the same region of the chicken GnRH-I cDNA sequence. The turkey cDNA starts at the last base of the decapeptide coding region.

Hemi-nested reverse transcription polymerase chain reaction (RTPCR) was used to obtain a partial cGnRH-I cDNA clone from turkey hypothalamic total RNA. The DNA sequence started after the decapeptide coding region and terminated at the polyadenylation site (Fig. 3). The nucleotide sequence of the partial cDNA amplified from the turkey hypothalamus is 90% identical to the equivalent chicken sequence. This high degree of homology is consistent with the Southern analysis (Fig. 2) which shows hybridisation of a chicken GnRH-I probe from this region of the GnRH gene with turkey genomic DNA.

### Comparison of the 5′ region of the chicken and rat GnRH genes

Comparison of the 5′ region of the chicken and rat GnRH genes indicates that there are stretches of AT repeats which are respectively, -778 bp to -723 bp and -676 bp to -621 bp from the rat and chicken transcription initiation sites (11, 16). The AT rich region has the potential to form areas of cruciform secondary structure which may facilitate the binding of transcription factors (18). The human and salmon 5′ regions do not contain any stretches of AT repeats (10, 14). The salmon GnRH 5′ region contains sequences which have the potential to form secondary structure and are at a similar distance from the transcription initiation site as the AT repeats in the rat and chicken GnRH genes (19).

No firm information is available on transcription factors which influence the expression of the GnRH gene. However, potential Pit-1-like binding sites occur in the 5′ region of the GnRH gene in the salmon, rat and chicken (11, 14, 16). The Pit-1 transcription factor is generally considered to be specific to the anterior pituitary gland (20) but this does not preclude a similar factor being involved in the regulation of the expression of the GnRH gene.

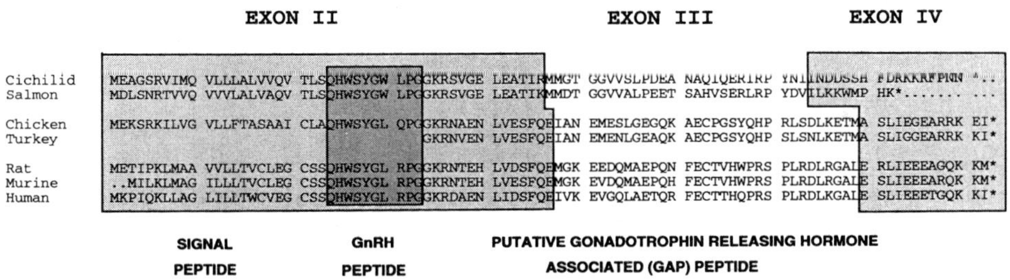


Fig. 4 A comparison of fish, avian and mammalian gonadotrophin-releasing hormone pre-pro-peptide sequences deduced from nucleotide sequences. The regions coded by exons II and IV are indicated by the boxed areas and exon III is between the boxed areas. The decapeptide sequence is indicated by the internal box and is followed by the C-terminal cleavage site, Gly-Lys-Arg. The functional domains of the peptide are named underneath the sequence.

## Predicted pro-peptide sequence

The chicken GnRH pre-pro-peptide deduced from the nucleotide sequence of the cGnRH cDNA is 92 amino acids long, comprising a 23 amino acid signal peptide, the 10 amino acid mature cGnRH-I, the Gly-Lys-Arg C-terminal cleavage site and the 56 amino acid cryptic peptide, GAP (Fig 4). The pre-pro-peptide sequence is similar in its overall length and in the length and amino acid composition of its functional domains to the human, rat and mouse pre-pro-GnRH. The decapeptide hormone is highly conserved, but the most highly conserved region is the Gly-Lys-Arg C-terminal cleavage domain which is the same in the GnRH pre-pro-peptide from several species and reflects the conservation of a common substrate for processing enzymes (21, 22). The conservation of the decapeptide hormone is probably dictated by the structure of its receptor (23). The degree of conservation in the signal peptide domain is low (30% between chicken and mammal, 26% between chicken and osteichthyes), but the hydrophobicity and length of the region is maintained. Of particular interest, in relation to the controversy which surrounds its potential as a prolactin secretion inhibiting factor, is the structure of GAP (24, 25). A GAP peptide similar to that in the mammals is present in both chicken and turkey pro-GnRH-Is. All of the 11 amino acids of chicken GAP coded in exon II are identical to at least one of the amino acids in the corresponding regions of the rat, mouse and human exon IIs while only one amino acid is common between salmon and chicken. In the region of GAP coded in exon III, 8 out of 33 amino acids are identical between rat and chicken with no identity between chicken and salmon (Fig. 4). Fifty percent of the 12 amino acids coded by exon IV are identical between rat and chicken but no identity is observed in this region between chicken and salmon (Fig. 4). The chicken GAP sequence has an overall identity with human, rat and mouse GAP sequences of between 38-46%. There is 88% identity between the chicken and turkey peptide sequence for the carboxyl cleavage region and GAP.

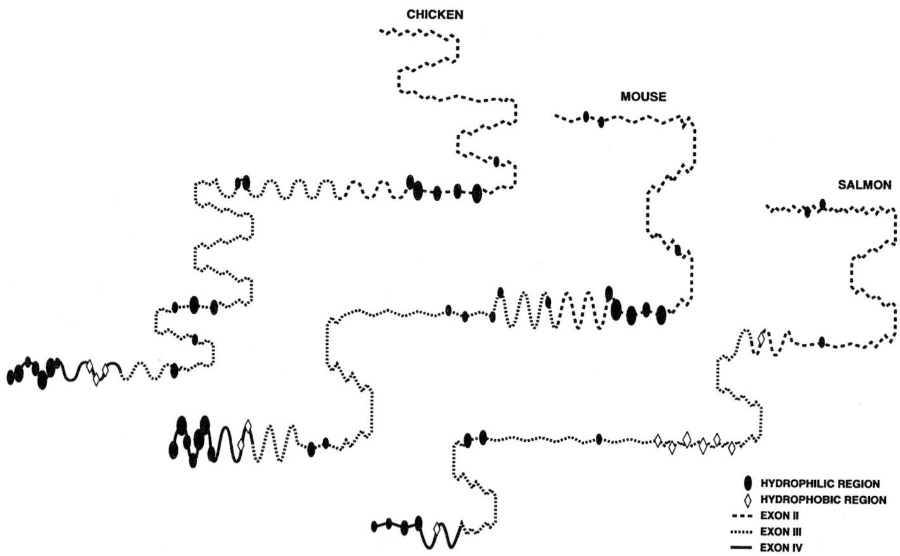


Fig. 5 Chou-Fasman secondary structure predictions for chicken, mouse and salmon GnRH pro-peptides. The predictions shown are from the first amino acid of the decapeptide to the last amino acid of GAP. The regions coded for by the different exons are represented by the continuous or broken lines (see code) as are the symbols for hydrophobicity and hydrophilicity, the size of the symbol indicates the degree of hydrophobicity and hydrophilicity. Helical structure is indicated by sine waves, β pleated sheets by saw teeth, β-turns by 180° bends and coils by blunt saw teeth.

The secondary structure of the mouse, chicken and salmon GnRH pro-peptide sequences, predicted using Chou-Fasman algorithms, contain regions of predicted alpha helix which are found in the areas of the GAP peptide coded by exon II and IV (Fig. 5). The carboxy terminals are hydrophilic in all three pro-peptides while the central region is relatively unstructured (Fig. 5). The strong conservation of the N- and C-terminals amino acid sequence in the mouse and chicken pro-peptide and the alpha helical structures in the mouse, chicken and salmon suggests that these parts of the sequence may have a common functional role in the different species.

The functional significance of the cryptic peptide, GAP, in the chicken pro-peptide remains to be established. The first stages of translocation of newly synthesised peptides across the membrane of the endoplasmic reticulum involves the signal peptide, but the later stages of translocation may be driven by ribosomal synthesis resulting in the extrusion of the peptide through the membrane. This process is thought to be assisted by the folding of the peptide in the endoplasmic reticulum. Since approximately 40 amino acid residues are intimately related to the ribosome and approximately 20 amino acids span the membrane, it would be impossible for a short peptide such as GnRH to pass through the membrane (26). The presence of GAP in the GnRH pro-peptide may therefore be required for translocation through the intracellular membranes. GAP may also play a role in the presentation of the C-terminal cleavage site for GnRH sequence specific endopeptidase, by producing loops on the surface of the pro-peptide (19, 21). It is also possible that GAP has an endocrine role in the chicken since it is reported to act as a prolactin release-inhibiting factor in the rat (7). However, the lack of homology between the amino acid sequences of GAPs from osteichthyes and those from chicken, turkey, mouse, rat and man suggests that, if GAP has an endocrine function, it may have evolved after the osteichthyes diverged from the higher vertebrate evolutionary line. The poor evolutionary conservation of exon III is surprising and it is possible that, in the osteichthyes, it may have a different origin from that in the chicken and mammals.

## References

1 King J A & Millar R P (1982) Structure of chicken hypothalamic luteinizing hormone-releasing hormone. I. Structural determination on partially purified material. *Journal of Biological Chemistry* **257** 10722-10728

2 King J A & Millar R P (1982) Structure of chicken hypothalamic luteinizing hormone-releasing hormone. II. Isolation and characterisation. *Journal of Biological Chemistry* **257** 10729-10732

3 Miyamoto K, Hasegawa Y, Igarishi M, Chino N, Sakakibra S, Kanagawa K & Matsuo H (1983) Evidence that chicken hypothalamic luteinizing hormone-releasing hormone is [Gln8]LH-RH. *Life Science* **32** 1341-1347

4 Miyamoto K, Hasegawa Y, Nomura M, Igarishi M, Kanagawa K & Matsuo H (1984) Identification of the second form of gonadotropin-releasing hormone in chicken hypothalamus; evidence that gonadotropin secretion is probably controlled by two distinct gonadotropin releasing hormones in avian species. *Proceedings of the National Academy of Sciences of the USA* **81** 3874-3878

5 Seeburg P H & Adelman J P (1984) Characterization of cDNA for precursor of human luteinizing hormone releasing hormone. *Nature* **311** 666-668

6 Adelman J P, Bond C T, Douglas J & Herbert E (1986) Two mammalian genes transcribed from opposite strands of the same DNA locus. *Science* **235** 1514-1517

7 Adelman J P, Mason A J, Hayflick J S & Seeburg P H (1986) Isolation of the gene and hypothalamic cDNA for the common precursor of gonadotrophin-releasing hormone and prolactin release-inhibiting factor in human and rat. *Proceedings of the National Academy of Sciences of the USA* **83** 179-183

8 Mason A J, Hayflick J S, Zoeller R T, Young W S, Phillips H S, Nikolics K & Seeburg P H (1986) A deletion truncating the gonadotrophin-releasing hormone gene is responsible for hypogonadism in the hpg mouse. *Science* **234** 1366-1371

9 Bond C T, Hayflick J S, Seeburg P H & Adelman J P (1989) The rat gonadotrophin releasing hormone : SH locus: structure and hypothalamic expression. *Molecular Endocrinology* **3** 1257-1262

10 Radovick S, Wondisford F E, Nakayama Y, Yamada M, Cutler G B & Weintraub B D (1990) Isolation and characterisation of the human gonadotropin-releasing hormone gene in the hypothalamus and placenta. *Molecular Endocrinology* **4** 476-480

11 Kepa J K, Wang C, Neeley C I, Raynolds M V, Gordon D F, Wood W M & Wierman M E (1992) Structure of the rat gonadotropin releasing hormone (rGnRH) gene promoter and functional analysis in hypothalamic cells. *Nucleic Acid Research* **20** 1393-1399

12 Hayflick J S, Adelman J P & Seeburg P H (1989) The complete nucleotide sequence of the human gonadotrophin-releasing hormone gene. *Nucleic Acid Research* **17** 6403-6404

13 Klungland H, Lorens J B, Anderson O, Kisen G O & Alestrom P (1992) The atlantic salmon pre-pro-gonadotropin releasing hormone gene and mRNA. *Molecular and Cellular Endocrinology* **84** 167-174

14 Bond C T, Francis R C, Fernald R D & Adelman J P (1991) Characterisation of complementary DNA encoding the precursor for gonadotropin-releasing hormone and its associated peptide from a teleost fish. *Molecular Endocrinology* **5** 931-937

15 Suzuki M, Hyodo S, Kobayashi M, Aida K & Urano A (1992) Characterisation and localisation of mRNA encoding the salmon-type gonadotrophin releasing hormone precursor of the masu salmon. *Journal of Molecular Endocrinology* **9** 73-82

16 Dunn I C, Chen Y, Hook C, Sharp P J & Sang H M (1993) Characterization of the chicken pre-pro-gonadotrophin-releasing hormone-I gene. *Journal of Molecular Endocrinology* **11** 19-29

17 Millar R P & King J A (1992) Evolution of gonadotropin-releasing hormones. *Trends in Endocrinology and Metabolism* **9** 339-346

18 Greaves D R, Patient R K & Lilley D M (1985) Facile cruciform formation by an (AT) 34 sequence from a Xenopus globin gene. *Journal of Molecular Biology* **185** 461-478

19 Andersen O & Klungland H (1993) The salmon GnRH encoding gene in teleost fish. *International Review of Cytology* (in press)

20 Karin M, Castrillo J L & Theill L E (1990) Growth hormone gene activation: a paradigm for cell type specific gene activation. *Trends in Genetics* **6** 92-96

21 Rangaraju N S, Xu J-F & Harris R (1991) Pro-gonadotropin-releasing hormone protein is processed within hypothalamic neurosecretory granules. *Neuroendocrinology* **53** 20-28

22 Rangaraju N S & Harris R (1991) Processing enzyme specificity is a consequence of pro-hormone precursor protein conformation. *Archives Biochemistry Biophysics* **290** 418-426

23 Davidson J S, Flanagan C, Wakefield I K, King J A, Del Milton R, Prescott R & Millar R P (1990) Gonadotropin-releasing hormone receptor interaction and intracellular signal transduction in the chicken gonadotrope. *Endocrinology of Birds: Molecular to Behavioral* pp 43-48. Eds M Wada, S Ishii & C G Scanes. Tokyo: Japanese Scientific Societies Press

24 Seeburg P H, Mason A J, Stewart T A & Nikolics K (1987) The mammalian gene and its pivotal role in reproduction. *Recent Progress in Hormone Research* **43** 69-98

25 Nikolics K, Mason A J, Szonyi E, Ramachandran J & Seeburg P H (1985) A prolactin inhibiting factor within the precursor for human gonadotropin-releasing hormone. *Nature* **316** 511-517

26 Wickner W T & Lodish H F (1985) Multiple mechanisms of protein insertion into and across membranes. *Science* **230** 400-407

Avian Endocrinology
Ed. P J Sharp, pp 11-28
Journal of Endocrinology Ltd, Bristol (1993)

# Hypophysiotrophic control of growth hormone secretion and its feedback regulation: an update

S Harvey
Department of Physiology, University of Alberta, Edmonton, T6G 2H7, Canada

## Introduction

The hypophysiotrophic regulation of growth hormone (GH) release and its modulation by some peripheral hormones was reviewed at the Fourth International Symposium on Avian Endocrinology (1, 2). A more comprehensive review of avian GH secretion has also been compiled since this meeting (3). This brief review will therefore only focus on recent literature and research related to GH secretion in birds.

## Neuropeptides regulating growth hormone secretion

As in other vertebrates, the hypothalamus inhibits or stimulates the synthesis and/or release of GH from the avian pituitary gland by neuronal factors (mostly peptides) that are released from the median eminence into hypophyseal circulation. The hypothalamus integrates neural and humeral information about the external and internal environments and in this way acts as a final common pathway in the neuroendocrine regulation of GH secretion. Numerous neuropeptides may act as chemical messengers in this hypothalamo-pituitary axis (Table 1), but only three: growth hormone releasing factor (GRF), thyrotrophin-releasing hormone (TRH) and somatostatin (somatotrophin release inhibiting hormone, SRIF), are probably of physiological importance.

### *Growth hormone releasing factor*

Growth hormone releasing factors with immunological similarities to those in mammalian species are present in the hypothalami of most lower vertebrates but have yet to be detected in birds (3, 4). Peptides with GH-releasing activity are present in the avian hypothalamus and presumably interact with the avian pituitary receptors stimulated by synthetic (mammalian) GRF preparations. Hybridization of mRNA moieties in the chicken hypothalamus with a fragment of the human GRF cDNA (3) provides evidence for the expression of a GRF-like gene in this tissue.

Recent studies indicate that the intracellular mechanisms involved in the induction of chicken GH release following GRF stimulation include an accumulation of cyclic adenosine 3′,5′-monophosphate (cAMP) and the activation of protein kinase A (PKA) and protein kinase C (PKC) (5). Calcium may normally participate in GRF-induced GH release (6, 7) but secretion can occur independently of extracellular calcium entry into pituitary cells (7, 8). The synergistic interaction between GRF and TRH in inducing GH release is also calcium-independent (6, 8). The interaction between TRH and GRF is also likely to occur via a pathway that arises prior to the activation of PKC, since phorbol esters have no synergistic interaction with either of these secretagogues. The synergistic interaction between GRF and TRH also contrasts with the additive interaction recently described for GRF and interleukin 6 (IL-6), another GH secretagogue (9).

Table 1 Neural regulation of GH secretion in birds

| **Regulatory factor** | **Effect** | **Reference** |
|---|---|---|
| Neuropeptide | | |
| GRF | ↑ | 72 |
| SRIF | ↓ | 73 |
| TRH | ↑ | 74 |
| pGlu-Glu-Pro$NH_2$ | ↑↓ | 16 |
| Galanin | ↑ | 75 |
| VIP | ↑ | 76 |
| Dynorphin | ↑/0 | 77 |
| Bombesin | ↑/0 | 78 |
| Substance P | ↑/0 | 76 |
| GnRH | ↑/0 | 79 |
| Met-enkephalin | 0 | 80 |
| Leu-enkephalin | 0 | 80 |
| Secretin | 0 | 81 |
| Cholestokinin | 0 | 81 |
| APP | 0 | 81 |
| Neurotransmitter | | |
| DA | ↑ | 82 |
| NA | ↑ | 83 |
| Histamine | ↑ | 84 |
| GABA | ↑ | 84 |
| 5HT | ↓ | 85 |
| ACh | ↓ | 86 |

Abbreviations: GH, growth hormone; GRF, growth hormone releasing factor; SRIF, somatostatin; TRH, thyrotrophin-releasing hormone; VIP, vasoactive intestinal polypeptide; GnRH, gonadotrophin-releasing hormone; APP, avian pancreatic polypeptide; DA, dopamine; NA, noradrenaline; GABA, gamma aminobutyric acid; 5HT, serotonin; Ach, acetylcholine; ↑, increase; ↓, decrease; 0, no effect.

The release of immunoreactive GH from GRF-stimulated chicken pituitary glands has recently been shown to include phosphorylated (10, 11) and non-phosphorylated (12) forms. *In vitro* stimulation with GRF depletes the total pituitary stores of immunoreactive GH, even though it dramatically increases a smaller pool of newly synthesized [$^{35}$S]-labelled GH (12). It is therefore likely that GRF initiates GH synthesis, although these results could indicate an inhibition of intracellular GH degradation. These results contrast with those of Aramburo *et al* (10), who failed to observe any stimulatory effect of GRF on the incorporation of [$^{32}$P] into chicken GH. This discrepancy may reflect the relatively small amount of phosphorylated GH in the total GH pool, or, alternatively, that the synthesis or phosphorylation of this moiety may be insensitive to GRF action. The failure of GRF to induce GH phosphorylation in the chicken pituitary is surprising. Protein kinase A mediates GRF action in birds (5) and this enhances the phosphorylation of chicken GH (11). Nevertheless, as the promotor region of the chicken GH gene has no overall homology with the corresponding regions of mammalian genes (13), the actions of mammalian GRFs on GH synthesis in birds may be dissimilar to those in mammals.

*Somatostatin*

The presence of SRIF moieties in the avian hypothalamus and the inhibitory effects of $SRIF_{14}$ on avian GH release are well established (3). However, while several novel forms of SRIF with inhibitory actions on GH release were reported also to be present in the chicken hypothalamus (14), more recent studies (J M Conlon, personal communication) have failed to confirm these findings.

$SRIF_{14}$ has previously been shown to inhibit basal GH release and GH release induced by the secretagogues GRF, TRH, galanin and vasoactive intestinal polypeptide (3). Recent studies also indicate that it inhibits IL-6-induced GH release in broiler fowl *in vivo* (9). These actions appear to be mediated by a blockade of cAMP accumulation and impaired PKA activity (15). Somatostatin does not influence GH release in the presence of PKC activators or calcium ionophores (15), suggesting selective inhibition of the adenylate-cyclase PKA system.

*Thyrotrophin-releasing hormone*

Authentic TRH (pGlu-His-Pro$NH_2$) has recently been identified in the chicken hypothalamo-pituitary axis (16). The concentration of TRH in the hypothalamus is far less (<10%) than that in the rat hypothalamus, but higher (by 2-3 fold) than the concentration in the chicken pituitary gland. Another tripeptide (pGlu-Glu-Pro$NH_2$) with structural and immunological similarities to TRH is also found in these tissues. This peptide, in contrast with TRH, is present at higher concentration in broiler fowl pituitary glands than in the broiler hypothalamus, suggesting a pituitary site of synthesis. The pGlu-Glu-Pro$NH_2$:TRH ratios in the broiler tissues are, however, much lower than those in mammalian species (17-19). This peptide is, moreover, undetectable in the hypothalamo-pituitary axis of White Leghorn chickens (16). Birds would thus appear to be relatively deficient in this newly discovered neuropeptide.

It is well established that TRH is a potent GH-releasing factor in juvenile birds (3, 20) and that it stimulates the release of stored and newly-synthesised GH (12). Thyrotrophin releasing hormone does not appear to stimulate GH synthesis *de novo*, in contrast with GRF (12). The GH response of broilers to TRH challenge is greater in birds reared at a high ambient temperature (33 °C) than in those raised at moderate temperatures, although the GH response is of shorter duration (21). The GH responsiveness of broilers selected for body weight gain is comparable with birds selected for food efficiency. The GH response of broiler fowl to TRH is, however, less than in egg-laying White Leghorns (S Harvey, unpublished observations).

The GH-releasing activity of TRH declines during ontogeny. This age-related diminution of GH responsiveness is probably due to a decline in GH secretory granule volume, somatotroph responsiveness, and the number of somatotrophs and TRH-binding sites (3). A suppression of TRH-induced cAMP, increased metabolic clearance of plasma GH and TRH, and increased sensitivity to thyroidal and SRIF inhibition are also implicated in these ontogenic changes. In old birds, GH is still released in response to exogenous TRH, particularly in birds in which endogenous triiodothyronine ($T_3$) or SRIF are suppressed and GH secretion enhanced by starvation or passive immunoneutralization, respectively (3, 22).

Although pGlu-Glu-Pro$NH_2$ is similar to TRH, it has no direct effects on the release of GH from incubated broiler fowl pituitary glands (16, 23). Unlike TRH, this peptide is also unable to downregulate TRH receptors on pituitary membranes. Thyrotrophin releasing hormone induced GH release is inhibited by pGlu-Glu Pro$NH_2$ *in vitro* and pGlu-GluPro$NH_2$ also competitively displaces TRH from membrane binding sites (16). This peptide may therefore be an endogenous antagonist of TRH receptors, since its actions *in vitro* are identical to those of benzodiazepine drugs that antagonize TRH-induced GH release by receptor occupancy (24). Benzodiazepines (24) and pGlu-Glu-Pro$NH_2$ (16) also similarly act *in vivo* to lower basal circulating GH concentrations, possibly by antagonism of tonic TRH induced GH release. Paradoxically, both agents potentiate the *in vivo* GH response of chickens to TRH challenge (16, 24). Since TRH modulates GH release at pituitary and central nervous sites (25), pGlu-Glu-Pro$NH_2$ and benzodiazepine drugs may therefore antagonize inhibitory, extra-pituitary actions of TRH that are normally masked by its direct stimulation of GH release.

*Neuropeptide interactions*

GRF, TRH and SRIF have post-receptor interactions within the pituitary gland to regulate GH release (5, 6, 15). Recent studies also indicate actions and interactions of these peptides within the central nervous system (CNS). Somatostatin (SRIF), for instance, enhances TRH-induced GH release when injected intracerebroventricularly (icv) (26). Since SRIF inhibits its own secretion (3) and icv SRIF depletes the SRIF content of the hypothalamus (26), this may reflect a decrease in inhibitory hypothalamic tone, even though icv SRIF concomitantly reduces the content of hypothalamic TRH (3). Tonic inhibitory actions of SRIF on the hypothalamic factors stimulating GH release are also indicated by the increased GH response to TRH challenge in birds icv injected with high-titre SRIF antisera (26). Interactions between SRIF, GRF and TRH in the CNS are also suggested by the depletion of hypothalamic SRIF stores in birds icv injected with GRF or TRH (3, 27). The inhibitory effect of centrally administered GRF on the GH response to systemic TRH challenge (27) is therefore likely to be mediated by increased SRIF secretion, since GRF induces SRIF release from hypothalami incubated *in vitro* (27). This inhibitory action of GRF is accompanied by a decrease in the dopamine (DA) content of the chicken hypothalamus and resembles the decrease in hypothalamic DA induced by centrally administered GH (28). A central dopaminergic mechanism may therefore normally stimulate GH release in chickens, even though DA acts at pituitary sites to inhibit secretagogue-induced GH release (29). This possibility is supported by the increased content of hypothalamic DA in birds injected centrally or peripherally with TRH which has overall actions at CNS sites to stimulate GH release (30). The inhibitory action of icv TRH on the content of hypothalamic TRH (3) may also indicate an autoregulatory component in the regulation of TRH neurones, although high-titre TRH antisera reduces the stores of hypothalamic TRH when injected icv (3). These homologous and heterologous interactions within the chicken CNS may therefore represent fine-tuning feedback or autoregulatory mechanisms to regulate GH secretion when the activity of one stimulatory or inhibitory releasing factor is enhanced.

**Autoregulation of GH secretion**

Stimulated GH secretion in domestic fowl is blunted by repeated or chronic administration of GH secretagogues (31, 32) or exogenous GH (33, 34). Since GH secretion is regulated by a complex interaction between stimulatory and inhibitory hypothalamic factors and is modulated by a variety of peripheral endocrine and metabolic factors responsive to GH action, GH may exert negative control on its own secretion via ultrashort- (actions at pituitary sites), short- (actions at hypothalamic sites) or long- (actions at peripheral sites) feedback loops.

*Ultrashort-loop feedback*

Basal and TRH-induced GH release from chicken hemipituitary glands is unaffected by exposure to 1 or 10 µg ovine GH/ml for periods of up to 200 min (33), suggesting the absence of ultrashort-loop feedback in the regulation of GH release. Growth hormone also has no direct effect on pituitary receptors for TRH (33) or SRIF (35). The expression of GH receptor (GHR) gene in the somatotroph-rich caudal lobe of the chicken pituitary gland (36) nevertheless suggests actions of GH in GH-secreting cells. This possibility is supported by the demonstration of GH-binding sites on pituitary membranes (37), indicating translation of the GHR mRNA. Alternatively, it is possible that these receptors mediate GH actions unrelated to GH autoregulation or that they are only present in cells secreting other pituitary hormones. Indeed the GHR message is also present in the cephalic lobe of the pituitary gland, which is lacking in somatotrophs.

Table 2 Effects of GH administration on hypothalamic and pituitary factors regulating the secretion of GH: evidence for autoregulation

| **Mechanism** | **Response** | **Reference**[+] |
|---|---|---|
| *In vivo* | | |
| Hypothalamic GHRA | ↓ | |
| Hypothalamic SRIF | ↑ | 34 |
| Hypothalamic TRH | 0 | 3 |
| Hypothalamic DA | ↓ | 47 |
| Hypothalamic DOPAC | ↓ | 47 |
| Hypothalamic A | ↓ | 47 |
| Hypothalamic NA | ↓ | 47 |
| Hypothalamic 5HT | 0 | 47 |
| Hypothalamic 5HIAA | 0 | 47 |
| Hypothalamic GH receptor mRNA | 0 | |
| Pituitary TRH receptors | ↓ | 34 |
| TRH-induced GH | ↓ | 33 |
| Pituitary GH content | ↓ | 45 |
| Pituitary GH receptor mRNA | 0 | |
| *In vitro* | | |
| Basal GH release | 0 | 33 |
| Pituitary TRH receptors | 0 | 33 |
| TRH-induced GH | 0 | 33 |
| Pituitary SRIF receptors | 0 | 35 |

[+] where published; if no reference given, personal observation.
Abbreviations: GH, growth hormone; GHRA, GH releasing activity; SRIF, somatostatin; TRH, thyrotrophin-releasing hormone; DA, dopamine; DOPAC, 3,4 dihydroxyphenylacetic acid; A, adrenaline; NA, noradrenaline; 5HT, serotonin; 5HIAA, 5-hydroxyindoleacetic acid; ↓, decrease; ↑, increase; 0, no effect.

*Short-loop feedback*

The detection of GHR mRNA in hypothalamic and extra-hypothalamic regions of the chicken brain (38, 39) and specific, saturable, high-affinity, low-capacity binding sites for GH (40) suggests GH actions within the CNS. Since the expression of the GHR gene in the chicken brain is inversely related to circulating GH concentrations during ontogeny (38), these receptors may mediate autoregulatory actions of GH on its own release. Although the blood-brain barrier is thought to be impermeable to GH (41) carrier-mediated transport systems (42, 43) and retrograde hypophyseal flow (44) are likely to facilitate GH interactions within circumventricular organs, particularly the median eminence.

Autoregulatory actions of GH within the CNS (Table 2) are indicated by the suppression of basal and stimulated GH secretion in birds receiving icv GH injections (28, 34). The blunting of GH release is likely to reflect an increased inhibitory hypothalamic tone and a reduction in stimulatory control. Indeed, increased SRIF concentrations are found in the medial basal hypothalamus (MBH) 24 h after a bolus icv injection of GH (34). However, as SRIF is not thought to inhibit GH synthesis, the diminution of pituitary GH stores in chickens bearing intracranial implants of GH-secreting adenohypophysial cells (45) is probably due to a suppression of stimulatory hypophysiotropic factors. A reduction in the size of the pituitary GH pool is therefore likely to be partially responsible for the impaired GH secretion in birds injected with GH centrally (28, 34). This may also partially account for the inability of TRH to release GH from the pituitary glands of broiler fowl injected with GH (Fig. 1). This finding may, however, also reflect inhibitory *in vivo* effects of GH (e.g. an increase in plasma $T_3$) on the abundance of pituitary TRH receptors (1, 33).

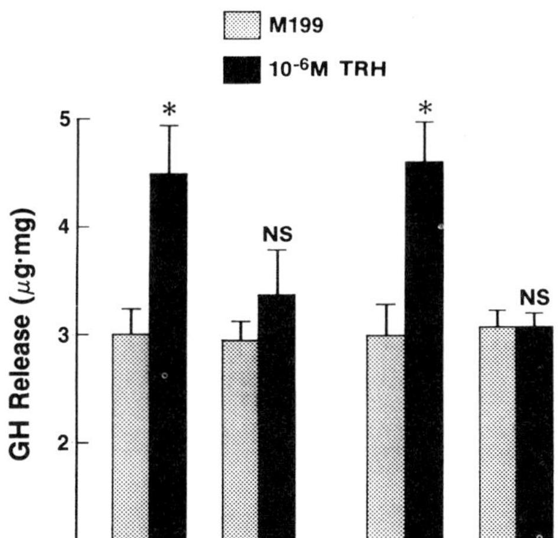


Fig. 1 Growth hormone (GH) release from chicken hemipituitary glands incubated for 2 h at 39 °C in the presence (solid bars) or absence (stippled bars) of thyrotrophin-releasing hormone ($10^{-6}$ M). The pituitary glands were collected from 6 week old broiler fowl that had been acutely (1 day) or chronically (once each day for 7 days) injected intramuscularly with recombinant DNA-derived chicken GH (200 µg/kg per day) or the 0.9% (w/v) NaCl vehicle (0.5 ml/kg per day). The birds were killed 4 h after the last injection of GH or the vehicle and preincubated for 60 min prior to incubation in fresh incubation media (Medium 199, M199; Gibco Grand Island, New York) or TRH. Values are shown as means ±s.e. (n=8). Values significantly different are indicated by asterisks ($P<0.05$ Student's *t* test). NS=not significant.

Although endogenous GH concentrations may not be reduced by systemic injections of exogenous GH (28, 33, 34), the pituitary GH response to TRH challenge is impaired between 2 and 12 h after a bolus administration of GH (33). This indicates that GH in peripheral plasma has central nervous sites of action. The reduction in the GH-releasing activity of the pigeon hypothalamus after the chronic stimulation of endogenous GH secretion (46) and the diminished GH releasing activity of the chicken hypothalamus after the systemic administration of exogenous GH (Fig. 2) also suggests central actions of peripheral GH. This possibility is also supported by the rapid (within 2 h) and prolonged (for 24 h) reduction in the turnover and content of catecholamines (dopamine, DA and noradrenaline, NA) in the hypothalami of laying hens given systemic GH injections (47). The icv administration of GH similarly lowers the catecholamine content of the chicken hypothalamus, without impairing the content or turnover of serotonin (47). The incubation of the chicken medial basal hypothalamus with recombinant DNA-derived chicken GH (10 μg/ml) for 3, 6, 12 or 24 h also lowers the tissue DA and DOPAC contents by approximately 50% (unpublished observations). Since DA may act centrally to stimulate GH release (30), a reduction in hypothalamic DA might be expected to inhibit GH release.

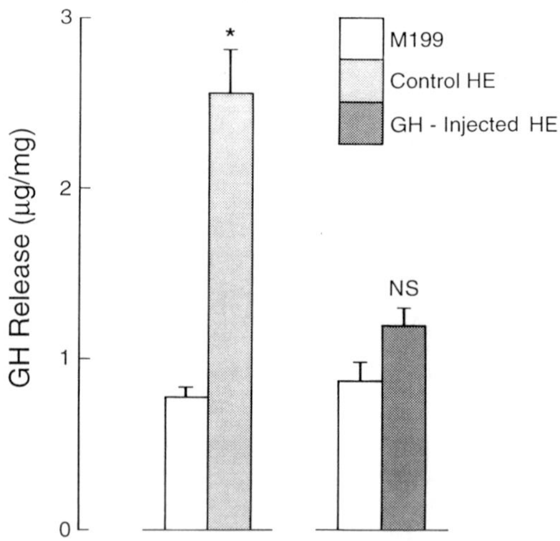


Fig. 2 Growth hormone (GH) release from broiler fowl hemipituitary glands incubated for 2 h at 39 °C in the presence (shaded symbols) or absence (open symbols) of acid (0.1 N HCl) extracts (1 hypothalamic equivalent) made from the medial basal hypothalami of 6 week old broilers (n=15) that had been intravenously injected for 7 days with recombinant DNA-derived chicken GH (200 μg/kg per day) or the 0.9% (w/v) NaCl vehicle (0.5 ml/kg per day). The birds were killed 4 h after the last injection of GH or vehicle and the hypothalami were pooled prior to extraction and pituitary incubation. Values are shown as mean ±s.e. (n=8). Values significantly different are indicated by asterisks ($P<0.05$, Student's *t* test). NS=not significant. The concentrations of immunoreactive somatostatin in the extracts from vehicle and GH-injected birds were 27.4 ng/mg protein and 116.3ng/mg protein, respectively.

These aminergic actions of exogenous GH occur in the preoptic hypothalamus (POA) and the MBH, suggesting actions on the perikarya and axon terminals of the peptidergic neurones regulating GH release. Catecholamine interneurones may therefore mediate the feedback action of GH on its own release, as catecholamine neurones make synaptic contact with GRF- and SRF- secreting neurones in the rat hypothalamus (48, 49). However, as tyrosine hydroxylase and DA are co-localized in rat GH-releasing neurones (50), GH may also autoregulate its secretion by direct actions on hypophysiotrophic neurones. The presence of GHR mRNA in SRIF-cell bodies in rat periventricular nuclei and in GRF-perikarya in MBH nuclei (51, 52), supports this view.

*Long-loop feedback*

Growth hormone receptors are almost ubiquitous and the functions of most organs and tissues are influenced by GH action. Numerous factors in peripheral circulation may therefore feedback at hypothalamic or pituitary sites to regulate GH secretion. In birds, hormones produced by the liver, thyroid, adrenal, pancreas and gonads in response to GH stimulation may contribute to feedback inhibition (Table 3). However, unlike mammalian species the hyperglycemic and lipolytic actions of GH (53) appear to be unrelated to feedback regulation in birds, even though nutritional deprivation provokes GH release (22). The stimulatory action of GH on the avian immune system (54, 55) may, conversely, provide positive feedback, since cytokines have recently been shown to stimulate GH release in poultry (9).

Table 3 Inhibition of GH secretion by peripheral stimuli: *long-loop* feedback?

| **Stimuli** | **Effect** | **Reference** |
|---|---|---|
| $T_3$ | ↓ | 66 |
| $T_4$ | ↓ | 66 |
| Insulin | ↓ | 87 |
| Glucagon | ↓ | 61 |
| Corticoids | ↓ | 88 |
| Catecholamines | ↓ | 89 |
| IGF-I | ↓ | 58 |
| Androgens | ↓ | 90 |
| Estrogens | ↓ | 91 |
| Glucose | 0 | 92 |
| FFA | 0 | 93 |

Abbreviations: GH, growth hormone; $T_3$, triiodothyronine; $T_4$, thyroxine; IGF-I, insulin-like growth factor-I; FFA, free fatty acids; ↓, decrease; 0, no effect.

Negative feedback in mammals is largely mediated by insulin-like growth factors (IGFs) produced in the liver. Hepatic IGFs are also GH-dependent in birds (56) and exogenous IGF-I can inhibit basal, TRH-and GRF- induced GH release in chickens (57, 58). Birds are, however, relatively insensitive to IGF-I (59, 60). The physiological importance of IGF-I in the control of GH secretion in poultry is therefore uncertain, especially as endogenous GH concentrations may fluctuate independently of plasma IGF-I in normal birds (61). The low circulating IGF concentrations in sex-linked dwarf birds has, however, been considered a causal factor in their hypersecretion of GH (62).

In birds, a physiological role for thyroid hormones in the feedback inhibition of GH secretion has been proposed (1), contrary to the stimulatory effects of these hormones on GH release in mammalian species. In recent studies $T_3$ treatment *in vivo* (100 μg/kg for 10 days) has been shown to reduce the immunoreactive GH content of the chicken pituitary and to block the *in vitro* GH response to TRH (12). Dietary supplementation with $T_3$ was also found to decrease basal and TRH-induced GH secretion in normal chickens and in $T_3$-deficient sex-linked dwarfs (62). The ability of dietary $T_3$ to inhibit the GH responses to TRH or GRF in birds in which basal GH secretion is not suppressed (63) indicates that basal and stimulated GH release might be differentially regulated or have different sensitivities to thyroidal inhibition. In recent studies the acute exposure of pituitary glands to $T_3$ *in vitro* has also been found to reduce the intracellular content of stored GH (64) and the accumulation of newly synthesised [$^{35}$S] labelled GH (12). The synthesis and release of GH induced by TRH, GRF and by activators of the PKA and PKC second messenger systems is also directly reduced by $T_3$ *in vitro* (12, 29, 64, 65). The chicken pituitary gland would, however, appear to be differentially responsive to thyroxine ($T_4$). Unlike $T_3$, $T_4$ does not downregulate pituitary TRH receptors nor inhibit TRH- or GRF- induced GH secretion *in vivo* or *in vitro* (63).

In the absence of thyroid hormones, GH synthesis is increased (12), pituitary TRH receptors are upregulated (63) and the GH responses to TRH/GRF are potentiated (3, 62, 63), indicating inhibitory thyroid actions at post-receptor sites. Thus, as GH increases the peripheral conversion of $T_4$ to $T_3$ (22), thyroid hormones appear to provide inhibitory long-loop stimuli to regulate the hypothalamo-somatotroph axis. They may also regulate circulating GH concentrations by modulating the rate of GH metabolism.

## GH metabolism

The metabolic clearance rate of GH is reduced in thyroidectomized fowl and restored to that in the controls after $T_4$ administration (66). The increased production of $T_3$ in GH treated (22) birds might therefore account for their increased rate of GH clearance (Fig. 3). The metabolic clearance rate of GH in mammals is, conversely, prolonged by exogenous GH administration (67). The recent demonstration of GH binding proteins in the peripheral plasma of chickens (68, 69) may, however, also modulate the rate of GH degradation, since protein-binding prolongs the metabolic clearance of GH in mammalian plasma (70). An increase in GH clearance and a feedback suppression of endogenous GH release may therefore partly account for the inability of exogenous GH to promote growth in rapidly growing chickens (71).

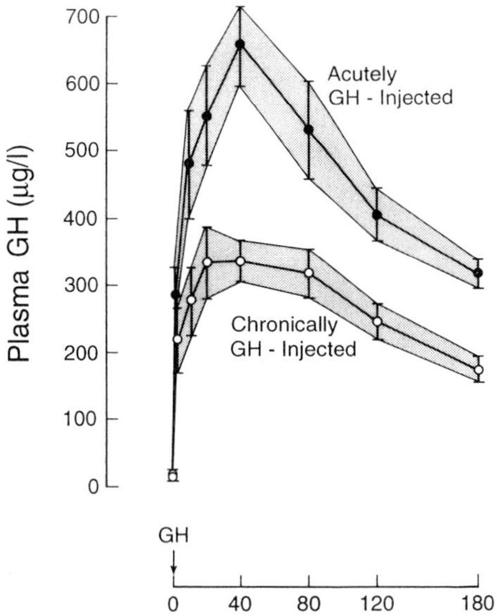


Fig. 3 Immunoreactive plasma growth hormone (GH) concentrations in 6 week old broiler fowl following a bolus intramuscular injection of recombinant DNA-derived chicken GH (200 µg/kg) in birds previously uninjected (open symbols) or chronically injected (once daily, for 7 days; solid symbols) with the same dose of GH. Values are shown as means ±s.e. (n=8).

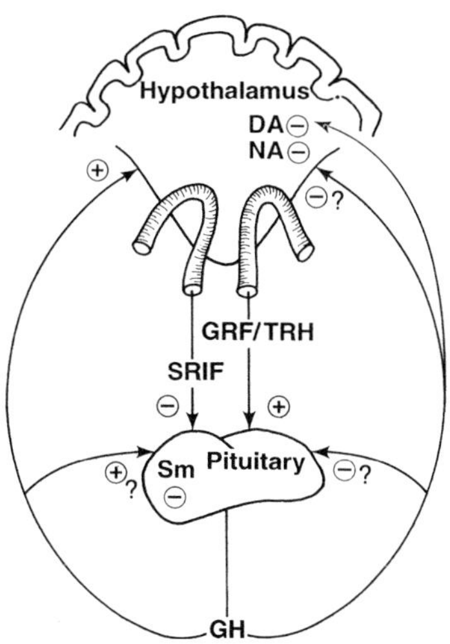


Fig. 4 Actions and putative actions of growth hormone (GH) in the avian hypothalamo-pituitary axis. Abbreviations: SRIF, somatostatin; GRF, GH-releasing factor; TRH, thyrotrophin-releasing hormone; Sm, somatomedin (insulin-like growth factor); DA, dopamine; NA, noradrenaline.

## Conclusion

Recent studies have provided new information on the regulation of the hypothalamo-somatotroph axis in birds. In particular, the multifactorial regulation of GH secretion has been shown to be characterised by feedback mechanisms and heterologous interactions at central nervous and peripheral sites (Fig. 4). These autoregulatory components should be further investigated in future studies, especially as the presence of GH receptors within the pituitary gland indicates hitherto unsuspected roles for GH in somatotroph regulation and pituitary function.

## Acknowledgements

The author would like to thank Dr L Souza (Amgen) for gifts of recombinant DNA-derived chicken GH, C M Josey and D Eichenlaub for typing the manuscript, and Imperial Chemical Industries and the National Sciences and Engineering Research Council of Canada for financial support of the author's research.

## References

1 Harvey S (1990) Thyroidal inhibition of growth hormone secretion: negative feedback? In *Endocrinology of Birds: Molecular to Behavioral* pp 11-127. Eds M Wada, S Ishii & C G Scanes. Tokyo: Japanese Scientific Societies Press

2 Scanes C G, Aramburo C & Campbell R M (1990) Hormonal involvement in avian growth and development: growth hormone and insulin-like growth factor 1. *Endocrinology of Birds: Molecular to Behavioral* pp 93-110. Eds M Wada, S Ishii & C G Scanes. Tokyo: Japan Scientific Societies Press

3 Harvey S, Fraser R A & Lea R W (1991) Growth hormone secretion in poultry. *Critical Reviews in Poultry Biology* **3** 239-282

4 Harvey S (1993) Growth hormone secretion in poikilotherms and homeotherms. *The Endocrinology of Growth, Development and Metabolism of Vertebrates* pp 151-182. Eds M Schriebman, C G & P K T Pang. Academic Press: New York

5 Perez F M, Malamed S & Scanes C G (1990) Stimulation of chicken growth hormone release by phorbol esters. *General and Comparative Endocrinology* **80** 181-188

6 Perez F M, Malamed S & Scanes C G (1989) Possible participation of calcium in growth hormone release and in thyrotropin-releasing hormone and human pancreatic growth hormone releasing factor synergy in a primary culture of chicken pituitary cells. *General and Comparative Endocrinology* **75** 481-491

7 Goodman H, Harvey S, Brooks D & Lea R W (1991) Extracellular calcium involvement in GRF-induced membrane depolarization of domestic fowl somatotrophs. *Journal of Endocrinology* **131** Supplement Abstract 43

8 Goodman H, Lea R W, Brooks D & Harvey S (1992) TRH/hpGRF synergy in domestic fowl is independent of changes in intracellular calcium. Abstract P28. *Proceedings of the Fifth International Symposium on Avian Endocrinology.* September 1992 Edinburgh: AFRC Roslin Institute

9 Buonomo F C (1992) Stimulation of somatotropin (GH) secretion following peripheral administration of interleukin-6 (IL-6). Abstract 461. *Program and Abstracts 75$^{th}$ Annual Meeting of the Endocrine Society.* San Antonio USA

10 Aramburo C, Donoghue D, Montiel J L, Berghman L R & Scanes C G (1990) Phosphorylation of chicken growth hormone. *Life Sciences* **47** 945-952

11 Aramburo C, Montiel J L, Proudman J A, Berghman L R & Scanes C G (1992) Phosphorylation of prolactin and growth hormone. *Journal of Molecular Endocrinology* **8** 183-191

12 Denver R J & Harvey S (1991) Thyroidal inhibition of chicken pituitary growth hormone: alterations in secretion and accumulation of newly synthesized hormone. *Journal of Endocrinology* **131** 39-48

13 Tanaka M, Hosokawa Y, Watahiki M & Nakashima K (1992) Structure of the chicken growth hormone-encoding gene and its promoter region. *Gene* **112** 235-239

14 Hasegawa Y, Miyamoto K, Nomura M, Igarashi M, Kangawa K & Matzuo H (1984) Isolation and amino acid compositions of four somatostatin-like substances in chicken hypothalamic extract. *Endocrinology* **105** 433-435

15 Donoghue D J & Scanes C G (1991) Possible involvement of adenylyl cyclase-cAMP-protein kinase A pathway in somatostatin inhibition of growth hormone release from chicken pituitary cells. *General and Comparative Endocrinology* **81** 113-119

16 Harvey S, Trudeau V L, Ashworth R J & Cockle S M (1993) pGlutamylglutamylprolineamide modulation of growth hormone secretion in domestic fowl: antagonism of TRH action? *Journal of Endocrinology* **138** 137-147

17 Ashworth R J & Cockle S M (1992) The TRH-related peptide pyroglutamylglutamylprolineamide: a possible autocrine or paracrine modulator of pituitary function? Abstract 677. *Program and Abstracts 74th Annual Meeting of the Endocrine Society*. San Antonio USA

18 Ashworth R J, Morell J M, Aitken A & Cockle S M (1991) Pyroglutamylglutamylprolineamide is present in rat anterior and posterior pituitary gland. *Journal of Endocrinology* **129** R1-R4

19 Ashworth R J, Visser T J & Cockle S M (1991) The TRH-like peptide pGlu-Glu-$ProNH_2$ is present in the porcine pituitary but not in reproductive tissues. *Biochemical and Biophysical Research Communications* **181** 1557-1563

20 Harvey S (1990) Thyrotrophin-releasing hormone: a growth hormone-releasing factor. *Journal of Endocrinology* **125** 345-358

21 Herremans M, Buyse J, Leenstra F R, Beuving G, Berghman L & Decuypere E (1992) Growth hormone response to TRH in male broiler chickens selected for body weight gain or food conversion and reared at either a moderate or a high ambient temperature. *Reproduction Nutrition Development* **37** 135-141

22 Kühn E R, Herremans M, Dewill E, Vanderpooten A, Rudas P, Bartha T, Verheyen G, Berghman L & Decuypere E (1991) Thyrotropin-releasing hormone (TRH) is not thyrotropic but somatotropic in fed and starved adult chickens. *Reproduction, Nutrition, Development* **31** 431-439

23 Trudeau V L, Harvey S, Ashworth R J & Cockle S M (1992) pGlutamylglutamylprolineamide (EEP) modulation of growth hormone secretion in the chicken. *Program and Abstracts Annual Neurosciences Meeting*. Anaheim USA

24 Harvey S (1993) Benzodiazepine antagonism of TRH receptors: biphasic actions on GH secretion in domestic fowl. *Journal of Endocrinology* **137** 35-42

25 Harvey S, Lea R W & Ahene C A (1990) Central action of thyrotrophin-releasing hormone on growth hormone secretion in domestic fowl. *Journal of Endocrinology* **126** 83-88

26 Ahene C A, Lea R W & Harvey S (1991) Central somatostatinergic regulation of growth hormone secretion in dwarf chickens. *Comparative Biochemistry and Physiology* **98A** 103-106

27 Ahene C A, Lea R W & Harvey S (1991) Thyrotrophin-releasing hormone-induced growth hormone (GH) secretion in anaesthetized chicken: inhibition by GH-releasing factor at central sites. *Journal of Endocrinology* **138** 13-19

28 Lea R W & Harvey S (1990) Feedback effects of growth hormone on anterior pituitary function in chickens. *Journal of Endocrinology* **125** 409-415

29 Donoghue D J, Perez F M, Diamante B S A, Malamed S & Scanes C G (1990) Influence of catecholamines, prostaglandins and thyroid hormones on growth hormone secretion by chicken pituitary cells *in vitro*. *Domestic Animal Endocrinology* **7** 35-42

30 Harvey S & Lea R W (1993) Thyrotrophin-releasing hormone-induced growth hormone secretion in domestic fowl: concomitant stimulation of dopamine turnover in the medial basal hypothalamus. *Journal of Endocrinology* (in press)

31 Cogburn L A, Liou S S, Alfonso C P, McGuiness M C & McMurtry J P (1989) Dietary thyrotropin releasing hormone stimulates growth rate and increases the insulin:glucagon molar ratio of broiler chickens. *Proceedings of the Society for Experimental Biology and Medicine* **192** 127-134

32 Harvey S, Scanes C G & Phillips J G (1985) Growth hormone secretion in anaesthetized fowl. 1. Refractoriness to repeated stimulation by human pancreatic growth hormone releasing factor (hpGRF) or thyrotrophin releasing hormone (TRH). *General and Comparative Endocrinology* **59** 1-9

33 Harvey S & Baidwan J S (1990) Feedback inhibition of growth hormone (GH) secretion in fowl: GH-induced downregulation of thyrotropin-releasing hormone binding to pituitary membranes. *Journal of Molecular Endocrinology* **4** 13-21

34 Lea R W, Ahene C A, Marsh J A & Harvey S (1990) Short-loop inhibition of TRH-induced GH secretion in fowl. *Journal of Endocrinology* **126** 237-244

35 Harvey S, Baidwan J S & Attardo D (1990) Homologous and heterologous regulation of somatostatin binding sites on chicken adenohypophysial membranes. *Journal of Endocrinology* **127** 417-425

36 Hull K L, Harvey S, Marsh J A, Boone F, McCann-Levorse L & Scanes C G (1993) Growth hormone receptor gene expression in sex-linked dwarf chickens: a model of Laron-type dwarfism? *Pediatric and Adolescent Endocrinology* **24** 294-309

37 Hull K L, Fraser, R A & Harvey S (1992) Expression and translation of the growth hormone receptor gene in chicken pituitary glands. *Journal of Endocrinology* **135** 459-468

38 Fraser R A, Attardo D & Harvey S (1990) Growth hormone receptors in hypothalamic and extra-hypothalamic tissues. *Journal of Molecular Endocrinology* **5** 231-238

39 Hull K L, Fraser R A, Marsh J A & Harvey S (1992) Growth hormone receptor gene expression in dwarf sex-linked chickens. *Canadian Journal of Pharmacology and Physiology* **70** Axi-Axii

40 Attardo D & Harvey S (1990) Growth-hormone binding sites in chicken hypothalamus. *Journal of Molecular Endocrinology* **4** 23-29

41 Belchetz P E, Ridley R M & Baker H F (1982) Studies on the accessibility of prolactin and growth hormone levels in serial, simultaneous plasma and cerebrospinal fluid samples in the rhesus monkey. *Brain Research* **239** 310-315

42 Banks W A & Kastin A J (1985) Aluminum alters the permeability of the blood-brain barrier to some non-peptides. *Neuropharmacology* **24** 407-412

43 Lai Z, Emtner M, Roos P & Nyberg F (1991) Characterization of putative growth hormone receptors in human choroid plexus. *Brain Research* **546** 222-226
44 Sato M, Chihara K, Kita T, Kashio Y, Okimura Y, Kitajima N & Fujita T (1989) Physiological role of somatostatin-mediated autofeedback regulation of growth hormone: importance of growth hormone in triggering somatostatin release during a trough period of pulsatile growth hormone release in conscious male rats. *Neuroendocrinology* **50** 139-151
45 Vasilatos-Younken R (1987) Hollow fibre-encapsulated pituitary cells for the study of adenohypophysial regulation of growth in poultry. 2. Recipient growth responses. *Poultry Science* **66** 899-903
46 Hall T R & Chadwick A (1984) Evidence of 'short-loop' feedback regulation for prolactin and growth hormone secretion in the pigeon. *General and Comparative Endocrinology* **58** 233-238
47 Lea R W & Harvey S (1993) Growth hormone suppression of catecholamine turnover in the chicken hypothalamus: implications for GH autoregulation. *Journal of Endocrinology* **136** 245-251
48 Bluet-Pajot M T, Mounier F, Durand D, Kordon C, Llorens-Cortes C, Videau C & Epelbaum J (1991) 6-hydroxydopamine lesions of the locus ceruleus induce a paradoxical increase in growth hormone secretion in male rats. *Journal of Neuroendocrinology* **4** 9-14
49 Nakai Y, Shioda S, Ochiai M & Kozaka K (1986) Catecholamine peptide interactions in the hypothalamus. *Current Topics in Neuroendocrinology, Morphology of Hypothalamus and its Connections* pp 135-160. Eds D Ganten & D W Pfaff. Springer: Verlag, Berlin
50 Niimi M, Takahara J, Sato M & Kawanishi K (1992) Identification of dopamine and growth hormone releasing factor-containing neurons projecting to the median eminence of the rat by combined retrograde tracing and immunohistochemistry. *Neuroendocrinology* **58** 92-96
51 Burton K A, Kabigting E B, Clifton D K & Steiner R A (1991) Distribution and regulation of growth hormone receptor mRNA in the adult male rat brain. Abstract 155. *Program and Abstracts 73rd Annual Meeting of the Endocrine Society.* Washington DC
52 Burton K A, Steiner R A & Clifton D K (1992) Double-label *in situ* hybridization confirms that the GRF gene and the GH receptor gene are co-expressed in subset of neurons in the arcuate nucleus. *Program and Abstracts 74th Annual Meeting of the Endocrine Society.* Abstract 1370. San Antonio Texas
53 Scanes C G (1992) Lipolytic and diabetogenic effects of native and biosynthetic growth hormone in the chicken: a re-evaluation. *Comparative Biochemistry and Physiology* **101** A 871-787
54 Haddad E E & Mashaly M M (1990) Effect of thyrotropin-releasing hormone, triiodothyronine, and chicken growth hormone on plasma concentrations of thyroxine triiodothyronine, growth hormone and growth of lymphoid organs and leukocyte populations in immature male chickens. *Poultry Science* **69** 1074-1102
55 Marsh J A, Johnson B E, Lillehoj H S & Scanes C G (1992) Effect of thyroxine and chicken growth hormone on immune function in autoimmune thyroiditis (obese) strain chicks. *Proceedings of the Society for Experimental Biology & Medicine* **199** 114-122

56 O'Neill I E, Houston B & Goddard C (1990) Stimulation of insulin-like growth factor-1 production in primary cultures of chicken hepatocytes by chicken growth hormone. *Molecular and Cellular Endocrinology* **70** 41-47

57 Perez F M, Malamed S & Scanes C G (1985) Biosynthetic human somatomedin-C inhibits hGRF (1-44 $NH_2$) -induced and TRH-induced GH release in a primary culture of chicken pituitary cells. *IRCS Medical Science* **13** 871-872

58 Buonomo F L, Lauterio T J, Baile C A & Daughaday W H (1987) Effects of insulin-like growth factor I (lGF-I) on growth hormone-releasing factor (GRF) and thyrotropin-releasing hormone (TRH) stimulation of growth hormone (GH) secretion in the domestic fowl (*Gallus domesticus*). *General and Comparative Endocrinology* **66** 274-279

59 McGuiness M C & Cogburn L A (1988) Plasma somatomedin levels during the growth cycle and unresponsiveness of broiler chickens to recombinant-derived human somatomedin-C. *Poultry Science* **67** 143

60 Huybrechts L M, Decuypere E, Buyse J, Kühn E R & Tixier-Boichard M (1992) Effect of recombinant human insulin-like growth factor-I on weight gain, fat content and hormone parameters in broiler chickens. *Poultry Science* **71** 181-187

61 Lazarus D D & Scanes C G (1988) Acute effects of hypophysectomy and administration of pancreatic and thyroid hormones on circulating concentrations of somatomedin-C in young chickens: relationship between growth hormone and somatomedin-C. *Domestic Animal Endocrinology* **5** 283-289

62 Tixier-Boichard M, Monvoisin J L, Decuypere E, Huybrechts L M & Kühn E R (1991) Effect of triiodothyronine supplementation on thyrotropin-releasing hormone-induced growth hormone secretion in sex-linked dwarf and normal chicks. *General and Comparative Endocrinology* **84** 147-154

63 Harvey S, Decuypere E, Darras V M & Berghman L (1991) Differential effects of $T_4$ and $T_3$ on TRH-and GRF-induced GH secretion in the domestic fowl. *Reproduction, Nutrition, Development* **31** 451-460

64 Donoghue D J & Scanes C G (1991) Triiodothyronine ($T_3$) inhibition of growth hormone secretion by chicken pituitary cells *in vitro*. *General and Comparative Endocrinology* **84** 344-354

65 Harvey S (1990) Trii-iodothyronine inhibition of TRH-induced growth hormone release from the chicken adenohypophysis *in vitro*. *Journal of Endocrinology* **126** 75-81

66 Harvey S, Klandorf H & Scanes C G (1990) Participation of triiodothyronine and metabolic clearance rate in the inhibition of growth hormone secretion in thyroxine-treated domestic fowl. *Journal of Endocrinology* **124** 215-223

67 Mullis P E, Pal B R, Mathews D R, Hindmarsh P C, Phillips P E & Dunger D B (1992) Half-life of exogenous growth hormone following suppression of endogenous growth hormone secretion with somatostatin in type 1 (insulin-dependent) diabetes mellitus. *Clinical Endocrinology* **36** 255-263

68 Vasilatos-Younken R, Gray K S, Bacon W L, Nestor K E, Long D W & Rosenberger J L (1990) Ontogeny of growth hormone (GH) binding in the domestic turkey: evidence of sexual dimorphism and developmental changes in relationship to plasma GH. *Journal of Endocrinology* **126** 131-139

69 Davis S L, Graf M, Morrison C A, Hall T R & Swift P J (1992) Identification and partial purification of serum growth hormone binding protein in domestic animal species. *Journal of Animal Science* **70** 773-780

70 Herington A C, Ymer S I & Tiong T S (1991) Does the serum binding protein for growth hormone have a functional role? *Acta Endocrinologica* **124** 14-20
71 Scanes C G, Peterla T A, Kantor S & Ricks C A (1990) *In vivo* effects of biosynthetic chicken growth hormone in broiler-strain chickens. *Growth Development and Aging* **54** 95-101
72 Scanes C G, Carsia R V, Lauterio T J, Huybrechts L, Rivier J & Vale W (1983) Synthetic human pancreatic growth hormone releasing factor (GRF) stimulates growth hormone secretion in the domestic fowl (*Gallus domesticus*). *Life Sciences* **34** 1127-1134
73 Scanes C G & Harvey S (1989) Somatostatin (SRIF) inhibition of thyrotropin-releasing hormone (TRH) and growth hormone (GH) releasing factor (GRF)-induced GH secretion in young and adult anaesthetized chickens. *General and Comparative Endocrinology* **75** 256-264
74 Klandorf H, Harvey S & Fraser H M (1985) Physiological control of growth hormone secretion by thyrotrophin-releasing hormone in the domestic fowl. *Journal of Endocrinology* **100** 351-355
75 Cheung A & Hall T R (1990) Direct stimulation of growth hormone secretion by galanin in the domestic fowl. *Journal of Neuroendocrinology* **2** 285-290
76 Hall T R, Cheung A & Griffiths E C (1988) Peptide effects on GH release in domestic fowl. *Program and Abstracts $8^{th}$ International Congress of Endocrinology.* Tokyo, Japan
77 Hall T R, Cheung A, Harvey S & Griffiths E C (1986) Opiatergic effects on hormone release from chicken pituitary glands incubated *in vitro.* *Regulatory Peptides* **15** 179
78 Hall T R, Harvey S & Scanes C G (1986) Control of growth hormone secretion in the vertebrates: a comparative survey. *Comparative Physiology and Biochemistry* **84A** 231-253
79 Harvey S (1977) Purification, radioimmunoassay, and physiology of chicken growth hormone. *Ph.D Thesis* University of Leeds
80 Harvey S, Chadwick A, Luck M, Scanes C G & Bolton N J (1978) Effects of enkephalins and analogues on the secretion of pituitary hormones in immature male chickens. *IRCS Medical Sciences* **6** 493
81 Leung F C (1986) Hormonal regulation of growth in chickens. *Control and Manipulation of Animal Growth* pp 233-243. Eds P J Buttery, D B Lindsay & N B Haynes. London: Butterworths
82 Buonomo F C, Zimmerman N G, Lauterio T J & Scanes C G (1984) Catecholamine involvement in the control of growth hormone secretion in the domestic fowl. *General and Comparative Endocrinology* **54** 360-371
83 Leung I C, Taylor J E & Ball C A (1985) Potent interaction between thyrotropin-releasing hormone (TRH) and human pancreatic growth hormone-releasing factor (hGRF) in stimulating chicken growth hormone (cGH) *in vivo*; hypothalamic noradrenergic mediation in TRH stimulation of cGH release. *Domestic Animal Endocrinology* **2** 183-190
84 Hall T R, Harvey S & Chadwick A (1984) Prolactin and growth hormone secretion in chickens; stimulation by histamine and inhibition by gamma-amino butyric acid. *Acta Endocrinologica* **107** 36-41

85 Hall T R, Harvey S & Chadwick A (1984) Relationship between hypothalamic serotoninergic activity and prolactin and growth hormone secretion in the domestic fowl. *Neuroendocrinology* **39** 206-213

86 Hall T R, Harvey S & Chadwick A (1984) Serotonin and acetylcholine affect the release of prolactin and growth hormone from pituitary glands of domestic fowl *in vitro* in the presence of hypothalamic tissue. *Acta Endocrinologica* **105** 455-462

87 Picarper G, Leclerq B, Saadoun A & Mongin P (1986) A radioimmunoassay of chicken growth hormone using growth hormone produced by recombinant DNA technology: validation and observations of plasma hormone variations in genetically fat and lean chickens. *Reproduction, Nutrition, Development* **25** 1105-1114

88 Harvey S & Scanes C G (1978) Plasma growth hormone levels in growth retarded corticosterone treated chickens. *British Poultry Science* **20** 331-335

89 Scanes C G & Harvey S (1980) Inhibition of thyrotrophin-releasing hormone induced growth hormone secretion in domestic fowl by adrenaline and prostaglandin $E_1$ and $E_2$. *Hormone and Metabolic Research* **12** 641-642

90 Fennell M J & Scanes C G (1992) Effects of androgen (testosterone, 5α-dihydrotestosterone, 19-nortestosterone) administration on growth in turkeys. *Poultry Science* **71** 539-547

91 Sommerville B A, Scanes C G, Swaninathan R, Care A O, Harvey S & Chadwick A (1989) Effect of estrogen on calcium homeostasis and pituitary hormones in the growing chick. *General and Comparative Endocrinology* **76** 261-266

92 Harvey S, Scanes C G, Chadwick A & Bolton N J (1978) Influence of fasting, glucose and insulin on the levels of growth hormone and prolactin in the plasma of the domestic fowl *(Gallus domesticus). Journal of Endocrinology* **76** 501-506

93 Foltzer C H, Strosser M T, Harvey S & Malhe P (1985) Control of plasma levels of growth hormone, glucagon and insulin in ducklings: roles of free fatty acids and somatostatin. *Journal of Endocrinology* **106** 21-25

Avian Endocrinology
Ed. P J Sharp, pp 29-46
Journal of Endocrinology Ltd, Bristol (1993)

# Hypothalamic and peripheral control of thyroid function during the life cycle of the chicken

E R Kühn[1], L R Berghman[2], L Moons[2], F Vandesande[2], E Decuypere[3] and V M Darras[1]
Catholic University of Leuven: [1]Laboratory of Comparative Endocrinology, Naamsestraat 61, B-3000 Leuven, Belgium; [2]Laboratory of Neuroendocrinology and Immunological Biotechnology, Naamsestraat 59, B-3000 Leuven, Belgium; [3]Laboratory of Physiology and Immunology of Domestic Animals, Kardinaal Mercierlaan 92, B-3001 Leuven, Belgium

## Introduction

The hypothalamic tripeptide thyrotrophin-releasing hormone (TRH) has been recognized for more than two decades as the major regulator of the biosynthesis and release of pituitary thyrotrophin (TSH) in vertebrates, with the possible exception of teleosts. This is consistent with the fact that thyroxine ($T_4$) is released from the thyroid gland following a TRH challenge. Most circulating triiodothyronine ($T_3$) is derived from $T_4$ by peripheral 5′-monodeiodination (5′D) (1). In sheep, and possibly other mammals, TSH also stimulates the intrathyroidal conversion of $T_4$ into $T_3$ and therefore may release $T_3$, prior to $T_4$ (2).

Apart from the release of TSH, TRH stimulates the release of other hypophyseal hormones. Under physiological conditions TRH releases prolactin in most mammals (3), whereas in birds the somatotrophs are stimulated by TRH (4, 5). In amphibians, TRH releases prolactin, growth hormone (GH) and α-melanophorotropin (α-MSH) from the hypophysis together with TSH (6). In teleosts no convincing data exist to show that TRH releases TSH, but some results indicate that TRH may be somatotropic (7).

In the chicken, peripheral $T_3$ concentrations are controlled by the simultaneous release of GH and TSH (8, 9), whereas TSH acting alone is purely thyrotrophic and has no direct effect on peripheral $T_4/T_3$ conversion (10). On the other hand peripheral fluctuations in thyroid hormones may exert a negative feedback control on both TSH and GH secretion (11).

This review deals with the hypothalamic and peripheral mechanisms controlling thyroid function and their evolution during the life cycle of the chicken.

## Control of thyroid function in the chick embryo

### *Hypothalamic control*

Studies on hypophysectomized chick embryos demonstrate that the pituitary gland is necessary for the normal increase in thyroidal activity that occurs between days 10 and 12 of incubation. During this period the sensitivity of the chick embryo to exogenous TRH increases markedly as measured by TRH-induced $T_4$ release. The number of immunocytochemically demonstrable TSH cells and the amount of immunoreactive material per cell increases also between days 10.5 and 11.5. It seems that during this period a transition from thyroidal autonomy to thyroidal adenohypophyseal dependency may occur since thyroids of embryos deprived of their pituitary glands develop normally until day 10-12. The maturation of a feedback loop has also been indirectly shown to occur during this time interval. However, the capacity of the pituitary and the thyroid glands to respond to TRH and TSH administration measured as an increase in the concentration of plasma $T_4$ is already present as early as day 6.5 of incubation (12).

Concentrations of plasma $T_4$ increase from day 10 and reach a maximum at day 20 of incubation prior to pipping whereas concentrations of plasma $T_3$ remain low (13, 14). The $T_3/T_4$ ratio in the thyroid glands equals that in the plasma before perforation of the air cell membrane. This can be explained by the observation that during this period, a low dose of TRH exerts its activity by a thyrotrophic stimulation of the hypophysis as shown by increased concentrations of plasma $T_4$ and reverse $T_3$ ($rT_3$), whereas the release of GH remains unaffected (15). This is further substantiated by the fact that plasma GH is hardly detectable until day 17 of incubation, but increases thereafter (14, 16). However, the peripheral mechanisms for $T_3$ production are already present since an injection of chicken GH (cGH) on day 14 of incubation, increases the concentration of plasma $T_3$ and the hepatic $T_3$ production (17). While plasma $T_4$ concentrations increase gradually up to pipping, the increase in $T_3$ is abrupt, between day 19 and hatching. The latter increase coincides with internal pipping and the transition from an allantoic to pulmonary respiration (14, 17, 18; Fig. 1).

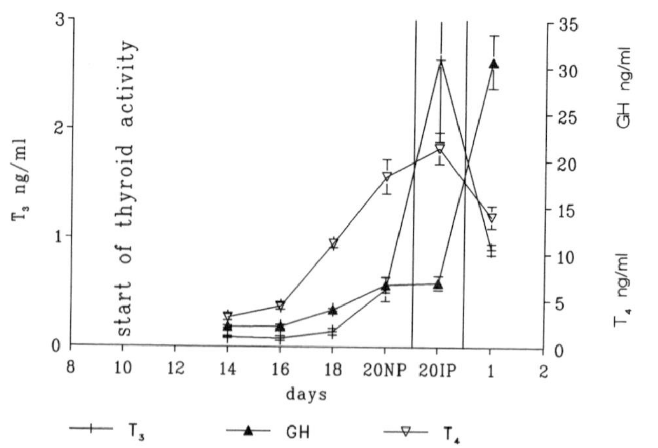


Fig. 1 Changes in concentrations of plasma thyroid hormones and cGH at the end of incubation in the domestic chicken (days 14-20) and the first day posthatch (day 1). The parallel vertical lines show the time when the chickens are pipping and hatching. Values are means ± s.e. (n=7). NP, non pipping, IP internal pipping.

On day 18 of incubation, several hypothalamic hormones influence circulating concentrations of thyroid hormones. An injection of 1 µg TRH into an allantoic blood vessel increases concentrations of plasma $T_3$ and $T_4$, whereas concentrations of $rT_3$ do not change. At the same time, the $T_3/T_4$ ratio increases together with an increase in capacity of liver homogenate to produce $T_3$ when incubated in the presence of $T_4$ and dithiothreitol. An intravenous (iv) injection of bovine TSH also increases plasma concentrations of $T_3$ and $T_4$ but the increase in $T_4$ is more pronounced, resulting in a decreased $T_3/T_4$ ratio. Interestingly, the increase in plasma $T_4$ is paralleled by an increase in plasma $rT_3$ (19).

Although the increase in concentrations of plasma $T_4$ following a TRH challenge is consistent with the view that TRH stimulates TSH release, no direct proof for TSH release has been presented up to now. A monoclonal antibody against the α-subunit of chicken glycoprotein hormones, in combination with the iodinated α-subunit and USDA chicken LH (K-3) as the standard, were used in order to measure concentrations of plasma α-subunit-like immunoreactivity (α-IR) by radioimmunoassay (RIA) (20). This would be predicted to measure TSH, LH and free α subunit combined. In critical samples, LH was assayed using the recently issued homologous USDA chicken LH RIA, to rule out LH changes as an explanation for the observed changes in α-IR components. Perifusion of pituitaries from 18-day-old embryos with 10 and 100 nM TRH stimulated the release of α-IR providing evidence for the release of TSH (Fig. 2). An iv injection of the anti-α-subunit (AA) alone or in combination with TRH in 18-day-old chick embryos decreased $T_4$ concentrations while TRH alone increased $T_4$ (Fig. 3). On the other hand, administration of AA had no effect on concentrations of plasma $T_3$, whereas a profound increase in $T_3$ is observed with TRH or TRH + AA (Fig. 3). These observations provide further evidence that TRH stimulates the release of TSH in the chick embryo (Fig. 4).

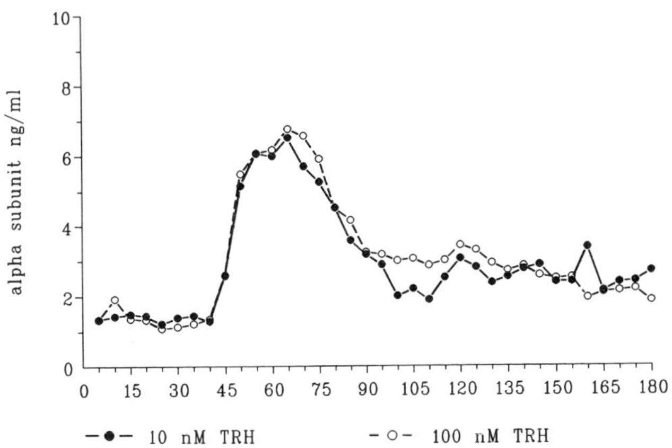


Fig. 2 Radioimmunoassay of the glycoprotein hormone α subunit in the medium used to perifuse pituitaries from 18-day-old chicken embryos with 10 and 100 nM TRH added between 30 and 60 min of perifusion. The radioimmunoassay is predicted to measure TSH together with LH and free glycoprotein hormone α subunit.

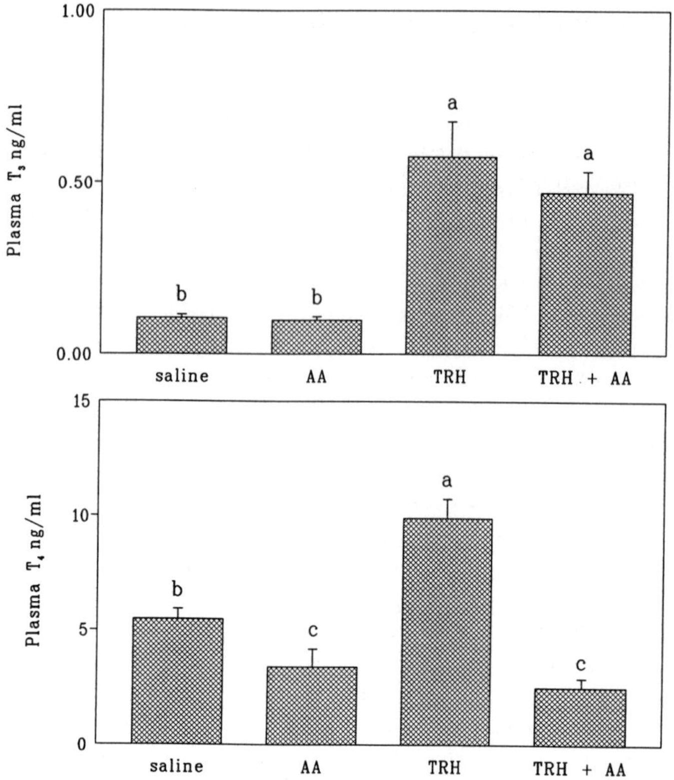


Fig. 3 The effect in 18-day-old chicken embryos of injections of saline, 5 μg anti-glycoprotein hormone α subunit monoclonal antibody (AA), 1 μg TRH and 1 μg TRH + 5 μg AA on concentrations of plasma $T_3$ and $T_4$. Blood samples were taken 4 h (saline and AA) and 2 h (TRH and TRH + AA) following injection. Values are means ± s.e. (n=20). Mean values with the same letter are not significantly different (ANOVA, $P < 0.05$).

In the chick embryo, concentrations of plasma corticosterone increase at the end of incubation (21) and iv injections of ovine corticotrophin-releasing hormone (oCRH) induce dose dependent increases in corticosterone concentration after 15 and 30 min and after 1 and 2 h, which may indicate both a stimulation of release and synthesis (22). Interestingly, injection of oCRH also increases concentrations of plasma $T_3$ and $T_4$, without affecting the $T_3/T_4$ ratio, indicating stimulation of the thyrotrophs rather than the peripheral conversion of $T_4$ into $T_3$ as was suggested by an earlier study involving injection of ACTH or glucocorticoids (23; Fig. 4).

Human growth hormone-releasing hormone (hGRH) increases the concentration of plasma GH in the 18-day-old chick embryo 15 min after iv injection. This increase is followed by an increase in plasma $T_3$ concentration after 2 h, whereas the concentration of plasma $T_4$ does not change (8).

*Peripheral control*

In chick embryos, the increase in plasma $T_3$ following TRH or hGRH injection is preceded by an increase in plasma GH and is accompanied by an increased $T_3$ recovery from liver homogenates incubated with $T_4$ (8). This observation suggests GH stimulates 5′D activity to increase the conversion of $T_4$ into $T_3$. Further support for this view comes from the observation that a hypophyseal extract devoid of GH following immunoabsorption is not effective in this regard (9, 24, 25).

The mechanism by which GH induces changes in peripheral thyroid hormone metabolism has only recently been established (Fig. 4). The liver of the chick embryo contains not only 6-propyl-2-thiouracil (PTU)-sensitive type I deiodinase, showing both outer ring deiodinating (5′D) and inner ring deiodinating (5D) activity, but, in contrast to the mammalian liver, contains considerable amounts of PTU-insensitive type III deiodinase, with only 5D activity (14, 26). The enzyme assays used to measure type I and type III activity are based on the deiodination of the $^{125}$I-labelled preferred substrate, namely $rT_3$ for type I (5′D-I) and $T_3$ for type III (5D-III) activity. There is no evidence in the chicken embryo liver for PTU-insensitive type II deiodinase, showing only 5′D activity (25, 26), although PTU-insensitive conversion of $T_4$ to $T_3$ has been reported in embryonic quail liver (27, 28). An ontogenetic study of hepatic 5′D-I activity in the chick embryo demonstrates a threefold increase up to the period of pipping and hatching and a slow decrease thereafter. Hepatic 5D-III activity increases threefold between days 14 and 17 of incubation, but decreases thereafter more than tenfold between day 17 and hatching. Moreover, during this period a positive correlation is found between plasma GH and $T_3$ and between plasma $T_4$ and 5′D-I activity, while a negative correlation is present between plasma GH and 5D-III activity and between plasma $T_3$ and 5D-III activity (14).

A single cGH injection increases plasma $T_3$ and decreases plasma $rT_3$ concentrations in all embryonic stages tested and in chicks immediately after hatch (17). Kinetic studies of the $T_4$ to $T_3$ conversion reveal, however, no increase in $K_m$ or $V_{max}$ and the specific tests described above, using liver microsomal fractions, reveal that GH does not affect the amount of 5′D-I, but decreases the amount of 5D-III. Growth hormone therefore does not stimulate the $T_3$ production *in vitro* by $T_4$ deiodination (5′D-I), but inhibits $T_3$ (and $T_4$) degrading activity (5D-III), resulting *in vivo* in increased concentrations of plasma $T_3$ and decreased concentrations of plasma $rT_3$ (17, 29).

The clear effect of an injection of GH on 5D-III inhibition, observed at the end of incubation, coincides with increased hepatic GH receptor capacity and a low concentration of plasma GH (30). A negative correlation has also been observed between plasma GH concentrations and specific binding of $^{125}$I-cGH to its hepatic receptor in growing chickens (31). These observations support the view that, as in mammals (32), high concentrations of plasma GH down-regulate GH receptors.

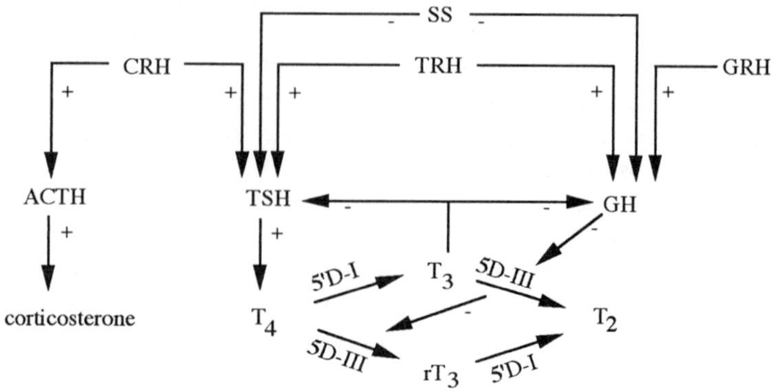


Fig. 4 Mechanisms controlling thyroid function in the chick embryo. CRH, corticotrophin releasing hormone; SS, somatostatin; TRH, thyrotrophin releasing hormone; GRH, growth hormone releasing-hormone; $rT_3$, reverse $T_3$; $T_4$, thyroxine; $T_3$ triiodothyronine; $T_2$, diiodothyronine; ACTH, adrenocorticotrophic hormone; 5′D-I, outer ring type I deiodinase; 5 D-III, inner ring type III deiodinase.

## Control of thyroid function in the growing chicken

After hatching, concentrations of plasma $T_3$ increase up to 2 weeks of age and decrease slowly thereafter, whereas $T_4$ may decrease during the first 2 weeks and increase thereafter (33). A positive correlation exists between relative growth of the chicks and plasma $T_3$, but there is no relationship with the concentration of plasma $T_4$. However, increasing body weight with age is positively correlated with plasma $T_4$ (33).

*Hypothalamic control*

The administration of TRH stimulates the incorporation of radioactive phosphate into the thyroid glands (34), the release of $^{131}I$ from the thyroid gland (35) and increases concentrations of plasma thyroid hormones in juvenile chickens (36). Increases in $T_4$ concentrations are observed in 3-, 5- and 7-week-old chickens 30 min after a single TRH injection (10 μg/kg body weight). An increase in plasma $T_3$ is not observed in 3-week-old chicks, but in 5- and 7-week-old birds $T_3$ is increased after 60 min. (37). In all animals the $T_3/T_4$ ratio in the serum is 10-to 30-times as high as in the thyroid glands during the first two weeks after hatching, suggesting that peripheral 5′D activity is predominantly responsible for the production of $T_3$.

Following injection of 0.1 or 0.25 I.U. bTSH/kg into 5-week-old chickens, a relatively small increase in $T_3$ is observed as compared to $T_4$. Serum $rT_3$ concentrations are also increased after an injection of 0.25 I.U. bTSH/kg but the maximal concentration (75 pg/ml) is low (38). Reverse $T_3$ is virtually absent in young posthatch birds, but increases in one to two year old chickens (39).

*Peripheral control*

As for TSH, GH is also released following TRH injection in growing chickens. In both conscious and anaesthetized immature cockerels the administration of 2.5 and 25 µg TRH/kg significantly increases the concentration of plasma GH (4).

Contrary to the situation in chicken embryos, in the posthatch chicken, plasma concentrations of $T_3$ do not increase and the liver does not show increased *in vitro* $T_3$ recovery from $T_4$ following injection of GH. Up to the age of 4 weeks cGH administration has no effect on thyroid hormone metabolism, but in 5-week-old chicks cGH is again able to increase the $T_3/T_4$ ratio, although it still has no significant effect on hepatic $T_3$ production (17).

Interestingly, nutritional status plays a role in mediating the action of GH on thyroid activity. Once chicks start to eat (±2 days after hatching) the stimulatory effect of an injection of GH on circulating $T_3$ and hepatic $T_3$ production is abolished. In contrast, food deprivation after a period of food intake in young chicks restores the stimulatory effect of GH on plasma $T_3$ but not on hepatic $T_3$ production (41).

The importance of GH in maintaining plasma concentrations of plasma $T_3$ during growth in the chicken has been investigated in several experiments designed to lower endogenous GH concentrations. An injection of anti-GH serum into 4-week-old chicks decreases concentrations of plasma $T_3$, compared to controls. Since an increase in plasma $T_4$ is also observed, a lack of substrate for the 5′D enzyme cannot be responsible for the decrease in $T_3$ (41). Hypophysectomy substantially reduces concentrations of plasma GH and $T_3$, whereas the effect on plasma $T_4$ is less pronounced. Following injection of GH, an increase in $T_3$ is observed in chicks hypophysectomized for 2 or 3 days, whereas $T_3$ concentrations in sham controls are not affected (41). These observations show that, even at an age when the chicks are normally not responsive to GH injection, the monodeiodination system remains GH-sensitive, and the endogenous GH levels are important in maintaining normal circulating $T_3$ concentrations (Fig. 5).

During growth 5D-III deiodinase activity is very low, indicating that the high circulating concentrations of endogenous GH may exert a maximal inhibitory effect on $T_3$ degradation. Hypophysectomy of growing birds increases 5D-III levels, an effect which is reversed by treatment with GH. These observations confirm that the regulation of hepatic type III, but not of type I deiodinase, is GH dependent (14, 29, 41; Fig. 5).

The question arises which factor might be involved in the control of 5′D-I activity and hence the peripheral production of $T_3$. Concentrations of plasma $T_4$ are maintained 3 and 7 days after hypophysectomy, whereas 5′D-I activity is slightly decreased after 3 but not after 7 days compared to sham operated controls. Therefore, hepatic 5′D-I seems not to be TSH-dependent. As mentioned previously, a positive correlation is found between concentrations of plasma $T_4$ and the hepatic 5′D-I activity during ontogeny (14). The stimulatory effect of $T_4$, and not of TSH, on hepatic 5′D-I activity in rat liver has been established (42, 43) and recently it has been shown that thyroid hormones regulate the transcription of 5′D-I mRNA in rat liver microsomes (44).

Hypophysectomy not only decreases concentrations of plasma GH but dramatically increases the number of GH receptors in the liver. Injection of cGH reduces the number of hepatic GH receptors in hypophysectomized chicks to the numbers measured in sham operated control birds (31). This observation explains why hypophysectomy restores the effect of GH injections on monodeiodination activity in growing chicks: the increased activity of 5D-III makes an effect of GH more easily detectable, while an increase in GH receptor sites further increases the effectiveness of the injected GH (Fig. 5).

*Negative feedback control by thyroid hormones*

Negative feedback control by thyroid hormones appears to be very complex, since the amount of TSH and GH released following a TRH challenge is modulated by concentrations of circulating thyroid hormones, particularly $T_3$. Therefore, any condition which influences $T_3$ production may affect both the synthesis and release of GH and TSH.

Not much is known about the negative feedback mechanisms controlling TSH release in birds. Thyroid hormones are reported in quail to decrease the TSH response to TRH by a down regulation of TRH receptors (45), and $T_3$ has been found to down regulate TRH binding sites on chicken pituitary membranes (46).

Thyroid hormones inhibit basal and both TRH-and GRH-induced GH release from pituitary tissue *in vitro* and *in vivo* (11). Thyroxine is less active in this respect than $T_3$ but might exert some of its effects via $T_3$-independent mechanisms, which implies that both thyroid hormones act at different sites in the hypothalamo-pituitary axis to control the synthesis and release of GH (47).

Growth hormone is required for a normal growth rate in the chicken since hypophysectomy in young chickens reduces growth while replacement therapy with mammalian GH partially restores growth rate (48). The absence of a clear stimulatory effect on growth by GH administration in rapidly growing broiler chicks or turkeys is believed to be due to the high concentrations of circulating GH found in growing birds already exerting a maximal effect on growth, and/or to the number of GH receptors being limiting (49). Present results seem to support this hypothesis since high concentrations of plasma GH maintain high concentrations of plasma $T_3$ by maximally inhibiting 5D-III activity (41). Therefore any additional effect of GH will only influence $T_3$ and presumably growth under those physiological conditions where type III monodeiodination is increased.

## Control of thyroid function in the adult chicken

Concentrations of plasma $T_3$ (33) and GH (49) are increased up to 4-8 weeks of age and reach low stable values after the onset of puberty. Concentrations of plasma $T_4$ increase steadily from the second week posthatch and reach their maximal values around 4-8 weeks of age (33). The adult chicken therefore has relatively low concentrations of plasma $T_3$ and GH and high concentrations of plasma $T_4$, with absolute values varying with the strain of bird.

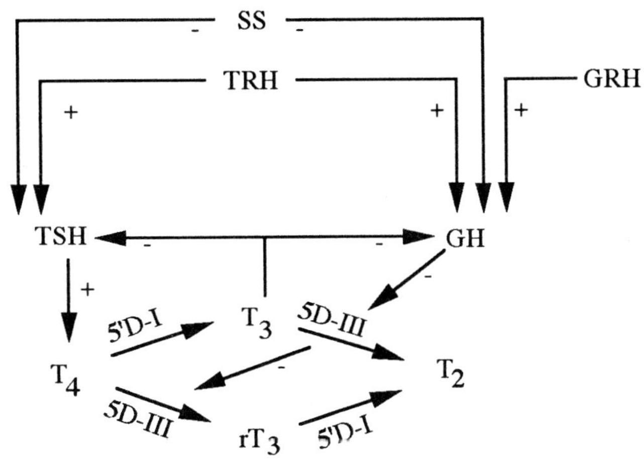


Fig. 5 Control mechanisms of thyroid function in juvenile, rapidly growing chickens. SS, somatostatin; TRH, thyrotrophin releasing hormone; GRH, growth hormone releasing-hormone; $rT_3$, reverse $T_3$; $T_4$, thyroxine; $T_3$ triiodothyronine; $T_2$, diiodothyronine; 5′D-I, outer ring type I deiodinase; 5 D-III, inner ring type III deiodinase.

*Hypothalamic control*

Contrary to the thyrotrophic effects of TRH described for chick embryos and growing chicks, where TRH increases both concentrations of plasma $T_3$ and $T_4$ (8, 19, 37, 38), doses of up to 13 μg TRH/kg are unable to increase plasma $T_4$ in the adult chicken, though the thyroid glands remain responsive to TSH injections. Moreover, $T_4$ concentrations even may decrease as a result of TRH injections. On the other hand, an increase in $T_3$ is observed with a dose as low as 0.6 μg TRH/kg, suggesting the existence of a very sensitive peripheral mechanism for the control of $T_3$ (19).

This aspect of thyroid physiology has been reinvestigated recently using adult fed and starved Warren chickens (2.5 kg) approaching the end of their second laying year. A comparison was made between the effects of injections of saline, 10 μg TRH, 100 μg bTSH and 100 μg oGH on concentrations of thyroid hormones and GH in the circulation. Starvation for 3 days decreased concentrations of plasma $T_3$ and increased plasma GH, a well known effect in most vertebrates studied. Starvation potentiated the effect of a TRH challenge on GH release as has also been observed in growing chicks fed a low protein diet (50). The increase in concentrations of plasma GH observed 15 min after TRH injection, was followed by an increase in plasma $T_3$ after 1 h, which was also more pronounced in starved than in fully fed birds. Concentrations of plasma $T_4$, however, remained unaffected or even decreased slightly, whereas increases in concentrations of plasma $T_4$ can readily be obtained with 100 μg TSH in normal fed or starved adult animals (10). These observations suggest that in the adult chicken, TRH is not thyrotrophic and does not release TSH.

Other observations also support this view:

1 An injection of 500 μg of anti-glycoprotein hormone α subunit (20) depresses the concentration of plasma $T_4$, but not of $T_3$ after 4 hours in 2-year-old Warren chickens (Fig. 6) indicating that endogenous TSH controls thyroidal output of $T_4$.

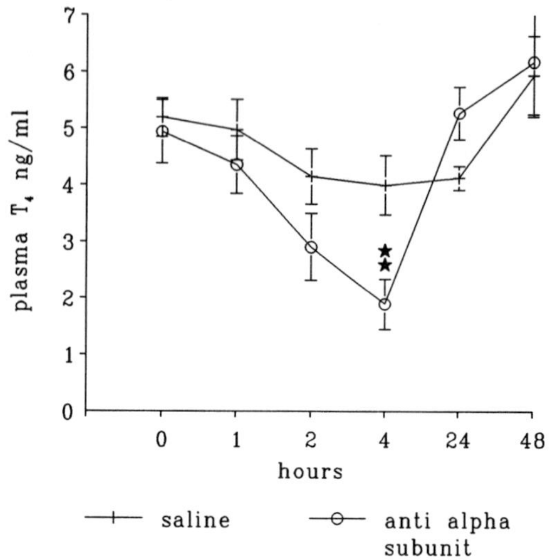


Fig. 6 Concentrations of plasma $T_4$ in 2-year-old Warren hens, after an intravenous injection of 500 μg anti-glycoprotein hormone α subunit serum or saline vehicle. Values are means ± s.e. (n = 8), $^{**}P < 0.01$ (ANOVA). The antibody is predicted to be capable of immunoneutralizing TSH in addition to the gonadotrophins and free glycoprotein hormone α subunit.

2 An injection of 5 μg TRH/kg or 50 μg oGH/kg increases the concentrations of plasma $T_3$ after 2 hours in 1-year-old broilers (body weight ~ 5 kg), whereas concentrations of plasma $T_4$ and α-IR remain unaffected. Moreover, using female broiler breeder chickens of the same age on a restricted feeding regime (body weight ~ 2.5 kg), an injection of 5 μg TRH/kg or 50 μg oGH/kg depresses the concentrations of plasma $T_4$ and α glycoprotein hormone subunit (Fig. 7), while the concentration of plasma $T_3$ is increased.

3 An autoradiographic study of the adenohypophysis reveals a decrease in TRH binding sites in the cephalic lobe - located most probably on the thyrotrophs - in adult chickens, as compared to chick embryos (L Moons, unpublished observation). This decrease in binding sites on the thyrotrophs is much more pronounced than the decrease in binding sites on the somatotrophs in the caudal lobe, which has been described previously (51).

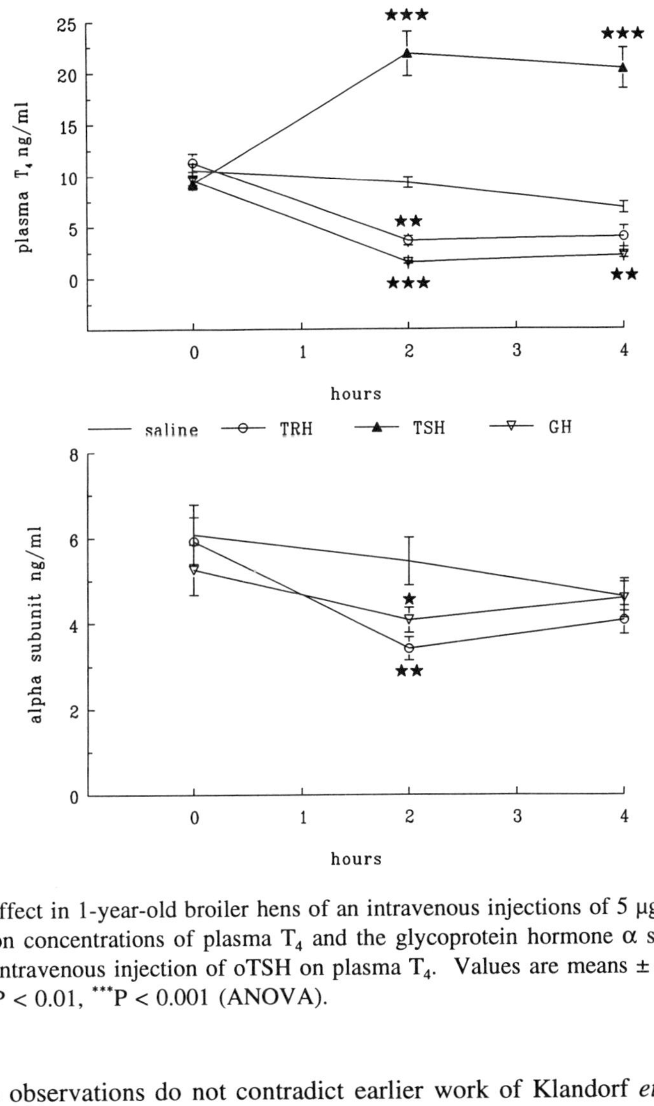


Fig. 7 The effect in 1-year-old broiler hens of an intravenous injections of 5 μg TRH/kg or 50 μg oGH/kg on concentrations of plasma $T_4$ and the glycoprotein hormone α subunit, and the effect of an intravenous injection of oTSH on plasma $T_4$. Values are means ± s.e. (n = 8), $^{*}P < 0.05$, $^{**}P < 0.01$, $^{***}P < 0.001$ (ANOVA).

These observations do not contradict earlier work of Klandorf *et al.* (52) who noted only a very small (1.3-fold) and short-lived increase in plasma $T_4$ after iv injection of a large dose of TRH (± 70 μg/kg) in a laying strain, whereas a 2.3-fold increase in plasma $T_3$ was observed which lasted for more than 3 h. Similarly, using 20-week-old Astra cocks, an injection of 10 μg TRH/kg resulted in a small increase in plasma $T_4$ in fed or fasted animals, which lasted only 1 hour whereas plasma $T_3$ was increased for at least 4 hours (53). There is apparently a contradiction, however, with an early *in vitro* assay where chicken anterior pituitary glands obtained from old laying hens were incubated with different doses of TRH (54). A significant *in vitro* release of thyrotrophic activity from the anterior pituitary was observed as measured by the ability of the incubation medium to stimulate the uptake of radioactive phosphate *in vivo* by the thyroids following i.v. injection. It remains, however, questionable whether those effects are due to TSH or to the simultaneous release of GH into the medium.

*Peripheral control*

As discussed earlier, TRH is a potent GH secretagogue in young chickens (4). It is however less potent in anaesthetized adults (01.-10 μg/kg) (55) and has no effect on GH release in conscious unrestrained fully fed adult chickens (56). On the other hand, similar doses of TRH in chickens deprived of food increase concentrations of plasma $T_3$ (19) and plasma GH (10). Age differences are also observed in the ability of $T_3$ and somatostatin (49) to suppress a TRH-induced GH release, with somatostatin being more effective in adults than young chicks. It is possible that the decrease in plasma $T_3$ during starvation (19) or severe protein restriction (50) removes somatostatin inhibition on GH release and therefore promotes TRH-induced GH release.

Injection of oGH increases concentrations of plasma $T_3$ and decreases $T_4$ plasma in fed and fasted chickens, whereas prolactin is ineffective in this regard (57). This effect has initially been explained by an activation of the liver 5′D activity, since hepatic $T_3$ recovery from $T_4$ is increased 1 and 2 h after injection of oGH. Careful analysis of hepatic monodeiodination activity reveals that in the microsomal fraction, 5′D-I activity is dominant, though less in adults than in growing chickens, and that GH injection does not affect this activity and therefore does not stimulate the conversion *in vitro* of $T_4$ into $T_3$. However, GH decreases 5D-III activity indicating that GH inhibits $T_3$ degradation in adults as it does in embryos and growing chicks (29, 41).

Circulating concentrations of $T_3$ control TRH- or GRH-stimulated GH secretion in young and adult chickens (10). Our results indicate that $T_3$ also controls the release of TSH. A summary of the mechanisms involved in the control of thyroid function in adult chickens is shown in Figure 8.

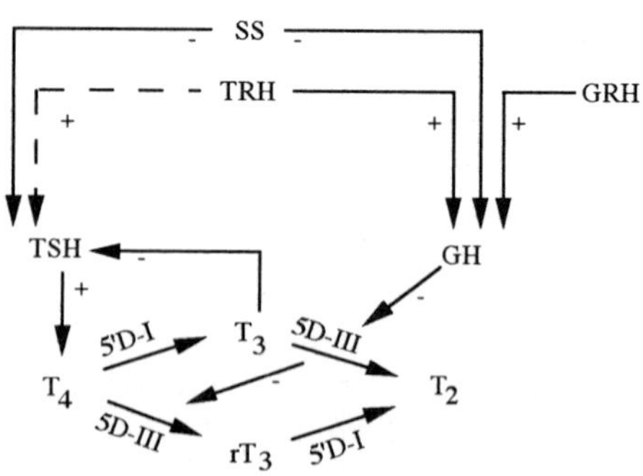


Fig. 8 Mechanisms controlling thyroid function in the adult chicken. Some thyrotrophic TSH-releasing activity may be present at high doses of TRH. SS, somatostatin; TRH, thyrotrophin releasing hormone; GRH, growth hormone releasing-hormone; $rT_3$, reverse $T_3$; $T_4$, thyroxine; $T_3$, triiodothyronine; $T_2$, diiodothyronine; 5′D-I, outer ring type I deiodinase; 5 D-III, inner ring type III deiodinase.

## Conclusion

During the entire life cycle of the chicken, the thyroid gland is responsive to TSH and mainly releases $T_4$ following stimulation. The mechanisms controlling the release of TSH and the peripheral production of $T_3$ change with age.

In the chick embryo the release of TSH may be controlled by CRH and TRH, resulting in increased concentrations of plasma $T_3$ and $T_4$. Growth hormone which may be released by TRH or GRH inhibits hepatic type III deiodination resulting in increased concentrations of plasma $T_3$ without altering plasma $T_4$.

Concentrations of plasma $T_3$ increase at the end of incubation and remain elevated for 2 weeks following hatching. During this period concentrations of plasma GH are high, while hepatic GH receptor availability is low, probably due to internalization of the GH-receptor complex. An injection of GH during this period, therefore, does not affect concentration of plasma $T_3$, but injection of anti-GH serum or hypophysectomy results in a decrease in plasma $T_3$. Type III deiodinase activity is very low in rapidly growing chickens but is increased after hypophysectomy, an effect which is reversed by treatment with GH. This suggests that in rapidly growing chickens endogenous GH exerts maximal inhibition of $T_3$ degradation.

Finally, in adult chickens, GH injection increases concentrations of plasma $T_3$ by regulating hepatic type III, but not, *in vitro*, type I deiodination. However, TRH does not increase plasma concentrations of $T_4$, but probably controls peripheral concentrations of $T_3$ by releasing GH, but not TSH. Concentrations of plasma $T_4$ as well as glycoprotein hormone $\alpha$ subunit immunoreactivity, may even be decreased following TRH injection, indicating that TSH release is controlled by circulating concentrations of $T_3$, and not directly by TRH.

## References

1 Kühn E (1990) Hormonal control of peripheral monodeiodination in vertebrates. *Progress in Comparative Endocrinology* pp 421-426. Eds A Epple, C G Scanes & M H Stetson. New York: Wiley-Liss

2 Peeters R, Buys N, Vanmontfort D, Van Isterdael J, Decuypere E & Kühn E R (1992) Preferential release of tri-iodothyronine following stimulation by thyrotrophin or thyrotrophin-releasing hormone in sheep of different ages. *Journal of Endocrinology* **132** 93-100

3 Fink G, Koch Y & Nurit Ben Aroya (1983) TRH in hypophyseal portal blood: characteristics of release and relationship to thyrotropin and prolactin secretion. *Thyrotropin-Releasing Hormone* pp 127-143. Eds E C Griffiths & G W Bennett. New York: Raven Press

4 Harvey S, Scanes C G, Chadwick A & Bolton N J (1978) The effect of thyrotropin-releasing hormone (TRH) and somatostatin (GHIH) on growth hormone and prolactin secretion *in vitro* and *in vivo* in the domestic fowl (*Gallus domesticus*). *Neuroendocrinology* **26** 249-260

5 Harvey S, Scanes C G & Klandorf H (1988) Thyrotrophin-releasing hormone induces growth hormone secretion in adult hypothyroid fowl. *General and Comparative Endocrinology* **69** 233-237

6 Kühn E R, Jacobs G & Vandorpe G (1990) Thyroid function and a possible thyroidal-gonadal interaction in reproduction and circannual rhythmicity. *Biology and Physiology of Amphibians* pp 3-27. Ed W Hanke. Stuttgart: Gustav Fischer Verlag

7 Wigham T & Batten T F C (1984) *In vitro* effects of thyrotropin-releasing hormone and somatostatin on prolactin and growth hormone release by the pituitary of *Poecilia latipinna. General and Comparative Endocrinology* **55** 444-449

8 Kühn E R, Vanderpooten A, Huybrechts L M, Decuypere E, Darras V & Sharp P J (1988) Hypothalamic hormones that release growth hormone stimulate hepatic 5′-monodeiodination activity in the chick embryo. *Journal of Endocrinology* **118** 233-236

9 Berghman L R, Darras V M, Huybrechts L M, Decuypere E, Vandesande F & Kühn E R (1988) Evidence for chicken GH as the only hypophyseal factor responsible for the stimulation of the hepatic 5′-monodeiodination activity in the chick embryo. *Reproduction, Nutrition, Development* **29** 197-202

10 Kühn E R, Herremans M, Dewil E, Vanderpooten A, Rudas P, Bartha T, Verheyen G, Berghman L & Decuypere E (1991) Thyrotropin-releasing hormone (TRH) is not thyrotropic but somatotropic in fed and starved adult chickens. *Reproduction, Nutrition, Development* **31** 431-439

11 Harvey S (1990) Thyroidal inhibition of growth hormone secretion: negative feedback? *Endocrinology of Birds: Molecular to Behavioral* pp 111-127. Eds M Wada, S Ishii & C G Scanes. Tokyo: Japan Scientific Societies Press. Berlin: Springer-Verlag

12 Thommes R C (1987) Ontogenesis of thyroid function and regulation in the developing chick embryo. *The Journal of Experimental Zoology* Suppl. **1** 273-279

13 Kühn E R, Decuypere E & Rudas P (1984) Hormonal and environmental interactions on thyroid function in the chick embryo and posthatching chicken. *Journal of Experimental Zoology* **232** 653-658

14 Darras V M, Visser T J, Berghman L R & Kühn E R (1992) Ontogeny of type I and type III deiodinase activity in embryonic and posthatch chicks: relationship with changes in plasma triiodothyronine and growth hormone levels. *Comparative Biochemistry and Physiology* **103A** 131-136

15 Decuypere E & Scanes C G (1983) Variation in the release of thyroxine, triiodothyronine and growth hormone in response to thyrotrophin releasing hormone during development of the domestic fowl. *Acta Endocrinologica* **102** 220-223

16 Harvey S, Davison T F & Chadwick A (1979) Ontogeny of growth hormone and prolactin secretion in the domestic fowl (*Gallus domesticus*). *General and Comparative Endocrinology* **39** 270-273

17 Darras V M, Huybrechts L M, Berghman L, Kühn E R & Decuypere E (1990) Ontogeny of the effect of purified chicken growth hormone on the liver 5′ monodeiodination activity in the chicken: reversal of the activity after hatching. *General and Comparative Endocrinology* **77** 212-220

18 Decuypere E, Nouwen E J, Kühn E R, Geers R & Michels H (1979) Differences in serum iodohormone concentration between chick embryos with and without the bill in the air chamber at different incubation temperatures. *General and Comparative Endocrinology* **37** 264-267

19 Kühn E R, Decuypere E, Iqbal A, Luysterborgh D & Michielsen R (1988) Thyrotropic and peripheral activities of thyrotrophin and thyrotrophin-releasing hormone in the chick embryo and adult chicken. *Hormone and Metabolic Research* **20** 158-162

20 Berghman L R, Proudman J A, Kempen S, Decuypere E, Kühn E R & Vandesande F (1992) Immunochemistry of chicken luteinizing hormone using homologous monoclonal antibodies with different specificities. Abstract P1 *Proceedings of the Fifth International Symposium on Avian Endocrinology* September 1992. Edinburgh: AFRC Roslin Institute

21 Marie C (1981) Ontogenesis of the adrenal glucocorticoids and of the target function of the enzymatic tyrosine transaminase activity in the chick embryo. *Journal of Endocrinology* **90** 193-200

22 Meeuwis R, Michielsen R, Decuypere E & Kühn E R (1989) Thyrotropic activity of the ovine corticotropin-releasing factor in the chick embryo. *General and Comparative Endocrinology* **76** 357-363

23 Decuypere E, Scanes C G & Kühn E R (1983) Effects of glucocorticoids on circulating concentrations of thyroxine ($T_4$) and triiodothyronine ($T_3$) and on peripheral monodeiodination in pre- and post-hatching chickens. *Hormone and Metabolic Research* **15** 233-236

24 Decuypere E, Dewil E & Kühn E R (1991) The hatching process and the role of hormones. *British Poultry Science Symposium* **22** 239-256. Ed. S G Tullett. London: Butterworth-Heinemann

25 Kühn E R, Decuypere E, Huybrechts L M & Darras V M (1990) Hormonal control of the hepatic 5′-monodeiodination activity and its relation to the somatomedin production in the chicken. *Endocrinology of Birds: Molecular to Behavioral* pp 129-142. Eds M Wada, S Ishii & C G Scanes. Tokyo: Japan Scientific Societies Press. Berlin: Springer-Verlag

26 Galton V A & Hiebert A (1989) The ontogeny of the enzyme system for the 5′- and 5 deiodination of thyroid hormones in chick embryo liver. *Endocrinology* **120** 2604-2610

27 McNabb F M A, Lyons L J & Hughes T E (1986) Avian hepatic $T_3$ generation by 5′-monodeiodination: characterization of two enzymatic pathways and the effects of goitrogens. *Comparative Biochemistry and Physiology* **85A** 249-255

28 Hughes T F & McNabb F M A (1986) Avian hepatic $T_3$ production by two pathways of 5′-monodeiodination: effects of fasting and patterns during development. *Journal of Experimental Zoology* **238** 393-399

29 Darras V M, Berghman L R, Vanderpooten A & Kühn E R (1992) Growth hormone acutely decreases type III iodothyronine deiodinase in chicken liver. *Federation of European Biochemical Societies* **300** 5-8

30 Vanderpooten A, Huybrechts L M, Decuypere E & Kühn E R (1991) Differences in hepatic growth hormone receptor binding during development of normal and dwarf chickens. *Reproduction, Nutrition, Development* **31** 47-55

31 Vanderpooten A, Darras V M, Huybrechts L M, Rudas P, Decuypere E & Kühn E R (1991b) Effect of hypophysectomy and acute administration of growth hormone (GH) on GH-receptor binding in chick liver membranes. *Journal of Endocrinology* **129** 275-281

32 Husman B, Gustafsson J A & Andersson G (1988) Receptor-mediated endocytosis and degradation of bovine growth hormone in rat liver. *Molecular and Cellular Endocrinology* **59** 13-25

33 Kühn E R, Decuypere E, Colen L M & Michels H (1982) Posthatch growth and development of a circadian rhythm for thyroid hormones in chicks incubated at different temperatures. *Poultry Science* **61** 540-549

34 Breneman W R & Rathkamp W (1973) Release of thyroid stimulating hormone from chick anterior pituitary glands by thyrotropin releasing hormone (TRH). *Biochemical and Biophysical Research Communications* **52(1)** 189-194

35 Newcomer W S & Huang F S (1974) Thyrotropin-releasing hormone in chicks. *Endocrinology* **95** 318-320

36 Kühn E R & Nouwen E J (1978) Serum levels of triiodothyronine and thyroxine in the domestic fowl following mild cold exposure and injection of synthetic thyrotropin-releasing hormone. *General and Comparative Endocrinology* **34** 336-342

37 Michels H, Decuypere E, Huybrechts L & Kühn E R (1986) Endocrinological effects of the sex-linked dwarf gene: Iv TRH sensitivity during growth. *Proceedings of the 7th European Poultry Conference* 24-28 August 1986 pp 970-974. France: Paris

38 Williamson R A & Davison T F (1985) The effect of a single injection of thyrotrophin on serum concentrations of thyroxine, triiodothyronine, and reverse triiodothyronine in the immature chicken (*Gallus domesticus*). *General and Comparative Endocrinology* **58** 109-113

39 Premachandra B N, Lang S, Andrada J A & Kite J H Jr (1977) Reverse triiodothyronine in the chicken. *Life Sciences* **21** 205-212

40 Darras V M, Vanderpooten A, Huybrechts L M, Berghman L R, Dewil E, Decuypere E & Kühn E R (1991) Food intake after hatching inhibits the growth hormone induced stimulation of the thyroxine to triiodothyronine conversion in the chicken. *Hormone and Metabolic Research* **23** 469-472

41 Darras V M, Rudas P, Visser T J, Hall T R, Huybrechts L M, Vanderpooten A, Berghman L R, Decuypere E & Kühn E R (1993) Endogenous growth hormone controls high plasma levels of 3,3′,5′-triiodothyronine ($T_3$) in growing chickens by decreasing the $T_3$-degrading type III deiodinase activity. *Domestic Animal Endocrinology* **10** 55-65

42 Kaplan M M & Utiger R D (1978) Iodothyronine metabolism in liver and kidney homogenates from hyperthyroid and hypothyroid rats. *Endocrinology* **103** 156-151

43 Kaplan M M (1979) Changes in the particulate cellular component of hepatic $T_4$ 5′ monodeiodinase in hyperthyroid and hypothyroid rat. *Endocrinology* **105** 548-554

44 Berry M J, Kates A-L & Larsen P R (1990) Thyroid hormone regulates type I deiodinase messenger RNA in rat liver. *Molecular Endocrinology* **4** 743-748

45 Parès-Herbuté N & Astier H (1985) The relative potencies of L-$T_4$ and L-$T_3$ on the pituitary-thyroid axes in the Japanese quail. *General and Comparative Endocrinology* **60** 298-305

46 Harvey S & Baidwan J S (1990) Thyroidal inhibition of growth hormone secretion in fowl: tri-iodothyronine-induced down-regulation of thyrotrophin-releasing hormone-binding sites on pituitary membranes. *Journal of Molecular Endocrinology* **4** 127-134

47 Harvey S, Decuypere E, Darras V M & Berghman L (1991) Differential effects of $T_4$ and $T_3$ on TRH- and GRF- induced GH secretion in the domestic fowl. *Reproduction, Nutrition, Development* **31** 451-460

48 King D B & Scanes C G (1986) Effect of mammalian growth hormone and prolactin on the growth of hypophysectomized chickens. *Proceedings of the Society for Experimental Biology and Medicine* **182** 201-207

49 Scanes C G, Aramburo C & Campbell R M (1990) Hormonal involvement in avian growth and development: growth hormone and insulin-like growth factor I. *Endocrinology of Birds: Molecular to Behavioral* pp 93-110. Eds M Wada, S Ishii & C G Scanes. Tokyo: Japan Scientific Societies Press. Berlin: Springer-Verlag

50 Lauterio T J & Scanes C G (1988) The role of thyroid hormones in the growth hormone response to protein restriction in the domestic fowl (*Gallus domesticus*). *Journal of Endocrinology* **117** 223-228

51 Harvey S & Baidwan J S (1989) Thyrotrophin-releasing hormone (TRH)-induced growth hormone secretion in fowl: binding of TRH to pituitary membranes. *Journal of Molecular Endocrinology* **3** 23-32

52 Klandorf H, Sharp P J & Sterling R (1978) Induction of thyroxine and triiodothyronine release by thyrotrophin-releasing hormone in the hen. *General and Comparative Endocrinology* **34** 377-379

53 Abdel-Fattah, Kamal I, Bobek S & Sechman A (1991) Serum pattern of thyroxine ($T_4$), 3,3′,5-triiodothyronine ($T_3$) and 3,3′,5′-triiodothyronine ($rT_3$) in fed and fasted cocks following TRH stimulation. *Journal of Veterinary Medicine* **38** 401-408

54 Scanes C G (1974) Some *in vitro* effects of synthetic thyrotrophin releasing factor on the secretion of thyroid stimulating hormone from the anterior pituitary gland of the domestic fowl. *Neuroendocrinology* **15** 1-9

55 Harvey S & Scanes C G (1984) Comparative stimulation of growth hormone secretion in anaesthetized chickens by human pancreatic growth hormone releasing factor (hpGRF) and thyrotrophin releasing factor (TRH). *Neuroendocrinology* **39** 314-320

56 Harvey S, Sterling R J & Phillips J G (1981) Diminution of thyrotrophin releasing hormone-induced growth hormone secretion in adult domestic fowl (*Gallus domesticus*). *Journal of Endocrinology* **89** 405-410

57 Kühn E R, Verheyen G, Chiasson R B, Huts C, Huybrechts L, Van den Steen P & Decuypere E (1987) Growth hormone stimulates the peripheral conversion of thyroxine into triiodothyronine by increasing the liver 5′-monodeiodinase activity in the fasted and normal fed chicken. *Hormone and Metabolic Research* **19** 304-308

Avian Endocrinology
Ed. P J Sharp, pp 47-60
Journal of Endocrinology Ltd, Bristol (1993)

# Neuroendocrine and neural changes associated with the photoperiodic control of reproduction

T S Juss
AFRC Research Group on Photoperiodism and Reproduction, Department of Biological Sciences, University of Bristol, Bristol BS8 1UG, UK

## Introduction

Breeding in most animals is restricted to the spring and summer so that 'ultimate factors' such as favourable conditions of temperature and food availability may be exploited to ensure a maximal chance of survival of the offspring. In temperate regions, it is the seasonal change in daylength or photoperiod which acts as a 'proximate factor' and determines the period of reproductive activity. Birds are highly photoperiodic and gonadal development occurs in response to increasing daylengths of spring. A phenomenon of further adaptive value is 'photorefractoriness' where reproductive activity switches off well in advance of a change in season/photoperiod in anticipation of a deterioration in environmental conditions. This is illustrated in Figure 1 which shows the stimulation followed by regression of testicular size in the European starling after transfer from 8h light:16h dark (8L:16D) to 18L:6D. Photosensitivity is not regained until the birds have experienced a period of short days. The notion that photostimulation and refractoriness may be two separate processes is supported by the observation that the latter is absent in 11L and that the former proceeds to a greater extent in 13L (significantly greater testicular growth attained than in 18L). These gonadal changes are driven by changes in pituitary gonadotrophin secretion (luteinizing hormone: LH, and follicle stimulating hormone: FSH). The secretion of prolactin, which has a role in incubation behaviour, is also highly photoperiodic. Note also in Figure 1 that photostimulation of these pituitary hormones is a graded response dependent upon the length of the day.

During photostimulation light is detected by extraretinal photoreceptors in the hypothalamus and this information is conveyed to a clock which measures the length of day accurately. The physiological nature of the clock remains unresolved, but functionally it involves a phase of photoinducibility which is controlled by the circadian system (1).

## Photoperiodic time measurement and an apparent lack of a role for melatonin

In mammals, seasonal changes in reproduction occur in response to seasonal changes in the duration of the night-time rise in melatonin secretion from the pineal gland. The rhythm in melatonin secretion is driven by a circadian clock in the suprachiasmatic nuclei (SCN) and its duration is truncated by light to reflect precisely the length of the scotophase. In birds, melatonin secretion also provides a calendar of seasonal photoperiodic information but they do not employ it to regulate reproduction (2, 3).

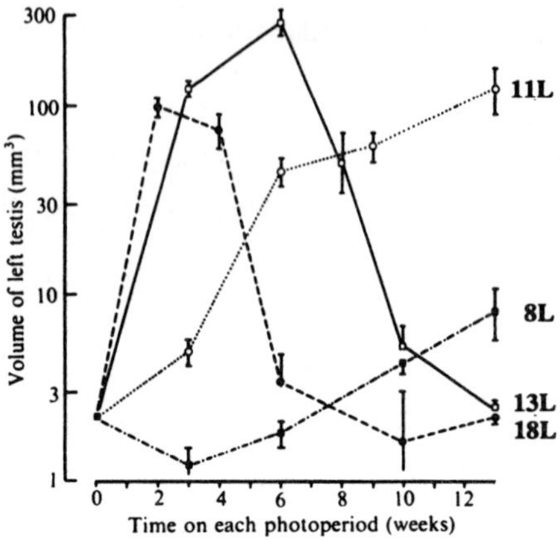


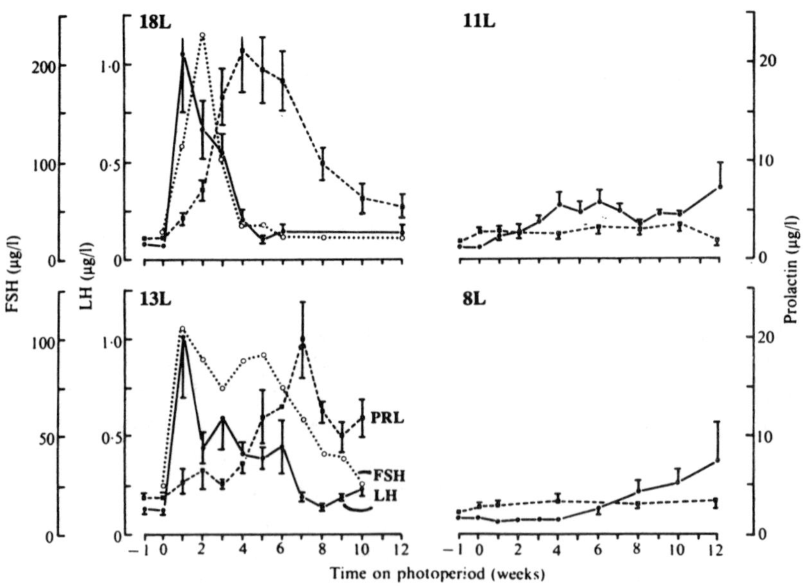


Fig. 1 Changes in gonadal volume and hormone secretion in male European starlings after transfer from 8h light:16h darkness (8L:16D) to 18L:6D (n=7), 13L:11D (n=8), 11L:13D (n=8) and 8L:16D (n=6, controls). Each point represents the mean ± s.e.

Top graph: Changes in volume of the left testis (note log scale). Bottom: Changes in LH, FSH and prolactin. Data from (14).

A recent experiment in male quail, mimicking the approach of Tamarkin *et al* (4) in the hamster, further reinforces this view. The duration of endogenous melatonin secretion was extended by two daily injections of melatonin (10 μg/bird) either around dusk (2h before, and at dusk) or dawn (2h before, and at dawn). To maximise the chances of obtaining a photoperiodic effect, birds were exposed to the critical daylength (12L) and daylengths just below and above (11L and 13L). In one experiment where melatonin was administered at dawn for 17 days (each group n=8), there was no effect on testicular growth in 11 and 13L. Because testicular growth was not prevented by the additional melatonin in the stimulatory photoperiods, as would have been predicted in the hamster, it was concluded that the melatonin duration hypothesis does not apply to quail. Paradoxically, there was a subtle but significant stimulation of testis mass in 12L (125±45 mg saline treated and 410±142 mg melatonin treated, *t*-test $P<0.05$), which may be the result of a pharmacological effect of melatonin on the position of the photoinducible phase. An advance of ~30 min in the photoinducible phase would account for this effect since it is known that in quail a shift from 12 to 12.5L causes a 30% increase in gonadal growth rate.

Data from these types of experiments lead to the hypothesis that in birds the readout of photoperiodic information is obtained via a neural link directly from the central clock, rather than from the melatonin rhythm which is merely a 'limb' of the clock.

**Hypothalamic releasing factors and refractoriness**

Environmental photoperiodic information processed into a long day (LD) or short day (SD) stimulus must finally impinge upon the neuroendocrine components of the reproductive system. In the case of gonadotrophin releasing hormone (GnRH), this is a group of about 600 cells located primarily in the preoptic and septal areas of the hypothalamus. In a parallel fashion, information for the control of prolactin secretion would be expected to target the vasoactive intestinal peptide (VIP) cells mainly situated in the infundibular nuclear complex (INC) of the mediobasal hypothalamus, since VIP is likely to be a prolactin releasing factor in birds (5).

In starlings, there are marked changes in the GnRH system that mirror the extent of reproductive activation. Using the techniques of brain immunocytochemistry and radioimmunoassay, Goldsmith *et al* (6) found that after exposure to LD, sexual maturation was associated with an increase in GnRH hypothalamic content as well as in cell size and density of staining. During refractoriness and following testicular collapse, there was a dramatic reduction in all these parameters such that GnRH cells became hardly visible. The correlation of testicular development and GnRH-I content is not a generalized phenomenon since it does not occur in quail (7) and visible changes in immunocytochemical staining for GnRH-I and GnRH-II are also absent (8). Unlike GnRH-I, in the starling, GnRH-II content in various brain areas remains constant during the course of photostimulation and refractoriness (A Goldsmith and P J Sharp, unpublished), and since GnRH-II cannot be detected immunocytochemically in the median eminence (domestic fowl and quail (9), starling and quail (8)) it has been suggested that GnRH-II is not involved in reproduction. Recent work in the domestic hen, however, shows that in addition to GnRH-I there are changes in brain GnRH-II content during the progesterone induced surge and diurnal rhythm of plasma LH (10, 11), and given the absence of a GnRH-II input to the median eminence it is possible that control of LH secretion by GnRH-II occurs via an interaction with the GnRH-I neural circuit. In support of this view we have found an intermingling of GnRH-I and GnRH-II fibres in the quail septal region and an extensive network of GnRH-II fibres in the lateral hypothalamic area (8).

As in the case of GnRH and LH control in the starling, there are changes in brain VIP immunoreactivity (12) and content (5) in the turkey during photostimulation and the hyperprolactinaemic state of incubation. During photostimulation, a rise in plasma prolactin was associated with the appearance of VIP immunoreactive cells in the INC, an increase in fibre intensity in the median eminence, and a doubling of hypothalamic VIP content, and further increases in immunostaining were found in the incubating bird.

Such studies of morphology and endocrine function have helped in defining the processes involved in LD stimulation and refractoriness. One idea put forward in the past has been that the GnRH cell itself could be the site of refractoriness resulting from the exhaustive depletion of GnRH in the face of constant stimulation of release. The data of Goldsmith *et al* (6) would not support this idea since GnRH immunoreactivity remained high at the onset of testicular collapse and diminished markedly only when the regression was complete suggesting that the depletion of immunoreactivity was a consequence rather than a cause of refractoriness. Parry and Goldsmith (13) have combined immunocytochemistry with electron microscopy to look for changes at a higher level outside the GnRH cell, and found a reduction in glial ensheathment and an increase in synaptic thickening around GnRH cells in deeply refractory starlings indicating an increased neural input. However, because once again these changes were absent during testicular regression the possibility remains that they are a consequence of refractoriness.

Another endocrine candidate for the cause of refractoriness is prolactin. Exogenous prolactin has long been known to inhibit LH secretion, probably at the level of the GnRH cell, and it has been shown in both wild and captive starlings (14) that the onset of photorefractoriness is associated with a marked increase in circulating levels of prolactin. Note how in Figure 1 the delay in refractoriness in starlings switched from 8 to 13L compared to 8 to 18L, is accompanied by a corresponding delay in the peak of prolactin secretion. To test the hypothesis that refractoriness is the result of increased prolactin secretion, intact and castrated male starlings with sustained reproductive stimulation on 11L were infused daily with a dose of intracerebroventricular ovine prolactin (1μg prolactin/2 μl vehicle) for two weeks. Within a few days plasma LH and FSH were reduced to SD values but upon termination of infusion and transfer to LD (18L), gonadotrophin levels rose rapidly. This indicates that the birds were still photosensitive and that they had failed to become refractory as a result of ovine prolactin administration (15).

A similar conclusion emerged from another study, where Japanese quail were transferred from 20L to 13 and 8L (24, Fig. 2). Quail maintained on LD do not spontaneously regress their gonads like starlings but they develop a 'relative refractory state'. This was first found by Robinson and Follett (16) who showed that when quail on LD experience a decrease in photoperiod to a daylength which is still longer than the critical daylength (and therefore strictly speaking is still a LD), reproduction switches off. By definition this is refractoriness because the quail have used the relatively small decrease in daylength as a prompt to *anticipate* the arrival of the season of short days and switched off reproduction even though the daylength is still stimulatory. Furthermore, if the initial drop in daylength is not followed by a progressive shortening of photoperiod as would happen in the natural daylengths of autumn, refractoriness eventually dissipates and the reproductive axis recrudesces in response to the daylength which is still longer than the critical daylength. The argument is further strengthened by the observation that the longer the quail are maintained on LD, deeper is the development of refractoriness, and the longer it takes for recrudescence to occur (17).

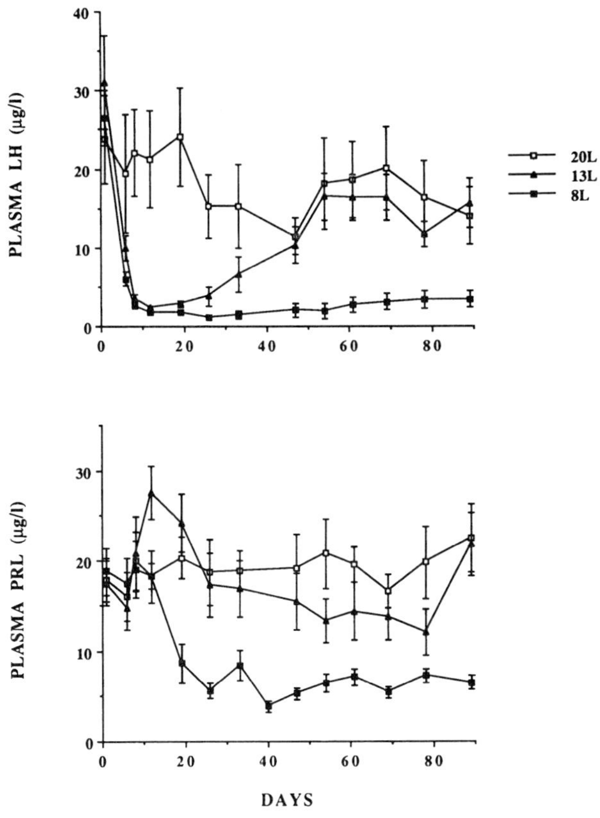


Fig. 2 Circulating LH and prolactin in gonadectomized Japanese quail after transfer from 20h light:4h darkness (20L:4D) to 8L:16D (n=10), 13L:11D (n=10), and 20L:4D (n=10, controls). Each point represents the mean±s.e.

Figure 2 shows that when quail on 20L were transferred to 13L, a photoperiod longer than the critical daylength, LH secretion fell sharply in exactly the same way as the 8L transfer, but then progressively rose to LD (20L) values; i.e. the LH profile showed relative refractoriness. There was no recrudescence on 8L. With respect to prolactin however, in the 13L transfer the prolactin response dissociated from the LH response because prolactin remained constantly high, and if prolactin secretion was responsible for the LH relative refractory profile through its ability to inhibit LH release, one might have expected to see an inverse relation between the prolactin and LH profile. This would suggest that prolactin is not the 'photorefractory agent'.

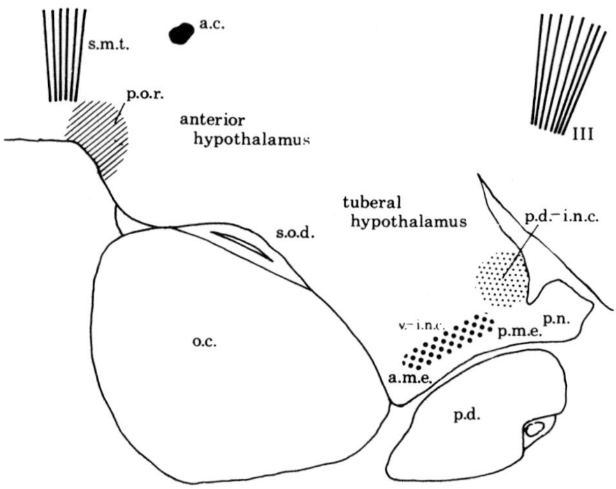


Fig. 3 A schematic diagram indicating regions in the hypothalamus of the Japanese quail where lesions block photoperiodically induced testicular or ovarian growth. These are the ventral (v.-i.n.c., large dots) or posterodorsal (p.d.-i.n.c., small dots) infundibular nuclear complex, and the preoptic region (p.o.r., dashed lines) of the anterior hypothalamus. Taken from Follett (18). a.c., anterior commissure; a.m.e., anterior median eminence; o.c., optic chiasma; p.d., pars distalis of the pituitary (adenohypophysis); p.m.e., posterior eminence; p.n., pars nervosa of the pituitary (neurohypophysis); s.m.t., septomesencephalic tract; s.o.d., supraoptic decussation; III, oculomotor nerve.

Nevertheless, the possibility cannot be ignored that even though prolactin may not be the cause of refractoriness, neural and neurochemical changes associated with LD and the onset of refractoriness may require elevated levels of prolactin in order to effect refractoriness.

**Involvement of other neural circuits**

The study of brain areas that may receive and convey information concerned with photostimulation of endocrine function has received great attention over a period of 30 years or so in several avian species, and a variety of techniques ranging from discrete local illumination to electrical stimulation/lesion and knife cut deafferentation has been employed. Two areas consistently highlighted throughout this period are the preoptic and tuberoinfundibular regions which when lesioned block photostimulated gonadal growth (18).

Local illumination of the tuberal hypothalamus using light guides, optic fibres or micropellets of radioluminous paint (duck (19), quail (20)) has induced maximal testicular growth and this would imply that extraretinal photoreception occurs at this site. Lesion work in quail has confirmed the requirement for the integrity of the preoptic (21) and tuberal areas (22, 23, Fig. 3). When this work was carried out, immunocytochemistry was not available and so the relation between lesions and the GnRH system was unknown. Two midline lesion sites were identified in the tuberal region. These were the ventral infundibular nuclear complex (v.inc.) which is close to and covers much of the antero-posterior extent of the median eminence, and a site only slightly more dorsal but located at the posterior-most extent of the tuberal structure - the posterodorsal infundibular nuclear complex (pd.inc.). Although the reason for the efficacy of the v.inc. lesions is still unknown, because of the proximity and spread of this site over the median eminence, it is possible that they damaged the GnRH input to the external layer. The more distal pd.inc. has a greater likelihood of being involved in a more fundamental role of processing photostimulatory information, but immunocytochemical studies (T S Juss, unpublished), show that in quail many diffuse GnRH fibres pass through the pd.inc. and the possibility that lesions here prevent photoinduction by preventing transport of GnRH to the median eminence has been re-investigated.

Quail maintained on 8L:16D were treated with radiofrequency lesions in the pd.inc. and then transferred to 18L:6D for 4-5 weeks. Figure 4 shows that the lesions had an all-or-none effect on reproductive activity. Whereas birds unaffected by the lesions showed the normal 5-fold increase in LH and cloacal gland growth after 2 weeks, birds reproductively blocked as a result of lesioning had basal LH and undeveloped cloacal glands throughout the experiment. There was a 240-fold difference in the terminal testicular weight between the two groups. Subsequent GnRH immunocytochemistry in the two groups revealed a distinct, but only minor difference in the site of lesion placement, implying that lesions in the posteroventral tuberal hypothalamus do not per se prevent photoinduced testicular development, but have to be very site-specific to be effective. On the other hand, it is known from previous immunocytochemical data that GnRH fibres passing through the pd.inc. do so in a diffuse, and not a discrete manner. From the point of view of the diffuse GnRH input, both sets of lesions would have been expected to transect the fibres to an equal extent.

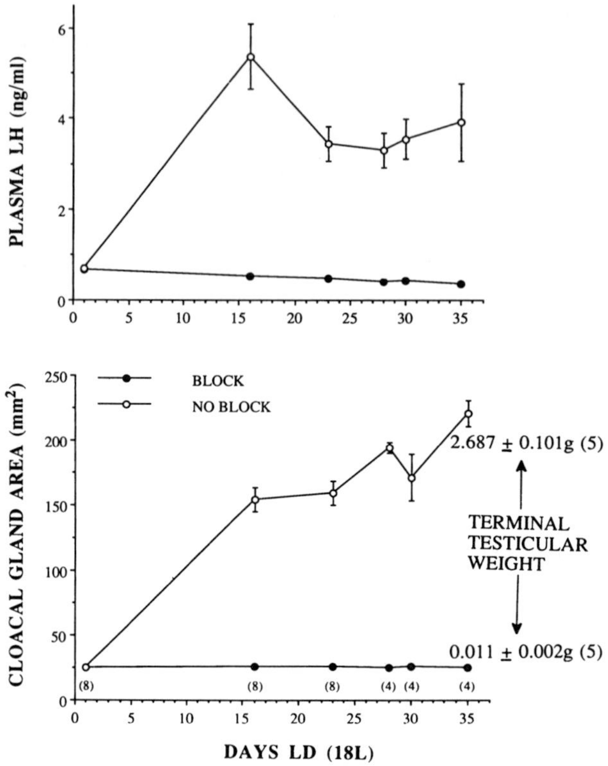


Fig. 4 Diagram showing the reproductive response (circulating LH, top graph; cloacal gland area, bottom graph) to long days in Japanese quail with radiofrequency lesions in the posterodorsal infundibular nuclear complex. Birds were either totally unresponsive to long days ('block' group) or showed a normal increase in reproductive activity ('no block' group). Each point represents the mean±s.e. Number of individuals for LH and cloacal gland measurements are given in brackets along the lower x-axis.

A thorough analysis of the GnRH immunostaining using the current criteria of image analysis confirmed that the GnRH systems in the two groups were similar. There was no GnRH deficit in the brains of blocked quail compared to the reproductively active quail at the level of the GnRH cell body. The total number of GnRH cell bodies in blocked quail (387±50) was not different from reproductively active birds (434±80). The cell areas of GnRH cells in the two groups were 102±6 and 98±5 μm sq. respectively and not significantly different. The optical density of GnRH immunoreactive product was the same in the perikarya and terminals in the median eminence in both groups. Figure 5 shows that staining in the median eminence was similar in one bird from each group throughout the rostro-caudal extent, indicating that transport of the releasing factor to the terminal site had not been interrupted. The conclusion from these data is that lesions in the pd.inc. do not block photoinduction of LH by interfering with the GnRH input to the median eminence, but that it has a more fundamental role as a 'release control centre' for GnRH. This control of release could be carried out by the pd.inc. as a photoreceptive site for daylength information, or as an area which projects more directly to the GnRH circuitry, perhaps to the level of the median eminence by means of axo-axonal contacts.

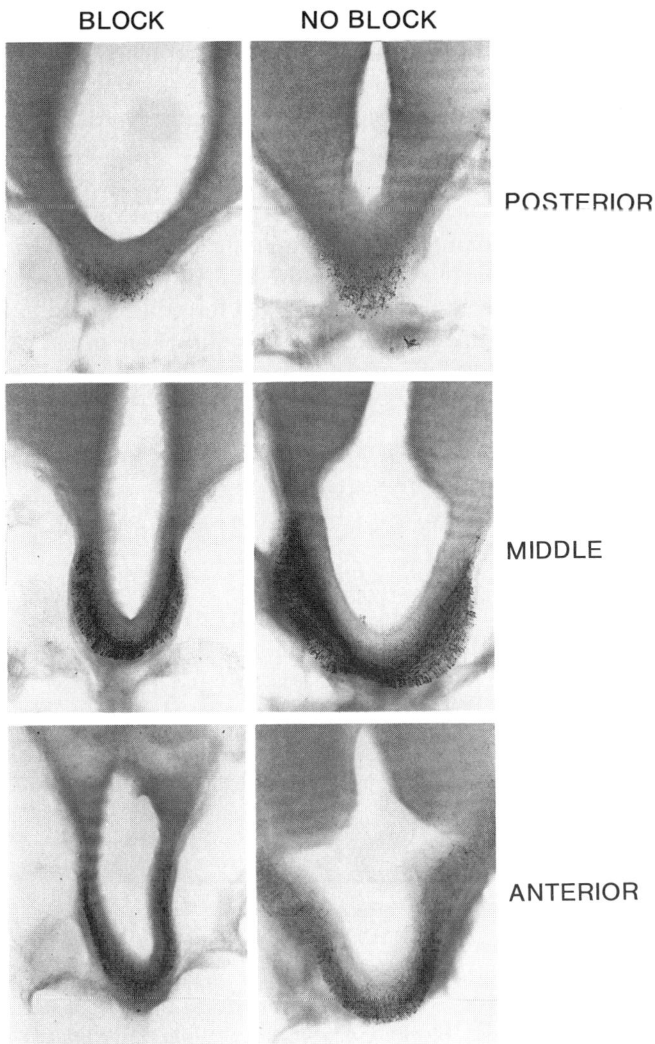


Fig. 5 Photomicrographs showing GnRH immunoreactivity in the anterior, middle and posterior part of the median eminence in one bird from the 'block' and 'no block' groups. The antero-posterior extent of staining in both birds appears to be the same.

To test whether the absence of photostimulation of reproductive activity in pd.inc. lesioned birds was due to the ablation of the tuberal photoreceptor implicated in the brain local illumination studies, the experiment was repeated to assess the function of the prolactin neuroendocrine mechanism in lesioned birds. Another batch of quail were lesioned and this also yielded a group of reproductively active and a group of reproductively blocked birds. As shown in Figure 6 the prolactin response to LD in both groups was the same; after the first 8 days in LD, plasma LH increased in the blocked birds from 8.5±0.4 (mean±s.e. n=7) to 18.1±2.4 ng/ml and in the non-blocked birds from 8.4±0.1 to 19.2±2.9 ng/ml. This indicated that blocked birds were able to read photoperiodic time in the same way as the non-blocked birds. The higher prolactin measurements taken later on in non-blocked birds are almost certainly due to testicular steroids, which would be minimal in the blocked birds having undeveloped gonads, since it is known that castrated quail have lower prolactin values compared to intact quail (24). The most conservative and obvious conclusion is that the pd.inc. is somehow involved in the GnRH (but not the prolactin) neurosecretory mechanism downstream of the daylength measuring apparatus, and is not involved in photoreception. However, it is possible that prolactin and GnRH pathways have separate photoperiodic systems and that the photoreceptors for prolactin are in the septal area (25), or the eye (26), while the photoreceptors for GnRH are in the pd.inc.

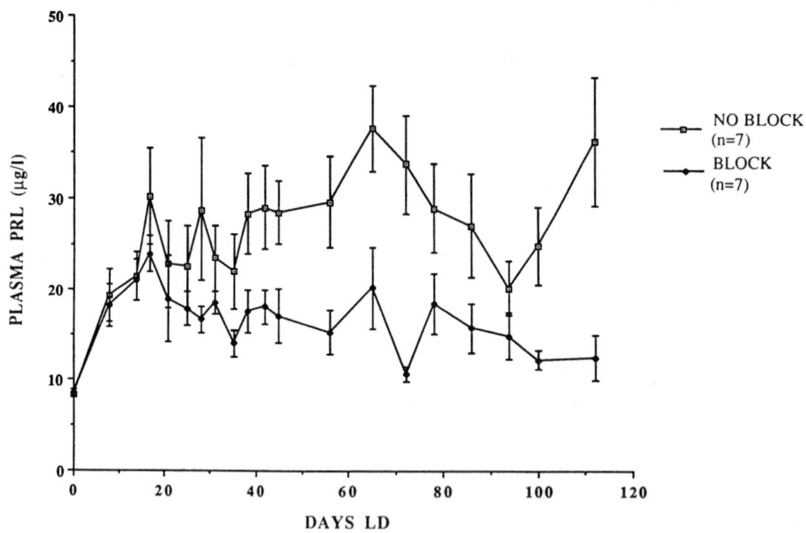


Fig. 6 Increase in circulating prolactin in response to long days in male Japanese quail treated with radiofrequency lesions in the posterodorsal infundibular nuclear complex. The lesioning resulted in two groups of birds; 'block' group where photoinduction of LH was completely blocked, and 'no block' group which showed a normal stimulation of LH and cloacal gland growth in long days.

Brain pathways for photoinduced prolactin secretion are not known, although partial involvement of the retino-hypothalamic-pineal tract is implied by the work of Siopes and El Halawani (26) who reported that enucleation and pinealectomy slowed the photoinduced rise of prolactin in yearling turkey hens and enucleation, but not pinealectomy, depressed prolactin in 12 week-old turkey males.

## Conclusion

The neural and neurochemical substrates for photoperiodic time measurement in birds are still a mystery. An advance in elucidating the locus of the deep brain photoreceptor came from the immunocytochemical detection of CSF contacting cells which contain an immunoreactive opsin-like material in the septal and tuberal regions of the ring dove brain (25), but physiological experiments to confirm this discovery have not yet been done. The structure and location of the avian homologue of the suprachiasmatic nucleus continues to be heavily debated. Investigators relying mainly upon cytoarchitecture have identified a nucleus in the medial hypothalamus as the SCN (e.g. 27) whereas those placing a greater emphasis on areas that receive an input from the retina favour a more lateral nucleus (e.g. 28). Consideration of the peptidergic content of the medial and lateral nuclei has not helped to resolve the issue since peptides found in the mammalian suprachiasmatic nucleus do not form an identifiable nucleus in birds. However, because of the effect of neuropeptide Y on circadian rhythms in mammals, Kuenzel and Blähser (29) consider a group of cells which surrounds and lies mainly on the medial border of the lateral anterior thalamic nucleus and which is immunoreactive for this peptide as a possible candidate for the avian suprachiasmatic nucleus. Whether the clock resides in the suprachiasmatic nucleus, and where the readout mechanisms of the clock lie, are separate unanswered questions. It is possible that thyroid hormones can affect the perception of daylength, because photorefractory starlings recover photosensitivity when thyroidectomized, and thyroxine treatment of refractory birds prevents recovery of photosensitivity (30), but the site and mechanism of action of thyroid hormones is unknown. The perennial question of the neural basis of photorefractoriness remains a feature of the study of photoperiodism.

## References

1 Hamner W H & Enright J T (1967) Relationship between photoperiodism and circadian rhythms of activity in the house finch. *Journal of Experimental Biology* **46** 43-61

2 Follett B K, Foster R G & Nicholls T J (1985) Photoperiodism in birds. *Photoperiodism, Melatonin and the Pineal* pp 93-105. (Ciba Foundation Symposium 117). London: Pitman

3 Kumar V, Juss T S & Follett B K (1992) Melatonin and the pineal gland melatonin secretion in quail provides a seasonal calendar but not one used for photoperiodic time measurement. *Melatonin and the Pineal Gland* pp 163-168. Eds Y Touitou, J Arendt and P Pévet. Amsterdam: Elsevier

4 Tamarkin L, Westrom W K, Hamill A I & Goldman B D (1976) Effect of melatonin on the reproductive systems of male and female hamsters: a diurnal rhythm in sensitivity to melatonin. *Endocrinology* **99** 1534-1541

5 Mauro L J, Youngren O M, Proudman J A, Phillips R E & El Halawani M E (1992) Effects of reproductive status, ovariectomy, and photoperiod on vasoactive intestinal peptide in the female turkey hypothalamus. *General and Comparative Endocrinology* **87** 481-493

6 Goldsmith A R, Ivings W E, Pearce-Kelly A S, Parry D M, Plowman G, Nicholls T J & Follett B K (1989) Photoperiodic control of the development of the LHRH neurosecretory system of European starlings (*Sturnus vulgaris*) during puberty and the onset of photorefractoriness. *Journal of Endocrinology* **122** 255-268

7 Creighton J A & Follett B K (1986) Changes in gonadotrophin-releasing hormone and LH in Japanese quail during the first few days of photostimulation. *Journal of Endocrinology* **113** 419-422

8 Juss T S, Ball G F & Parry D M (1992) Immunohistochemical localization of cGnRH I and cGnRH II in the brains of photosensitive and photorefractory European starlings and Japanese quail. Abstract P87 *Proceedings of the Fifth International Symposium on Avian Endocrinology* September 1992. Edinburgh: AFRC Roslin Institute

9 Mikami S, Yamada S, Hasegawa Y & Miyamoto K (1988) Localization of avian LHRH-immunoreactive neurons in the hypothalamus of the domestic fowl, *Gallus domesticus*, and the Japanese quail, *Coturnix coturnix*. *Cell and Tissue Research* **251** 51-58

10 Wilson S C, Chairil R A, Cunningham F J & Gladwell R T (1990) Changes in the hypothalamic contents of LHRH-I and -II and in pituitary responsiveness to synthetic chicken LHRH-I and -II during the progesterone-induced surge of LH in the laying hen. *Journal of Endocrinology* **127** 487-496

11 Wilson S C, Gladwell R T & Cunningham F J (1991) Diurnal changes in the plasma concentrations of LH and hypothalamic contents of LHRH-I and LHRH-II in the domestic hen. *Journal of Endocrinology* **130** 457-462

12 Mauro L J, Elde R P, Youngren O M, Phillips R E & El Halawani M E (1989) Alterations in hypothalamic vasoactive intestinal peptide-like immunoreactivity are associated with reproduction and prolactin release in the female turkey. *Endocrinology* **125** 1795-1804

13 Parry D M & Goldsmith A R (1992) Structural plasticity and synaptic input to the hypothalamic LHRH neurones in photosensitive and photorefractory starlings. Abstract P86 *Proceedings of the Fifth International Symposium on Avian Endocrinology* September 1992. Edinburgh: AFRC Roslin Institute

14 Dawson A & Goldsmith A R (1983) Plasma prolactin and gonadotrophins during gonadal development and the onset of photorefractoriness in male and female starlings (*Sturnus vulgaris*) on artificial photoperiods. *Journal of Endocrinology* **97** 253-260

15 Juss T S & Goldsmith A R (1992) Intracerebroventricular prolactin is potently gonadoinhibitory but does not induce photorefractoriness. Abstract P88 *Proceedings of the Fifth International Symposium on Avian Endocrinology* September 1992. Edinburgh: AFRC Roslin Institute

16 Robinson J E & Follett B K (1982) Photoperiodism in Japanese quail: the termination of seasonal breeding by photorefractoriness. *Proceedings of the Royal Society of London B* **215** 95-116

17 Follett B K & Pearce-Kelly A (1990) Photoperiodic control of the termination of reproduction in Japanese quail (*Coturnix coturnix japonica*) *Proceedings of the Royal Society of London B* **242** 225-230

18 Follett B K (1984) Birds. *Marshall's Physiology of Reproduction Vol. 1 Reproductive Cycles of Vertebrates* pp 283-350. Ed. G E Lamming. Edinburgh: Churchill Livingstone

19 Benoit J (1970) Étude de l'action des radiations visibles sur la gonadostimulation et leur pénétration intra-cranienne chez les oiseaux et les mammifères. *La photorégulation de la reproduction chez les oiseaux et les mammifères* pp 121-149. Eds J Benoit & I Assenmacher. Paris: CNRS

20 Oliver J & Baylé J D (1976) The involvement of the preoptico-suprachiasmatic region in the photosexual reflex of the quail: effects of selective lesions and photostimulation. *Journale de Physiologie Paris* **72** 627-637

21 Davies D T & Follett B K (1975) The neuroendocrine control of gonadotrophin release in the Japanese quail. II. The role of the anterior hypothalamus. *Proceedings of the Royal Society of London B* **191** 303-315

22 Sharp P J & Follett B K (1969) The effect of hypothalamic lesions on gonadotrophin release in Japanese quail, *Coturnix coturnix japonica. Neuroendocrinology* **5** 205-218

23 Davies D T & Follett B K (1975) The neuroendocrine control of gonadotrophin release in the Japanese quail. I. The role of the tuberal hypothalamus. *Proceedings of the Royal Society of London B* **191** 285-301

24 Juss T S & Sharp P J (1992) Description of photoperiodic and gonadal steoridal control of prolactin secretion in Japanese quail reveals a temporal dissocation between prolactin and luteinizing hormone secretion and hence photorefractoriness. Abstract P84 *Proceedings of the Fifth International Symposium on Avian Endocrinology* September 1992. Edinburgh: AFRC Roslin Institute

25 Silver R, Witkovsky P, Horvath P, Alones V, Barnstable C J & Lehman M N (1988) Co-expression of opsin- and VIP-like immunoreactivity in CSF-contacting neurons of the avian brain. *Cell and Tissue Research* **253** 189-198

26 Siopes T D & El Halawani M E (1989) Pineal and ocular influences on male and female turkeys; plasma luteinizing hormone and prolactin levels during gonadal development. *General and Comparative Endocrinology* **73** 284-292

27 Panzica G C (1988) Vasotocin-immunoreactive elements and neuronal typology in the suprachiasmatic nucleus of the chicken and Japanese quail. *Cell and Tissue Research* **242** 371-376

28 Cassone V M & Moore R Y (1987) Retinohypothalamic projection and suprachiasmatic nucleus of the house sparrow, *Passer domesticus. Journal of Comparative Neurology* **266** 171-182

29 Kuenzel W J and Blähser S (1991) The distribution of gonadotropin-releasing hormone (GnRH) neurons and fibres throughout the chick brain (*Gallus domesticus*). *Cell and Tissue Research* **264** 481-495

30 Boulakoud M S, Ivings W E & Goldsmith A R (1991) Thyroxine treatment prevents the development of photosensitivity in European starlings (*Sturnus vulgaris*). *Journal of Comparative Physiology* **161** 516-520

# Part Two
# Environmental and Behavioural Interactions

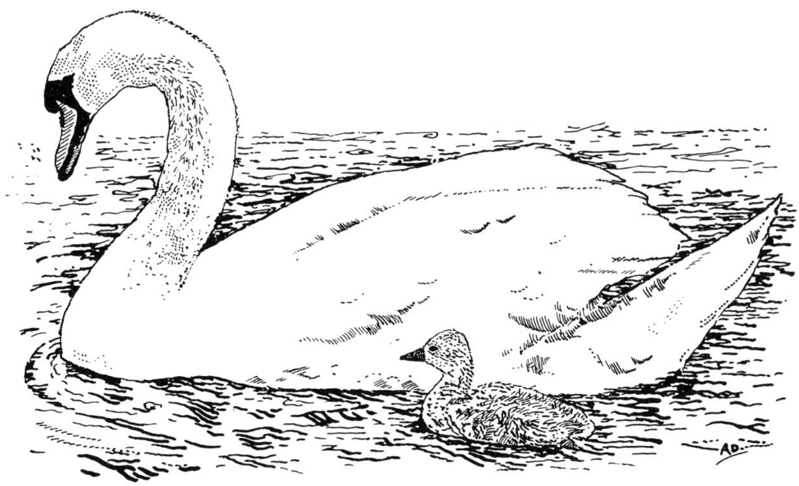

Avian Endocrinology
Ed. P J Sharp, pp 61-72
Journal of Endocrinology Ltd, Bristol (1993)

# Testosterone concentrations in males of an atypical species: the brood parasitic brown-headed cowbird

A M Dufty, Jr
Biology Department, Boise State University, 1910 University Drive, Boise, ID 83725, USA

## Introduction

The focus of this paper is on the relationship between concentrations of plasma testosterone and reproductive and accompanying social behaviours in the brown-headed cowbird (*Molothrus ater*). It will highlight some areas in which interesting comparative results have been obtained for cowbirds and other species, and will touch upon areas where potentially exciting work remains to be conducted.

The cowbird is notable for being an obligate brood parasite. Brood parasitism is an unusual avian breeding strategy characterized by the deposition of eggs in nests of heterospecifics and by the total absence of subsequent parental care by the parasite. Species with such unconventional breeding strategies provide 'natural experiments' that can be used to explore ideas about the role of hormones in the regulation of behaviour. That is, an understanding of the endocrine correlates of breeding and social behaviour in atypical species can provide insight into the role of hormones in birds in general. As outlined below, in some cases the hormonal profiles of parasitic and nonparasitic species are quite different, while in other cases their endocrine responses are strikingly similar. However, in all cases the physiological responses of the parasitic cowbird can be explained in terms of the behaviourial and ecological features that distinguish brood parasitism.

## Life history of the brown-headed cowbird

The brown-headed cowbird is a migratory passerine that breeds throughout the continental United States, as well as in southern Canada and northern Mexico. In spring, male cowbirds arrive on breeding grounds prior to females. Unlike males in most other passerine species, male cowbirds do not establish territories; rather, they develop dominance hierarchies through group displays and agonistic encounters. When females arrive, a dominant male pairs with the resident female and guards her from other males for the remainder of the breeding season. Males outnumber females in the adult population and unpaired males will court and copulate with an unguarded female if the dominant male is inattentive.

Female cowbirds in some populations defend large breeding territories, probably to ensure access to host nests (1). Parts of these areas may overlap with those of adjacent birds so females do not necessarily have exclusive access to all of the host nests in their territories (2). The breeding season lasts for approximately 10 weeks (May to mid-July). Females lay parasitic eggs throughout this period, and individual females may lay up to 50 eggs per season (3). This represents over three times their own body weight in egg production. Thus, cowbirds have adapted to their parasitic lifestyle by becoming highly proficient egg producers, leading to the sobriquet of 'passerine chicken' (4).

The extraordinary reproductive output by these birds is accompanied by the loss of all vestiges of parental behaviour. Nest building, incubation, and feeding young are all missing from the cowbird's behaviourial repertoire, and the attendant hormonal framework likewise is modified, as outlined below.

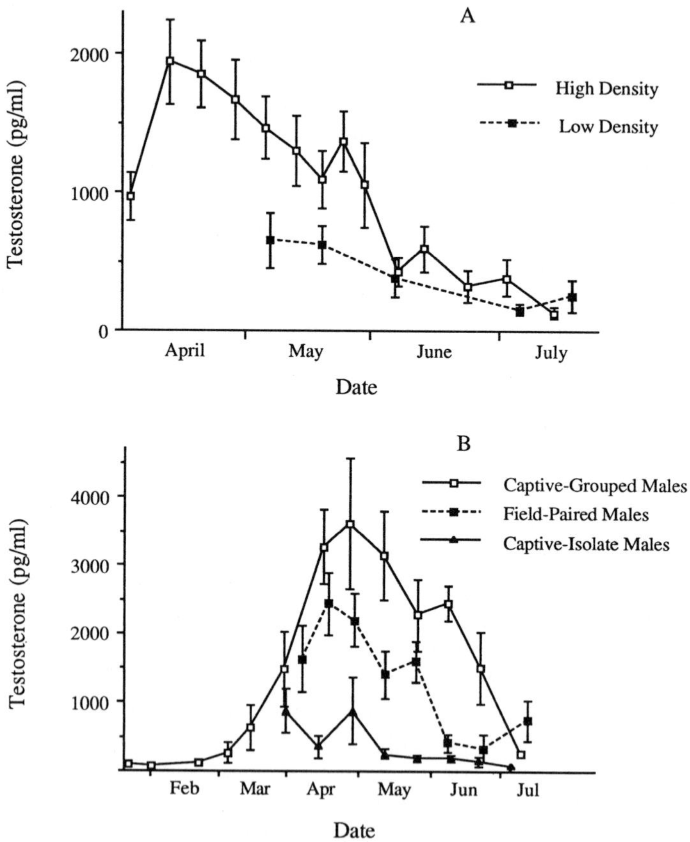


Fig. 1 Seasonal changes in concentrations of plasma testosterone in male brown-headed cowbirds (*Molothrus ater*) A) in populations with high and low densities (A M Dufty Jr, R C Fleischer, J Wingfield & S I Rothstein, unpublished observations); and B) in captivity and housed singly but with visual and vocal access to other males (open squares), or paired and free-living (closed squares) or in captivity and housed in visual and vocal isolation (closed triangles). From A M Dufty Jr (unpublished observations and 25).

## Seasonal testosterone patterns and brood parasitism

The past decade has seen publication of many investigations relating stage of the breeding cycle with patterns of testosterone secretion in free-living monogamous birds (5). From these studies a 'typical' pattern has emerged. Migratory males exhibit elevated concentrations of plasma testosterone when they arrive on breeding grounds and compete for territories and mates. Testosterone concentrations remain high until their mate is no longer receptive during egg laying, at which point testosterone levels decline, but not to basal values. Testosterone shows little or no increase in the event of a second clutch, although it will increase if the first clutch is destroyed or if the second clutch is accompanied by advertisement for a new mate. After the last clutch is completed, testosterone concentrations decrease to baseline and the birds prepare for moult. This basic pattern varies somewhat depending on the particular species, demographics, and environmental factors encountered (6).

The amplitude and extent of the testosterone peak at the beginning of the breeding season largely depends on two factors: the amount of male-male aggression in the population and the amount of parental care provided by the male (7). These two factors are related to testosterone, but in very different ways. First, male-male aggression is well known to be positively associated with testosterone (8). Recent studies indicate that this relationship is seen most clearly during periods of social instability, such as during territory procurement, establishment of dominance hierarchies, and acquisition and defence of mates. Once social systems stabilize, the relationship between male-male aggression and testosterone disappears (7). Second, male parental care has been shown to be incompatible with increased testosterone concentrations in plasma, and males reduce their testosterone concentrations when feeding young. Males given exogenous testosterone continue to be aggressive and to court females, but they provide no subsequent parental care, resulting in a high degree of nest failure (9). Thus, there is a cost to sustained elevated testosterone concentrations in nonparasitic species in terms of reproductive success.

Cowbirds take male-male aggression and parental care to opposite extremes. On the one hand, the extended breeding season exerts a selective force on males to maintain increased concentrations of plasma testosterone in order to remain sexually active, to court and defend females, and to increase their own reproductive success. The result is a prolonged period of agonistic interactions among males.

On the other hand, the total absence of parental care in cowbirds removes a potential constraint on the maintenance of increased plasma testosterone. Consequently, male cowbirds, unlike males in other apparently monogamous passerine species, exhibit elevated testosterone levels throughout most of the reproductive period (10 and Fig. 1A). Concentrations of plasma testosterone decline significantly only in the latter stages of breeding.

Interestingly, a nonparasitic passerine species that demonstrates a testosterone profile similar to that of the cowbird is the red-winged blackbird (*Agelaius phoeniceus*) (11). These birds are ecologically similar to cowbirds in many respects. For example, the males have access to a fertile mate throughout much of the breeding season. This is not a single female, as in cowbirds. Instead, red-winged blackbirds are polygynous, and at any given time at least one of several females may be receptive and must be guarded from other males. Similarly to cowbirds, red-winged blackbird males do not feed their young (at least in populations in western North America), and unpaired males solicit females. It is not surprising, therefore, that the pattern of testosterone secretion in male redwing blackbirds is similar to that of cowbirds.

Even apparently monogamous, nonparasitic birds like song sparrows (*Melospiza melodia*) have the capacity to exhibit 'cowbird-like' hormonal responses. Concentrations of plasma testosterone in male song sparrows typically decline after clutch completion when females incubate and are no longer sexually receptive, at a time when the birds prepare to feed young. Concentrations of plasma testosterone do not increase again when females prepare for their second clutch because males are still feeding fledglings (12), a behaviour that would be impeded by increased testosterone secretion.

However, should the first clutch fail, the stimulus for feeding behaviour is removed. Male song sparrows then encounter a situation that male cowbirds experience all season long: a fertile female and no parental duties. Under these conditions male song sparrows respond as do cowbirds (and red-winged blackbirds), with increased testosterone secretion. Males no longer must trade-off feeding young, which requires low testosterone levels, against the hormonally-incompatible behaviour of mate-guarding, which is facilitated by increased concentrations of plasma testosterone. They may respond fully in the direction of the latter behaviour (as they did with their first clutch) in an effort to maximize their genetic contribution to the second clutch.

It would be enlightening to use DNA fingerprinting and hormone analyses to examine the relationship between genetic contribution, testosterone concentration, and clutch number in nonparasitic birds. That is, a male feeding young from a successful first clutch will have lowered testosterone levels and, consequently, may be less likely to guard his mate as she lays the second clutch. This favours the success of the nearly-completed initial reproductive effort at the possible cost of reduced paternity in the second clutch. In contrast, a failed first clutch would enable the male to increase his androgen concentration and, presumably, his mate-guarding efficiency, and his genetic representation in the second clutch would be high.

In summary, although parasitic cowbirds and nonparasitic species like song sparrows seemingly have highly disparate breeding biologies, when ecological pressures are similar in the two groups the resulting patterns of plasma testosterone secretion likewise are similar.

### Testosterone and survival

If concentrations of plasma testosterone in male cowbirds are not restricted by physiological demands associated with parental care, then why do these concentrations fall at all prior to termination of breeding? One possibility is that through repeated interactions the birds come to recognize one another's dominance rank late in the season. Precedence can be exerted with a minimum of overt aggression and testosterone levels can be reduced (13). This implies that there is a cost to elevated testosterone concentrations, and that birds maintain these concentrations only as long as they must. Indeed, elevated testosterone values are known to be energetically expensive in birds (14). Prolonged high testosterone concentrations could affect survival if the cost was great enough, either directly via increased energy demands, or indirectly through testosterone-induced aggression.

To examine whether elevated testosterone concentrations affect survival in cowbirds, a group of males was implanted with testosterone-filled silastic tubing during the breeding season and their survival to the following breeding season was measured (15). Plasma testosterone concentrations in implanted birds were significantly higher than in unimplanted birds, but within the range of values normally encountered. Although cowbirds are highly philopatric (16), the number of implanted males that returned the following year (1/16) was significantly lower than for control males given empty implants (3/6) or for unimplanted males that had been banded in previous breeding seasons and recaptured the following spring (33/81). Furthermore, testosterone-implanted males bore signs of having engaged in severe fighting during the first few weeks after implantation: defeathering of the head, puncture wounds, and injuries to the wings or legs. A regression of testosterone values against days since implantation suggested that testosterone did not influence the occurrence or timing of prebasic moult.

These data indicate that male cowbirds with artificially augmented concentrations of testosterone suffer increased mortality through agonistic interactions with other males. This may be facilitated by the unusual social system of this brood parasite. The lack of territoriality by male cowbirds enhances their capacity to move about and to interact aggressively with many other males. Male cowbirds cannot retire to territorial refuges in which they are dominant; thus, they are more exposed to possible injury from conspecifics.

Territorial nonparasitic species with artificially elevated testosterone concentrations have not been shown to suffer reduced survival through agonistic interactions (17, 18), possibly because territories act as psychological buffers against intruding males, regardless of their hormonal state. However, if elevated testosterone values are sustained to the point where they interfere with the occurrence of moult, then survival is adversely affected in territorial species (18). Thus, the extent to which elevated testosterone concentrations are deleterious may be partly a function of the social system of the organism involved.

## Population density and plasma testosterone

Population and individual differences in seasonal patterns of testosterone secretion occur within species. Birds that contend for limited resources such as breeding sites, mates, or food, would be expected to compete more aggressively at high than at low population densities. The positive relationship between testosterone and aggression likewise should produce a strong correlation between testosterone and population density, at least during resource appropriation when male-male aggression (and social instability) would be most severe.

Cowbird populations discussed thus far have resided in favourable habitats: residential areas and pasture land with cattle that provide feeding areas, and field-forest edge that attracts potential host species. Cowbird population densities in these habitats are moderate to high (19). Conversely, the Sierra Nevada of California (USA) represents a marginal, high altitude habitat that has been invaded by cowbirds only within the last 50 years (20). Potential host species are plentiful, but colonization occurred only when human activity provided feeding sites that previously had been unavailable (i.e. campgrounds, horse corrals, grazing cattle). Even so, the birds move up to 7 km between breeding grounds that they occupy in the morning, and feeding areas where they spend the rest of the day (21). Females do not exhibit territoriality in this area, probably because the relatively low cowbird population density makes territorial behaviour energetically inefficient.

The seasonal testosterone profile of male cowbirds in this low density Sierran population is compared to that of males from a high density eastern population (Fig. 1A). The dense population shows much higher and more sustained concentrations of testosterone than in the low density population. Breeding in the latter population begins in earnest in mid-May and continues until mid-July (3), a reproductive season that is slightly delayed compared to its eastern counterpart. Despite the reduced amplitude of the testosterone profile in the low density population, the pattern of testosterone secretion is similar to that of the high density population. Both have maxima early in the breeding season.

Concentrations of plasma testosterone also vary as a function of population density in both starlings (*Sturnus vulgaris*) (22) and yellow-headed blackbirds (*Xanthocephalus xanthocephalus*) (23), although this is observed only during the early part of the breeding season. Thus, despite differences in seasonal patterns of testosterone secretion in parasitic and nonparasitic species, the effects of differing population densities on the amplitude of peak androgen levels is similar in both.

### Individual variation in testosterone secretion

The testosterone patterns illustrated in Figure 1A represent mean profiles averaged over numerous males. As in other species, the actual patterns exhibited by individual cowbirds reflect the effects of many environmental influences, including social cues (24). For example, in free-living cowbirds, paired males have significantly higher concentrations of plasma testosterone than do unpaired males or males whose mating status is unknown (25). This difference occurs early in spring, when males are competing with each other for dominance status and for access to females; i.e. at a time when the social hierarchy is particularly unstable. Once breeding has begun, profiles of paired and unpaired/unknown birds are indistinguishable.

### Captive versus free-living males

It is difficult to investigate the relative importance of social cues in the field. Laboratory studies have shown that both intrasexual and intersexual social cues affect testosterone secretion in males, including male cowbirds. However, one problem with captive studies is that captive nonparasitic males often have testosterone titres that are only a fraction of those of their free-living counterparts (26). Cowbirds are unusual in this regard because plasma androgen concentrations in captive males not only equal, but can even exceed those of free-living males (Fig. 1B).

The reasons for this difference have not been investigated, but they may have to do with differences in the importance of non-photoperiodic environmental cues in stimulating hormonal secretion in parasitic and nonparasitic birds. Habitat requirements for nonparasitic species can be complex, and include a mate, a territory, nest locations, areas that will provide food for young, and sentinel sites. If one or more of these environmental cues is inadequate or absent (as they often are in captivity), then it could reduce the stimulatory effect of the habitat, impair hormonal secretion and, ultimately, lower reproductive success (24). Conversely, cowbirds are widely distributed and have adapted to a broad range of habitats, from open prairie to woodlands. They can be bred in captivity, needing little more than long days, a mate, adequate food, and a supply of host nests to stimulate reproduction. In other words, cowbirds may be less sensitive to the specific environmental conditions of their surroundings than are nonparasitic species.

Vernal photoperiodic stimulation alone elicits physiological androgen concentrations in the low-normal range in isolate male cowbirds (Fig. 1B). However, vocal and visual cues provided by conspecific males housed in nearby cages stimulate maximal testosterone levels in captive male cowbirds. These values are higher than those found in free-living paired males, birds that might be expected to exhibit the highest testosterone titres because of the comprehensive set of stimuli available to them. These results suggest that the ability of male cowbirds to interact physically with male conspecifics may actually dampen the maximal hormonal response stimulated by the latter's mere presence.

An interesting behaviourial response lends credence to this argument. Male cowbirds raised in isolation from other males, but with access to females, develop songs with amplitude features that normally characterize songs of dominant males (27). Males housed with male conspecifics sing songs that are significantly less 'potent'. When isolate and group-housed birds are placed in aviaries with additional males, the isolate (potent-sounding) males are severely attacked, whereas the socially-housed males are not.

Hence, social interactions seem to reduce the potency of cowbird songs, just as they reduce maximal testosterone concentrations. This suggests that the ability to interact physically with conspecifics is an important part of both the behaviourial and hormonal development of cowbirds, and that such interactions limit the extent of behavioural and endocrine responses below maximum levels. Given the well-known relationship between testosterone and avian vocalizations, there may even be a direct link between socially-induced suppression of testosterone and vocal potency in male cowbirds.

**Intrasexual cues affecting testosterone secretion**

One of the most obvious behaviours of male cowbirds during the breeding season is their group bouts of song and display. Dominance rank is determined during these bouts, and it is not unusual to see 2 - 6 males interacting in a tree, all within a few metres of each other. Birds jockey for position in these groups, with threat displays and posturing. They also sing to one another while performing characteristic bow displays with wings spread. Thus, auditory and/or visual sensory pathways potentially could transmit important social information.

Cues from same-sexed conspecifics can modify the pattern of hormone secretion in nonparasitic species (28) and in parasitic cowbirds (29). The relative contribution of visual and/or auditory cues to this process in cowbirds was investigated by housing male cowbirds in individual cages in one of four conditions: 1) isolate males (birds in visual + vocal isolation), 2) tutored males (birds in visual isolation and played tape-recorded cowbird songs for two hours per day), 3) muted-companion males (birds in vocal isolation and housed next to a temporarily devocalized male), and 4) grouped males (birds given visual + vocal access to nearby males).

Grouped males and muted-companion males exhibit the greatest peak testosterone values and maintain elevated testosterone titres longer than do males in the other two groups. Tutored males show a mixed response: their peak testosterone concentrations are no higher than those of isolate males, but they maintain these values longer than do isolate males. Thus, both visual and vocal cues from other males affect hormonal secretion in male cowbirds. Of the two, visual stimulation appears to elicit a stronger response, but such a conclusion may be premature. Vocal stimulation was provided for only two hours per day and consisted of exposure to the same tape-recorded songs each day. It is possible that this procedure provided sub-optimal stimulation that resulted in a diminished endocrine response. Interestingly, song sparrows require both vocal and visual intrasexual cues for maximal endocrine titres to develop (28). The contrast with cowbirds may represent another case in which cowbirds require less extensive environmental stimulation to elicit a complete hormonal response.

**Intersexual cues affecting testosterone secretion**

Male cowbirds are also very responsive to vocal and postural behaviours of conspecific females. For example, males rapidly approach tape-recorded broadcasts of female vocalizations, and sing, display, and occasionally, attempt to copulate with accompanying mounts of females (1). Further, recent studies indicate that females use subtle displays to influence the quality of song development in males (30).

In other avian species, the presence of a mate affects hormone secretion in males (31). A similar phenomenon has been demonstrated in cowbirds (25). Male cowbirds housed with females and exposed to long days (16L:8D) show a more rapid increase in plasma testosterone and maintain elevated concentrations of testosterone longer than do unpaired males on long days. However, the peak concentrations of testosterone are the same in both groups. The presence or absence of a female has no effect on short day (8L:16D) males. Additional effects of the presence of a female may have been masked by the abrupt transfer of the birds to long days. Such transfers are known to stimulate maximally gonadal development and hormonal secretion (32). A gradual increase in photostimulation may have revealed more subtle intersexual effects on the rate or amplitude of endocrine responses in males. The relative importance of intersexual visual and vocal cues in eliciting hormonal responses in male cowbirds is currently unknown.

**Conclusion**

Studies of the reproductive and behaviourial endocrinology of brood parasitic brown-headed cowbirds reveal both similarities and contrasts with nonparasitic species. The pattern of testosterone secretion in cowbirds is consistent with current theory that recognizes the contribution of both physiological and ecological factors to the evolution of hormonal responses. Modulation of testosterone secretion in individual male cowbirds may depend largely on social interactions rather than on other environmental features.

**Acknowledgement**

This paper was written with support from Idaho State Board of Education Grant number 89-077.

**References**

1 Dufty A M Jr (1982) Response of brown-headed cowbirds to simulated conspecific intruders. *Animal Behaviour* **30** 1043-1052

2 Dufty A M Jr (1982) Movements and activities of radio tracked brown-headed cowbirds. *Auk* **99** 316-327

3 Fleischer R C, Smyth A P & Rothstein S I (1987) Temporal and age-related variation in the laying rate of the parasitic brown-headed cowbird *Canadian Journal of Zoology* **65** 2724-2760

4 Scott D M & Ankney C D (1983) The laying cycle of brown-headed cowbirds: passerine chickens? *Auk* **100** 583-592

5 Wingfield J C, Ball G F, Dufty A M Jr, Hegner R E & Ramenofsky M (1987) Testosterone and aggression in birds. *American Scientist* **75** 602-608

6 Wingfield J C & Moore M C (1987) Hormonal, social, and environmental factors in the reproductive biology of free-living male birds. *Psychobiology of Reproductive Behaviour: an Evolutionary Perspective* pp 149-175. Ed. D C Crews. Englewood Cliffs, USA: Prentice-Hall

7 Wingfield J C, Hegner R E, Dufty A M Jr & Ball G F (1990) The "Challenge Hypothesis": theoretical implications for patterns of testosterone secretion, mating systems, and breeding strategies. *American Naturalist* **136** 829-846

8 Balthazart J (1983) Hormonal correlates of behaviour. *Avian Biology, Vol. 7* pp 221-365. Eds D S Farner & J R King. New York: Academic Press

9 Silverin B (1980) Effects of long-acting testosterone treatment on free-living pied flycatchers, *Ficedula hypoleuca*, during the breeding period. *Animal Behaviour* **28** 823-840

10 Dufty A M Jr & Wingfield J C (1986) Temporal patterns of circulating LH and steroid hormones in a brood parasite, the brown-headed cowbird, *Molothrus ater*. I. Males. *Journal of Zoology* **208** 191-203

11 Beletsky L D, Orians G H & Wingfield J C (1989) Relationships of steroid hormones and polygyny to territorial status, breeding experience, and reproductive success in male red-winged blackbirds. *Auk* **106** 107-117

12 Wingfield, J C (1984) Environmental and endocrine control of reproduction in the song sparrows, *Melospiza melodia* I. Temporal organization of the breeding cycle. *General and Comparative Endocrinology* **56** 406-416

13 Ramenofsky M (1984) Agonistic behaviour and endogenous plasma hormones in male Japanese quail. *Animal Behaviour* **32** 698-708

14 Hännsler I & Prinzinger R (1979) The influence of the sex hormone testosterone on body temperature and metabolism of Japanese quail. *Experientia* **35** 509-510

15 Dufty A M Jr (1989) Testosterone and survival: a cost of aggressiveness? *Hormones and Behavior* **23** 185-193

16 Dufty A M Jr (1985) Song sharing in the brown-headed cowbird (*Molothrus ater*). *Zeitschrift für Tierpsychologie* **69** 177-190

17 Wingfield J C (1984) Environmental and endocrine control of reproduction in the song sparrow, *Melospiza melodia* II. Agonistic interactions as environmental information stimulating secretion of testosterone. *General and Comparative Endocrinology* **56** 417-424

18 Nolan V Jr, Ketterson E D, Ziegenfus C, Cullen D P & Chandler C R (1992) Testosterone and avian life histories: effects of experimentally elevated testosterone on prebasic moult and survival in male dark-eyed juncos. *Condor* **94** 364-370

19 Van Velzen W T (1972) Distribution and abundance of the brown-headed cowbird. *Jack-Pine Warbler* **50** 110-113

20 Rothstein S I, Verner J & Stevens E (1980) Range expansion and diurnal changes in dispersion of the brown-headed cowbird in the Sierra Nevada. *Auk* **97** 253-267

21 Rothstein S I, Verner J & Stevens E (1984) Radio-tracking confirms a unique diurnal pattern of spatial occurrence in the parasitic brown-headed cowbird. *Ecology* **65** 77-88

22 Ball G F & Wingfield J C (1987) Changes in plasma levels of sex steroids in relation to multiple broodedness and nest site density in male starlings. *Physiological Zoology* **60** 191-199

23 Beletsky L D, Orians G H & Wingfield J C (1990) Steroid hormones in relation to territoriality, breeding density, and parental behaviour in male yellow-headed blackbirds. *Auk* **107** 60-68

24 Wingfield J C (1983) Environmental and endocrine control of reproduction: an ecological approach. *Avian Endocrinology: Environmental and Ecological Aspects* pp 265-288. Eds S-I Mikami, S Ishii & M Wada. Tokyo: Japanese Scientific Societies Press. Berlin:Springer-Verlag

25 Dufty A M Jr & Wingfield J C (1986) The influence of social cues on the reproductive endocrinology of male brown-headed cowbirds: field and laboratory studies. *Hormones and Behavior* **20** 222-234

26 Wingfield J C & Farner D S (1980) Control of seasonal reproduction in temperate-zone birds. *Progress in Reproductive Biology* pp 62-101. Ed. P O Hubinont. Basel: Karger

27 West M J & King A P (1980) Enriching cowbird song by social deprivation. *Journal of Comparative and Physiological Psychology* **94** 263-270

28 Wingfield J C & Wada M (1989) Changes in plasma levels of testosterone during male-male interactions in the song sparrow, *Melospiza melodia*: time course and specificity of response. *Journal of Comparative Physiology A* **166** 189-194

29 Dufty A M Jr & Wingfield J C (1990) Endocrine response of captive male brown-headed cowbirds to intrasexual cues. *Condor* **92** 613-620

30 King A P & West M J (1983) Epigenesis of cowbird song - a joint endeavour of males and females. *Nature* **305** 704-706

31 Moore M C (1983) Effect of female sexual displays on the endocrine physiology and behaviour of male white-crowned sparrows, *Zonotrichia leucophrys. Journal of Zoology (London)* **199** 137-148

32 Dawson A & Goldsmith A R (1983) Plasma prolactin and gonadotrophins during gonadal development and the onset of photorefractoriness in male and female starlings (*Sturnus vulgaris*) on artificial photoperiods. *Journal of Endocrinology* **97** 253-260

Avian Endocrinology
Ed. P J Sharp, pp 73-86
Journal of Endocrinology Ltd, Bristol (1993)

# Hormones, reproduction, and behaviour in birds of the Sonoran desert

C M Vleck
Department of Ecology and Evolutionary Biology, University of Arizona, Tucson, AZ 85721, USA

## Introduction

Deserts are one of the major terrestrial ecosystems and birds are a highly visible component of their faunal make-up. Although there is not a well-accepted set of adaptations that uniquely characterize desert-dwelling birds (1, 2; but see 3), the physical characteristics of desert habitats differ sufficiently from those of other habitats such as mesic and temperate-zone woodland that it is of interest to investigate the hormonal organization of reproduction in birds that breed in deserts. In this habitat it is generally accepted that breeding tends to be dissociated from photoperiodic control and is driven by other environmental variables such as rainfall.

## Desert characteristics

Desert areas are found in both hemispheres, occupying over one-fourth of the world's land surface, usually within a band between approximately latitudes 15° and 35° (4). Deserts, as an ecosystem type, are not easy to describe in a few quantitative terms (5), but in general we think of deserts as hot and arid with relatively low primary productivity that is often unpredictable in time and space. For an endothermic desert-resident this has the potential to compound problems with thermoregulation, water balance, and nutritional state, all of which can impinge on successful reproduction.

Over recent decades, several investigators have discussed the ability of some desert birds to breed opportunistically in response to unpredictable rainfall and associated effects on vegetation and insect densities (6-11). In particular, in very arid areas such as central Australia, and to a lesser extent in southwest Africa, rainfall has been proposed as an important exogenous cue controlling breeding (9, 12, 13), similar to the situation in tropical regions (14, 15). Other suggested adaptations of birds for breeding in deserts have included nomadism, early sexual maturation, continuous breeding as long as suitable conditions prevail, participation in nest-building by both sexes, and simultaneous molt and breeding (1). Flexibility in breeding schedules however does not operate uniformly over all arid regions (1) and even within different populations of the same species, there may be differences in breeding periodicity, suggesting the potential evolutionary flexibility of this characteristic (10, 16, 17, 18). Schodde (18) has suggested that breeding opportunism may be characteristic of only some nomadic species, and even then may be superimposed on a more basic breeding seasonality.

Deserts vary tremendously in seasonality, productivity, and predictability. There is no single optimal adaptive strategy for desert organisms. For instance, Weins (19) recently investigated the avifaunas of shrub-desert habitats in Australia and North America. Despite nearly identical vegetation structures in the two localities, the breeding bird assemblages differed significantly in densities and social structure, and several life history parameters including length of breeding season and number of breeding attempts per season. Weins suggested these differences were related to the greater variance in precipitation in the Australian desert than in the North American desert (the average yearly rainfalls were nearly identical). Two major directions of life history adaptation have been discussed for desert organisms including plants, invertebrates, and vertebrates (20). In deserts where rich resources are short-lived and unreliable in time and space (episodic), one would predict breeding adaptations for a rapid activation of reproductive behaviour and a high reproductive rate. This is characteristic of at least some of the desert birds of Australia and Africa (see above). In deserts where food and water resources are scarce, but relatively predictable, efficient use of these resources should be mediated by mechanisms that activate reproduction seasonally, but with low reproductive output, coupled with adaptations to increase longevity of individuals.

Quantitative data to support such generalizations in desert birds are meagre at best and the physiological underpinnings are almost totally unexplored. I review here recent work on the endocrine control of reproduction in several birds of the Sonoran desert. The Sonoran desert is one division of the North American desert that covers most of southwestern Arizona and northwestern Mexico including the peninsula of Baja at a latitude between 27° and 34° N. It is the warmest of the North American deserts with a long hot season from April to October during which time maximum temperatures over 38 °C often occur for 90 consecutive days. Winter temperatures rarely fall below freezing (21). In the Sonoran desert around Tucson the vegetation is mostly brush, cacti and small trees. Near Tucson the annual rainfall averages about 29 cm (US National Climatic Data Center) with extremes ranging from 16 to 56 cm over the last 30 years (coefficient of variation = 30%). Rainfall is bimodal with more than half of the annual precipitation usually falling during summer months between July and September in brief, torrential and local thunderstorms. Winter rains usually occur between December and March as prolonged rainstorms. There is considerable variance about these dates for both winter and summer rains.

## Breeding seasons and gonadal cycles of Sonoran desert birds

Approximately 56 species commonly breed in the desert scrub habitat of the Sonoran desert (22). Of those for whom breeding seasons are known, five species nest primarily in the summer, 25 nest primarily in the spring and 19 species nest in both spring and summer. For many but not all of these species, the potential breeding season determined by breeding records tends to be long, extending for 6 to 12 months of the year (Table 1). Sixteen of the 56 species nesting in Arizona desert scrub habitat have nesting seasons lasting 6 months or longer (22) presumably because temperature by itself is rarely low enough to preclude successful reproduction. Although active nests have been found throughout these potential breeding seasons, this does not mean that birds routinely breed throughout the entire period. When conditions are favourable, however, Sonoran desert birds appear to be reproductively capable over a longer period than birds from higher latitudes or regions with more severe winters. Multiple nesting events are common for these desert birds. Curve-billed thrashers (*Toxostoma curvirostre*) and cactus wrens (*Campylorhynchus brunneicapillus*) can raise four broods in one year although two or three per year is more common (personal observation). Breeding in some of these species is associated with favourable local conditions, particularly rainfall. For example, nesting of rufous-winged sparrows (*Aimophila carpalis*) usually occurs after the summer rains in July, following a slow gonadal recrudescence between March and June. Following heavy winter rains, however, there is rapid gonadal recrudescence and nesting occurs in spring as well as summer (23, 24). The breeding season for black-throated sparrows (*Amphispiza bilineata*) in Baja, where summer rains are much reduced, coincides with winter rains, whereas in Arizona with a bimodal rain distribution, birds nest in the spring and summer (25, 26).

Table 1 Months of the year when active nests have been found for several birds breeding in the Sonoran desert.

| Species | Months | Reference |
|---|---|---|
| *Zenaida macroura*<br>Mourning Dove | all year,<br>especially Apr-Aug. | 26 |
| *Geococcyx californianus*<br>Roadrunner | Apr-Sept | 45 |
| *Campylorhynchus brunneicapillus*<br>Cactus Wren | Feb-Sept | 46, pers. obs. |
| *Toxostoma curvirostre*<br>Curve-billed Thrasher | Jan-Aug | 47, 48, pers. obs. |
| *Pipilo fuscus*<br>Brown Towhee | Feb- Sept | 45, 48 |
| *Pipilo aberti*<br>Abert's Towhee | Feb-Sept | 49 |
| *Aimophila carpalis*<br>Rufous-winged Sparrow | Apr-Sept | 23, 24 |
| *Amphispiza bilineata*<br>Black-throated Sparrow | Mar-Sept (Arizona)<br>Oct-June (Baja) | 26<br>25 |

Wyndham (27) has suggested that the long duration of the potential breeding season generally associated with deserts is better correlated with the general effects of latitude rather than local physical characteristics of habitats. His analysis was based on calculating the number of equally good months for breeding at several localities around the world. Wyndham's definition of breeding season provides a measure of the length and evenness of the distribution of the breeding seasons for all species at a given locality and as such has ecological relevance for understanding community structure. It does not necessarily indicate the interval of breeding competency for individuals of a species, which reflects the underlying physiological mechanisms of reproductive control.

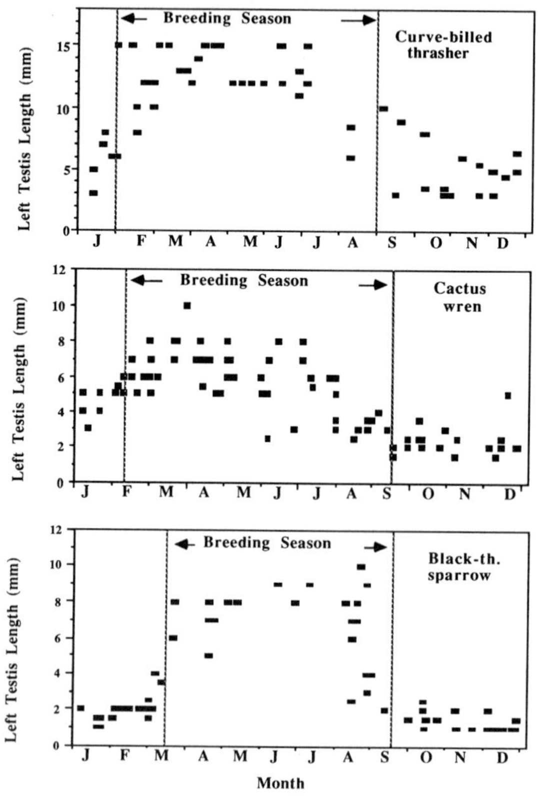


Fig. 1 Testis length versus time of the year in field-caught males of three species of birds (curve-billed thrasher, cactus wren and black-throated sparrow) living in the Sonoran desert near Tucson, Arizona. The area between the hatched lines indicates the reported breeding season for each species.

For three species of birds that I have studied in the Sonoran desert the gonads, measured by laparotomy after capture in the field (Fig. 1), progress through an annual cycle that coincides well with the reported breeding season (Table 1), although regression of gonads in most individuals tends to occur before published dates for the latest nests. Testes are small and inactive during the fall and early winter season, but are enlarged and producing sperm (determined by histological examination) throughout the nesting season. The environmental cues for gonadal recrudescence in the winter in these species have not been experimentally investigated, but are likely to involve day length since gonadal growth always occurs over the course of several weeks preceding the breeding season. In curve-billed thrashers, gonadal recrudescence is concurrent with the winter solstice, so short days, rather than increasing daylength, must activate recrudescence if photoperiod serves as the cue. Daylengths are just beginning to increase when gonadal recrudescence commences in the black-throated sparrow and cactus wren.

Thus, gonadal cycles of these Sonoran desert birds are not like those of central Australian zebra finch *(Taeniopygia guttata)* which are not photoperiodic and can have continuously active gonads as long as birds are not severely dehydrated (28-31). Rather, the gonad cycles resemble those of Australian budgerigars *(Melopsittacus undulatus)* in which the gonads become inactive when the birds are not breeding, which occurs at different times in different parts of their range (32). Fall breeding in the Sonoran desert is rare because food resources needed to raise young are depressed as summer rainfall lapses and temperatures decline. Conditions for successful reproduction improve again following winter rains and the accompanying increased food availability. These annual changes in the Sonoran desert are predictable and photoperiod provides a reliable Zeitgeber for gonadal cycles. In North American deserts in general, some form of photoperiodic timing probably prevails for most plant and animal annual cycles. Breeding is seasonal but variation in population numbers of many species is correlated quite tightly with variation in rainfall (33).

**Concentrations of plasma reproductive hormones in Sonoran desert birds**

The prolonged breeding season of desert birds provides a special opportunity to examine the role of reproductive hormones in maintenance, as well as initiation, of physiological competency and behavioural correlates of reproduction. In three species of birds in the Sonoran desert, there is a rise in the plasma concentrations of the gonadotrophin, luteinizing hormone (LH) and testosterone in males at the same time that gonadal recrudescence is occurring (Figs 2 and 3). The highest concentrations of these hormones, however, are associated only with the early part of the annual breeding cycle. Maximum measured testosterone concentrations decrease after the beginning of the season even though the gonads remain active and multiple nesting is common. The asynchrony of nesting between pairs in these species is such that a comparison of individuals based on the stage of their own annual reproductive cycle provides more power than making comparisons based on calendar date (Table 2). In curve-billed thrashers, actively breeding males in their first nesting cycle of the year have higher concentrations of both plasma testosterone and LH than during subsequent nesting cycles (Table 2).

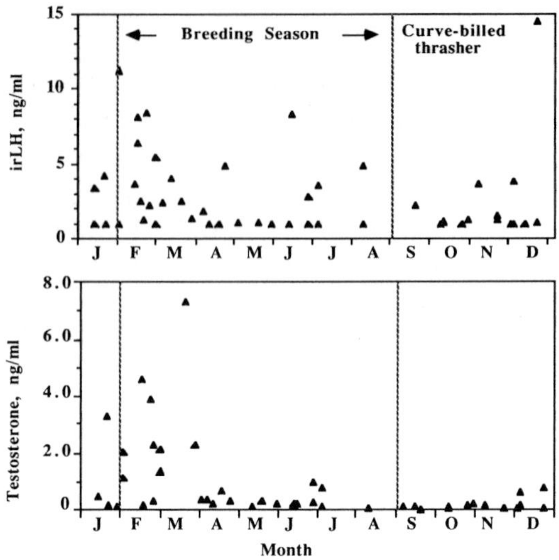


Fig. 2 Plasma concentrations of luteinizing hormone and testosterone versus time of the year in male curve-billed thrashers captured in the field. The area between the hatched lines indicates the reported breeding season.

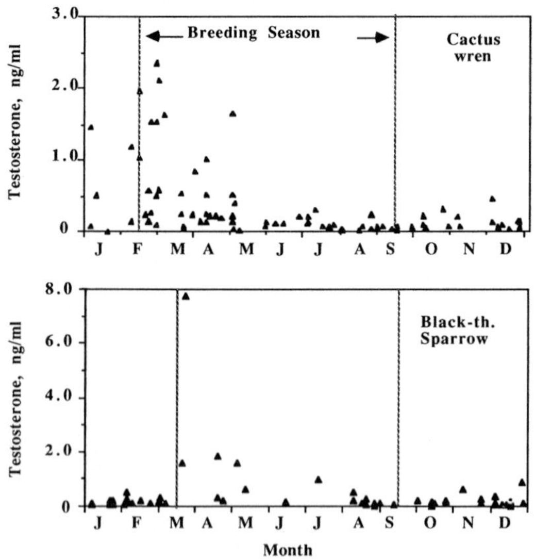


Fig. 3 Plasma concentrations of testosterone versus time of the year in male cactus wrens and black-throated sparrows captured in the field. The area between the hatched lines indicates the reported breeding seasons.

Table 2 Plasma hormone concentrations in male curve-billed thrashers during the annual reproductive cycle (means ±SD).[1]

| | Testosterone (ng/ml) | | Luteinizing hormone (ng/ml) | |
|---|---|---|---|---|
| *First Nest* (January-March) | | | | |
| Nest building | 3.19 ± 1.81 | (3) | 8.64 ± 2.39 | (3) |
| Incubation | 1.92 ± 0.48 | (3) | 2.26 ± 2.13 | (4) |
| Care of Nestlings | 7.33 | (1) | 2.50 | (1) |
| *Subsequent Nests* (March-July)[2] | | | | |
| Nest building | 0.66 | (1) | 1.03 ± 0.04 | (2) |
| Incubation | 0.18 ± 0.03 | (3) | 3.43 ± 4.21 | (3) |
| Care of Nestlings | 0.18 ± 0.08 | (4) | 1.04 ± 0.07 | (3) |
| *Nonbreeding* (October-November)[2] | 0.11 ± 0.06 | (11) | 1.52 ± 1.00 | (13) |

[1]Both sexes incubate the eggs and care for nestlings.
[2]Testosterone but not LH concentration, significantly lower ($t=3.87$, $P=0.002$) than during first nesting cycle.

This pattern of testosterone secretion in breeding curve-billed thrashers supports Wingfield *et al.* (34) who suggested that the highest concentrations of plasma testosterone in animals are associated primarily with male-male interactions and with heightened levels of aggression. The highest plasma levels of testosterone do not appear necessary for spermatogenesis, maintenance of secondary sex characteristics, or for display of sexual behaviours. In curve-billed thrashers, the early part of the nesting season is likely to be a time of relative social instability as nesting territories and pair-bonds are being established for the first time or re-established after the winter. It is during this time that there is the greatest likelihood of a challenge by a conspecific male and that testosterone levels are high and variable between individuals. Later in the season, when territories are well established, male-male interactions become less frequent and testosterone levels decline, even though breeding continues.

Desert habitats may furnish systems that provide a good test of another aspect of the hypotheses of Wingfield *et al.* (34). They suggested that in polygynous species of birds, male should have relatively high and invariant concentrations of plasma testosterone. This is because high testosterone concentrations and the outcome of aggressive interactions with competing males have a greater impact on fitness than does contribution to care of young which appears not to require relatively high concentrations of plasma testosterone. Males of monogamous species that care for young, in contrast, should have on average low testosterone concentrations, but with high variance depending on their previous encounters with other males in the population. Most desert species tend to be monogamous, possibly in part because resource densities are such that usually the female alone could not raise the young successfully. The cooperatively breeding Harris' hawk *(Parabuteo unicinctus)* whose reproductive endocrinology in the Sonoran desert has been well-worked out (35, 36) provides an opportunity to extend these hypotheses to birds with another type of social system.

Harris' hawks live in groups of two to seven individuals. These groups consist of a breeding pair and usually a number of related or unrelated auxiliary birds that help defend the group-held territory and bring food from group kills to the nest. Harris' hawks have a well-defined breeding season that begins in late January or early February with nest building and extends until late summer when fledglings become essentially independent of adult aid. Those male helpers in the group that are related to the breeders do not have any seasonal increase in plasma LH or testosterone and are not reproductively active. Elevated levels of testosterone and LH are found in both breeding males and unrelated male helpers in the group. These reproductively competent helpers do occasionally copulate with the female or inherit the breeding slot if the breeding male disappears. The testosterone concentrations in these primary and secondary breeding males, even at the height of the breeding season, are very low, averaging only about 300 pg/ml (compare to Figs. 2 and 3). The maximum levels of testosterone ever found in Harris' hawk males (~1500-2000 pg/ml) are about six times higher than the average (35). This is about the expected rise in testosterone predicted for monogamous males following a behavioural challenge by another male (34). Interactions between male Harris' hawks from different groups are often aggressive, although behavioural interactions between males within a group are surprisingly nonaggressive, usually consisting only of supplanting behaviour, even when the sexually active helpers are courting the breeding female (37). Thus the group as a whole tends to resemble the pattern expected for a monogamous pair; harmonious interactions within the group and low testosterone levels, but with the potential to interact aggressively with hawks outside the group, correlated with transient increases in testosterone. The selection pressure leading to this type of cooperative social behaviour may involve increased foraging success that results from group hunting (38). Group hunting allows hawks to capture more and larger prey items which may increase the nesting success for this hawk in this desert habitat. The low testosterone concentrations may be a necessary correlate of this kind of social cooperation.

## Summer breeding

In the Sonoran desert most nesting occurs in the spring months (March-May). Some notable exceptions are birds that breed only or primarily in the summer, including the rufous-winged sparrow mentioned above and the white-winged dove (*Zenaida asiatica)* which is a summer resident, arriving from its southern wintering range in March-April. Many birds that begin breeding in the late winter and spring do have a second, but shorter period of nesting in the summer following the summer rains. For instance during one year of bi-weekly field work, I located 30 cactus wren nests containing eggs. Eighty-seven percent of these were found in February-May. No active nests were found in June, the hottest and driest month, but four additional nests were found in July, after summer rains began. For 31 curve-billed thrasher nests, 71% had eggs in January through April, none were started in May and the other 29% were started in mid-June or early July. The reason why only a minority of birds renest in the summer is not clear - it may depend on the previous nesting success in the spring and the presence of offspring on the territory.

The early part of the summer before the summer rains commence is potentially the most physiologically stressful time of year in the Sonoran desert with ambient temperature reaching the seasonal maximum in June. May and June are on average the driest months; daytime relative humidities are usually less than 15%, and free water is scarce. Birds are well-equipped to cope with such conditions with body temperatures which are usually around 40 °C. Since this is higher than ambient temperature for most of the time, birds can usually lose any excess body heat down a thermal gradient without spending water to evaporatively cool by remaining inactive and out of direct sunlight (39). Panting or gular fluttering can increase evaporative heat loss at the expense of water, but birds lose relatively less water in the process of nitrogen excretion than mammals do because they excrete a semi-solid uric acid paste instead of urea (40).

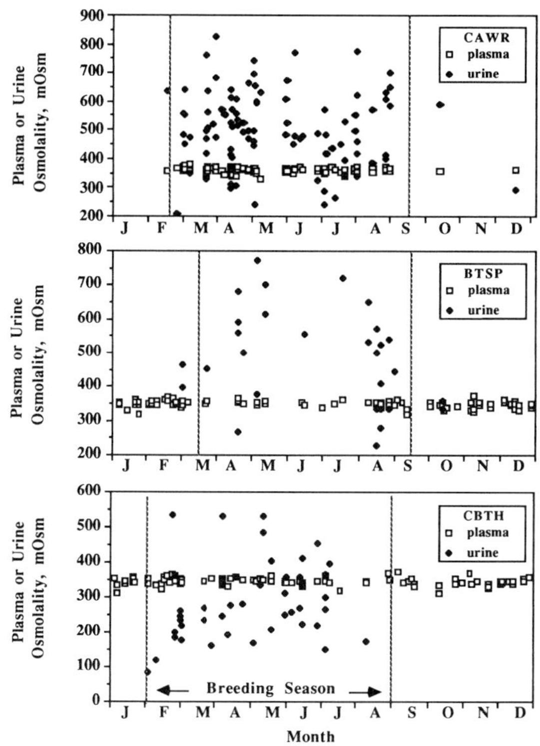


Fig. 4 The osmolality measured with a Wescor osmometer versus time of the year of plasma (open symbol) and urine (closed symbol) in three species of birds living in the Sonoran desert. The area between the hatched lines indicates the reported breeding seasons.

One indication of water stress in organisms is the solute concentration of body fluids which tends to rise with dehydration. In the primarily insectivorous cactus wren and primarily granivorous black-throated sparrow, the osmolality of urine concentrations during the breeding season tends to be quite variable, but is usually higher than that of plasma ranging to over 800 mOsm (Fig 4). The highest osmolality of urine measured in these species are close to the maximum reported for several other birds that were experimentally dehydrated (41). This suggests that these desert birds may sometime be approaching their maximum physiological ability to maintain water balance. In the omnivorous curve-billed thrasher, urine osmolality can be higher than that of the plasma but tends to be much lower than in the sparrow and cactus wren, even in the hottest months of the year. In May and June, thrashers are eating large amounts of cactus fruit which probably provides a more abundant and reliable source of water than either seeds or insects.

The osmolality of plasma in all three species (Fig. 4) is tightly regulated (average values range from 346 to 359 mOsm; S.D. = 10 mOsm for the three species) although they tend to be higher than in mammals that maintain plasma osmolality at around 300 mOsm. They are also higher than the plasma osmolality in the Australian emu (*Dromaius novaehollandiae*) (42), but within the range reported for several other birds, including desert species (41). These data support the hypothesis that Sonoran desert birds are not usually water stressed, even in the hottest, driest months. Concentrations of the plasma adrenocorticosteroid hormone that typically rise in birds following stress, including dehydration (43), support this view (44). Basal concentrations of plasma corticosterone measured in the summer in several species of Sonoran desert birds, including the three analyzed here, are either no different from winter values or are actually lower! During times of intense heat or failure of food supplies, though, it is probable that desert species would face problems with maintaining water balance. Wingfield *et al.* (44) collected a few plasma samples in black-throated sparrows, in June of 1990 near Phoenix, Arizona during a period in which the highest-ever temperatures for the area were recorded (over 50 °C). During this interval the basal concentrations of corticosterone in these individuals were over four times higher than basal concentrations measured in other birds during the previous summer.

Wingfield *et al.* (44) also measured the rate at which corticosterone rises following capture and handling in several birds of the Sonoran desert. There is the expected stress-induced increase in the secretion of corticosterone, but the response is markedly reduced in the summer months (June-August) compared to that in the winter months (December-early January) when the birds are not breeding. The extent of attenuation of this stress response, however, varies with the local habitat in which the birds reside. For curve-billed thrashers, black-throated sparrows, and cactus wrens, the reduction in stress response in the summer is highly significant, but is less so for Abert's towhee (*Pipilo aberti*) and Inca dove (*Scardafella inca*), species that are normally found in mesic habitats within the desert ecosystem. It is not known whether desert birds breeding in the spring also show this reduced adrenal response. The significance of summertime resistance to stress-induced rises of corticosterone in the most arid-adapted species is not clear. Reduced adrenal response to acute stress may have evolved to allow reproduction to proceed despite the severe summertime conditions that can prevail. Drought inhibition of reproductive systems involving elevated corticosterone has been suggested for another arid-zone species, the Texas bobwhite quail (*Colinus virgianus*) (43). It would be interesting to know whether those individual birds in the Sonoran desert species that continue to nest into the summer differ in their response to environmental stress compared to those that terminate nesting in the spring or early summer.

## Conclusion

While the basic hormonal control of breeding cycles probably does not differ in fundamental ways between birds living in the desert and in other habitats, adjustments of breeding cycles to local conditions are probably as varied as are the local conditions themselves. The evolutionary flexibility of reproductive patterns can be demonstrated by examining breeding records across geographical and climatological clines, but the precise physiological mechanisms underlying these differences are mostly unknown, and await extensive comparative and experimental investigation. In the three bird species of the Sonoran desert reviewed here, reproduction is seasonal with activation of gonads associated with rises in the typical reproductive steroids and gonadotrophins. Hormone levels do not remain at maximal levels even though the potential breeding season continues for several months. The duration and intensity of the reproductive effort vary from year to year depending on local conditions and temporal variation.

## Acknowledgements

The work was supported by the National Science Foundation (R11-8310562). Rick Bowers provided unfailing assistance in the field. Luteinizing hormone levels were measured in the laboratory of Dr John Wingfield and I gratefully acknowledge his assistance and support. I thank David Vleck for his insightful comments on all aspects of this work.

## References

1 Serventy D L (1971) Biology of desert birds. *Avian Biology* pp 287-339. Eds D S Farner, J R King & K C Parkes. New York: Academic Press

2 Weins J A (1991) The ecology of desert birds. *The Ecology of Desert Communities* pp 278-310. Ed G A Polis. Tucson: University of Arizona

3 Williams J B, Withers P C, Bradshaw S D & Nagy K A (1991) Metabolism and water flux of captive and free-living Australian parrots. *Australian Journal of Zoology* **39** 131-142

4 Polis G A (1991) Desert communities: an overview of patterns and processes. *The Ecology of Desert Communities* pp 1-26. Ed G A Polis. Tucson: University of Arizona

5 Evenari M (1985) The Desert Environment. *Hot Deserts and Arid Shrublands, A* pp 1-22. Eds M Evenari, I Noy-Meir & D W Goodall. Amsterdam: Elsevier

6 Marshall A J & Disney H J de S (1957) Experimental induction of the breeding season in a xerophilous bird. *Nature* **180** 647-649

7 Serventy D L & Marshall A J (1957) Breeding periodicity in Western Australian birds: with an account of unseasonal nesting in 1953 and 1955. *Emu* **57** 99-126
8 Keast A (1959) Australian birds: their zoogeography and adaptations to an arid continent. *Biogeography and Ecology in Australia* pp 89-114. Eds A Keast, R L Crocker & C S Christian. Hague: Junk
9 Immelmann K (1963) Drought adaptations in Australian desert birds. *Proceedings XIII International Ornithological Congress* **13** 5549-657
10 Immelmann K (1971) Ecological aspects of periodic reproduction. *Avian Biology* pp 341-389. Eds D S Farner, J R King & K C Parkes. New York: Academic Press
11 Maclean G L (1970) The breeding seasons of birds in the south-western Kalahari. *Ostrich Supplement* **8** 179-192
12 Marshall A J (1970) Environmental factors other than light involved in the control of sexual cycles in birds and mammals. *La Photorégulation de la Réproduction chez les Oiseaux et les Mammifères* pp 53-64. Eds J Benoit & I Assenmacher. Paris: CNRS
13 Lofts B & Murton R K (1968) Photoperiodic and physiological adaptations regulating avian breeding cycles and their ecological significance. *Journal of Zoology* **155** 327-394
14 Dittami J P & Gwinner E (1985) Annual cycles in the African stonechat and their relationship to environmental factors. *Journal of Zoology London (A)* **207** 1-14
15 Dittami J P (1986) Seasonal reproduction, moult and their endocrine correlates in two tropical *Ploceidae* species. *Journal of Comparative Physiology B* **156** 641-647
16 Davies S J J F (1977) The timing of breeding by the zebra finch *Taeniopygia castanotis* at Mileura, Western Australia. *Ibis* **119** 369-372
17 Kikkawa J (1980) Seasonality of nesting by zebra finches at Armidale, NSW. *Emu* **80** 13-20
18 Schodde R (1982) Origin, adaptation and evolution of birds in arid Australia. *Evolution of the Flora and Fauna of Arid Australia* pp 191-224. Eds W R Barker & P J M Greenslade. Frewville, South Australia: Peacock
19 Weins J A (1991) Ecological similarity of shrub-desert avifaunas of Australia and North America. *Ecology* **72** 479-495
20 Shmida A, Evenari M & Noy-meir I (1986) Hot desert ecosystems: an integrated view. *Hot Deserts and Arid Shrublands, B* pp 379-387. Eds M Evenari, I Noy-Meir & D W Goodall. Amsterdam: Elsevier
21 McGinnies W G (1981) *Discovering the Desert.* Tucson: University of Arizona
22 Davis W A & Russell S M (1984) *Birds in Southeastern Arizona.* Tucson: Tucson Audubon Society
23 Phillips A R (1968) Rufous-winged sparrow. *US National Museum Bulletin* **237** 902-219
24 Ohmart R D (1969) Physiological and ethological adaptations of the rufous-winged sparrow (*Aimophila carpalis*) to a desert environment. PhD Dissertation University of Arizona
25 Miller A H (1960) Adaptation of breeding schedule to latitude. *Proceedings XII International Ornithological Congress* **12** 513-522
26 Phillips A R, Marshall J & Monson G (1964) *The Birds of Arizona.* Tucson: University of Arizona
27 Wyndham E (1986) Lengths of birds' breeding seasons. *American Naturalist* **128** 155-164

28 Marshall A J & Serventy D L (1958) The internal rhythm of reproduction in xerophilous birds under conditions of illumination and darkness. *Journal of Experimental Biology* **35** 666-671

29 Oksche A, Farner D S, Serventy D L, Wolff F & Nicholls C A (1963) The hypothalamo-hypophysial neurosecretory system of the zebra finch, *Taeniopygia castanotis*. *Zeitshrift für Zellforschung* **58** 846-914

30 Sossinka R (1974) Der Einfluss von Durstperioden auf die Schilddrüsen- and Gonaden-aktivität und ihre Bedeutung für die Brutperiodik des Zebrafinken (*Taeniopygia castanotis* Gould). *Journal für Ornithologie* **115** 128-141

31 Vleck C M & Priedkalns J (1985) Reproduction in zebra finches: hormone levels and effect of dehydration. *Condor* **87** 37-46

32 Pohl-Apel G, Sossinka R & Wyndham E (1982) Gonadal cycles of wild budgerigars *Melopsittacus undulatus* (Psittaciformes: Platycercidae). *Australian Journal of Zoology* **30** 791-797

33 Macmahon J A & Wagner F (1985) The Mojave, Sonoran and Chihuahuan deserts of North America. *Hot Deserts and Arid Shrublands, A* pp 105-202. Eds M Evenari, I Noy-Meir & D W Goodall. Amsterdam: Elsevier

34 Wingfield J C, Hegner R E, Dufty A M Jr & Ball G F (1990) The "challenge hypothesis": theoretical implications for patterns of testosterone secretion, mating systems and breeding strategies. *American Naturalist* **136** 829-846

35 Mays N A, Vleck C M & Dawson J (1991) Plasma luteinizing hormone, steroid hormones, behavioural role, and nest stage in cooperatively breeding Harris' hawks (*Parabuteo unicinctus*). *Auk* **108** 619-637

36 Vleck C M, Mays N A, Dawson J W & Goldsmith A R (1991) Hormonal correlates of parental and helping behaviour in cooperatively breeding Harris' hawks (*Parabuteo unicinctus*). *Auk* **108** 638-648

37 Dawson J W & Mannan R W (1991) The role of territoriality in the social organization of Harris' hawks. *Auk* **108** 661-672

38 Bednarz J C (1988) Cooperative hunting in Harris' hawks (*Parabutea unicinctus*). *Science* **239** 1525-1527

39 Davies S J J F (1982) Behavioural adaptations of birds to environments where evaporation is high and water is in short supply. *Comparative Biochemistry and Physiology* **71A** 557-566

40 Braun E J (1982) Renal function. *Comparative Biochemistry and Physiology* **71A** 511-517

41 Goldstein D L & Braun E J (1989) Structure and concentrating ability in the avian kidney. *American Journal of Physiology* **256** R501-R509

42 Skadhauge E, Maloney S K & Dawson T J (1991) Osmotic adaptation of the emu (*Dromaius novaehollandiae*). *Journal of Comparative Physiology B* **161** 173-178

43 Cain J R & Lien R J (1985) A model for drought inhibition of bobwhite quail (*Colinus virginianus*) reproductive systems. *Comparative Biochemistry and Physiology* **82A** 925-930

44 Wingfield J C, Vleck C M & Moore M C (1992) Seasonal changes of the adrenocortical response to stress in birds of the Sonoran desert. *Journal of Experimental Zoology* **264** 419-428

45 Ohmart R D (1973) Observations on the breeding adaptations of the roadrunner. *Condor* **75** 140-149

46 Ricklefs R E & Hainsworth F R (1969) Temperature regulation in nestling cactus wrens. The nest environment. *Condor* **71** 32-37
47 Bent A C (1948) Life histories of North American nuthatches, wrens, thrashers, and their allies. *US National Museum Bulletin* **195** 351-402
48 Ricklefs R E (1965) Brood reduction in the curve-billed thrasher. *Condor* **67** 505-510
49 Marshall J T (1963) Rainy season nesting in Arizona. *Proceedings XIII International Ornithological Congress* **13** 620-622

Avian Endocrinology
Ed. P J Sharp, pp 87-98
Journal of Endocrinology Ltd, Bristol (1993)

# Mast cells containing GnRH-like immunoreactivity in the CNS during courtship in doves

R Silver[1], X Zhuang[2], R Millar[3] and A-J Silverman[4]
[1]Barnard College of Columbia University, New York, NY 10027, USA; [2]Psychology Department, Columbia University, New York, NY 10027, USA; [3]MRC Regulatory Peptides Research Unit, University of Capetown Medical School, Capetown, South Africa; [4]Department of Anatomy and Cell Biology, Columbia University, New York, NY 10032, USA

## Introduction

Studies of neural and endocrine interactions are well documented, but the ways in which elements of the immune system participate in such interactions during normal behavioural and physiological responses remain largely unexplored (1, 2). The goal of this paper is to review recent observations showing that an interaction occurs among the immune, nervous and endocrine systems during courtship.

In his initial studies with ring doves, Lehrman (3) discovered the ways in which external stimuli, such as those presented by the courting male, induce ovarian development in the female dove. This work established the principle that cues external to the animal reach the central nervous system and result ultimately in an endocrine response. The work of Green and Harris (4) demonstrated a functional link between the brain and pituitary gland, and proved that the flow of information was from the nervous to the endocrine system via the hypophyseal portal capillaries in the median eminence. This neuroendocrine link provided a mediating mechanism for the phenomena discussed by Lehrman (3). Since that time, observations in many species have confirmed the principle that environmental stimuli, including those from the sexually active partner, can produce changes in behavioural responses (5) and in endocrine secretions (6). Despite the solidity and antiquity of these observations, the loci reached by such afferent information, and the site(s) of neural-endocrine signal transduction are not known.

## The effect of behavioural and endocrine state on the gonadotrophin releasing hormone (GnRH)-like immunoreactive cells in the central nervous system

Since GnRH controls gonadotrophin secretion, the GnRH-neuronal system might be expected to respond to external stimuli during courtship. The distribution of cells containing GnRH-like immunoreactivity (GnRH-ir) was compared in the brains of courted versus isolated animals using an antiserum (LR-1) directed against the mammalian form of the decapeptide (9). Subjects were adult ring doves (*Streptopelia*

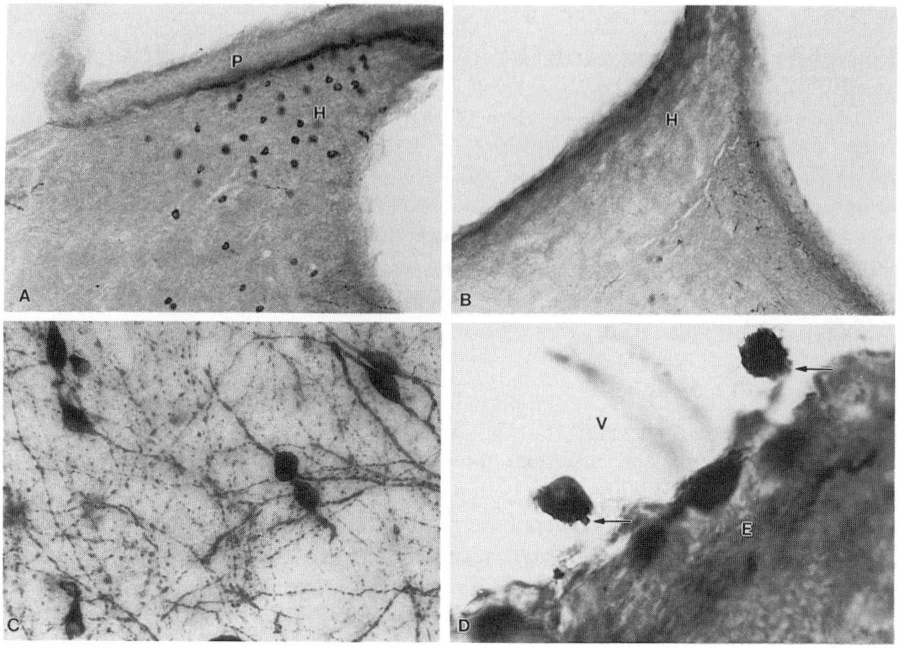


Fig. 1 Light micrographs of GnRH-like immunoreactive cells in the dove brain. A and B are both of the medial habenula region (H) and illustrate the differences in GnRH-ir mast cell number between courting (A) and isolated (B) males. Note the axonal fibres near the third ventricle in B (arrows). In C, GnRH neurones and their processes in the preoptic area are shown. The typical fusiform morphology of these cells is evident. Mast cells as shown in D can be identified on the ventricular surface. These cells have numerous fine filamentous processes (arrow). Other cells are intercalated into the ependymal (E) wall.

*roseogrisea*) with previous breeding experience, placed together in a cage provided with straw and a nest bowl and permitted to court. GnRH-ir neurones and fibres had a similar distribution in the two groups and there was no obvious difference produced by mating in the staining intensity of neuronal elements. However, a new non-neuronal immunoreactive cell type was observed within the medial habenular region of courted animals. In isolated animals, a few small (8-10 µm in diameter) round GnRH-positive cells were seen in the choroid plexus, on pial and ventricular surfaces in the medial habenular region, intercalated into the ependymal wall, and around circumventricular organs (organum vasculosum lamina termalis and organum subseptale) and blood vessels (Fig. 1B; see also 8). After a 1 hour period of courtship with a sexually active partner, this cell type was observed in the grey matter at the tip of the medial habenula (9). Within two hours of pairing, these GnRH-immunoreactive cells became more numerous, and occupied deeper regions of the grey matter of the medial habenula as well being present on the pial and ventricular surfaces in this region (Fig. 1A, D). These cells differed from GnRH-ir neurones (Fig. 1C) in morphology, size, and cytoplasmic staining pattern. Initially, the small size and unusual shape of these newly appearing cells suggested that they were an unknown artifact; but the consistent finding of a difference in their number and distribution between isolated and courted animals provoked us to examine them more closely. The effects of behavioural and endocrine state on the number of GnRH-like immunoreactive cells was explored further in four groups of male birds (8). These were, 1) paired with a female, and displaying courtship behaviour; 2) paired with a squab, and displaying aggressive behaviour; 3) visually isolated from other doves; and 4) long-term castrates housed communally. In the first three groups, GnRH-like immunoreactive non-neuronal cells were observed in all the areas described above, with differences among groups in their numbers in the medial habenula region. As shown in Table 1 there were significantly fewer GnRH-like immunoreactive cells in the medial habenula of isolated than in courted animals, while doves placed with a squab and displaying aggressive behaviour were intermediate between these two groups. Long-term castration had the most profound effect in reducing the number of such cells in the brain.

Table 1 A comparison of the number of GnRH-like immunoreactive cells in male ring doves in different behavioural states (means ± DS).

| **Behavioural group** | **Number of habenular mast cells** | **Number of birds** |
|---|---|---|
| Courting | $1170 \pm 210^{a}$ | 7 |
| Given squab | $700 \pm 130^{b}$ | 6 |
| Isolated | $350 \pm 110^{c}$ | 6 |
| Castrate | $220 \pm 120^{d}$ | 4 |

The following differences among groups are tested by one-sided *t*-test: a vs b, p=.06; b vs c, p=.05; c vs d, p=.26; a vs c, p=.01; a vs d, p=.02; b vs d, p=.04. (Data from 8).

## Identification of the non-neuronal GnRH-like immunoreactive cell type

Ultrastructural observations (see below) suggest that the GnRH-like immunoreactive non-neuronal cells have either a monocyte or mast cell lineage. Two different markers for macrophage/microglial cells were used: the monoclonal antibody OX42 (Serotec, 1:100), which recognises an antigen associated with complement receptor function in macrophages (10) and the *Bandeiraea simplicifolia* isolectin B4 conjugated to biotin or FITC (11) known to be associated with both resting and active microglia. Neither the OX42 antibody nor the *Bandeiraea simplicifolia* isolectin labelled cells in the medial habenula of courted animals. In alternate sections, GnRH-like immunoreactivity was demonstrable in both small round non-neuronal cells and in axonal fibres. In the ring dove, the isolectin outlined all the capillaries in the brain and adjacent choroid plexus. In the latter, a few cells were also labelled. The OX42 antibody did not react with any structures in these sections, suggesting that the GnRH-like immunoreactive non-neuronal cells were not microglia.

Of the small molecules found in mast cells, histamine and serotonin are amongst those most frequently encountered across species. The presence of these potential markers for mast cells was investigated using highly specific polyclonal antibodies to histamine (12) and to serotonin (13) to determine if one or both amines were contained within non-neuronal cells of the brain and other organs. In addition, the metachromatic response of the newly appearing cells to acidic toluidine blue was also tested. Mast cells are known to contain proteoglycans, which give them their metachromatic properties, and a variety of small molecules which are bound by the proteoglycan in the granule matrix (14). Staining with acidic toluidine blue (15) resulted in the visualisation of numerous small, round vacuolated cells with pink/purple cytoplasm and deep blue nuclei. This colour change distinguished them from surrounding neurones which had blue cytoplasm and clear nuclei containing deep blue nucleoli.

GnRH-like immunoreactivity was identified in small round cells in the brain, lung and liver but was absent from the gut (gastric gland). Alternate serial sections from the medial habenula, lung and liver were reacted with antibodies to GnRH and histamine. Cells immunoreactive for the peptide and amine with a similar morphology and distribution were observed in all three organs. Double label immunocytochemical analysis was then carried out on brain and lung from male doves that had courted for 2 h. In the medial habenula and in the lung, under these physiological conditions, all GnRH-immunoreactive cells also contained histamine. Antibodies to both substances resulted in punctate fluorescence and most puncta appeared to fluoresce at both emission spectra when appropriately excited, suggesting that the amine and peptide were in the same granules. Control sections in which the second primary antibody was replaced with normal rabbit serum contained only single labelled cells indicating that the double labelling was not due to binding of the second fluorophore to the initial layer of reagents.

Serotonin immunoreactivity in the region of the medial habenula was confined to axons and nerve terminals. The antiserum did not react with any elements within the lung or liver but stained enteroendocrine cells in the gut.

These studies strongly suggest that the non-neuronal GnRH-like immunoreactive elements are mast cells. Mast cells which possess Fc receptors specific for immunoglobulin E (IgE) have been known for years in connective and mucosal tissue where they play an important role in allergic reactions (16, 17). During allergic reactions, mast cells are coated by IgE molecules produced and released by B lymphocytes. When allergens bind to these IgE molecules and cross-link two of them, a series of reactions are triggered that cause mast cells to release their granular contents, including histamine and neuropeptides (18, 19). However, IgE-allergen binding is not the only potent stimulus for mast cell degranulation. In many regions, including the CNS, mast cells appear to be innervated or in close approximation to nerve terminals (20) and can be stimulated to release their granular contents. A number of neuropeptides including GnRH and GnRH antagonists have been found to be mast cell secretogogues (21-24). Amine release can also be triggered directly by progesterone (25) and induced release can be augmented by oestradiol (26). Oestrogen receptors have been demonstrated in CNS mast cells of rats (27).

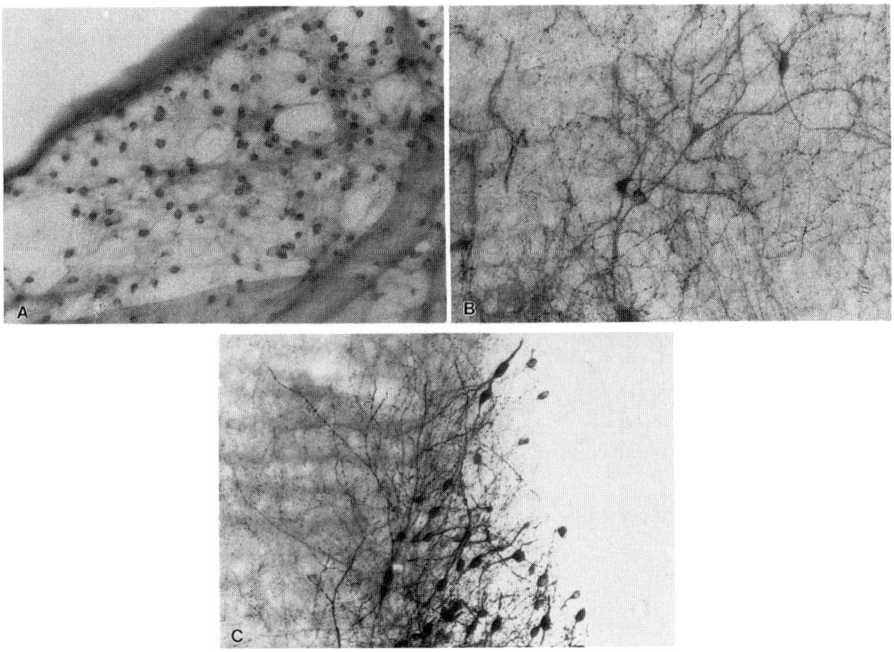


Fig. 2 Light microscopic identification of substance P positive mast cells (A) and neuronal elements (B) in the pia and CNS, respectively, of the ring dove. VIP positive neurones were found (C) but no mast cells.

## Identification of GnRH-like material and other peptides in GnRH-like immunoreactive (ir) mast cells

The avian brain contains more than one form of GnRH decapeptide. Chicken GnRH-I (cGnRH-I), which predominates, differs from mammalian GnRH by the substitution of glutamine for arginine in position 8. The chicken GnRH-II (cGnRH-II) sequence contains amino acids $His^5$-$Trp^7$-$Tyr^8$. Staining in GnRH-like immunoreactive mast cells was abolished when the LR-1 antiserum was absorbed with either cGnRH-I or cGnRH-II while the staining in neurones and axons was abolished only by absorption with cGnRH-I (9). This suggests that the peptide recognised by the antiserum differs between neuronal and non-neuronal cells. Immunocytochemical studies have also been done using GnRH monoclonal antibodies including EL-14 which is directed against the N- and C- termini of mGnRH, and Rb 1076 which is directed against the middle amino acids and recognises mGnRH precursor and mGnRH (28), and other polyclonal GnRH antisera including #3-3 and #6DL9/4 supplied by P J Sharp, #IV-H10 supplied by L Jennes and HU-11B supplied by H Urbanski. None of these antisera appear to react with the GnRH-like immunoreactive mast cells identified with antiserum LR-1, but all show immunocytochemical reactions with neurones in the preoptic area and fibres in the median eminence. The identity of the GnRH-like immunoreactive peptide present in the non-neuronal cells therefore remains unknown.

Previous reports indicate that mast cells contain vasoactive intestinal polypeptide (VIP) and Substance P (SP) (18, 29). In fact, SP is one of the most potent secretagogues of mast cells (21). Using an SP antiserum (Incstar) we observed positive round-shaped cells with a mean cell diameter of 8.5±0.71 μm (X±SE) (n=10) characteristic of mast cells in the pia mater, in a location similar to GnRH-ir mast cells (Fig. 3A). Although SP immunopositive neurones (Fig. 2B) were also observed, no SP-containing mast cells were present in the medial habenula proper in courted males. Antibodies against VIP (Incstar), did not stain any cells with a mast cell morphology in any of the brain regions examined, though labelled neurones were found (Fig. 2C), as previously described for the ring dove brain (30). While double-label experiments for histamine and SP remain to be done, the results suggest that distinct subpopulations of peptide containing mast cells occur in different microenvironments of the brain.

**Ultrastructural characteristics of GnRH-like immunoreactive mast cells**

Ultrastructurally, the GnRH-like immunoreactive mast cells contain large vacuoles, a highly condensed nucleus filled with heterochromatin (Fig. 3A) and numerous filamentous processes (Fig. 1D). There is little rough endoplasmic reticulum and a very reduced Golgi apparatus. In some cells, the secretory granules are dilated and partially electron lucent with dense reaction product deposited on the inner surface of the limiting membrane (Fig. 3A). Many such vacuoles appear fused and reach toward the surface of the cell. Other cells contain considerably less immunoreactive deposits (Fig. 3B, C). In these cells, the secretory vacuoles were almost completely electron lucent. It is possible that the degranulation observed is an artifact of preparation or of the immunocytochemical staining (e.g. the mild detergent treatment). The morphology of these medial habenular GnRH cells is quite different from ultrastructure of GnRH neurones or GnRH secretory granules in synaptic terminals, and differs from adjacent ependyma or glial elements of the central nervous system (9). This is consistent with the cells being of a non-neuronal lineage.

**GnRH-like immunoreactive mast cells do not endocytose**

The tracer substance Fluorogold (FG) is a stilbene which can insert into the plasma membrane of cells and then be internalised by endocytosis. It is widely used as a retrograde neuronal tracer for this reason. We reasoned that administration for some days before and at the time of pairing would label the mast cell population if it

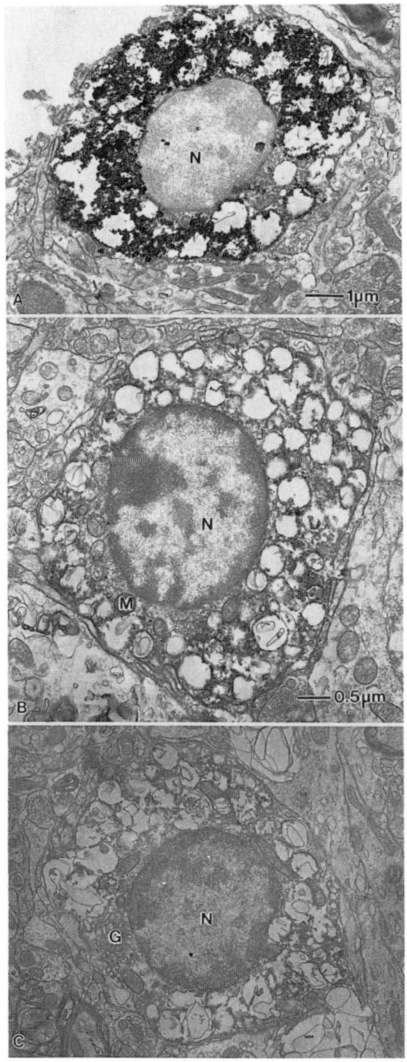


Fig. 3 Three electron micrographs depicting LR-1 positive cells in the medial habenula of a male dove sacrificed two hours after initiation of courtship. In A, the cell is actually intercalated just below the ependymal layer. The secretory granules are somewhat dilated. The silver-gold deposits used to intensify the DAB reaction product are intense throughout the cell and frequently obscure the boundaries of the granules. In B and C, GnRH-ir cells located several cell diameters deep into the parenchyma contain noticeably less reaction product. The cell in C is almost completely devoid of any reaction product and its granules are fused with each other. The loss of reaction product may be due to the immunocytochemical procedure. In all three cells the nucleus (N) is condensed and filled with heterochromatin. These are a few mitochondria (M) and reduced Golgi apparatus (G). The micron bar in B also applies to C.

were capable of active internalisation of substances from the blood or CSF. If so, these cells might belong to the macrophage rather than mast cell lineage. Doves were given intraventricular injections of 3μl of 0.08% FG in saline on 3 consecutive days, paired immediately after the third injection, and sacrificed after 2 hours of courtship. Higher doses (e.g. 3μl of 1.6%) made the animals ill and were discontinued. Fluorogold was also given intraperitoneally to males on 5-11 consecutive days, in doses ranging from 1.5-4.4 mg FG/animal. As with cerebro-intraventricular injections, animals were paired after the last injection and were sacrificed after 2 hr courtship. Injection of FG into the ventricle resulted in intense labelling of the choroid plexus. However, no FG containing cells were observed in the medial habenula, suggesting that the cells did not take up this tracer. Neurosecretory cells were labelled after intra-peritoneal injection, but again, no mast cells were labelled.

### Origins of GnRH-like material in mast cells

Immunocytochemical studies show that there are numerous GnRH axons which appear to terminate in or near the ventricular surface of the medial habenula (9). It seems possible that these axons might release GnRH into the CSF which is then endocytosed by the non-neuronal population. To test for this possibility, a $^{125}$I-labelled GnRH agonist ([DAla$^6$, NMeLeu$^7$, Pro$^9$ NEt] GnRH) was prepared, purified and iodinated as previously reported (31), and administered into the lateral cerebroventricle via an indwelling cannula at the time of pairing. Tissue was then prepared for autoradiography. The results indicate that radioactive neuropeptide did not accumulate in any cells within the brain although abundant silver grains were found in the ventricular space. This result suggests that the epitope recognised by the GnRH antiserum is not endocytosed by a receptor-mediated mechanism from the cerebrospinal fluid, but is produced by the mast cell. The presence of the peptide epitope in both lung and liver recognised by LR-1 (unpublished data from the lab), [where there are no other GnRH-positive elements and where circulating GnRH is likely to be very low], also strongly supports the hypothesis of prior synthesis of the peptide within the mast cells.

### GnRH in the cerebrospinal fluid does not induce the appearance of immunoreactive mast cells

The presence of numerous GnRH fibres that terminate in or near ventricles in the medial habenular region suggest the possibility that these terminals secrete into the CSF at the time of courtship and that local high concentrations of GnRH might be the stimulus causing the appearance of these non-neuronal cells into the medial habenular region. To test this hypothesis, synthetic chicken GnRH-I or -II (10μg) was injected into the lateral ventricle of isolated females. Two hours later animals (n=4/group) were sacrificed and tissue was prepared for immunocytochemistry. This treatment did not result in any increase in the number or distribution of GnRH positive non-neuronal cells in the injected birds. This is consistent with the conclusion that GnRH-like immunoreactive material within the mast cells is not endocytosed.

**Occurrence of GnRH-like immunoreactive mast cells in other species**

Cells with the appearance of mast cells, immunostained with LR-1, have been observed in hamster habenula (material prepared by T Nunez) and in the pia arachnoid and olfactory bulb of shrew (material prepared by T Dellovade and E Rissman), and in cells isolated from the rat peritoneum (J Nunez and R Silver, unpublished data). Finally, C Sisk and collaborators have detected GnRH mRNA with *in situ* hybridization in cells of the choroid plexus of prepubertal ferrets using an oligonucleotide probe to GnRH (48 base pairs 102-149 of human cDNA) (personal communication).

**Conclusions**

Mast cells occur in the brain but are usually associated with the blood vessels and lie just inside the blood brain barrier (20, 32-37). Following insult to the brain or damage of brain tissue caused by disease as in multiple sclerosis (2), mast cells enter deeper into CNS. While changes in mast cell distribution during ontogeny and in disease have been well documented, our data is, to the best of our knowledge, the first demonstration of a behaviourally triggered redistribution of mast cells in the normal brain parenchyma.

The specific challenge is to understand the function of mast cells containing GnRH-like immunoreactivity in the medial habenula and elsewhere in the brain. Our ultrastructural studies suggest that release of granule contents occurs within the medial habenula and presumably makes histamine and the GnRH-like peptide available in this region as courtship proceeds. Since axons immunoreactive for both substances are also found in this area, it is reasonable to assume that histamine and GnRH receptors are present. It should be noted that mast cells can contain up to 20 mediators (19) including interleukins 3-6, granulocyte/macrophage colony stimulating factor (38) and a variety of proteoglycans, proteases and neuropeptides (39). Studies must now be undertaken to determine the release properties of the avian brain mast cells and the alteration in both release and synthesis of mediator substances with behavioural state.

The present results suggest that mast cells, a component of the immune system, are involved in nervous and endocrine interactions during normal behavioural activity. Future studies will explore the nature of the signal that elicits the appearance of mast cells in the brain and the role that the mast cells might play in altering nervous system function during reproduction. Mast cells in the brain may represent a novel delivery system for biologically active substances. While the preliminary data suggest that mast cells containing GnRH-like immunoreactivity occur in other species, the generality of our observation that such cells can increase in number in the brain under specific behavioural and endocrine conditions, remains to be examined.

**Acknowledgements**

The Fluorogold studies were carried out by Ann Wilensky. We wish to thank Robert Benoit for the LR-1 antibody, P J Sharp (anti-chicken GnRHI raised in rabbit, #3-3; and anti-chicken GnRHI raised in sheep, #6DL9-4; P Panula for the histamine antiserum and H Tamir for the serotonin antiserum. This work was supported by NIMH grant 02983 (RS) and HD 10665 (AJS).

## References


1 Besedovsky H O & del Ray A (1992) Immuno-neuroendocrine circuits: integrative role of cytokines. *Frontiers in Neuroendocrinology* **13** 61-94

2 Theoharides T C (1990) Mast cells: the immune gate to the brain. *Life Sciences* **46** 607-617

3 Lehrman D S (1959) Hormonal responses to external stimuli in birds. *Ibis* **101** 478-496

4 Green J D & Harris G W (1949) Observations on the hypophysio-portal vessels on the living rat. *Journal of Physiology London* **108** 359-361

5 Cheng M-F, Desiderio C, Havens M & Johnson A (1988) Behavioral stimulation of ovarian growth. *Hormones and Behavior* **22** 388-401

6 Silver R (1992) Environmental factors influencing hormone secretion. *Behavioral Endocrinology* pp 401-422. Eds J Becker, S M Breedlove & D Crews. Cambridge, MA: MIT Press

7 Silver R, Ramos C, Machuca H & Silverin B (1992) Immunocytochemical distribution of GnRH in the brain of adult and posthatching great tit (*Parus major*) and ring dove (*Streptopelia roseogrisea*). *Ornis Scandanavica* **23** 222-232

8 Zhuang X, Silverman A-J & Silver R (1993) Reproductive behavior, endocrine state, and the distribution of GnRH-like immunoreactive mast cells in dove brain. *Hormones and Behavior* (in press)

9 Silver R, Ramos C L & Silverman A-J (1992) Sexual behavior triggers the appearance of non-neuronal cells containing Gonadotropin-Releasing Hormone-like immunoreactivity. *Journal of Neuroendocrinology* **4** 207-210

10 Robinson A P, White T M & Mason D W (1986) Macrophage heterogeneity in the rat as delineated by two monoclonal antibodies MRC OX-41 and MRC OX-42, the latter recognizing complement receptor type 3. *Immunology* **57** 239-247

11 Streit W J & Kreutzberg G W (1987) Lectin binding by resting and reactive microglia. *Journal of Neurocytology* **16** 249-260

12 Panula P, Happola O, Airaksinen M S, Auvinen S & Virkamaki A (1988) Carbodiimide as a tissue fixative in histamine immunohistochemistry and its application in developmental neurobiology. *Journal of Histochemistry and Cytochemistry* **36** 259-269

13 Krichgessner A, Gershon M D, Lu K P & Tamir H J (1988) Co-storage of serotonin binding protein with serotonin in the rat CNS. *Neuroscience* **8** 3879-3890

14 Enerback L (1986) Mast cell heterogeneity: the evolution of the concept of a specific mucosal mast cell. *Mast Cell Differentiation and Heterogeneity* pp 1-25. Eds A D Befus, J Bienenstock & J A Denburg. New York: Raven Press

15 Humason G L (1979) *Animal Tissue Techniques*. San Francisco: Freeman

16 Galli S J, Dvorak A M, and Dvorak H F (1984) Basophils and mast cells: morphologic insights into their biology, secretory patterns and function. *Progress in Allergy* **34** 1-141

17 Serafim W E & Austen K F (1987) Mediators of immediate hypersensitivity reactions. *New England Journal of Medicine* **317** 30-34

18 Cutz E, Chan W, Track N S, Goth A & Said S I (1978) Release of VIP in mast cells by histamine liberators. *Nature* **275** 661-662

19 Marom Z & Casale D T B (1983) Mast cells and their mediators. *Annals of Allergy* **50** 367-370

20 Dimitriadou V, Aubineau P, Taxi J & Seylaz J (1987) Ultrastructural evidence for a functional unit between nerve fibers and type II cerebral mast cells in the cerebral vascular wall. *Neuroscience* **22** 621-630
21 Fewtrell C M S, Foreman J C, Jordan C C, Oehme P, Renner H & Stewart J M (1982) The effects of substance P on histamine and 5-hydroxytryptamine release in the rat. *Journal of Physiology* **330** 393-411
22 Hook W A, Karten M & Siragania R P (1985) Histamine release by structural analogs of LHRH. *Federation Proceedings* **44** 1323
23 Phillips A, Hahn D W, McGuire J L, Ritchie D, Capetola R J, Bowers C & Folkers K (1988) Evaluation of the anaphylactoid activity of a new LHRH antagonist. *Life Sciences* **43** 883-888
24 Yano H, Wershel B K, Arizono N & Galli S J (1989) Substance P-induced augmentation of cutaneous vascular permeability and granulocyte infiltration in mice is mast cell dependent. *Journal of Clinical Investigation* **84** 1276-1286
25 Vliagoftis H, Dimitriadou V & Theoharides T C (1990) Progesterone triggers selective mast cell secretion of serotonin. *International Archives of Allergy and Applied Immunology* **93** 113-119
26 Cocchiara R, Albeggiani G, DiTrapani G, Azzolina A, Lampiasi N, Rizzo F & Geraci D (1990) Modulation of rat peritoneal mast cell and human basophil histamine release by estrogens. *International Archives of Allergy and Applied Immunology* **93** 192-197
27 Vliagoftis H, Dimitriadou V, Boucher W, Rozniecki J J, Correia I, Raam S & Theoharides T C (1992) Estradiol augments while tamoxifen inhibits rat mast cell secretion. *International Archives of Allergy and Immunology* **98** 398-409
28 Silverman A-J, Witkin J W & Millar R P (1990) Light and electron microscopic immunocytochemical analysis of antibodies directed against GnRH and its precursor in hypothalamic neurons. *Journal of Histochemistry and Cytochemistry* **38** 803-813
29 Montoya G A, Villena F, Jofre A & Pezo A A (1989) Substance P-like immunoreactivity in sympathetic ganglion from toad. *Cellular and Molecular Biology* **35** 325-331
30 Cloues R, Ramos C & Silver R (1990) Vasoactive intestinal polypeptide-like immunoreactivity during reproduction in doves: influence of experience and number of offspring. *Hormones and Behavior* **24** 215-231
31 Millar R P & King J A (1983) Synthesis, luteinizing hormone-releasing activity, and receptor binding of chicken hypothalamic luteinizing hormone-releasing hormone. *Endocrinology* **113** 1364-1369
32 Cammermeyer J (1966) Reaction of mast cells identified in the rabbit brain. *Journal of Neuropathology and Experimental Neurology* **25** 130
33 Campbell D J & Kiernan J A (1966) Mast cells in the nervous system. *Nature* **210** 756-757
34 Dimitriadou V, Lambracht-Hall M, Reichler J & Theoharides T C (1990) Histochemical and ultrastructural characteristics of rat brain perivascular mast cells stimulated with compound 48/80 and carbechol. *Neuroscience* **39** 209-224
35 Dropp J J (1972) Mast cells in the central nervous system of several rodents. *Anatomical Record* **174** 227-238
36 Ibrahim M Z M (1974) The mast cells of the mammalian central nervous system - Part 1. Morphology, distribution and histochemistry. *Journal of Neurological Science* **21** 431-478

37 Ibrahim M Z M, Al-Wirr M E & Bahuth N (1979) The mast cells of the mammalian central nervous system. III. Ultrastructural characteristics in the adult rat brain. *Acta Anatomica* **104** 134-154

38 Plaut M, Pierce J H, Watson C J, Hanley-Hydes J, Nordan R P & Paul W E (1989) Mast cell lines produce lymphokines in response to cross-linkage of Fc-episilon RI or to calcium ionophores. *Nature* **339** 64-67

39 Gleich G J, Ackerman S J, Hisamatsu K & Leiferman K M (1986) Mast cell heterogeneity: biological and clinical significance. *Mast Cell Differentiation and Heterogeneity* pp 347-366. Eds A D Befus, J Bienenstock & J A Denburg. New York: Raven Press

Avian Endocrinology
Ed. P J Sharp, pp 99-110
Journal of Endocrinology Ltd, Bristol (1993)

# Incubation behaviour in the turkey: molecular and endocrinological implications

M E El Halawani and I Rozenboim
Division of Animal Physiology, Department of Animal Science, University of Minnesota, St Paul, MN 55108, USA

## Introduction

The onset of incubation behaviour in turkey hens is correlated with decreased concentrations of plasma lutcinizing hormone (LH), the ovarian steroids, progesterone (P) and oestradiol (E), and increased concentrations of plasma prolactin. Incubation behaviour is also associated with cessation of ovulation and ovarian regression. The mechanisms underlying these changes in the secretion of reproductive hormones in incubating birds are far from clear. This review is concerned with factors in the incubating turkey which are responsible for the suppressed concentrations of plasma LH and ovarian steroids and for the very high concentrations of circulating prolactin.

## Mechanisms regulating prolactin secretion

Studies from our laboratory are consistent with the view that increased prolactin secretion is responsible for the shift from the egg laying to the incubation phase of the reproductive cycle in the turkey (1, 2). Thus, intracranial perfusion of prolactin in laying turkey hens reversibly induces characteristics of incubation behaviour: increased nesting and decreased egg production (Fig. 1).

The mechanisms responsible for the increase in prolactin secretion in the incubating hen appear to operate at the levels of both the hypothalamus and the anterior pituitary gland. Acute electrical stimulation of the medio-basal hypothalamus increases the concentration of plasma prolactin in female turkeys, with the increase being much greater in incubating than in laying birds (3, 4; Fig. 2). The central nervous contribution to the increase in prolactin secretion may be related, in part, to the activity of basal hypothalamic neuroendocrine cells containing vasoactive intestinal peptide (VIP), a hypothalamic prolactin-releasing neuropeptide (5). The increase in circulating prolactin during incubation behaviour is accompanied by a dramatic increase in the number of VIP-like immunoreactive (VIP-IR) cells in the infundibular nuclear complex of the hypothalamus and in VIP immunoreactivity in the median eminence, whereas disruption of incubation results in a decrease in both plasma prolactin and the number of hypothalamic VIP-IR cells (6).

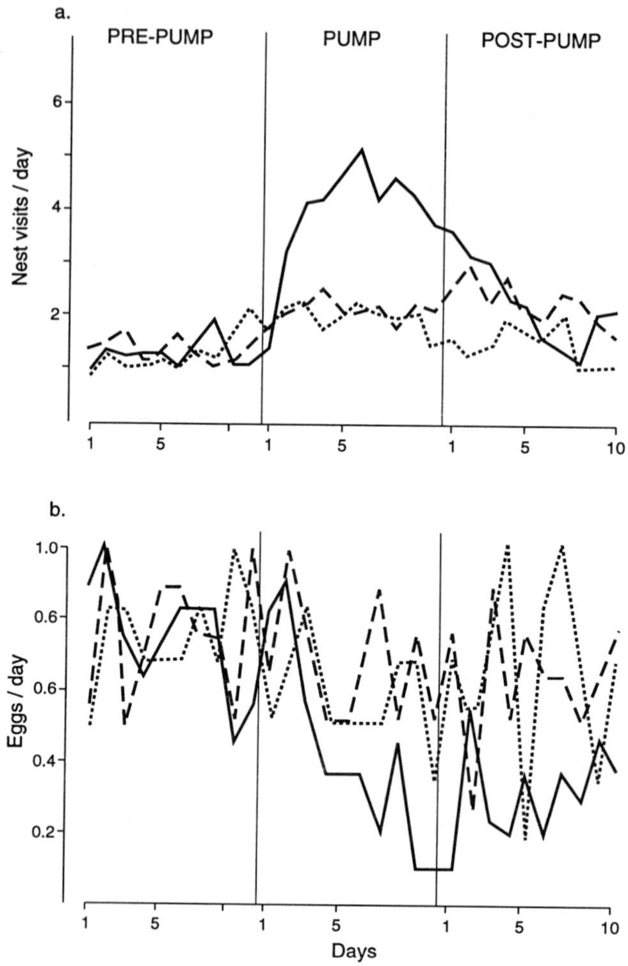


Fig. 1 The effect of intracranial perfusion of ovine prolactin or vehicle using an osmotic pump in laying turkey hens on A) nest visits and B) egg laying. The experiment demonstrates that prolactin, acting at a central nervous site, induces incubation behaviour in laying hens. Solid line, prolactin birds; dashed line, vehicle birds; dotted line, control birds. Modified from (2).

Other studies have also shown increases in the number and size of VIP-IR neurones within the medio-basal hypothalamus when concentrations of plasma prolactin are increased, including in the incubating bantam hen, and in the domesticated pigeon and ring dove following the initiation of crop milk secretion and feeding of offspring (7-9).

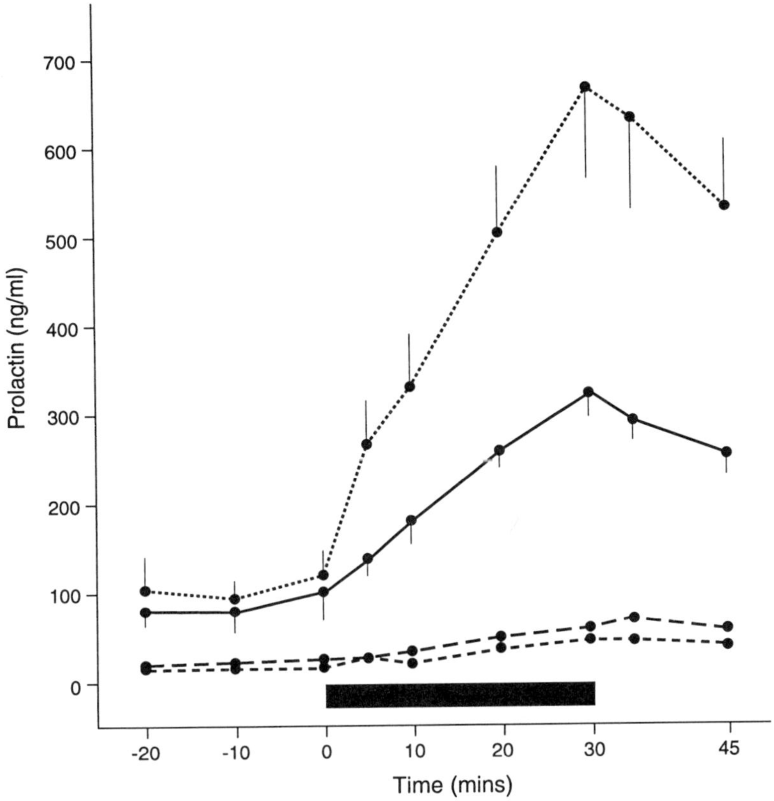


Fig. 2 The effect on concentrations of plasma prolactin of electrical stimulation of the median eminence in turkey hens which were held on short days and were not laying (- - -), or were held on long days and were photorefractory (– – –), laying (——) or incubating (· · ·). The bar indicates the 30 minute stimulation period. Data from (14).

Quantification of hypothalamic VIP reveals changes in content throughout the reproductive cycle exclusively within the median eminence/infundibular nuclear complex region (10). These changes in VIP content are directly related to concentrations of circulating prolactin except at the onset of incubation and in photorefractory birds. The increase in plasma prolactin associated with the onset of incubation in the absence of changes in hypothalamic VIP concentration might indicate that VIP turnover is maximum and that an increase in pituitary responsiveness to VIP might account for the elevations in circulating prolactin. Vasoactive intestinal polypeptide-induced release of prolactin *in vitro*, by pituitary cells from turkey hens at various stages of the reproductive cycle, reflects the circulating concentrations of prolactin at these stages (11). In support of this view, anterior pituitary cells from laying hens have lower spontaneous and VIP-stimulated rates of prolactin secretion than pituitary cells from incubating hens. In addition, pituitary cells from laying hens have greater spontaneous and VIP-stimulated rates of prolactin secretion than cells from photorefractory or non-photostimulated hens. The release of prolactin induced by VIP *in vitro* also reflects the number of VIP binding sites in the anterior pituitary gland. Reproductively quiescent non-photostimulated birds have the lowest number of anterior pituitary VIP binding sites. These increase following photostimulation to reach maximal values in incubating hens and then decrease during the development of photorefractoriness. Further evidence for the central role played by an increase in hypothalamic VIP in maintaining increased prolactin secretion during incubation is provided by the reduction in plasma prolactin and termination of nesting behaviour which is observed after passive immunization of incubating chickens with anti-chicken VIP serum (7, 12).

The apparent dissociation between the high concentration of hypothalamic VIP and low prolactin secretion in photorefractory birds might be caused by a decrease in pituitary responsiveness to VIP, since pituitaries from photorefractory birds have fewer VIP binding sites and are less responsive to VIP than pituitaries from laying or incubating birds (11, 13). In addition, electrical stimulation of the median eminence results in a lower increase in prolactin secretion in photorefractory birds than in laying or incubating hens (14; Fig. 2). Thus, a decrease in pituitary responsiveness to VIP without a change in hypothalamic VIP concentration may explain the apparent uncoupling of hypothalamic VIP and prolactin secretion following termination of incubation and subsequent photorefractory state.

The high VIP content of the hypothalamus in the photorefractory birds is reduced following transfer to short days (10) suggesting that the synthesis/secretion of this neuropeptide is controlled by the photoperiod. Indeed, hypothalamic immunoreactive VIP content increases following photostimulation (10) confirming our previous immunocytochemical studies (6).

Recent findings from our laboratory (15) demonstrate that the posterior pituitary of the turkey contains a potent prolactin-releasing factor activity capable of causing a rapid, dose-dependent stimulation of prolactin secretion by anterior pituitary cells *in vitro*. A marked increase in the prolactin-releasing factor activity in the posterior pituitary occurs only as hens shift from the laying to the incubation phase of the reproductive cycle (between 2.9-fold and 11.1-fold). These results, taken together with the finding that there is a dramatic increase in anterior pituitary prolactin and prolactin mRNA content in incubating turkeys (16), suggest that the increased prolactin-releasing factor activity in the posterior pituitary promotes stimulation of prolactin synthesis and release in incubating hens.

An increase in prolactin gene expression in the anterior pituitary gland might be expected to be associated with increased prolactin secretion. This appears to be the case since anterior pituitary gland prolactin mRNA and pituitary prolactin contents are highest during the incubation phase of the reproductive cycle (12, 16; Fig. 3). Thus, when hens are incubating, the expression of the prolactin gene is increased, prolactin secretion increases, and the pituitary shows an enhanced secretory responsiveness to prolactin releasing factors.

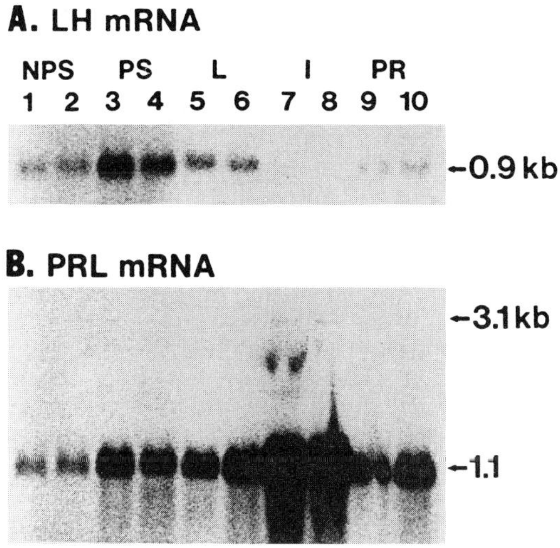


Fig. 3 Northern blots of A) LH β subunit mRNA and B) Prolactin mRNA in anterior pituitary glands from turkey hens which were non-photostimulated (NPS), photostimulated (PSD), laying (L) incubating (I) or photorefractory (PR). Data from (25).

**Mechanisms regulating luteinizing hormone secretion**

Luteinizing hormone secretion is suppressed during incubation in gallinaceous birds (17). The mechanism underlying this suppression remains to be clarified. A suppressive effect of prolactin on LH secretion is suggested by the association between high concentrations of plasma prolactin and low concentrations of LH in incubating birds (17). This is supported by the findings that passive immunization of incubating chickens with anti-prolactin serum increases circulating LH (18), and exogenous prolactin suppresses concentrations of plasma LH and induces ovarian regression (1, 19).

Persistent nesting activity, a behaviour displayed by incubating hens, may also be important in the suppression of LH secretion, since the administration of ovine prolactin diminishes but does not prevent the rise in plasma LH when incubating chickens are deprived of their nests (19). In the turkey, the administration of ovine prolactin does not prevent the increase in LH secretion which occurs after ovariectomy (13). Further, in incubating bantam hens, the increase in plasma LH after injection of anti-cVIP serum, which reduces circulating prolactin, occurs only after the hens desert their nests (7). These observations taken together support the concept that the neuronal stimuli associated with nesting play a role in suppressing LH secretion in incubating birds (1).

Ovarian hormones, increased concentrations of plasma prolactin, and neuronal stimuli associated with nesting may act synergistically to inhibit LH secretion during incubation. In the incubating turkey, removal of nesting stimuli, which reduces prolactin secretion, has no effect on concentrations of plasma LH (20). Only when incubating hens are also ovariectomized and deprived of their nests does LH secretion increase (21, 22). Thus, increased plasma prolactin and ovarian hormones may act synergistically in incubating birds to suppress gonadotrophin secretion.

The mechanisms responsible for the depressed concentrations of plasma gonadotrophins appear to involve the hypothalamic-pituitary system. The reduced concentration of plasma LH in the incubating hen is apparently not due to an inadequate releasable pool of LH in the anterior pituitary gland since pituitaries of incubating birds release more LH in response to luteinizing hormone-releasing hormone (LHRH) challenge than pituitaries of laying birds (14, 23, 24; Fig. 3). Moreover, intramuscular injection of LHRH in turkey hens at weekly intervals, as the hens progress from laying to incubating, elicits a progressively larger LH response (24). It is possible that the enhanced pituitary response of the incubating hen to LHRH challenge results from diminished ovarian steroid feedback inhibition on the pituitary (22). Alternatively, the pituitary may become more, rather than less, sensitive to hypothalamic input, by compensatory up-regulation of LHRH receptors on pituitary cells induced by low levels of LHRH release. It is not known whether a reduced release of LHRH from the hypothalamus and/or an increase in anterior pituitary LHRH binding sites occurs as the turkey hen begins to incubate.

The amount of pituitary mRNA transcript specific for the LH-β subunit is lowest in hyperprolactinemic, incubating hens (25; Fig. 3), implying that prolactin and/or the nesting stimulus may have a suppressive effect on LH-β mRNA. Indeed, recent findings, utilizing anterior pituitary cells in culture, indicate that ovine prolactin (2 μg) suppresses steady state amounts of LHβ-mRNA and LHRH-stimulated LHβ-mRNA production, without altering LH release (26). LHRH-induced LH release *in vitro* is unaffected by ovine prolactin. Thus, reduced amounts of LHβ-mRNA in the anterior pituitary may be partially responsible for the reduced concentrations of plasma LH in incubating hens. This is substantiated by earlier findings that cultures of anterior pituitary cells from laying or incubating turkeys release LH in proportion to the circulating concentrations of plasma LH (27).

Electrical stimulation of the hypothalamus of incubating turkeys does not release LH as effectively as intravenous injection of LHRH (14; Fig. 4). This suggests the presence of an inhibitory mechanism(s) within the hypothalamus that restricts pituitary LH secretion. This may account for the finding in turkeys and bantam hens that hypothalamic LHRH-I content is lower in incubating than in laying birds (22, 28).

Deprivation of incubating turkeys of their nests results in reduced concentrations of plasma prolactin, increased hypothalamic LHRH content, and increased pituitary LH and LHβ-mRNA without increasing the suppressed concentrations of plasma LH. This observation suggests that increased plasma prolactin and/or increased nesting activity suppresses hypothalamic LHRH content, and LH and LHβ-mRNA in the anterior pituitary gland. Thus, when ovine prolactin is administered to laying turkeys, there is a reduction of hypothalamic LHRH-I and II content (29). These findings further implicate a suppression of hypothalamic LHRH release as being the reason why LH secretion does not increase in incubating turkeys in response to nest-deprivation. Concentrations of plasma LH in incubating turkeys increase after nest-deprivation only if the birds are also ovariectomized. Therefore, it appears that in the incubating turkey, nesting stimulus and/or prolactin act synergistically with ovarian hormones to inhibit LHRH release, resulting in low concentrations of plasma LH.

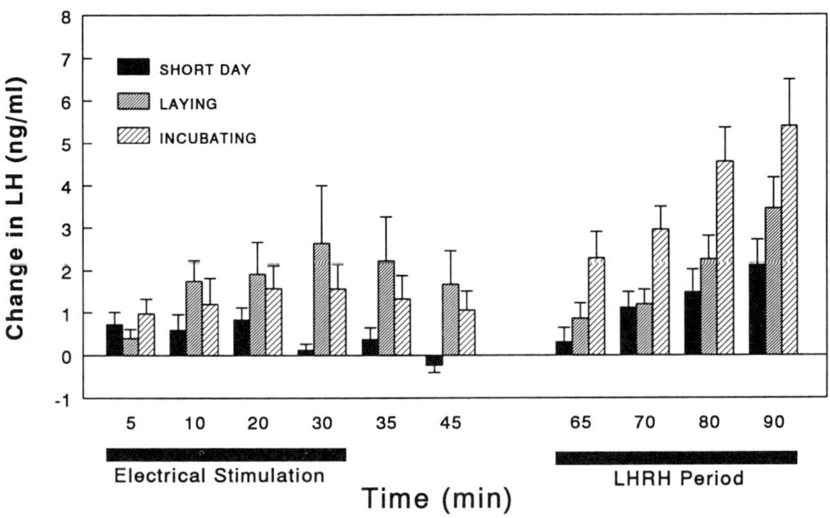


Fig. 4 A comparison of changes in plasma LH concentrations in turkey hens held on short days to prevent egg laying, and on long days to induce laying or incubation behaviour, after electrical stimulation of the hypothalamus or after an intravenous injection of 4 μg luteinizing-hormone releasing-hormone (LHRH)/kg body weight.

## Possible mechanisms for ovarian regression in incubating hens

At the onset of incubation the ovary, which contains a full complement of large preovulatory follicles, suddenly undergoes follicular atresia (30). Follicular atresia can also be induced in turkey hens by the administration of large amounts of prolactin (31). Follicular atresia begins with the largest preovulatory follicles and proceeds down the follicular hierarchy until all the preovulatory follicles have become atretic. During ovarian regression associated with the onset of incubation, follicular atresia is extensive five days after the last egg is laid (30). Circulating concentrations of LH decrease concurrently with the increase in nesting behaviour which precedes the full expression of incubation behaviour and prior to any visible evidence of follicular atresia. Thus, ovarian collapse may result from the decreased LH secretion. The sequential nature of the collapse may reflect a continuum within the follicular hierarchy of decreasing follicular sensitivity to LH.

The mechanisms underlying the decrease in concentrations of plasma ovarian steroids and subsequent ovarian regression at the onset of incubation are not known, but may involve the associated decline in plasma gonadotrophins and the increase in plasma prolactin. Prolactin is a prime candidate because it induces ovarian regression when administered exogenously (1, 31), and inhibits the ability of injections of LH to stimulate secretion of testosterone and oestrogen (32). In addition, stimulation of steroidogenesis in ovarian follicular theca cells by LH *in vitro* is suppressed by prior treatment of turkey hens with ovine prolactin (33). However, LH-induced progesterone and testosterone secretion by turkey granulosa and theca cells obtained from pre-ovulatory follicles from hens at the onset of incubation is greater than that by theca cells from follicles of laying hens (34). Therefore, it appears that granulosa and theca cells remain responsive to LH until well after the initiation of ovarian regression associated with incubation behaviour. The enhanced LH-induced progesterone and testosterone responses may reflect compensatory up-regulation of granulosa and theca cell LH receptors, since depressed concentrations of plasma LH, progesterone and testosterone are observed in turkeys well before the final oviposition and before the onset of incubation behaviour.

Unlike concentrations of progesterone and testosterone, depressed concentrations of plasma oestrogen in hens at the onset of incubation may be due, in part, to a limited potential for ovarian oestrogen production. The small white ovarian follicles are the major source of oestrogen in the chicken (35, 36) and turkeys (37). Both basal and stimulated oestrogen production by the small white follicles are depressed in hens at the onset of incubation (M El Halawani, unpublished observation). Thus, both a decrease in circulating LH and diminished oestrogen production by small white follicles occur concomitantly at the onset of incubation behaviour. It is possible that decreased oestrogen secretion by the small white follicles may interrupt the positive feedback of ovarian steroids on pituitary LH release (38, 39), thereby leading to reduced circulating LH concentrations and ovarian regression. This contention is supported by the finding that ovine prolactin inhibits LH-stimulated production of oestrogen by 'super-small follicles' (<1 mm diameter) in chickens (40). It therefore follows that an increase in circulating prolactin at the onset of incubation behaviour may induce ovarian regression by directly inhibiting ovarian oestrogen production. Alternatively, depressed oestrogen production by the small white follicles may be due to the decrease in concentrations of plasma LH.

## References

1 El Halawani M E, Silsby J L, Youngren O M & Phillips R E (1991) Exogenous prolactin delays photo-induced sexual maturity and suppresses ovariectomy-induced luteinizing hormone secretion in the turkey (*Meleagris gallopavo*). *Biology of Reproduction* **44** 420-424

2 Youngren O M, El Halawani M E, Silsby J L & Phillips R E (1991) Intracranial prolactin perfusion induces incubation behavior in turkey hens. *Biology of Reproduction* **44** 425-431

3 El Halawani M E, Youngren O M, Silsby J L & Phillips R E (1987) Prolactin release following electrical stimulation of the brain in female turkeys (*Meleagris gallopavo*). *General and Comparative Endocrinology* **65** 1-8

4 El Halawani M E, Youngren O M, Silsby J L & Phillips R E (1988) Involvement of serotonin in prolactin release induced by electrical stimulation of the hypothalamus of the turkey (*Meleagris gallopavo*). *General and Comparative Endocrinology* **72** 323-328

5 El Halawani M E, Silsby J L & Mauro L J (1990) Vasoactive intestinal peptide is a hypothalamic prolactin releasing neuropeptide in the turkey (*Meleagris gallopavo*). *General and Comparative Endocrinology* **78** 66-73

6 Mauro L J, Elde R P, Youngren O M, Phillips R E & El Halawani M E (1989) Alterations in hypothalamic vasoactive intestinal peptide-like immunoreactivity are associated with reproduction and prolactin release in the female turkey. *Endocrinology* **125** 1795-1804

7 Sharp P J, Sterling R J, Talbot R T & Huskisson N S (1989) The role of hypothalamic vasoactive intestinal polypeptide in the maintenance of prolactin secretion in incubating bantam hens: observations using passive immunization, radioimmunoassay and immunohistochemistry. *Journal of Endocrinology* **122** 5-13

8 Peczely P & Kiss J Z (1988) Immunoreactivity to vasoactive intestinal polypeptide (VIP) and thyrotropin-releasing hormone (TRH) in hypothalamic neurons of domesticated pigeon (*Columbia livia*). Alterations following lactation and exposure to cold. *Cell Tissue Research* **251** 485-494

9 Cloues R, Ramos C & Silver R (1990) Vasoactive intestinal polypeptide-like immunoreactivity during reproduction in doves: influence of experience and number of offspring. *Hormones and Behavior* **24** 215-231

10 Mauro L J, Youngren O M, Proudman J A, Phillips R E & El Halawani M E (1992) Effects of reproductive status, ovariectomy and photoperiod on vasoactive intestinal peptide in the female turkey hypothalamus. *General and Comparative Endocrinology* **87** 481-493

11 El Halawani M E, Silsby J L & Mauro L J (1990) Enhanced vasoactive intestinal peptide-induced prolactin secretion from anterior pituitary cells of incubating turkeys (*Meleagris gallopavo*). *General and Comparative Endocrinology* **80** 138-145

12 Talbot R T, Hanks M C, Sterling R J, Sang, H M & Sharp P J (1991) Pituitary prolactin messenger ribonucleic acid levels in incubating and laying hens: effects of manipulating plasma levels of vasoactive intestinal polypeptide. *Endocrinology* **129** 496-502

13 Rozenboim I & El Halawani M E (1992) Characterization of vasoactive intestinal peptide pituitary membrane receptors in turkey hens during different stages of reproduction. *Biology of Reproduction* **46** Supplement 1 96

14 El Halawani M E, Youngren O M, Silsby J L & Phillips R E (1991) Effect of reproductive condition on luteinizing hormone (LH) and prolactin (prolactin) release induced by electrical stimulation (ES) of the turkey hypothalamus. *Poultry Science* **70** Supplement 1 39

15 El Halawani M E, Silsby J L, Koike T I & Robinzon B (1992) Evidence of a role for the turkey posterior pituitary in prolactin release. *General and Comparative Endocrinology* **87** 436-442

16 Wong E A, Ferrin N H, Silsby J L & El Halawani M E (1991) Cloning of a turkey prolactin cDNA: expression of prolactin mRNA throughout the reproductive cycle of the domestic turkey (*Meleagris gallopavo*). *General and Comparative Endocrinology* **83** 18-26

17 El Halawani M E, Fehrer S, Hargis B M & Porter T E (1988) Incubation behaviour in the domestic turkey: physiological correlates. *CRC Critical Reviews in Poultry Biology* **1** 285-314

18 Lea R W, Dods A S M, Sharp P J & Chadwick A (1981) The possible role of prolactin in the regulation of nesting behaviour and the secretion of luteinizing hormone in broody bantams. *Journal of Endocrinology* **91** 89-97

19 Sharp P J, Macnamee M C, Sterling R J, Lea R W & Pederson H C (1988) Relationships between prolactin, LH and broody behaviour in bantam hens. *Journal of Endocrinology* **118** 279-286

20 El Halawani M E, Silsby J L, Fehrer S C & Behnke E J (1983) Reinitiation of ovulatory cycles in incubating female turkeys by an inhibitor of serotonin synthesis, p-chlorophenylalanine. *Biology of Reproduction* **28** 221-228

21 Zadworny D & Etches R J (1987) Effects of ovariectomy or forced feeding on the plasma concentrations of prolactin and luteinizing hormone in incubating turkey hens. *Biology of Reproduction* **36** 81-88

22 El Halawani M E, Silsby J L, Foster L K, Rozenboim I & Foster D N (1992) Prolactin-ovarian steroid synergism in the suppression of luteinizing hormone (LH) in the incubating turkey (*Meleagris gallopavo*). *Poultry Science* **71** Supplement 1 22

23 Sharp P J, Sterling R J, Milton R C & Millar R P (1986) Effect of luteinizing hormone releasing hormone and its analogues on plasma luteinizing hormone concentrations in incubating bantam hens. *British Poultry Science* **27** 124-135

24 El Halawani M E, Fehrer S C & Silsby J L (1987) Enhanced luteinizing hormone release by luteinizing hormone-releasing hormone in incubating female turkeys (*Meleagris gallopavo*). *Biology of Reproduction* **36** 884-889

25 Wong E A, Silsby J L, Ishii S & El Halawani M E (1992) Pituitary luteinizing hormone and prolactin mRNA levels are inversely related in laying and incubating turkey hens. *Biology of Reproduction* **47** 598-602

26 You S, Foster L K, Youngren O M, El Halawani M E & Foster D N (1992) Turkey luteinizing hormone β subunit complimentary DNA nucleotide sequence. *Poultry Science* **71** Supplement 1 57

27 El Halawani M E, Silsby J L & Fehrer S C (1988) Basal and hypothalamic extract-induced luteinizing hormone and prolactin secretion by cultured anterior pituitary cells from female turkeys in various stages of the reproductive cycle. *General and Comparative Endocrinology* **71** 45-54

28 Burke W H, Kamiyoshi M & Arbtan K D (1990) Hypothalamic neuropeptides in incubating and non-incubating Black Wyandotte bantam chickens. *Neuroendocrinology* **52** Supplement 1 51

29 Rozenboim I, Tabibzadeh C H, Silsby J L & El Halawani M E (1992) The effect of ovine prolactin administration on hypothalamic vasoactive intestinal peptide (VIP) and anterior pituitary VIP receptors in laying turkey hens. *Poultry Science* **71** Supplement 1 21

30 Porter T E (1988) Steroidogenesis during follicular growth and atresia in the domestic turkey. PhD Thesis, University of Minnesota

31 Hargis B M (1987) Anti-ovarian activity of prolactin in the domestic turkey hen. PhD Thesis, University of Minnesota

32 Morrison T O, Lein R J & El Halawani M E (1990) Influence of ovine prolactin administration on *in vivo* steroidogenic responses of laying turkeys to ovine LH (oLH). *Poultry Science* **69** Supplement 1 96

33 Lien R J, Pusztai A & El Halawani M E (1989) Influence of the administration of ovine prolactin to laying turkeys on subsequent ovarian steroidogenesis *in vitro*. *Poultry Science* **68** Supplement 1 86

34 Porter T E, Silsby J L, Behnke E J, Knapp T R & El Halawani M E (1991) Ovarian steroid production *in vitro* during gonadal regression in the turkey. 1. Changes associated with incubation behavior. *Biology of Reproduction* **45** 581-586

35 Robinson F E & Etches R J (1986) Ovarian steroidogenesis during follicular maturation in the domestic fowl (*Gallus domesticus*). *Biology of Reproduction* **35** 1096-1105

36 Armstrong D G (1984) Ovarian aromatase activity in the domestic fowl (*Gallus domesticus*). *Journal of Endocrinology* **100** 81-86

37 Porter T E, Hargis B M & El Halawani M E (1991) Characterization of dissimilar steroid productions by granulosa, theca interna and theca externa cells during follicular maturation in the turkey (*Meleagris gallopavo*). *General and Comparative Endocrinology* **84** 1-8

38 El Halawani M E, Silsby J L, Fehrer S C & Behnke E J (1983) Effects of estrogen and progesterone on serum prolactin and luteinizing hormone levels in ovariectomized turkeys (*Meleagris gallopavo*). *General and Comparative Endocrinology* **52** 67-78

39 Wilson S C & Sharp P J (1976) Induction of luteinizing hormone release by gonadal steroids in the ovariectomized domestic hen. *Journal of Endocrinology* **71** 87-98

40 Zadworny D, Shimada K, Ishida H & Sata K (1989) Gonadotropin-stimulated estradiol production in small ovarian follicles of the hen is suppressed by physiological concentrations of prolactin *in vitro*. *General and Comparative Endocrinology* **74** 468-473

Avian Endocrinology
Ed. P J Sharp, pp 111-122
Journal of Endocrinology Ltd, Bristol (1993)

# Integration of environmental factors regulating transitions of physiological state, morphology and behaviour

J C Wingfield[1], T P Hahn[1] and D Doak[2]
[1]Department of Zoology, NJ-15, University of Washington, Seattle, WA 98195, USA; [2]Department of Biology, University of California, Santa Cruz, CA 95604, USA

## Introduction

Virtually all organisms livc in fluctuating environments: night and day; high tide/low tide; spring, summer, autumn and winter; dry season and rainy season. Accordingly, they must make adjustments for each environmental condition so that fitness is maximized. In vertebrates, as in many other organisms, endocrine secretions regulate morphological, physiological and behavioural adjustments in anticipation of these predictable changes. The individual thus prepares for variations in its environment by changing phenotype.

Phenotypic changes are orchestrated by neural and hormonal responses to environmental signals that provide information predictive of future conditions in the habitat. Neural responses allow very rapid changes to an external signal, but phenotypic change in anticipation of environmental fluctuations depends more upon neuroendocrine and endocrine secretions. It is here that the field of endocrine physiology interfaces with ecology and behaviour, forming a continuum from purely organismal biology to cell and molecular aspects of hormone action, transduction of environmental signals in the brain, cell communication systems and secretion. The emerging field of 'environmental endocrinology' has the potential to integrate many fields of biology at a time when polarization and reductionism is the norm.

There is a vast literature on the responses of birds to environmental signals. However, there appear to be two major types of environmental signal that provide predictive information regulating transitions in morphology, physiology and behaviour (1). Initial predictive information, such as photoperiod, provides very reliable long term predictive cues so that an individual can begin preparing for a future event several weeks or even months in advance. Such long-term predictive signals are then integrated with the second group of environmental factors, called supplementary information (local temperature, food availability etc.), that provide short-term predictive cues. In most habitats there is variation in predictable events. For example, some springs are early and warm, others are late and cool; rainy seasons can fail or be prolonged. Therefore the degree to which individuals of a given population integrate these types of environmental information is critical and varies depending upon how accurately timed the future event must be (e.g. onset of breeding, migration, moult).

The extent to which environmental signals are integrated also has profound implications for the mechanisms by which the central nervous system perceives those signals, and transduces them into neuroendocrine and endocrine secretions that then regulate phenotypic change. It has been shown mathematically that if a future event is very predictable, then only a very few, or even one, reliable environmental cue(s) should be required to prepare for and time that future event (2). The more variable the future event becomes, the more environmental signals need to be processed to coordinate phenotypic change accurately with environmental fluctuations (2, 3). In other words, the degree to which initial predictive information is important, and the extent to which it is integrated with supplementary information, depends upon predictability of phenological progression. Incorporation of mathematical techniques such as log-linear analysis (4) and information theory (5) can be used to model when and if populations should integrate predictive environmental signals, or rely on one or the other. Simple demographic data such as egg-laying dates, timing of moults or migrations will provide bases from which to test these models. Mechanisms by which the brain transduces such information can then be compared in populations that represent different extremes of responsiveness, so that the pathways by which this information is then passed on to specific tissues, and how those tissues respond, can be determined.

## Predictability of future events

It is well known that different species of birds differ in their annual cycles as indicated by timing of migrations, moult, egg-laying dates and so on. Some have very consistent cycles with little variation from year to year, some are continuous breeders, and yet others are highly variable or even opportunistic (1, 3). Here we will use breeding seasons as examples of how mathematical techniques can provide an indication of predictability of a future event, and adjustments of phenotypic change (i.e. gonadal development), in anticipation of that event. Egg-laying dates from several breeding seasons (years) can be tabulated in a matrix with, for example, month as times within a cycle for each year that data are available. Actual numbers of clutches initiated can be entered for each month or, if such precise data are not available, an indication of whether eggs were laid or not within any month is sufficient. Log-linear analysis (4) then provides an indication of variability in egg-laying dates that can be partitioned among seasonality (variability from month to month), year to year variation, and the interaction of month and year. The degree of month-year interaction indicates unpredictability of the breeding season over the years in which data are available. Clearly the more years of information collected, the greater precision the values for percent variations will have, but even a few years (e.g. 3-5) are still very useful and indicative of how variability is partitioned.

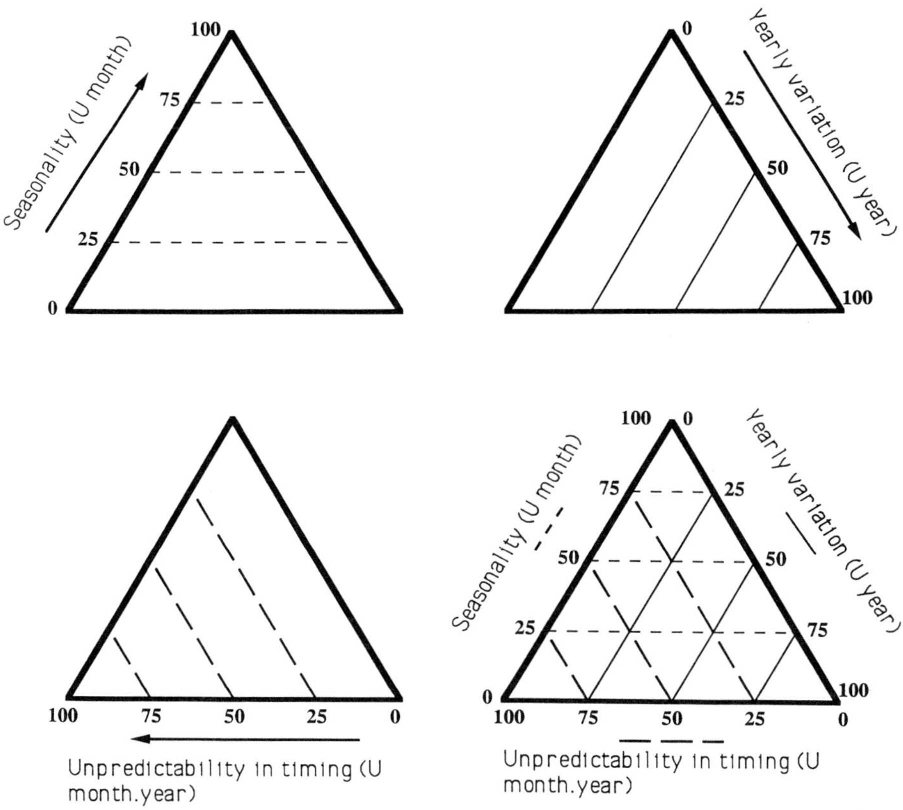


Fig. 1 Representation of the percent variation due to seasonality (U month), yearly variation (U year) and unpredictability in timing (U month.year). Percent contribution by each variable can be plotted along the lines indicated within the triangle to give a plot of predictability.

These variables can be presented along the axes of a triangle (Fig. 1). Percent variability due to seasonality (U month) is plotted on one axis along a line projecting to the next axis. Percent yearly variation (U year) is plotted on the next axis along a line projecting to the third axis. Finally, percent variation due to an interaction of month and year (U month.year, or unpredictability) is plotted along the third axis on a line projecting to the first axis. These plots and the combination of all three components are outlined in the lower right hand panel of Figure 1. This triangle is a visual representation of the relative contributions of each component to timing of breeding. Note that predictability is the inverse of unpredictability (U month.year) and can be obtained by subtracting the value for U month.year from 100%.

**Maximum month to month variation, no contribution by year**

| Year | Jan | Feb | Mar | Apr | May | Jun | Jul | Aug | Sep | Oct | Nov | Dec |
|---|---|---|---|---|---|---|---|---|---|---|---|---|
| 1 | 0 | 0 | 0 | 0 | 0 | 10 | 10 | 0 | 0 | 0 | 0 | 0 |
| 2 | 0 | 0 | 0 | 0 | 0 | 10 | 10 | 0 | 0 | 0 | 0 | 0 |
| 3 | 0 | 0 | 0 | 0 | 0 | 10 | 10 | 0 | 0 | 0 | 0 | 0 |
| 4 | 0 | 0 | 0 | 0 | 0 | 10 | 10 | 0 | 0 | 0 | 0 | 0 |

**Seasonality (U month) = 1.0; Yearly variation (U year) = 0; Unpredictability in timing (U month.year = 0)**

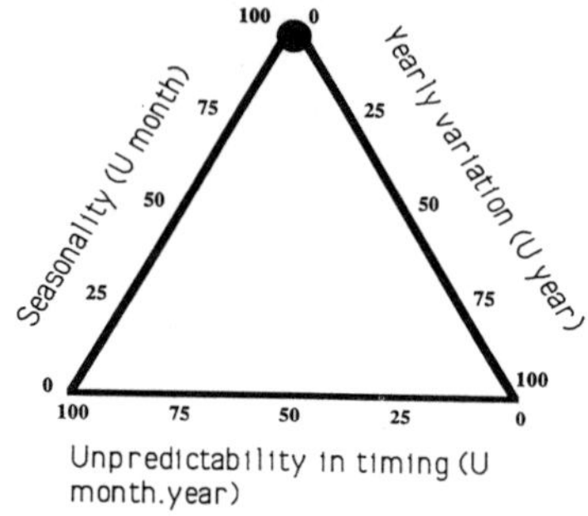


**Maximum year to year variation, no contribution by months**

| Year | Jan | Feb | Mar | Apr | May | Jun | Jul | Aug | Sep | Oct | Nov | Dec |
|---|---|---|---|---|---|---|---|---|---|---|---|---|
| 1 | 10 | 10 | 10 | 10 | 10 | 10 | 10 | 10 | 10 | 10 | 10 | 10 |
| 2 | 0 | 0 | 0 | 0 | 0 | 0 | 0 | 0 | 0 | 0 | 0 | 0 |
| 3 | 0 | 0 | 0 | 0 | 0 | 0 | 0 | 0 | 0 | 0 | 0 | 0 |
| 4 | 10 | 10 | 10 | 10 | 10 | 10 | 10 | 10 | 10 | 10 | 10 | 10 |

**Seasonality (U month) = 0; Yearly variation (U month) = 1.0; Unpredictability in timing (U month.year) = 0**

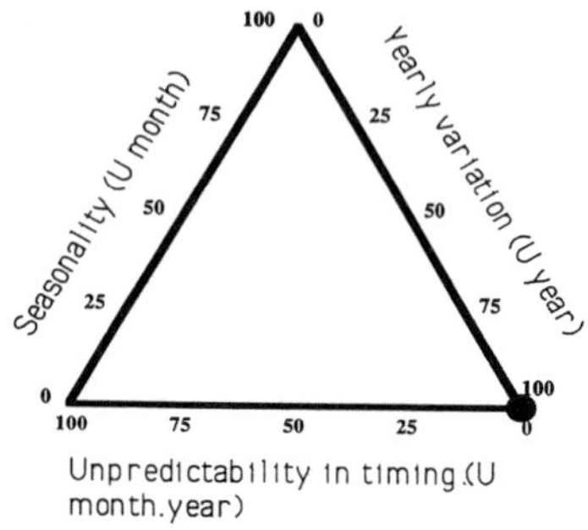


**No year to year variation, strong contribution of year x month**

| Year | Jan | Feb | Mar | Apr | May | Jun | Jul | Aug | Sep | Oct | Nov | Dec |
|---|---|---|---|---|---|---|---|---|---|---|---|---|
| 1 | 10 | 10 | 10 | 0 | 0 | 0 | 0 | 0 | 0 | 0 | 0 | 0 |
| 2 | 0 | 0 | 0 | 10 | 10 | 10 | 0 | 0 | 0 | 0 | 0 | 0 |
| 3 | 0 | 0 | 0 | 0 | 0 | 0 | 10 | 10 | 10 | 0 | 0 | 0 |
| 4 | 0 | 0 | 0 | 0 | 0 | 0 | 0 | 0 | 0 | 10 | 10 | 10 |

**Seasonality (U month) = 0; Yearly variation (U year) = 0; Unpredictability in timing (U month.year) = 1.0**

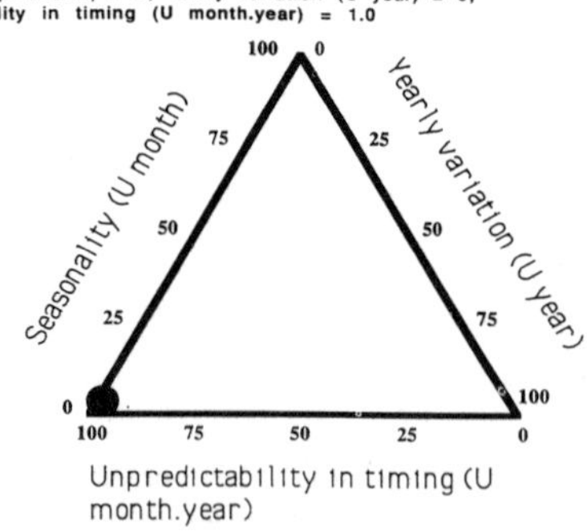


**Combination of variation by month, year, and month x year**

| Year | Jan | Feb | Mar | Apr | May | Jun | Jul | Aug | Sep | Oct | Nov | Dec |
|---|---|---|---|---|---|---|---|---|---|---|---|---|
| 1 | 0 | 0 | 0 | 0 | 10 | 10 | 0 | 0 | 0 | 0 | 0 | 0 |
| 2 | 0 | 0 | 10 | 10 | 10 | 10 | 10 | 0 | 0 | 0 | 0 | 0 |
| 3 | 0 | 0 | 0 | 0 | 10 | 10 | 10 | 10 | 10 | 10 | 0 | 0 |
| 4 | 0 | 0 | 0 | 0 | 0 | 10 | 0 | 0 | 0 | 0 | 0 | 0 |

**Seasonality (U month) = 0.47; Yearly variation (U year) = 0.15; Unpredictability in timing (U month.year) = 0.38**

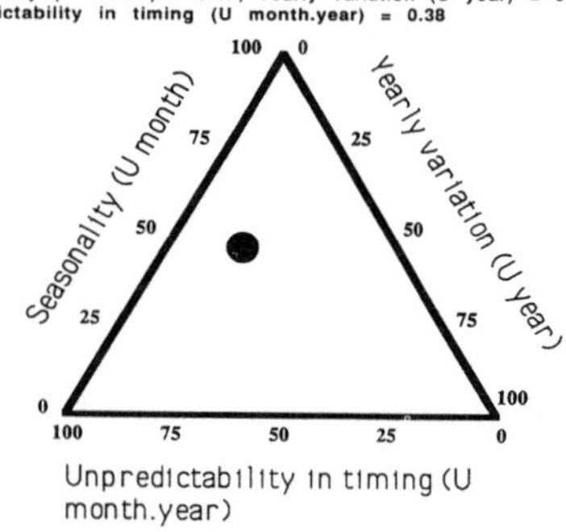


Fig. 2 Examples of matrices representing hypothetical extremes of egg laying periods. In the matrix the value 10 indicates that eggs were laid in that month and 0 indicates that no eggs were laid in that month. The percent variation for each axis is given beneath the matrix. The large black spot indicates where the point representing each percent variation should be placed on the triangle. See text for further explanation.

To demonstrate how the model works we have constructed 4 hypothetical matrices in Figure 2. At the upper left, eggs are laid in June and July of every year (as indicated by presence of a 10 - eggs present, or 0 - no eggs present). In this case, all variation is due to U month and none to U year, or interaction between month and year (U month.year). A point can be placed at the upper apex as a visual representation of the contribution of all three components. Note that predictability is thus 100%. In the upper right hand example, eggs are laid in every month in two years, but no laying occurred in the other two years. These data reveal that all variation is due to U year and none to monthly variation (seasonality, U month) or interaction of month and year (U month.year), thus placing the point at the right hand apex. In this extreme case it should be pointed out that although breeding is possible in all months, each year can be completely different (i.e. boom or bust). Again predictability is 100%, but note that this refers to timing of breeding within a year rather than to variation from year to year. In other words, when conditions are favourable, breeding is possible in all months; when conditions are unfavourable, no breeding can occur that year.

The lower left panel of Figure 2 shows a matrix in which there is no contribution of U month or U year. All variation is explained by an interaction of month and year (U month.year) thus placing the point at the left hand apex of the triangle. In other words the month in which breeding occurs varies from year to year. In this example predictability is zero. Finally, in the lower right example of Figure 2, there is a combination of variation by month, year, and the interaction of month and year placing the point toward the centre of the triangle. This hypothetical case is the most complex. We have generated a matrix in which year to year variation and seasonality (U month) is significant, and in which the months when breeding can occur vary from year to year. Here we wish to demonstrate that all three components can contribute to variation in breeding times. In this example predictability is about 62%.

Obviously, such artificial matrices given above may not exist in nature, but they do serve to illustrate the flexibility of this type of analysis and the information it can generate. According to Cohen (2), if a future event is predictable then there should be selection for sensitivity to a few or even one very reliable environmental signal, and other factors could be ignored. It thus follows that if predictability is high, then initial predictive information (that provides reliable long term cues) would be important. Going back to Figure 2, if U month is high then predictability is high due to seasonality. The time of year in which breeding occurs is similar in all years and initial predictive information would be very important. As U month.year increases then predictability declines because there is variation in when breeding can be timed from year to year. In this case it follows that, although initial predictive information may still be important, other (e.g. supplementary) information should also be integrated in order to time breeding accurately. The contribution made by U year indicates whether or not breeding occurs equally in all years, but does not provide any information about the timing within years.

Log-linear analysis is a method that can provide a measure of predictability and thus indicate to what extent an individual should use initial predictive information (such as photoperiod) to time a future event. However, we should be cautious in concluding how this model provides a measure of whether initial predictive information should be integrated with supplementary factors (i.e. when U month.year makes a significant contribution). For example, if U month.year is high, then low predictability suggests that initial predictive information is less important. On one hand it could be argued that low predictability from this model requires integration with supplementary information because the environment is truly uncertain, or conversely that no information is required because the environment is certain and favourable virtually at all times (e.g. continuous breeders, see below). Log-linear analysis does not distinguish between these equally plausible interpretations. We suggest that although log-linear analysis may be useful for assessing the relative importance of initial predictive information, it is less useful for determining the extent to which supplementary cues should also be used. Other mathematical methods are sensitive to this issue.

**Integration of environmental cues**

Using the same matrix in a slightly different form (month as times in a cycle and two states - eggs present or not present - summed over all the years of study) we have applied a model based on information theory (5). In this method, predictability has two components: constancy (C) and contingency (M). An environment can be predictable because it is always the same (constant) or because there are maximal changes in conditions from favourable to unfavourable within each year (contingency) (5). Predictability (Pr) is then equal to constancy plus contingency ($Pr = M + C$). It is very important to point out here that the predictability value from this model is not the same as that determined by log-linear analysis. From log-linear analysis we infer predictability of timing from U month and U month.year. In Colwell's method (5), predictability can be high due to contingency or constancy and thus has little bearing on whether initial predictive information is important.

Wingfield *et al.* (3) have applied this model to several avian breeding systems. They suggest that the combination of predictability and its components, constancy and contingency, gives an indication of the degree to which environmental information should be integrated. This reduces to give a ratio M/C or the 'environmental information factor', Ie (3). Matrices of egg-laying dates for several species indicate that continuous breeders (a high degree of constancy) have a low Ie value and thus probably use few if any environmental cues to time breeding (they can always breed), whereas periodic breeders have a wide range of Ie values indicating great variation in the amount of information that must be integrated to time reproduction (3). Data on species that have been tested experimentally for a correlation of Ie with responsiveness to environmental cues support this idea (3).

It is proposed here that the application of log-linear analysis and the contingency/constancy model of predictability to the same data can provide an indication of firstly, whether initial predictive information is important (a high predictability value from log-linear analysis) and secondly, to what extent integration with supplementary cues is important (the Ie value from the contingency/constancy model of predictability). Simple information gleaned from demographic data collected in the field could thus provide a potentially in-depth assessment of how an individual may transduce environmental signals that in turn regulate neuroendocrine and endocrine secretions. Several examples are given below.

Breeding seasons are well studied for several taxa of *Zonotrichia*, and matrices can easily be constructed (3). Briefly, *Z. leucophrys gambelii* has a highly predictable and brief breeding season at high latitudes whereas the other taxa have varying lengths of breeding season at mid to low latitudes.

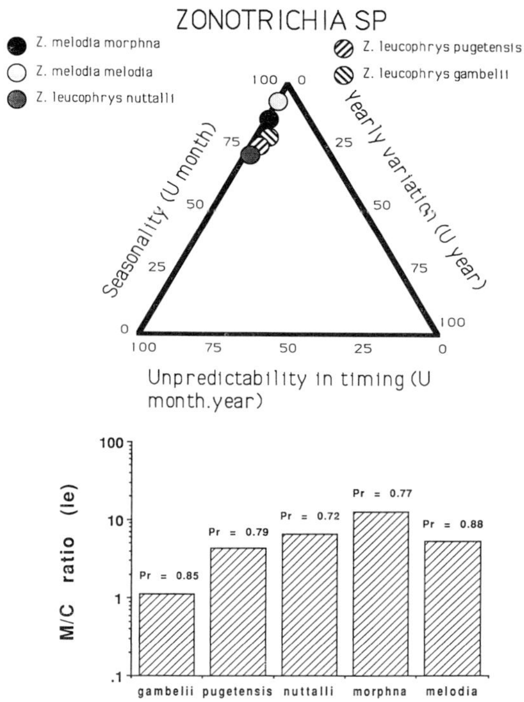


Fig. 3 Log-linear analysis of breeding data for five taxa of *Zonotrichia*. Based on the same matrices, the M/C ratio, (Ie) values are plotted for the same taxa (lower panel). The Ie values are a measure of the degree to which environmental information factors are integrated to time seasonal breeding. See reference (3) for details on matrices and breeding biology.

Log-linear analysis of matrices made up of egg laying dates are summarized in the triangle in Fig 3. All species show high seasonality (U month) with little yearly variation (U year) or interaction (U month.year). Thus predictability is high in all cases, suggesting that initial predictive information is important in timing gonadal development in all these populations. Existing data indeed show that these species are highly photoperiodic (reviewed in 1, 3). However, log-linear analysis may be less sensitive in distinguishing ability to integrate environmental cues that fine tune onset of breeding among the taxa. We know that these populations of *Zonotrichia* show wide variation in breeding periodicity (especially individuals within populations). If we apply the contingency/constancy model of predictability and calculate Ie values (3) we see considerable variation, perhaps indicating better resolution of the degree to which integration of environmental signals is favoured (lower panel of Fig. 3). *Z. l. gambelii* has the lowest Ie factor and thus probably relies mostly on photoperiod to regulate gonadal development and ignores other cues (3). In contrast, others such as *Z. melodia morphna* has an over ten-fold higher Ie value, suggesting that it integrates photoperiodic cues with many other factors. Field data support this view insofar as gonadal maturation in *gambelii* is constant from year to year, whereas it varies (probably with local temperature) at least in *Z. m. melodia* (3).

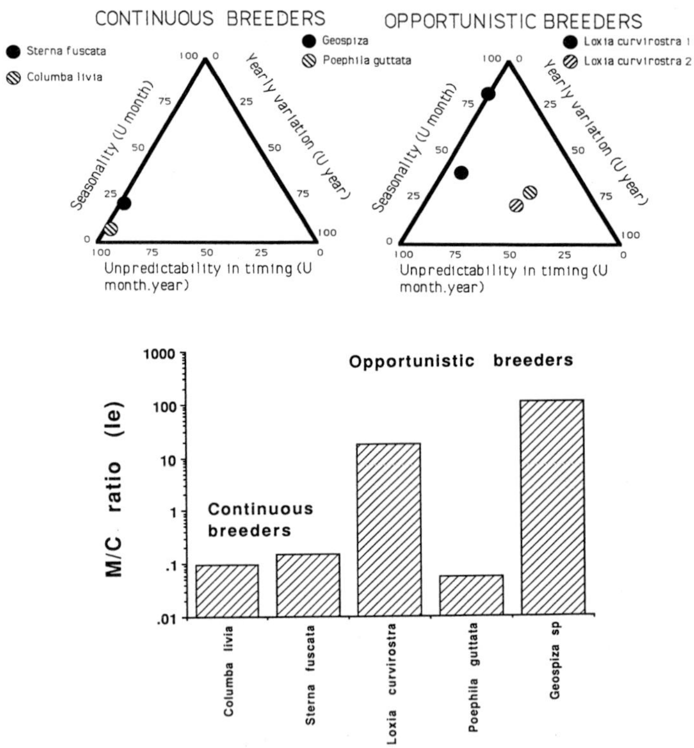


Fig. 4 Log-linear analysis of breeding data for continuous and opportunistic breeders. The lower panel shows corresponding M/C ratio (Ie) values for these taxa. The Ie values are a measure of the degree to which environmental information factors are integrated to time seasonal breeding. See reference (3) for matrices and breeding biology.

In Figure 4 we compare different species with widely different breeding seasons. Two continuous breeders - the sooty tern, *Sterna fuscata*, and rock dove, *Columba livia* - show very little seasonality (U month) or yearly variation (U year). However, the interaction between month and year is high (U month.year) suggesting very low predictability (left hand triangle of Fig. 4). In this case we would expect initial predictive information to be important in neither species. Indeed, it has been shown in rock doves that gonadal development is not influenced by photoperiod (6). It also appears that the Ie values for continuous breeders are extremely low (ca 0.1). We interpret this as an indication that continuous breeders need very little environmental information (if any) to time gonadal development and onset of breeding (i.e. they can always breed). Note that the predictability value using the contingency/constancy model is high because of constancy (3). This should not be confused with the predictability value from log-linear analysis which does not take constancy and contingency into account. In contrast to the continuous breeders, several opportunistic species (right hand triangle of Fig. 4) show varying degrees of seasonality (U month), year to year variation (U year) and unpredictability (U month.year). Some show high predictability in which an initial predictive cue, such as endogenous rhythm, may be important (e.g. a Galapagos finch, *Geospiza sp.*) and others apparently rely less on initial predictive cues (e.g. red crossbill, *Loxia curvirostra* and zebra finch, *Poephila guttata*). Furthermore, there is also great variation in Ie values (lower panel of Fig. 4). The zebra finch has a very low Ie value, indicating that integration of environmental cues may be negligible. Data do suggest that some truly opportunistic populations of this species may need little environmental information to time breeding. They are thought to maintain gonads in a near functional condition at all times and then respond to a short-term predictive cue, such as rainfall, to trigger actual onset of breeding (7). On the other hand, red crossbills and Galapagos finches show one and two orders of magnitude greater Ie values (respectively) indicating that they may integrate more cues to time reproductive development and onset of breeding. It is known that both these taxa show marked variation in gonadal development from year to year, and timing of breeding may have both seasonal and opportunistic components (3, 8). Thus, we might expect these species to integrate initial predictive and supplementary cues to a much greater extent than the zebra finch (3).

Although the examples given thus far focus on breeding seasons, it should be noted that these kinds of analyses can be applied to any periodic process in any organism. As an example of the versatility of these methods we show how they can be applied to several events within the annual cycle of one species, *Z. l. gambelii* (Fig. 5). Log-linear analysis indicates a high degree of seasonality (U month) for both moults and migratory periods, as well as the breeding and non-breeding seasons. Year to year variation (U year) and unpredictability (U month.year) are very low. This suggests that initial predictive information is very important for regulation of these processes. In support of this conclusion, a number of investigations indicate that they are all regulated by photoperiod (e.g. 9-11). On the other hand, Ie values (lower panel of Fig. 5) vary by almost 20-fold, suggesting that at least autumn migration and the winter period may require much greater integration of initial predictive and supplementary cues than the other processes. The data also suggest that moults and spring migration, in addition to gonadal development (see above), are regulated primarily by photoperiod and less by other supplementary cues.

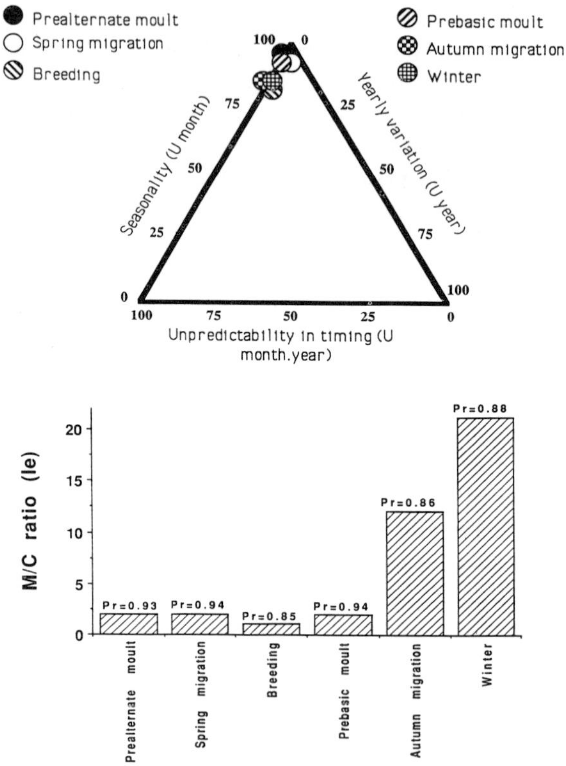


Fig. 5 Log-linear analysis of various events in the annual cycle of *Z. l. gambelii*. The lower panel shows corresponding M/C ratio (Ie) values for the same processes. The Ie values are a measure of the degree to which environmental information is integrated to time the various seasonal events. See reference (3) for further details.

## Tests of predictability models

The matrices and their analysis outlined above generate specific hypotheses concerning the integration of environmental information in the control of periodic phenotypic change. The models apply equally well to morphological, physiological and behavioural adjustments, but here we will continue to focus on control of reproductive cycles as an example. By far the most well-known initial predictive factor is the annual change in day length. The mechanisms by which day length influences gonadotrophin-releasing hormone (GnRH) synthesis and release, and subsequently its effects on luteinizing hormone (LH), follicle-stimulating hormone (FSH) and gonadal development have been investigated in depth (12). However, the degree to which other environmental factors influence photoperiodically-induced gonadal development have been neglected (1). Investigations show that low environmental temperature can inhibit photoperiodically-induced testicular development in some species but not others (1, 3). In *Z. l. gambelii*, that has a low Ie value and thus should be less sensitive to supplementary factors, low environmental temperature had only slight inhibitory effects on gonadal development following transfer from short to long days (13, 14). Other species in which Ie values are probably higher, tend to show marked effects of low temperature on photoperiodically-induced gonadal development (3). Effects on females are much less well known but deserve further study.

The effects of other supplementary factors have been even less well investigated. Although there is field evidence for effects of several environmental cues, both to inhibit and accelerate gonadal development (1), the mechanisms remain largely unknown. It is well documented that food deprivation can suppress gonadal recrudescence and inhibit ovulation, possibly through stress-induced increases in corticosterone (15). However, controlled experiments in which food is restricted, but not to a starvation level, should throw more meaningful light on this issue. In mammals, several investigators have reported effects of quality of diet (essential amino acids, free fatty acids, calcium etc.) on secretion of gonadotrophins and their control (1, 16), but none have related these effects to ecological and behavioural aspects of an organism's reproductive biology. The models described here suggest a series of experiments comparing populations that are potentially sensitive or insensitive to environmental cues. These could provide valuable comparisons for elucidating brain pathways for processing of environmental signals in general.

The genus *Zonotrichia*, with several closely related species showing a wide range of Ie values, but similar predictabilities as derived from log-linear analysis, would be ideal models to test the hypotheses generated from Figure 3. They would also be ideal for further tests of the models concerning processes of phenotypic change other than those involving the reproductive system (Fig. 5). Of particular interest would be to determine the interaction of photoperiod and other cues in the regulation of moult, metabolism, migratory behaviour and wintering strategies. Other processes are also candidates for these types of analyses.

## Acknowledgements

These investigations were supported by grant number DCB-9005081 from the National Science Foundation to JCW.

## References

1 Wingfield J C & Kenagy G J (1991) Natural regulation of reproductive cycles. *Vertebrate Endocrinology: Fundamentals and Biomedical Implications* pp 181-241. Eds M Schreibmann & R E Jones. New York: Academic

2 Cohen D (1967) Optimizing reproduction in a varying environment. *Journal of Theoretical Biology* **16** 1-14

3 Wingfield J C, Hahn T P, Levin R & Honey P (1991) Environmental predictability and the control of gonadal cycles in birds. *Journal of Experimental Zoology* **261** 214-231

4 Sokal R R & Rohlf F J (1981) *Biometry* 859 pp. New York: Freeman

5 Colwell R K (1974) Predictability, constancy and contingency of periodic phenomena. *Ecology* **55** 1148-1153

6 Murton R K, Thearle R J P & Thompson J (1972) Ecological studies of the feral pigeon, *Columba livia*, var. 1. Population, breeding biology and methods of control. *Journal of Applied Ecology* **9** 835-874

7 Vleck C M & Priedkalns J (1985) Reproduction in zebra finches: hormone levels and effect of dehydration. *Condor* **87** 37-46

8 Gibbs H L & Grant P R (1987) Ecological consequences of an exceptionally strong El Niño event on Darwin's finches. *Ecology* **68** 1735-1746

9 Farner D S & Gwinner E (1980) Photoperiodicity, circannual and reproductive cycles. *Avian Endocrinology* pp 331-366. Eds A Epple and M H Stetson. New York: Academic

10 Moore M C, Farner D S, Donham R S & Matt K S (1985) Endocrine and photoperiodic relationships during photorefractoriness, postnuptial moult, and onset of migration. *Acta XVIII Congressus Internationalis Ornithologici* pp 946-952. Eds V O Ilyichev & V M Gavrilov. Moscow: Nauka

11 Wingfield J C & Farner D S (1980) Control of seasonal reproduction in temperate-zone birds. *Progress in Reproductive Biology* **5** 62-101

12 Nicholls T J, Goldsmith A R & Dawson A (1988) Photorefractoriness in birds and comparison with mammals. *Physiological Reviews* **68** 133-176

13 Farner D S & Mewaldt L R (1952) The relative roles of photoperiod and temperature in gonadal recrudescence in male *Zonotrichia leucophrys gambelii. Anatomical Record* **113** 106

14 Lewis R A & Farner D S (1973) Temperature modulation of photoperiodically-induced vernal phenomena in white-crowned sparrows (*Zonotrichia leucophrys*). *Condor* **75** 279-286

15 Wingfield J C (1988) Changes in reproductive function of free-living birds in direct response to environmental perturbations. *Processing of Environmental Information in Vertebrates* pp 121-148. Ed. M H Stetson. Berlin: Springer-Verlag

16 Bronson F H (1988) Seasonal regulation of reproduction in mammals. *The Physiology of Reproduction* Vol 2 pp 1831-1871. Eds E Knobil & J Neill. New York: Raven

# Part Three
# Pituitary Hormones and their Receptors

Avian Endocrinology
Ed. P J Sharp, pp 123-134
Journal of Endocrinology Ltd, Bristol (1993)

# Molecular biology of gonadotrophins

S Ishii[1], H Ando[1], H Wako[2] and Y Kubota[3]
[1]School of Education, Waseda University and [2]School of Social Science, Waseda University, Tokyo 169-50, Japan; [3]Biosystems Laboratory, Nova Inc, Shinjuku-ku, Tokyo 160, Japan

## Introduction

Information on the chemical structure of avian gonadotrophins is limited, although a number of studies have been published on the chemical structure of mammalian and fish gonadotrophins (1). The primary structures of the LH beta subunit (2-4) or the gonadotrophin alpha subunit (5, 6) have been reported for three birds: the chicken (2, 5), Japanese quail (3, 6) and turkey (4). No study has been published on the primary structure of follicle-stimulating hormone (FSH) of any bird. In mammals and fish, the primary structure of gonadotrophins has been determined directly by chemical analysis of purified hormones and/or indirectly by deduction from the nucleotide sequence of cloned hormone genes. In birds, however, only the latter method has been used for determination of the primary structure of the gonadotrophin subunit molecules.

In this review, the primary structures of avian gonadotrophin subunits are compared with the structures of subunits from other vertebrate gonadotrophins and are discussed in relation to their evolution and the significance of their secondary structure for biological activity.

## Cloning and nucleotide sequence of cDNA for the LH beta subunit precursor molecule in birds

The primary structure of an avian LH beta subunit was first reported at the 4th International Symposium on Avian Endocrinology in Tokyo in 1989 (2, 3). The cDNA encoding the precursor molecule of the chicken LH beta subunit was cloned, and the primary structure of the LH beta precursor molecule was deduced from the nucleotide sequence.

The authors of the study constructed a cDNA expression library in a vector, lamda gt11, from poly(A)+RNA from broiler chicken pituitary glands. After screening the library with an antiserum against chicken LH beta (provided by Professor K Wakabayashi of Gunma University), a cDNA clone (L12) was obtained containing a 436 bp insert. Using a subclone (pL12) of L12 in pUC19 as the hybridization probe, another cDNA clone (LF127) with a 533 bp insert was isolated. Clone LF127 contained the full length cDNA encoding the putative chicken LH beta subunit (Fig. 1). Using LF127 as the hybridization probe, Ando *et al.* (3) cloned the cDNA for the LH beta subunit precursor molecule from a cDNA library from the pituitary of Japanese quail. The nucleotide sequence of the protein coding regions of this quail cDNA showed strong homology with the corresponding cDNA from the chicken, showing 68 and 88% homology in the signal peptide and apoprotein coding regions respectively.

```
Chicken LH beta
TL=533 or 436  NT5=24   LS=117   AP=357 or 354  NT3=35 or 83

----+----10---+----20---+----30---+----40---+----50---+----60
GGTGACCTGCGGCCCCATAGAGCCATGGGGGAGCGCAGGTGTTGGTGCTGATGACCCTT
                        MetGlyGlyAlaGlnValLeuValLeuMetThrLeu
----+----70---+----80---+----90---+---100---+---110---+---120
TTGGGGACCCCCCCGGCGACAACGGGGAACCCCCCCGTGGCTGTGGACCCCCCCCTGGCC
LeuGlyThrProProAlaThrThrGlyAsnProProValAlaValAspProProLeuAla
----+---130---+---140---+---150---+---160---+---170---+---180
GTCGTGGGACCCCCGATGGGGTTGGGGGGGGGCGGCCGCCCCCCGTGTCGCCCCATAAAC
ValValGlyProProMetGlyLeuGlyGlyGlyGlyArgProProCysArgProIleAsn
----+---190---+---200---+---210---+---220---+---230---+---240
GTAACGGTGGCGGTGGAGAAGGACGGATGCCCCCAATGTATGGCTGTGACCACCACGGCC
ValThrValAlaValGluLysAspGlyCysProGlnCysMetAlaValThrThrThrAla
----+---250---+---260---+---270---+---280---+---290---+---300
TGCGGGGGGTACTGCAGGACGCGGGAGCCGGTGTATCGCAGCCCTTTGGGCCCCCCCCCC
CysGlyGlyTyrCysArgThrArgGluProValTyrArgSerProLeuGlyProProPro
----+---310---+---320---+---330---+---340---+---350---+---360
CAGTCGGCGTGCACTTATGGGGCGCTGCGCTACGAGCGTTGGGCGCTGTGGGGCTGCCCC
GlnSerAlaCysThrTyrGlyAlaLeuArgTyrGluArgTrpAlaLeuTrpGlyCysPro
----+---370---+---380---+---390---+---400---+---410---+---420
ATAGGGAGCGACCCCCGCGTCCTCCTCCCCGTGGCTCTGAGCTGCCGCTGCGCCCGCTGC
IleGlySerAspProArgValLeuLeuProValAlaLeuSerCysArgCysAlaArgCys
----+---430---+---440---+---450---+---460---+---470---+---480
CCCATGGCGACCTCCGACTGCACCGTGCAGGGCTTGGGGCCGGCCTTCTGTGGGGCGCCA
ProMetAlaThrSerAspCysThrValGlnGlyLeuGlyProAlaPheCysGlyAlaPro
----+---490---+---500---+---510---+---520---+---530---+---540
GGGGGGTTCGGGGGGGAATAAATAGGACCGGGACCCCCCCCGGACCCCAAAAG-poly(A)
GGGGGGTTCGGGGGGGAATAAATAGGACCGGGACCCCCCCCGGACCCCAAAAGACCCCCA
GlyGlyPheGlyGlyGlu
----+---550---+---560---+---570---+---580
1-533 (LF127)
TATGTCCATTACAGTTACCTTATAGAGGGGCTCTGCCCCAT-poly(A) 146-581(L12)
```

Fig. 1 Nucleotide sequences of cDNAs encoding the chicken LH beta subunit precursor molecule and predicted amino acid sequence of the protein coding region. The translation initiation sequence and polyadenylation signal are underlined with thick continuous lines. The initiation and termination codons are indicated in bold letters. The codon for the predicted amino terminus of the apoprotein is underlined with a thin broken line. The nucleotide sequence between position 146 and 480 of the clone L12 is not shown, since the sequence of this region is identical to the corresponding region of clone LF127. (Data from 2).

You *et al.* (4) reported the nucleotide sequence of the cDNA for the turkey LH beta subunit precursor molecule which was cloned by using a chicken LH beta cDNA (LF127) as the hybridization probe. The turkey cDNA shows 92% homology with the chicken cDNA.

The avian LH beta subunit precursor cDNAs have longer leader sequences than those of any other vertebrate species reported: 117, 141 and 117 bases in chicken, quail and turkey respectively, compared with 60 bases in man and rat. The relative incidence of codons ending in G or C is high in the LH beta subunit precursor cDNA of chicken and quail (85 and 72%, respectively). Similarly high incidences of codons ending in G or C are found in cDNAs for beta subunits of mammalian FSH, LH and hCG (68-83%) but not in cDNAs for beta subunits of mammalian thyrotrophin and alpha subunits of mammalian pituitary glycoprotein hormones (52-58%). In the chicken LH beta subunit precursor cDNA, the polyadenylation signal sequence (AATAAA) overlaps with a part of the last codon of the apoprotein coding region and the termination codon (Fig. 1). Similar overlaps also occur in the cDNAs encoding the beta subunits of quail (6) and turkey LH (4), human chorionic gonadotrophin (7) and bovine FSH (8).

In the 3′ untranslated region of the quail LH beta cDNA, two series of completely or partially repetitive sequences are observed: one consisting of two repeated sets of 72 bases and the other consisting of 5 repeated sets of 30 bases. Such repeated sequences are not found in chicken LH beta cDNA. In the quail cDNA, there are also two sets of polyadenylation signal sequences.

```
Chicken alpha TL=709  NT5=81  LS=72  AP=288  NT3=265
----+----10---+----20---+----30---+----40---+----50---+----60---+----70---+
CCAAGGACAGCTCACATTTGAACATCGTCCTGCATTTTCTCATCTTTCTGATTTATTCTCATTGAACAAGGGAGA
----80---+----90---+---100---+---110---+---120---+---130---+---140---+---150
AAGATCATGGATTGCTACAGGAAGTATGCAGCTGTCACTTTGACCATTTTGTCTGTGTTTCTGCATCTTCTTCAT
      MetAspCysTyrArgLysTyrAlaAlaValThrLeuThrIleLeuSerValPheLeuHisLeuLeuHis
----+---160---+---170---+---180---+---190---+---200---+---210---+---220---+
ACTTTCCCAGATGGAGAATTTCTCATGCAGGGTTGTCCAGAGTGCAAGCTAGGGGAGAACAGGTTCTTTTCAAAA
ThrPheProAspGlyGluPheLeuMetGlnGlyCysProGluCysLysLeuGlyGluAsnArgPhePheSerLys
---230---+---240---+---250---+---260---+---270---+---280---+---290---+---300
CCAGGAGCCCCCATTTACCAGTGCACTGGGTGCTGTTTCTCCCGGGCCTATCCTACTCCAATGAGATCCAAGAAG
ProGlyAlaProIleTyrGlnCysThrGlyCysCysPheSerArgAlaTyrProThrProMetArgSerLysLys
----+---310---+---320---+---330---+---340---+---350---+---360---+---370---+
ACCATGCTTGTTCCAAAGAACATTACATCGGAAGCAACGTGCTGTGTAGCAAAGGCTTTCACCAAGATTACCCTT
TheMetLeuValProLysAsnIleThrSerGluAlaThrCysCysValAlaLysAlaPheThrLysIleThrLeu
---380---+---390---+---400---+---410---+---420---+---430---+---440---+---450
AAGGACAATGTGAAGATAGAGAACCACACAGACTGTCACTGCAGTACCTGCTACTATCATAAATCTTAAAGCCTG
LysAspAsnValLysIleGluAsnHisThrAspCysHisCysSerThrCysTyrTyrHisLysSer
----+---460---+---470---+---480---+---490---+---500---+---510---+---520---+
TCCCTTTGCTAATGATCAAGAACAACGGTGAATGAAATATTTGTTGTTCAGCTTTTACAGCACCGCTGTGTATAA
---530---+---540---+---550---+---560---+---570---+---580---+---590---+---600
TCTTGTGTTTTCTGGTCAAGACACCGAGTAGACTTTTGAATGAGATGGATGGCTGTTTTATTTCCTCTTTGCTTC
----+---610---+---620---+---630---+---640---+---650---+---660---+---670---+
TTCATGCATTTAAGTAAGTTTAACTATTTCCATTAGGGATTAGATGTAGCCCTTGCATGACAACCATAAGCTTGA
---680---+---690---+---700---+---710
TCTGTTTTTAAAATAAACTGCTAGATAAATTGTC-poly(A)      709
```

Fig. 2 Nucleotide sequence of a cDNA encoding the chicken alpha subunit precursor molecule and predicted amino acid sequence of the protein coding region. The translation initiation sequence and the polyadenylation signal are underlined with thick continuous lines. The initiation and termination codons are indicated in bold letters. The codon for predicted amino terminus of the apoprotein is underlined with a thin broken line. (Data from 5).

## Cloning and nucleotide sequences of cDNAs encoding avian glycoprotein hormone alpha subunit precursor molecules

Foster *et al.* (5) cloned the cDNA which encodes the entire region of the common alpha subunit precursor molecule of chicken pituitary glycoprotein hormones (Fig. 2). They constructed a cDNA library in a vector, lamda gt10, from chicken pituitaries treated with GnRH-I. For screening, they used a cDNA encoding the apoprotein region of the chicken alpha subunit molecule which was initially cloned by screening with mammalian alpha subunit cDNA (9). Foster *et al.* (5) reported that the GnRH-I treatment increased the proportion of positive clones more than 200 times.

Using this chicken alpha subunit precursor cDNA as the hybridization probe, Ando and Ishii (6) cloned cDNA encoding the glycoprotein alpha subunit precursor molecule from the Japanese quail. This cDNA shares 98% homology with the chicken alpha subunit precursor cDNA (Fig. 3).

```
Chicken alpha & Quail alpha
    +   *   +   *   +   *   +   *   +   *   +   *   +   *   +
CCAAGGACAGCTCACATTTGAACATCGTCCTGCATTTTCTCATCTTTCTGATTTATTCTCATTGAACAAGGGAGA
            CACATTTGAACATCGTCCTACGTTTTCTCATCTTTCTGATTTATTCTCATTGAACAAGGGAGA
    *   +   *   +   *   +   *   +   *   +   *   +   *   +   *
AAGATCATGGATTGCTACAGGAAGTATGCAGCTGTCACTTTGACCATTTTGTCTGTGTTTCTGCATCTTCTTCAT
AAGATCATGGATTGCTACAGGAAGTATGCAGCTGTCACTTTGACCATTTTGTCTGTATTTCTGCATCTTCTTCAT
    +   *   +   *   +   *   +   *   +   *   +   *   +   *   +
ACTTTCCCAGATGGAGAATTTCTCATGCAGGGTTGTCCAGAGTGCAAGCTAGGGGAGAACAGGTTCTTTTCAAAA
ACTTTCCCAGATGGAGAATTTCTCATGCAGGGTTGTCCAGAGTGCAAGCTAGGGGAGAACAGGTTCTTTTCAAAA
    *   +   *   +   *   +   *   +   *   +   *   +   *   +   *
CCAGGAGCCCCCATTTACCAGTGCACTGGGTGCTGTTTCTCCCGGGCCTATCCTACTCCAATGAGATCCAAGAAG
CCAGGAGCCCCCATTTACCAGTGCACTGGGTGCTGTTTCTCCCGGGCCTATCCCACTCCAATGAGATCCAAGAAG
    +   *   +   *   +   *   +   *   +   *   +   *   +   *   +
ACCATGCTTGTTCCAAAGAACATTACATCGGAAGCAACGTGCTGTGTAGCAAAGGCTTTCACCAAGATTACCCTT
ACCATGCTTGTTCCAAAGAACATTACATCGGAAGCAACGTGCTGCGTAGCAAAGGCTTTCACCAAGATTACCCTT
    *   +   *   +   *   +   *   +   *   +   *   +   *   +   *
AAGGACAATGTGAAGATAGAGAACCACACAGACTGTCACTGCAGTACCTGCTACTATCATAAATCTTAAAGCCTG
AAGGACAATGTGAAGATAGAGAACCACACAGACTGTCACTGCAGTACCTGCTACTATCATAAATCTTAAAGCCTG
    +   *   +   *   +   *   +   *   +   *   +   *   +   *   +
TCCCTTTGCTAATGATCAAGAACAACGGTGAATGAAATATTTGTTGTTCAGCTTTTACAGCACCGCTGTGTATAA
TCTGTTTGTTAATGATCAAGGACAACGGTGAATGGAATATTTGTTGTTCAGCTTTTATAGCACCGCCGTGTATAA
    *   +   *   +   *   +   *   +   *   +   *   +   *   +   *
TCTTGTGTTTTCTGGTCAAGACACCGAGTAGACTTTTGAATGAGATGGATGGCTGTTTTATTTCCTCTTTGCTTC
TCTTGTGTTTTCTGGGCAAGACACCGAGTAGACTTGTGAATGAGATGGATGGCTGTTTTATTTCCTCTTTGCTTC
    +   *   +   *   +   *   +   *   +   *   +   *   +   *   +
TTCATGCATTTAAGTAAGTTTAACTATTTCCATTAGGGATTAGATGTAGCCCTTGCATGACAACCATAAGCTTGA
TTCATGCATTTAAGTAAGTTTAACTATTTCCATTAGGGATTAGATGCAGCACTTGCATGACAACCATAAGCTTGA
    *   +   *   +   *   +   *   +   *   +   *   +   *   +   *
TCTGTTTTTAAAATAAACTGCTAGATAAATTGTC
TCTGTTTTTAAAATAAACTGTCAGATAAAAAG
```

Fig. 3 Comparison of nucleotide sequences of cDNAs of the glycoprotein hormone alpha subunit precursor molecules from chicken and Japanese quail. Non-identical positions are indicated in bold letters. Positions of the initiation codon, termination codon and the codon for the amino terminus of the apoprotein are underlined. (Data from 5, 6).

## The primary structures of avian LH beta subunit precursor molecules

The deduced amino acid sequences of the chicken (Fig. 4) and Japanese quail LH beta subunit apoprotein molecules are 92% homologous. You *et al.* (4) reported an even higher homology (99%) between the amino acid sequences of the LH beta apoproteins from turkey and chicken. Twelve Cys residues at positions 9, 23, 26, 34, 38, 57, 72, 88, 90, 93, 100, and 110 are common not only between chicken and quail LH beta subunits but also in the LH beta subunits of gonadotrophins from all the vertebrate species for which the sequences are known. Similarly, amino acid residues at positions 36-38, 71-72, 84-85 and 99-100 are also the same in the LH beta subunits from all vertebrate species. It should be noted that position 13, Asn, which is an attachment site for the carbohydrate moiety, is similarly conserved. The sequence between 84 and 88 (Val-Ala-Leu-Ser-Cys) is conserved in LH beta subunits in all the tetrapod vertebrates and in most of the gonadotrophin beta subunits in teleosts. Accordingly, this sequence seems to be necessary for LH or steroidogenic activity (2). In contrast, positions 42, 79, 89 and the sequence from 114 to 119 are extremely variable in beta subunits from different animal species.

Chicken LH beta
Quail LH beta

1 2 3 4 5 6 7 8 9 10 11 12 13 14 15 16 17 18 19 20
**Leu**-Gly-Gly-Gly-Gly-Arg-Pro-Pro-Cys-Arg-Pro-Ile-Asn-Val-Thr-Val-Ala-Val-Glu-Lys-
**Met**-Gly-Gly-Ser-Gly-Arg-Pro-Pro-Cys-Arg-Pro-Ile-Asn-Val-Thr-Val-Ala-Val-Glu-Lys-
21 22 23 24 25 26 27 28 29 30 31 32 33 34 35 36 37 38 39 40
**Asp-Gly**-Cys-Pro-Gln-Cys-Met-Ala-Val-Thr-Thr-Thr-Ala-Cys-Gly-Gly-Tyr-Cys-Arg-Thr-
**Glu-Glu**-Cys-Pro-Gln-Cys-Met-Ala-Val-Thr-Thr-Thr-Ala-Cys-Gly-Gly-Tyr-Cys-Arg-Thr-
41 42 43 44 45 46 47 48 49 50 51 52 53 54 55 56 57 58 59 60
Arg-Glu-Pro-Val-Tyr-Arg-Ser-Pro-Leu-Gly-Pro-Pro-Pro-Gln-Ser-**Ala**-Cys-Thr-Tyr-Gly-
Arg-Glu-Pro-Val-Tyr-Arg-Ser-Pro-Leu-Gly-Pro-Pro-Pro-Gln-Ser-**Ser**-Cys-Thr-Tyr-Gly-
61 62 63 64 65 66 67 68 69 70 71 72 73 74 75 76 77 78 79 80
Ala-Leu-Arg-Tyr-Glu-Arg-Trp-**Ala**-Leu-Trp-Gly-Cys-Pro-Ile-Gly-Ser-Asp-Pro-**Arg**-Val-
Ala-Leu-Arg-Tyr-Glu-Arg-Trp-**Asp**-Leu-Trp-Gly-Cys-Pro-Ile-Gly-Ser-Asp-Pro-**Lys**-Val-
81 82 83 84 85 86 87 88 89 90 91 92 93 94 95 96 97 98 99 100
**Leu**-Leu-Pro-Val-Ala-Leu-Ser-Cys-Arg-Cys-Ala-Arg-Cys-Pro-**Met**-Ala-Thr-Ser-Asp-Cys-
**Ile**-Leu-Pro-Val-Ala-Leu-Ser-Cys-Arg-Cys-Ala-Arg-Cys-Pro-**Ile**-Ala-Thr-Ser-Asp-Cys-
101 102 103 104 105 106 107 108 109 110 111 112 113 114 115 116 117 118 119
Thr-Val-Gln-Gly-Leu-Gly-Pro-Ala-Phe-Cys-Gly-Ala-Pro-Gly-Gly-Phe-Gly-Gly-**Glu**
Thr-Val-Gln-Gly-Leu-Gly-Pro-Ala-Phe-Cys-Gly-Ala-Pro-Gly-Gly-Phe-Gly-Gly-**Gln**

Homology: 91.6 %

Fig. 4 Amino acid sequences of the chicken and Japanese quail LH beta subunit apoprotein molecules. Non-identical amino acid residues are indicated in bold letters.

Likewise, all the half-cystine residues at positions 11, 14, 32, 35, 36, 63, 64, 86, 88, and 91 and two Asn residues at position 19 and 82 in alpha subunit apoproteins are common to all vertebrate species. The amino acid sequences of the alpha subunits from chicken (Fig. 2) and Japanese quail are identical. The amino acid sequence of the alpha subunit is more highly conserved in different species than is the sequence of the LH beta subunit.

**Homology of the amino acid sequences of the signal peptides and apoproteins of the LH alpha and beta subunit precursor molecules**

An analysis of the homology of the amino acid sequences of gonadotrophin subunit molecules among various vertebrates provides an insight into the evolution of gonadotrophin molecules in vertebrates, and especially of the avian gonadotrophins. The homologies between the signal peptides and apoproteins of the alpha subunits from various animal species are shown plotted in matrices for different vertebrate classes in Figures 5 and 6.

a) Signal peptide of the alpha subunit precursor molecule

| | Porcine | Quail | Chicken | Carp |
|---|---|---|---|---|
| Bovine | 92 | 85 | 85 | 26 |
| Porcine | | 84 | 84 | 74 |
| Quail | | | 100 | 17 |
| Chicken | | | | 17 |

b) Apoprotein of the alpha subunit precursor molecule

| | Porcine | Quail | Chicken | Carp | Salmon |
|---|---|---|---|---|---|
| Bovine | 97 | 85 | 85 | 72 | 66 |
| Porcine | | 84 | 84 | 74 | 64 |
| Quail | | | 100 | 71 | 66 |
| Chicken | | | | 71 | 66 |
| Carp | | | | | 74 |

Fig. 5 Homology matrices of the amino acid sequence of the signal peptides (a) and apoproteins (b) of the glycoprotein hormone alpha subunit precursor molecule in various vertebrates.

a) Signal peptide of the LH beta subunit precursor molecule

| | Porcine | Quail | Chicken | Salmon | Carp |
|---|---|---|---|---|---|
| Bovine | 85 | 45 | 35 | 15 | 15 |
| Porcine | | 35 | 25 | 20 | 15 |
| Quail | | | 54 | 26 | 15 |
| Chicken | | | | 22 | 7 |
| Salmon | | | | | 4 |

b) Apoprotein of the beta subunit precursor molecule

| | Porcine | Quail | Chicken | Bullfrog | Salmon | Carp |
|---|---|---|---|---|---|---|
| Bovine | 85 | 44 | 43 | 44 | 39 | 42 |
| Porcine | | 45 | 44 | 44 | 40 | 43 |
| Quail | | | 92 | 39 | 45 | 47 |
| Chicken | | | | 41 | 45 | 48 |
| Bullfrog | | | | | 48 | 50 |
| Salmon | | | | | | 77 |

Fig. 6 Homology matrices of the amino acid sequence of the signal peptides (a) and apoproteins (b) of the LH beta subunit precursor molecule in various vertebrates.

Signal peptides of the alpha subunit of chicken and Japanese quail show complete homology (Fig. 5a). The signal peptide of the alpha subunit is highly homologous among mammalian species, ranging between 67 and 92%, while homology is lower between mammalian and avian species (67 to 75%) and lowest between mammalian and teleostean species (17 to 26%).

The apoproteins of the alpha subunits of chicken and Japanese quail, as well as their signal peptides, have identical amino acid sequences. Unlike the signal peptides, the apoproteins of the alpha subunits share high homology even between phylogenically distant groups, homology values between mammals and teleosts being 59-66% (Fig. 5b). However, homology of the alpha subunit apoproteins between phylogenically distantly related groups is less than that between more closely related groups (71-97% within mammals, 69-85% between mammals and birds, 59-74% between mammals and teleosts), although the overall homology value is high.

Signal peptides of the LH beta subunit of chicken and Japanese quail are only 54% homologous, although homology values within mammals are relatively high, 70-85% (Fig. 6a). The values between mammals and birds and between mammals and teleosts were 30-45% and 10-20%, respectively. The signal peptide of the carp shows exceptionally low homology when compared with chicken and salmon. Generally, the sequence shows less similarity as the phylogenic relationship becomes more distant, as observed for the signal peptide and apoprotein of the alpha subunit.

The homology matrices of apoproteins of the LH beta subunits show a different pattern from that of the apoproteins of the alpha subunit or of the signal peptides of both alpha and LH beta subunits. As in the homology matrices of the other peptides, high homology values are observed within each vertebrate class (65-89% within mammals, 92-99% within birds, and 77-97% within fishes) (Fig. 6b). In the other hand, interclass homology values are small (37 to 50%), with no correlation with the phylogenic distance between classes (Fig. 6b).

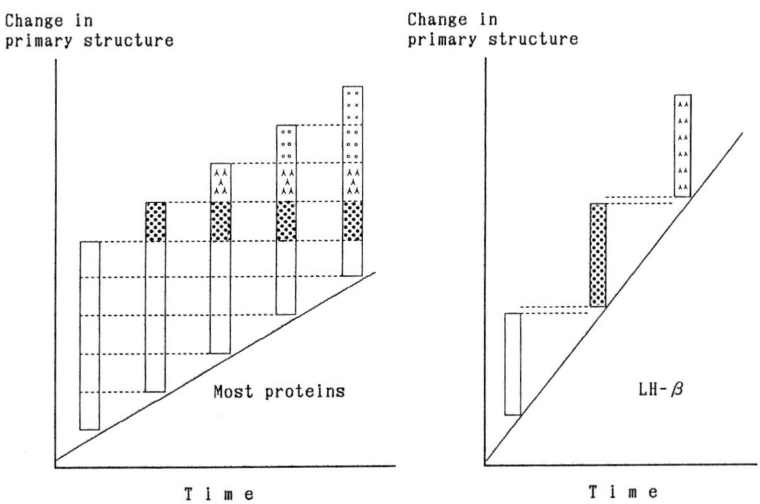


Fig. 7 Models to show difference in the evolution of the LH beta subunit molecule (right) and most of protein molecules (left). Common sequences are expressed with common patterns in each column which shows the known portion of the molecule.

The almost constant interclass homology values (about 45%) suggest that the amino acid sequence that occupies 45% of the peptide chain is indispensable for expression of hormone activity and has been conserved through the course of vertebrate evolution. In contrast, the rest of the peptide chain that occupies about 55% of the molecule is diversified among vertebrate classes. The LH beta apoprotein consists of only these two different types of the sequence, i.e. conserved and diversified sequences. This can be explained by assuming that the speed of evolution of the variable sequence of the LH beta subunit has been rapid or that the incidence of mutation in the variable sequence has been high (Fig. 7). These observations led Ando and Ishii (6) to propose that, in higher vertebrates, the specificity of the LH-LH receptor interaction is due to the highly variable sequence in the LH beta subunit molecule.

## The secondary structure of the chicken LH beta subunit molecule

The secondary structure of avian gonadotrophin subunit molecules is largely unknown although some information is available for the secondary structure of mammalian gonadotrophins (1). A computer analysis predicting the secondary structure of the chicken LH beta subunit molecule is shown in Figure 8.

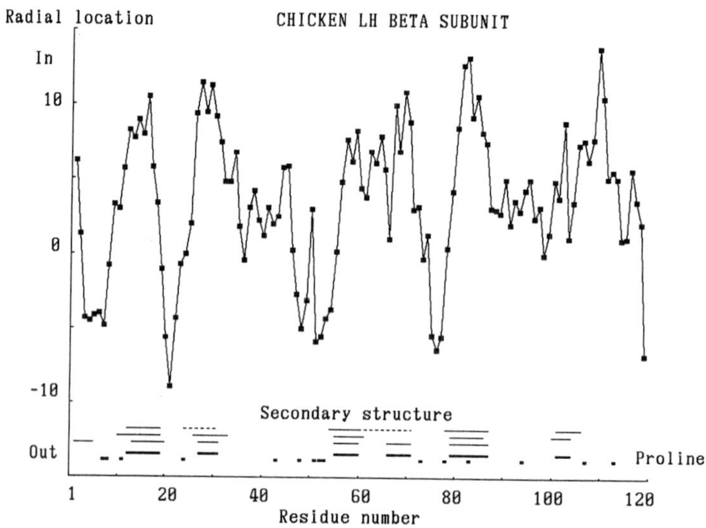


Fig. 8 Radial location (or exposure) of amino acid residues and positions of the elements of the secondary structure and proline residues in the LH beta molecule of chicken. The radial location of each residue is given with an arbitrary value of relative probability on the vertical axis. The larger the value the more internally it is located. Beta strands are indicated with thin horizontal lines and alpha helices with dotted horizontal lines. The methods used to predict the secondary structures shown at the bottom of the figure are from top to bottom, based on Chou and Fasman (12), Robson (13) and Nishioka and Ooi (14). The prediction using a combination of all methods is indicated by the thick horizontal lines. Locations of Pro residues are indicated as squares near the horizontal axis.

The position of Pro residues in the secondary structure of the chicken LH beta subunit is of special interest since Ishii (10) observed that the number of Pro residues in the LH beta subunit is greater in the more recently evolved than in older vertebrate classes, while the number of Pro residues in the FSH beta subunit is similar between vertebrate classes. This suggests that the presence of large numbers of Pro residues in the LH beta subunit in birds (15 residues) and mammals (about 20 residues) may be the basis for the high species (Table 1) and hormone specificity.

Table 1 Species specificity of specific binding of radioiodinated FSH and LH of rat, chicken and bullfrog to the testicular receptors of various vertebrates.

| **Donor of receptors** | **Radioiodinated hormones** | | | | | |
|---|---|---|---|---|---|---|
| | **rFSH** | **cFSH** | **fFSH** | **rLH** | **cLH** | **fLH** |
| Rat | + | + | - | + | - | - |
| Rabbit | + | + | | | | |
| Lonchura | + | | | | | |
| Melopsittacus | + | + | | | | |
| Pigeon | + | | | | | |
| Chicken | + | + | | | | |
| Quail | + | + | | - | + | - |
| Turtle | + | + | - | - | - | - |
| Lizard | + | + | + | - | + | + |
| Bufo | - | - | + | - | - | - |
| Rana | - | - | + | - | - | + |
| Xenopus | + | | - | - | - | +- |
| Newt | + | + | - | - | - | - |
| Salamander | + | | | - | - | |
| Teleosts | - | - | - | - | - | - |

\+ High specific binding. +- Low but significant specific binding. - Slight or not detectable specific binding.

Proline residues have less conformational freedom in unfolded structures of the peptide chain than any other residues, since the proline side chain is fixed by a covalent bond to the main chain (11). Thus, Pro residues stabilize the native structure of protein or make the native structure more rigid than any other residues. High rigidity, in turn, increases the specificity of protein-receptor interactions. The substantial number of Pro residues in the avian and mammalian LH beta subunit are thus considered to enhance the specificity of the binding of LH to its receptor.

In view of the potential importance of Pro residues in receptor binding it is important to know the positions of Pro residues in the secondary structure of the LH beta subunit molecule. Direct determination of the secondary structure of protein involves the use of the X-ray crystallography or, more recently, nuclear magnetic resonance. However, because it is difficult to obtain appropriate samples of avian gonadotrophin for these methods, a computer assisted indirect empirical/theoretical approach has been used to predict the secondary structure. The accuracy of indirect methods may be as low as about 50% or even less (11). Three different indirect methods have been used based on the Chou and Fasman (12), Robson (13) and the homology (14) methods. By combining the these methods, accuracy may be increased to about 70%. In addition, the radial location or exposure of each amino acid residue in the molecule has also been predicted using the method of Nishikawa and Ooi (14).

All the three methods predict the presence of beta strands and/or alpha helices at positions 13-19, 27-31, 79-86, and 101-104. Alpha helices were predicted by Chou and Fasman's method but not by Robson's method; the alpha helix cannot be predicted using the homology method. Furthermore, Chou and Fasman's method predicts additional beta strands at positions 54-60 and 101-106, as does Robson's method at positions 55-61 and 101-104. Only the homology method shows the presence of a beta strand at positions 1-5. Using a combination of these methods, the chicken LH beta subunit is predicted to contain 6 beta strands while the positions 1-11, 20-26, 32-54, 61-65, 72-78, 88-99, and 105-119 are predicted to be loop regions. The positions of the 15 proline residues in the chicken LH beta subunit are 7, 8, 11, 24, 43, 48, 51, 52, 53, 73, 78, 83, 94, 107, and 113. It can therefore be deduced that 14 of the 15 Pro residues occur in the predicted loop regions of the secondary structure. The presence of such a large number of Pro residues in the loop regions makes the LH beta subunit molecule highly rigid, and consequently highly specific for its receptors.

The prediction of the radial positions of amino acid residues in the LH beta subunit molecule also show that most Pro residues locate on or near the surface of the molecule, especially those which occur in a cluster at positions 48, 51, 52, 53.

## Conclusion

Avian LH shows greater similarity to mammalian LH in terms of specificity than to amphibian LH and probably reptilian LH. These specificities of avian and mammalian LH can be accounted for by 1) the lack of homology of the amino acid sequence of about 55% of the peptide chain of the LH beta molecule between different vertebrate classes, and 2) the increase in number of Pro residues in the LH beta subunit in the evolutionary more advanced vertebrate classes. These residues are located mostly in the predicted loop regions of the secondary structure of the molecule. The highly diversified sequence in the primary structure and localization of Pro residues in the predicted loop of the secondary structure may be responsible for the high specificity of LH in higher vertebrate (avian and mammalian) classes.

## References

1 Ishii S (1991) Gonadotropins. *Vertebrate Endocrinology: Fundamentals and Biomedical Implications* Vol. 4 Part B pp 33-66. Eds P K T Pang & M P Schreibman. New York: Academic Press

2 Noce T H, Ando H, Ueda T, Kubokawa K, Higashinakagawa T & Ishii S (1989) Molecular cloning and nucleotide sequence of the putative cDNA for the precursor molecule of the chicken LHβ subunit. *Journal of Molecular Endocrinology* **3** 129-137

3 Ando H, Noce T, Kubokawa K, Ueda T, Higashinakagawa T & Ishii S (1989) Molecular cloning and nucleotide sequence analyses of the chicken and quail luteinizing hormone beta-subunit complimentary DNAs. *Hormones and Environment* pp 17-168. Ed. D K O Chan. Hong Kong: University of Hong Kong

4 You S, Foster L K, Youngren O M, El Halawani M E & Foster D N (1992) Turkey luteinizing hormone beta subunit complimentary DNA nucleotide sequence. *Poultry Science* **71** (Suppl 1) 57

5 Foster D N, Galehouse D, Giordano T, Min B, Lam I C, Porter D A, Intehr H J & Bacon W L (1991) Nucleotide sequence of the cDNA encoding the common alpha subunit of the chicken pituitary glycoprotein hormones. *Journal of Molecular Endocrinology* **8** 21-27

6 Ando H & Ishii S (1990) Molecular cloning of complementary deoxyribonucleic acids for the quail luteinizing hormone alpha and beta subunits. *Zoological Science* **7** 1133

7 Fiddes J C & Goodman H M (1980) The cDNA for the beta subunit of human chorionic gonadotropin suggests evolution of a gene by readthrough into 3′ untranslated region. *Nature* **286** 684-687

8 Esch F S, Mason A J, Cooksy K, Mercado M & Shimasaki S (1986) Cloning and dDNA sequence analysis of the cDNA for the precursor of the beta chain of bovine follicle stimulating hormone. *Proceedings of the National Academy of Sciences of the USA* **83** 6618-6621

9 Foster D N & Galehouse D (1987) Molecular cloning and nucleotide sequence of a chicken pituitary alpha subunit cDNA. *Poultry Science* **66** (Suppl 1) 100

10 Ishii S (1988) Evolution of gonadotropin receptors. *Progress in Endocrinology* pp 233-238. Eds H Imura, K Shizume & S Yoshida. Amsterdam: Excerpta Medica

11 Branden C & Tooze J (1991) *Introduction to Protein Structure.* New York: Garland

12 Chou P Y & Fasman G D (1978) Prediction of the secondary structure of proteins from their amino acid sequence. *Advances in Enzymology* **47** 45-148

13 Garnet J, Osguthorpe D J & Robson B (1978) Analysis of the accuracy and implication of simple methods for predicting the secondary structure of globular protein. *Journal of Molecular Biology* **120** 97-120

14 Nishioka K & Ooi T (1986) Amino acid sequence homology applied to the prediction of protein secondary structures, and joint prediction with existing methods. *Biochimica et Biophysica Acta* **871** 45-54

Avian Endocrinology
Ed. P J Sharp, pp 135-148
Journal of Endocrinology Ltd, Bristol (1993)

# The molecular biology of prolactin

K Shimada[1], T Ohkubo[1], N Saito[1], R T Talbot[2] and P J Sharp[2]
[1]Department of Animal Physiology, Faculty of Agricultural Sciences, University of Nagoya, Nagoya 464-01, Japan; [2]AFRC Roslin Institute (Edinburgh), Roslin, Midlothian EH25 9PS, UK

## Introduction

Prolactin is a single chain polypeptide hormone produced in the anterior pituitary gland and is a member of a family of hormones including growth hormones, placental lactogens and proliferins (1). Its functions in birds are very diverse and include the control of Columbiform crop-sac development (2), brood patch formation (3), incubation behaviour (e.g. 4-6), ovarian function (7), gonadotrophin secretion (5, 8, 9), the production of 1,25-dihydroxycholecalciferol in the kidney (10), adrenal corticosteroidogenesis (11), water and electrolyte balance (12-14), pineal function (15) and immune responsiveness (16). Just how a single molecule can control so many biological functions is unknown.

Research on avian prolactins is assisted by the availability of bioassays including the pigeon crop sac assay (2), the local lactogenic response in the rabbit (17) and the proliferation of Nb2 lymphoma cells (18). Radioligand receptor assays using membranes from the rabbit mammary gland (19) or chicken or turkey kidneys (20) are also available. These assays are useful in monitoring the purification and assessing the biological specificity of avian prolactins. The development of radioimmunoassays for native turkey and chicken prolactins (21-24) and of a heterologous radioimmunoassay suitable for the measurement of prolactin in a wide range of avian species (25) has resulted in a large literature of measurements of concentrations of plasma prolactin in relation to incubation (e.g. 5, 21-24, 26) seasonal breeding (e.g. 27, 28), development (29) and osmoregulation (e.g. 13).

During the last three years, further advances in understanding of the biology of avian prolactin have been made by the application of molecular biological techniques to clone, sequence and measure the expression of avian prolactin genes. Advanced chromatographic procedures and Western blotting have been used to identify isoforms of prolactin. This brief review will focus on these recent developments.

## Molecular cloning of chicken and turkey prolactin cDNAs

Nucleotide sequences of chicken prolactin cDNAs were published by three groups in 1989 (30-32) and these have been used to isolate turkey prolactin cDNAs (32-34). The open reading frames in both chicken and turkey prolactin cDNAs contain 687bp beginning with a methionine codon (ATG) which is part of the sequence 5′ ACC ATG

A3′ for translational initiation by eukaryotic ribosomes. Polyadenylation signals (AATAAA) were present in some of the isolated chicken (31) and turkey (33) prolactin cDNAs. No information has been published on the structure of an avian genomic prolactin gene but the similarities between avian and mammalian cDNAs suggest that as in mammals, (35) it is likely to consist of five exons and four introns. As has been found in mammals, the chicken genome contains one copy of the prolactin gene (30).

| | |
|---|---|
| | 1 10 |
| chicken(28) | LeuProIle***<u>Cys</u>***ProIleGlySerValAsn***<u>Cys</u>***GlnValSerLeuGlyGluLeu |
| chicken (29) | - - - - - - - - - - - - - - - - - - - |
| turkey (32) | - - - - SerSer - - - - - - - - - - - - - |
| | 20 30 |
| chicken | PheAspArgAlaValLysLeuSerHisTyrIleHisTyrLeuSerSerGluIle |
| chicken | - - - - - - - - - - - - - - - - - - |
| turkey | - - - - - Arg - - - - - - Phe - - - - - |
| | 40 50 |
| chicken | PheAsnGluPheAspGluArgTyrAlaGlnGlyArgGlyPheIleTyrLysAla |
| chicken | - - - - - - - - - - - - - - - - - - |
| turkey | - - - - - - - - - - - - - - - - - - |
| | 60 70 |
| chicken | ValAsnGly***<u>Cys</u>***HisThrSerSerLeuThrThrProGluAspLysGluGlnAla |
| chicken | - - - - - - - - - - - - - - - - - - |
| turkey | - - - - - - - - - - - - - - - - - Thr |
| | 80 90 |
| chicken | GlnGlnIleHisHisGluAspLeuLeuAsnLeuValValGlyValLeuArgSer |
| chicken | - - - - - - - - - - - - - - - - - - |
| turkey | - - - - - - Glu - - - Ile Leu - - - - - |
| | 100 |
| chicken | TrpAsnAspProLeuIleHisLeuAlaSerGluValGlnArgIleLysGluAla |
| chicken | - - - - - - - - - - - - - - - - - - |
| turkey | - - - - - - - - - - - - - - - - - - |
| | 110 120 |
| chicken | ProAspThrIleLeuTrpLysAlaValGluIleGluGluGlnAsnLysArgLeu |
| chicken | - - - - - - - - - - - - - - - - - - |
| turkey | - - - - - - - - - - - - - - - - - Arg |
| | 130 140 |
| chicken | LeuGluGlyMetGluLysIleValGlyArgValHisSerGlyAspAlaGlyAsn |
| chicken | - - - - - - - - - - - - - - His - - - |
| turkey | - - - - - - - - - - Ile - - - - - - - |
| | 150 160 |
| chicken | GluIleTyrSerHisTrpAspGlyLeuProSerLeuGlnLeuAlaAspGluAsp |
| chicken | - - - - - Ser - - - - - - - - - - - - |
| turkey | - ValPhe - Gln - - - - - - - - - - - - - |
| | 170 180 |
| chicken | SerArgLeuPheAlaPheTyrAsnLeuLeuHis***<u>Cys</u>***LeuArgArgAspSerHis |
| chicken | - - - - - - - - - - - His - - - - - - |
| turkey | - - - - - - - - - - - - - - - - - - |
| | 190 |
| chicken | LysIleAspAsnTyrLeuLysValLeuLys***<u>Cys</u>***ArgLeuIleHisAspSerAsn |
| chicken | - - - - - - - - - - - - - - - - - - |
| turkey | - - - - - - - - - - - - - - - - Asn - |
| chicken | ***<u>Cys</u>*** |
| chicken | - |
| turkey | - |

Fig. 1 A comparison of the amino acid sequences of mature chicken and turkey prolactins as deduced from their cDNA nucleotide sequences. The two chicken cDNAs were cloned and sequenced by Hanks *et al.* (30) and Watakiki *et al.* (31) and the turkey cDNA by Wang *et al.* (32). The 6 cysteine residues are shown in bold italics with underlining.

Proof that the isolated cDNAs encode the amino acid sequence for prolactin has been obtained by expressing a putative chicken prolactin cDNA in *E. coli* and testing for biological activity (36). A protein extract of *E. coli* containing the expressed cDNA and injected locally into the ring dove crop sac stimulated a significant increase in mucosal weight compared with the control injection. The control injection was an extract of *E. coli* containing a plasmid with the putative prolactin cDNA inserted in the reverse orientation. Western blotting of the recombinant-derived protein using a polyclonal antiserum to native chicken prolactin revealed a single 23kDa protein band. This corresponds to the size of mature prolactin predicted from the amino acid sequence deduced from the nucleotide sequence of the prolactin cDNA.

Further proof that the isolated cDNAs encode prolactin comes from immunocytochemical studies using a monoclonal antibody to a synthetic decapeptide, [Lys]-His-Thr-Ser-Ser-Leu-Thr-Thr-Pro-Glu-COOH, corresponding to amino acids 59-67 of chicken and turkey prolactins (Fig. 1). Using this antibody, cells corresponding to the known distribution of prolactin cells (37) were localized in the cephalic lobe of the anterior pituitary glands of chickens and quail. Immunocytochemical double staining experiments showed that the prolactin monoclonal antibody did not recognize thyrotrophes, gonadotrophes, somatotrophes or adrenocorticotrophes (38).

**Amino acid sequences of avian prolactins deduced from their cDNAs**

The open reading frames of chicken and turkey prolactin cDNAs translate into a pre-prolactin composed of 229 amino acids. These comprise a signal peptide of 30 amino acids following by a mature prolactin of 199 amino acids (Fig. 1). This has a calculated molecular weight of 23kDa. The amino-terminal structure, Leu-Prl-Ile-Cys, in both turkey and chicken mature prolactins, is the same as in the crocodile, sea turtle, pig, human, mouse and sei whale (39).

The six cysteine residues at positions 4, 11, 58, 174, 191 and 199 are in the same positions in prolactins from other species, representing all classes of tetrapods (39). They are responsible for the secondary structure of the molecule forming disulphide loops by linking residues 4-11, 58-174, 191-199 (Fig. 2). Prolactin from teleosts has the same structure except for an absence of the disulphide loop across residues 4-11 at the N-terminal. This is because teleost prolactins lack the 12 amino acids found at the N-terminal of tetrapod prolactins.

A comparative analysis of the amino acid structures of prolactins from a range of vertebrate species shows that conserved residues cluster in four distinct domains in the prolactin molecules (31). These correspond, approximately, to residues between positions 17-32, 58-72, 83-99 and 160-199 (Fig. 1). It is likely that these domains are separated by three exon-exon boundaries as shown for the human and rat prolactin genes by genomic sequence analyses (40, 41). They are probably important for the binding of prolactin to its receptor.

Analyses of sequence homology between chicken prolactin and other species demonstrates the phylogenetic relationship between birds and other vertebrate classes. The highest homology is with reptiles (alligator, 92% sea turtle, 86%), high with amphibians (bullfrog, 72%) and the lowest with teleosts (salmon 28%) (39, 42).

Amongst mammalian prolactins, homologies with chicken prolactin range between 79% for the pig and 54% for the mouse (30). As might be expected the homology between chicken and turkey prolactin is greater than 90% representing a difference of 13-16 amino acids (Fig. 1). It is of interest to note that the amino acid sequences of chicken prolactins deduced from cDNAs derived from broiler chicken or broody bantam hen pituitary glands differ at positions 141, 150, 175 (Fig. 1). Both cDNAs have been used to produce recombinant-derived proteins which have been shown to be biologically active prolactin (36,43). The different amino acid substitution between broiler chickens and bantam hens possibly reflects their different ancestries.

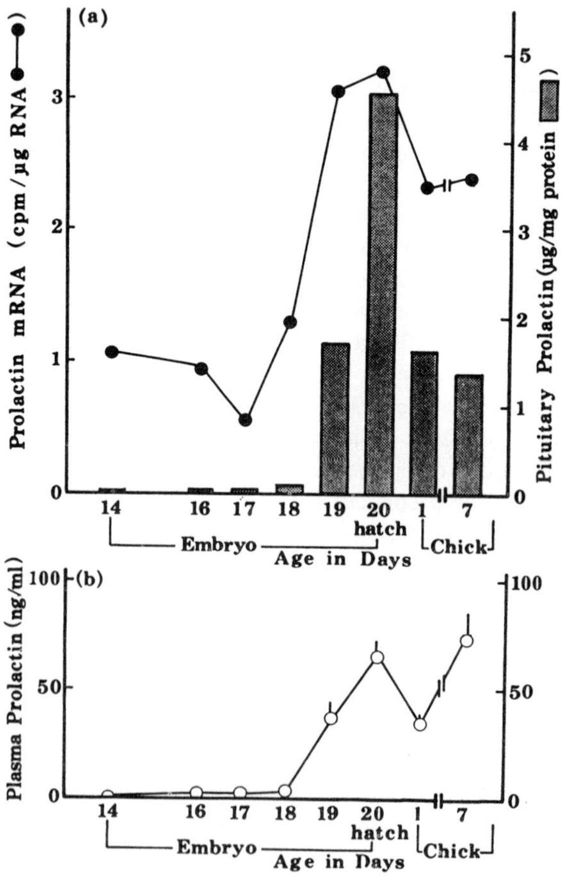


Fig. 2 Changes in concentrations of pituitary prolactin mRNA, pituitary prolactin and plasma prolactin during embryogenesis and after hatch in Gifujidori bantams. (Data from 46).

**Expression of prolactin mRNA in different physiological states**

The availability of chicken and turkey cDNAs has made it possible to investigate the relationships between steady-state amounts of prolactin mRNA in the pituitary gland and concentrations of pituitary and plasma prolactin. *In situ* hybridization using a chicken cDNA demonstrated that prolactin is expressed in the cephalic lobe of the anterior pituitary gland and is expressed more markedly in incubating than in laying hens (44).

Three papers published in 1991 showed that, during incubation, increased pituitary and plasma prolactin are associated with a marked increase in steady-state amounts of pituitary prolactin mRNA in the bantam hen (44, 45) and turkey (33). The decrease in plasma prolactin observed in incubating bantam hens following nest deprivation is associated with an immediate decrease in pituitary prolactin mRNA, although the total pituitary prolactin increased (44). This indicates that, following nest deprivation, mRNA decreases immediately while prolactin synthesis continues for a few days, but is not released.

Northern hybridization analysis shows that the major prolactin mRNA transcript in the pituitary glands of incubating Gifujidori bantam hens is 1.38kb (45) and, in incubating bantam hens of predominantly Silky/Wyandotte ancestry, is 0.88kb (44). This breed difference may arise from different lengths of untranslated regions at the 5′ and 3′ ends of the transcript or from differences in use of the alternative polyadenylation sizes.

Steady-state amounts of pituitary prolactin mRNA have also been shown to increase after photostimulation in bantam hens (45) and turkeys (33) and to decrease in turkeys after the development of photorefractoriness (33). These changes are directly related to changes in concentrations of pituitary and plasma prolactin.

An analysis of changes in the concentrations of pituitary prolactin mRNA during embryogenesis in the Gifujidori bantam hen shows they are low until day 18 of incubation, increase on day 19 and reach maximum values on the day of hatch (Fig. 2). The concentration of prolactin mRNA decreases on the day following hatch. These changes in prolactin mRNA are directly related to changes in the concentrations of prolactin in the pituitary gland and plasma. Since prolactin is considered to play an important role in osomoregulation during chick embryogenesis (47, 48), the increase in the expression of the prolactin gene 1-2 days before hatch may be associated with changes in osmoregulatory or respiratory functions at the time of hatching.

**Regulation of the expression of the prolactin gene**

Interest in factors controlling the expression of the prolactin gene has focused on the role of hypothalamic prolactin releasing factors. In mammals, prolactin biosynthesis is stimulated by thyrotrophin releasing hormone (TRH) by altering gene transcription (49, 50), although in the chicken, injection of TRH does not stimulate prolactin secretion (51). The possibility that TRH may stimulate prolactin gene expression has been investigated in photostimulated pre-laying Gifujidori bantam hens (Fig. 3). Injections of 2.5μg, but not 0.25μg, TRH/day for 4 days increased the amount of prolactin mRNA in the anterior pituitary gland. This increase was not associated with increases in concentrations of prolactin in the pituitary gland or plasma. This suggests that TRH stimulates transcription, but not the translation, of the prolactin gene and that pituitary prolactin concentrations are regulated both translationally and transcriptionally (52).

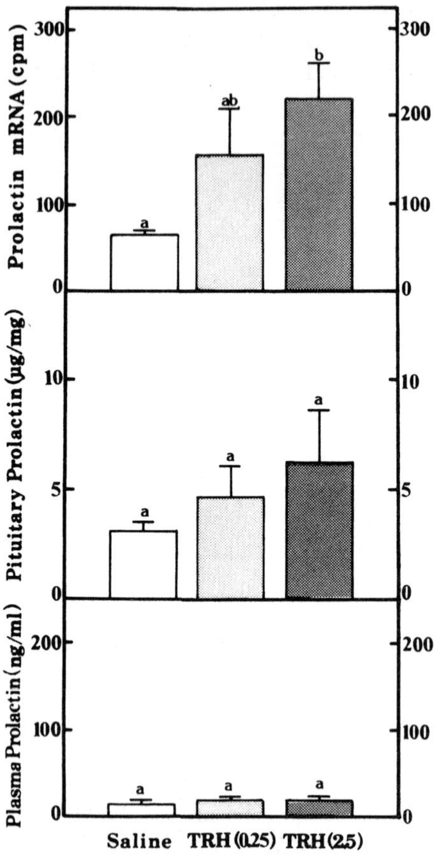


Fig. 3 Amount of pituitary prolactin mRNA, pituitary and plasma prolactin concentrations in photostimulated pre-laying hens treated with 0.25 or 2.5 µg thyrotrophin releasing hormone/day for 4 days. Values are means ± sem for 5 hens/group. Different superscripts indicate significant difference (P<0.05). (Data from 45).

Although the physiological role of TRH in the regulation of prolactin synthesis in birds is uncertain, it is now clearly established that vasoactive intestinal polypeptide (VIP) is a major physiologically significant avian prolactin releasing hormone (53, 54, 55). Some of the best evidence for this comes from the findings that passive immunization of incubating bantam hens with anti-chicken VIP serum results in an immediate decrease in concentration of plasma prolactin to below levels of detection (53) and that passive immunization of incubating ring doves with the same antiserum prevents the development of the crop sac (56). There is also evidence that in addition to stimulating prolactin release, VIP also stimulates the expression of the prolactin gene (53). A single injection of VIP in a laying hen stimulates prolactin secretion and increases the amount of prolactin mRNA in the anterior pituitary gland (Fig. 4). Conversely, the passive immunization of incubating bantam hens with anti chicken VIP serum results in a marked decrease in the amount of pituitary prolactin mRNA (53).

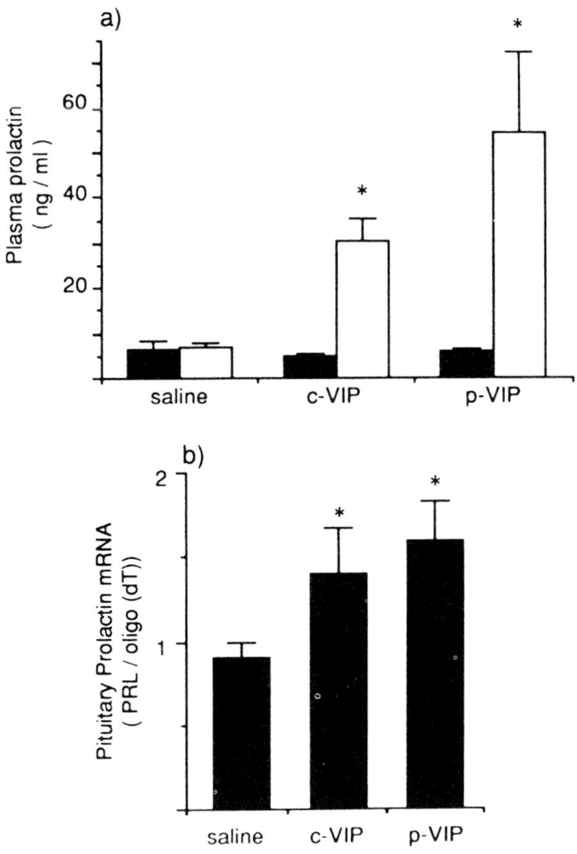


Fig. 4 Effect of an injection of 75μg porcine (p) or chicken (c) VIP/kg or saline vehicle in laying bantam hens on a) concentrations of plasma prolactin immediately before (solid box) and 10 min after (open bar) injection of VIP *P<0.05 versus the value immediately before injection and b) the amount of pituitary prolactin mRNA 90 min after injection of VIP *P<0.05 versus saline injection. The values are means ± sem (n=6) (Data from 44).

## Molecular forms of prolactin

The diverse biological actions of prolactin may be mediated by prolactin variants. These could be generated by alternative splicing of the primary mRNA transcript or by post-translational modifications including glycosylation, phosphorylation, deamidation, sulphation or the formation of high molecular weight polymers of prolactin (57). Alternative forms of prolactin, produced by alternative splicing may be smaller because of the removal of an exon, or larger because an intron is not removed (Fig. 5).

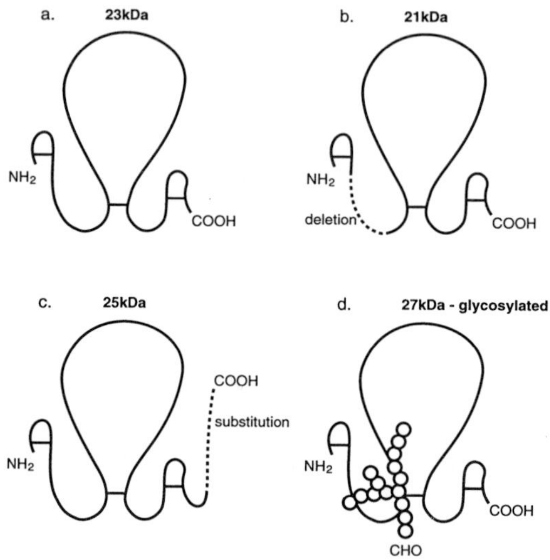


Fig. 5 Diagrammatic representations of possible structural variants of avian prolactin including a) the primary form with 3 disulphide loops, b) a smaller form produced by alternative splicing, involving amino acid deletion, c) a large form produced by the non-removal of an intron and d) a glycosylated form with a carbohydrate moiety attached to an Asn-X-Cys site.

The largest number of forms of an avian prolactin has been observed in the bantam hen using a sodium dodecyl sulphate-polyacrylamide gel electrophoresis, isoelectric focusing and two dimensional polyacrylamide gel electrophoresis and Western blotting (58). More than 10 forms occur ranging between 75 and 23kDa with pIs of between 6.57 and 4.6. Using less complex forms of electrophoresis, other authors find fewer forms of turkey (59-62) or chicken prolactin (37, 61). The possibility that some forms may be phosphorylated has been demonstrated by incubating chicken and turkey prolactin *in vitro* with the catalytic subunit of protein kinase A and $\gamma$-$^{32}$PATP (61). All forms of prolactin were intensely phosphorylated by protein kinase A.

Most interest in molecular forms of avian prolactins has focused on glycosylated variants (37, 59). In the turkey two such variants occur with a molecular size of

24.5kDa, one of which contains only *0*-linked carbohydrate. In the glycosylated forms of prolactins in whale, horse, man, pigs and sheep, the carbohydrate moiety is attached to the asparagine sequence (Asn-X-Ser) for *N*-linked glycosylation (57). Since this sequence does not occur at this position in either the chicken or turkey (Fig. 1) it has been proposed that *N*-linked glycosylation may occur in avian prolactin at Asn-X-Cys sites (59). The biological significance of glycosylated forms of prolactin is not fully understood but turkey (59) and mammalian forms are less biologically active that their non-glycosylated variants (57). There is evidence that at least one form of turkey prolactin, which has a molecular size of 25kDa and may be glycosylated, is present in reduced amounts in the pituitary gland of the incubating turkey (62). In the incubating bantam hen additional forms of prolactin have been observed in the anterior pituitary gland (58).

## Conclusion

Rapid advances in knowledge of the molecular structure of avian prolactins and their variants have been made during the past 3 years. The cloning of the chicken prolactin receptor, which is expressed in many tissues (63), opens up new opportunities for establishing the ways in which avian prolactins exert their many biological effects.

## References


1 Nicholl C S, Mayer G L & Russell S M (1986) Structural features of prolactins and growth hormones that can be related to their biological properties. *Endocrine Reviews* **7** 169-203

2 Nicoll C S (1967) Bioassay of prolactin. Analysis of the pigeon crop sac response to local prolactin injection by an objective and quantitative method. *Endocrinology* **80** 641-655

3 Hutchison R E, Hind R A & Steel E (1967) The effects of oestrogen, progesterone and prolactin on brood patch formation in ovariectomized canaries. *Journal of Endocrinology* **39** 379-385

4 Janik D S & Buntin J D (1985) Behavioural and physiological effect of prolactin in incubating ring doves. *Journal of Endocrinology* **105** 201-209

5 Sharp P J, Macnamee M C, Sterling R J, Lea R W & Pedersen H C (1988) Relationship between prolactin, LH and broody behaviour in bantam hens. *Journal of Endocrinology* **118** 279-286

6 Youngren O M, El Halawani M E, Silsby J L & Phillips R E (1991) Intracranial prolactin perifusion induces incubation behaviour in turkey hens. *Biology of Reproduction* **44** 425-431

7 Zadworny D, Shimada K, Ishida H & Sato K (1989) Gonadotropin-stimulated estradiol production in small ovarian follicles of the hen is suppressed by physiological concentrations of prolactin *in vitro*. *General and Comparative Endocrinology* **74** 468-473

8 Buntin J D, Lea R W & Figge G R (1998) Reductions in plasma LH concentration and testicular weight in ring doves following intracranial injection of prolactin or growth hormone. *Journal of Endocrinology* **118** 33-40

9 El Halawani M E, Silsby J L, Youngren O M & Phillips R E (1991) Exogenous prolactin delays photoinduces sexual maturity and suppresses ovariectomy-induced luteinizing hormone secretion in the turkey *(Meleagris gallopavo). Biology of Reproduction* **44** 420-424

10 Bikle D D, Spencer E M, Burke W H & Post C R (1980) Prolactin but not growth hormone stimulates 1,25 dehydroxyvitamin $D_3$ production by chick renal preparations *in vitro. Endocrinology* **107** 81-84

11 Carsia R V, Scanes C G & Malamed S (1987) Polyhormonal regulation of avian and mammalian corticosteroidogenesis *in vitro. Comparative Biochemistry and Physiology* **88A** 131-140

12 Phillips J G & Harvey S (1982) A re-appraisal of the role of prolactin in osmoregulation. *Aspects of Avian Endocrinology* pp 30-327. Eds C G Scanes, M A Ottinger, A D Kenney, J Balthazart, J Cronshaw & I Chester-Jones. Lubbock: Graduate Studies Texas Technical University

13 Harvey S, Hall T R & Chadwick A (1984) Growth hormone and prolactin secretion in water-deprived chickens *General and Comparative Endocrinology* **54** 46-50

14 Murphy M J, Brown P S & Brown S C (1986) Osmoregulatory effects of prolactin and growth hormone in embryonic chicks. *General and Comparative Endocrinology* **62** 485-492

15 Chakraborty S & Maiti B R (1981) Influence of prolactin on pineal gland activity in chicks. *General and Comparative Endocrinology* **44** 226-229

16 Skwarło-Sońta, K (1992) Prolactin as an immunoregulatory hormone in mammals and birds. *Immunology Letters* **33** 105-122

17 Scanes C G, Bolton N J & Chadwick A (1975) Purification and properties of an avian prolactin. *General and Comparative Endocrinology* **27** 371-379

18 Soares M J & Proudman J A (1990) Turkey and chicken prolactins stimulate the proliferation of rat Nb2 lymphoma cells. *Proceedings of the Society for Experimental Biology and Medicine* **197** 384-386

19 Shiu R P C & Friesen H G (1974) Properties of a prolactin receptor from the rabbit mammary gland. *Biochemical Journal* **140** 301-311

20 Krishnan K A, Proudman J A & Bahr J M (1991) Radioligand receptor assay for prolactin using chicken and turkey kidney membranes. *Comparative Biochemistry and Physiology* **100B** 769-774

21 Proudman J A & Opel H (1981) Turkey prolactin: validation of a radioimmunoassay and measurement of changes associated with broodiness. *Biology of Reproduction* **25** 537-580

22 Burke W H & Papkoff H (1980) Purification of turkey prolactin and the development of a homologous radioimmunoassay for its measurement. *General and Comparative Endocrinology* **40** 297-307

23 Etches R J & Cheng K W (1982) A homologous radioimmunoassay for turkey prolactin: changes during the reproductive and ovulatory cycle. *Poultry Science* **61** 1354-1364

24 Lea R W, Sharp P J & Chadwick A (1982) Daily variations in the concentrations of plasma prolactin in broody bantams. *General and Comparative Endocrinology* **48** 254-275

25 McNeilly A S, Etches R J & Friesen H G (1978) A heterologous radioimmunoassay for avian prolactin: application to the measurement of prolactin in the turkey. *Acta Endocrinologica* **89** 66-69

26 Zadworny D, Shimada K, Ishida H, Sumi C & Sato K (1988) Changes in plasma levels of prolactin and estradiol, nutrient intake, and time spent nesting during the incubation phase of broodiness in the Chabo hen. *General and Comparative Endocrinology* **71** 406-412

27 Nicholls T J, Goldsmith A R & Dawson A (1984) Photorefractoriness in European starlings: associated hypothalamic changes and the involvement of thyroid hormones and prolactin. *Journal of Experimental Zoology* **232** 567-572

28 Sharp P J, Massa R, Bottoni L, Lucini V, Lea R W, Dunn I C & Trocchi V (1986) Photoperiodic and endocrine control of seasonal breeding in Grey partridge *(Perdix perdix). Journal of Zoology* **209** 187-200

29 Harvey S, Davison T F & Chadwick A (1979) Ontogeny of growth hormone and prolactins secretion in the domestic fowl *(Gallus domesticus). General and Comparative Endocrinology* **39** 270-273

30 Hanks M C, Alonzi J A, Sharp P J & Sang H M (1989) Molecular cloning and sequence analysis of putative chicken prolactin cDNA. *Journal of Molecular Endocrinology* **2** 21-30

31 Watahiki M, Tanaka M, Masuda N, Sugisaki K, Yamamoto M, Yamakawa M, Nagai J & Nakashima K (1989) Primary structure of chicken prolactin deduced from the cDNA sequence. *Journal of Biological Chemistry* **264** 5535-5539

32 Kumar M V, Burke W H & Denny T P (1989) Isolation, cloning, and sequencing of a cDNA coding of chicken prolactin. *Poultry Science* **68** 81

33 Wong, E A, Ferrin N H, Silsby J L & El Halawani M E (1991) Cloning of a turkey prolactin cDNA: expression of prolactin mRNA throughout the reproductive cycle of the domestic turkey *(Meleagris gallopavo). General and Comparative Endocrinology* **83** 18-26

34 Karatzas K, Zadworny D & Kuhnlein U (1990) Nucleotide sequence of turkey prolactin. *Nucleic Acids Research* **18** 3071

35 Di Mattia G E, Gellersen B, Duckworth M L & Friesen H G (1990) Human prolactin gene expression. *Journal of Biological Chemistry* **16** 16412-16421

36 Hanks M C, Talbot R T & Sang H M (1989) Expression of biologically active recombinant-derived chicken prolactin in *Escherichia coli. Journal of Molecular Endocrinology* **3** 15-21

37 Mikami S (1980) Comparative anatomy and evolution of the hypothalamus hypophysial system in higher vertebrates. *Hormones, Adaptation and Evolution* pp 57-76. Eds S Ishii, T Hirano & M Wada. Tokyo: Japanese Scientific Societies Press

38 Berghman L R, Grauwels L, Vanhamne L, Proudman J A, Foidart A, Balthazart J & Vandesande F (1992) Immunocytochemistry and immunoblotting of avian prolactins using polyclonal and monoclonal antibodies toward a synthetic fragment of chicken prolactin. *General and Comparative Endocrinology* **85** 346-357

39 Noso T, Swanson P, Lance V A & Kawauchi H (1992) Isolation and characterization of glycosylated and non-glycosylated prolactins from alligator and crocodile. *International Journal of Peptide and Protein Research* **39** 250-257

40 Truong A T, Duez C, Belayeu A, Renard A, Picket R, Bell G I & Martial J A (1984) Isolation and characterization of the human prolactin gene. *EMBO Journal* **3** 429-439

41 Cook N & Baxter J D (1982) Structural analysis of the prolactin gene suggests a separate origin for its 5′ end *Nature* **297** 603-606

42 Yasuda A, Yamaguchi K, Kobayashi T, Yamamoto K, Kikuyana S & Kawauchi H (1991) The complete amino acid sequence of prolactin from the bullfrog *(Rana catesbeiana). General and Comparative Endocrinology* **83** 218-226

43 Ohkubo T, Tanaka M, Nakashima K, Shimada K, Saito N & Sato K (1991) High-level expression of biologically active prolactin in *E coli. Prolactin Conference Monograph* pp 1-12. Ed. K Hoshino. Kyoto: Shinko Printing Company

44 Talbot R T, Hanks M C, Sterling R J, Sang H M & Sharp P J (1991) Pituitary prolactin messenger ribonucleic acid levels in incubating and laying hens: effects of manipulating plasma levels of vasoactive intestinal polypeptide. *Endocrinology* **129** 496-502

45 Shimada K, Ishida H, Sato K, Seo H & Matsui N (1991) Expression of prolactin gene in incubating hens. *Journal of Reproduction and Fertility* **91** 147-154

46 Ishida H, Shimada K, Sato K, Seo H, Murata Y, Matsui N & Zadworny D (1991) Developmental expression of the prolactin gene in the chicken. *General and Comparative Endocrinology* **83** 463-467

47 Doneen B A & Smith T E (1982) Ontogeny of endocrine control of osmoregulation in chick embryo I. Role of pituitary gland in the distribution of water and ions among embryonic and extraembryonic compartments. *General and Comparative Endocrinology* **48** 300-309

48 Donnen B A & Smith T E (1982) Ontogeny of endocrine control of osmoregulation in chick embryo II. Actions of prolactin, arginine vasopressin and aldosterone. *General and Comparative Endocrinology* **48** 310-318

49 Murdock G H, Waterman M, Evans R M & Rosenfeld M G (1985) Molecular mechanisms of phorbol ester, thyrotropin-releasing hormone and growth factor stimulation of prolactin gene transcription. *Journal of Biological Chemistry* **200** 11852-11858

50 Laverière J N, Tixier-Vidal A, Busson N, Morin A, Martial J A & Gourdji D (1988) Preferential role of calcium in the regulation of prolactin gene transcription by thyrotropin-releasing hormone in GH3 pituitary cells. *Endocrinology* **122** 337-340

51 Harvey S, Scanes C G, Chadwick A & Bolton N J (1978) The effect of thyrotropin-releasing hormone (TRH) and somatostatin (GHRIH) on growth hormone and prolactin secretion *in vitro* and *in vivo* in the domestic fowl *(Gallus domesticus). Neuroendocrinology* **26** 247-260

52 Framley L S & Miller H A III (1989) Ontogeny of prolactin secretion in the neonatal rat is regulated post-transcriptionally. *Endocrinology* **124** 3-6

53 Sharp P J, Sterling R J, Talbot R T & Huskisson N S (1989) The role of hypothalamic vasoactive intestinal polypeptide in the maintenance of prolactin secretion in incubating bantam hens: observations using passive immunisation, radioimmunoassay and immunohistochemistry. *Journal of Endocrinology* **27** 5-13

54 Mauro L J, Elde R P, Youngren O M, Phillips R E & El Halawani M E (1989) Alterations in hypothalamic vasoactive intestinal peptide-like immuno-reactivity are associated with reproduction and prolactin release in the female turkey. *Endocrinology* **125** 1795-1804

55 Cloues R, Ramos C & Silver R (1990) VIP-like immunoreactivity during reproduction in doves: influence of experience and number of offspring. *Hormones and Behaviour* **33** 215-231

56 Lea R W, Talbot R T & Sharp P J (1991) Passive immunization against chicken vasoactive intestinal polypeptide suppresses plasma prolactin and crop sac development in incubating ring doves. *Hormones and Behaviour* **25** 287-294
57 Sinha Y N (1992) Prolactin variants. *Trends in Endocrinology and Metabolism* **3** 100-106
58 Talbot R T, Burke W H, Scougall R K & Sharp P J (1992) Multiple forms of prolactin in the chicken pituitary in different reproductive states. Abstract P5 *Proceedings of the Fifth International Symposium on Avian Endocrinology.* Edinburgh: AFRC Roslin Institute
59 Proudman J A & Corcoran D H (1981) Turkey prolactin: purification by isotachophoresis and partial characterization. *Biology of Reproduction* **25** 375-384
60 Corcoran D H & Proudman J A (1991) Isoforms of turkey prolactin: evidence for differences in glycosylation and in tryptic peptide mapping. *Comparative Biochemistry and Physiology* **99B** 563-570
61 Arámburo C, Montiel J L, Proudman J A, Berghman L R & Scanes C G (1992) Phosphorylation of prolactin and growth hormone. *Journal of Molecular Endocrinology* **8** 183-191
62 Karatzas C N, Guémené D, Zadworny D & Kuhnlein U (1992) The ratio of prolactin isoforms in the pituitary gland: changes during the reproductive cycle of the turkey. Abstract 13 *Proceedings of the Fifth International Symposium on Avian Endocrinology.* Edinburgh: AFRC Roslin Institute
63 Tanaka M, Maeda K, Okubo T & Nakashima K (1992) Double antenna structure of chicken prolactin receptor deduced from the cDNA sequence. *Biochemical and Biophysical Research Communications* **188** 490-496

Avian Endocrinology
Ed. P J Sharp, pp 149-160
Journal of Endocrinology Ltd, Bristol (1993)

# The avian neurohypophysial hormone-neurophysin precursors: structure, post-translational processing and evolution

R Acher, J Chauvet, M T Chauvet and G Michel
Laboratory of Biological Chemistry, University of Paris VI, 96 Boulevard Raspail, 75006 Paris, France

## Introduction

Hormonal activities of the neurohypophysis were discovered at the beginning of the century by investigating the effects of extracts of mammalian posterior pituitaries on mammalian organs *in vivo* or *in vitro* (1). Four biological or pharmacological responses were used for bioassays. These were uterotonic activity measured as the contractile response of the isolated rat uterus, the milk ejection reflex in the lactating rabbit, pressor activity measured as an increase in rat blood pressure, and antidiuretic activity measured as a decrease in urine volume. An additional bioassay for measuring neurohypophysial hormones, avian depressor activity, measured as a decrease in chicken blood pressure, was adopted as the standard bioassay for oxytocin (2). Whereas the mammalian neurohypophysial hormones, oxytocin and vasopressin, and their functions have received much attention, studies on birds are relatively rare (3). Purification and chemical characterization of chicken, goose and turkey neurohypophysial hormones show that in birds, in common with other non-mammalian tetrapods, they comprise mesotocin ([Ile$^8$]-oxytocin) and vasotocin ([Ile$^3$]-vasopressin) (4). In common with the 12 known vertebrate neurohypophysial hormones (5), these four peptides are amidated nonapeptides with a disulphide bridge 1-6, differing from each other only by substitutions in positions 3 and/or 8 (5). It can therefore be expected that their receptors should have similar hormone-binding domains although receptor specificities allow distinct physiological functions.

Oxytocin and vasopressin are cleaved from protein precursors 105-150-residues long (6) which contain specific neurophysins termed MSEL- and VLDV-neurophysins depending on the amino acids in positions 2, 3, 6 and 7 (7). Mesotocin and vasotocin precursors identified in amphibians and birds have a similar organization, with a VLDV-neurophysin associated with mesotocin and a MSEL-neurophysin associated with vasotocin (8). Such precursors also occur in fish but the duplication of the oxytocin-like hormone in dogfish (9, 10) and the great diversity of this hormone in cartilaginous fish makes it difficult to establish phylogenic relationships.

## Avian neurohormones

Mesotocin and vasotocin have been identified in birds that fly (4) as well as in the flightless birds (Ratites) such as the ostrich (11).

*Structure*

The structures of mesotocin, vasotocin, oxytocin and vasopressin are shown in Figure 1. Although mesotocin and vasotocin seem to be the evolutionary counterparts, respectively, of oxytocin and vasopressin, they display different pharmacological profiles in bioassays using mammalian and avian organs (12) (Table 1).

**Oxytocin-like hormones**

| | 1 | 2 | 3 | 4 | 5 | 6 | 7 | 8 | 9 | |
|---|---|---|---|---|---|---|---|---|---|---|
| Oxytocin | Cys. | Tyr. | Ile. | Gln. | Asn. | Cys. | Pro. | Leu. | Gly($NH_2$) | Mammals |
| Mesotocin | Cys. | Tyr. | Ile. | Gln. | Asn. | Cys. | Pro. | Ile. | Gly($NH_2$) | Non-mammalian tetrapods, Marsupials |

**Vasopressin-like hormones**

| | 1 | 2 | 3 | 4 | 5 | 6 | 7 | 8 | 9 | |
|---|---|---|---|---|---|---|---|---|---|---|
| Vasopressin | Cys. | Tyr. | Phe. | Gln. | Asn. | Cys. | Pro. | Arg. | Gly($NH_2$) | Mammals |
| Lysipressin | Cys. | Tyr. | Phe. | Gln. | Asn. | Cys. | Pro. | Lys. | Gly($NH_2$) | Pigs, Macropodids, Didelphids |
| Phenypressin | Cys. | Phe. | Phe. | Gln. | Asn. | Cys. | Pro. | Arg. | Gly($NH_2$) | Macropodids |
| Vasotocin | Cys. | Tyr. | Ile. | Gln. | Asn. | Cys. | Pro. | Arg. | Gly($NH_2$) | Non-mammalian vertebrates |

Fig. 1 Amino acid sequences of tetrapod neurohypophysial hormones. Underlined amino acids indicate differences from the sequence in oxytocin.

Table 1 Pharmacological activities of neurohypophysial hormones[1]

| | **Oxytocin-like activities** | | | **Vasopressin-like activities** | |
|---|---|---|---|---|---|
| | **Rat, uterus contraction** | **Chicken, decrease in blood pressure** | **Rabbit mammary gland, milk ejection** | **Rat, increase in blood pressure** | **Rat, antidiuresis** |
| Oxytocin | 450 | 450 | 450 | 5 | 5 |
| Mesotocin | 291 | 502 | 330 | 6.3 | 1.1 |
| Vasopressin | 17 | 62 | 69 | 412 | 465 |
| Vasotocin | 120 | 300 | 220 | 225 | 260 |

[1] Values are given in International Units per micromole (from 12).

*Functions*

The physiological function of vasotocin in birds is antidiuresis (3). During flight, metabolic rate increases dramatically, and the associated increase in oxygen consumption is accompanied by a proportionate increase in water loss. Since birds are uricotelic, they expend a minimum of urinary water in excreting catabolic nitrogen and the ability to form hypertonic urine saves water. Vasotocin exerts its antidiuretic effects in the domestic fowl by increasing water resorption from renal tubules rather than by reducing the glomerular filtration rate (3, 13).

Vasotocin induces contractions of the oviduct of the domestic fowl *in vitro* (3) which is consistent with its role in the control of oviposition (14, 15). Oviposition in chickens is associated with a transient five-fold increase in concentrations of plasma vasotocin whereas concentrations of plasma mesotocin remain unchanged (15). Since neurohypophysectomized fowls with diabetes insipidus lay eggs (3), it is possible that the increase in plasma vasotocin is not the prime cause of oviposition.

The hormonal function of mesotocin in birds and other non-mammalian tetrapods remains to be established. Interestingly, mesotocin is found in Australian marsupials instead of oxytocin (16) and, in these species, it controls milk ejection which is a function of oxytocin in placental mammals.

Vasotocin and mesotocin may also act as neurotransmitters, since in the chicken there is evidence that vasotocinergic and mesotocinergic neurones project to diverse sites in the central nervous system (17).

**Neurophysins**

Mesotocin and vasotocin are linked, respectively, to a VLDV-neurophysin and a MSEL-neurophysin in their precursors. These precursors are subjected to post-translational proteolytic processing which probably occurs in neurosecretory granules (18, 19).

Homologues of two mammalian neurophysins have been recognized in chicken, goose and ostrich (23) and their complete amino acid sequences have been determined in the goose (21, 22) and ostrich (24, 25).

*Structure of VLDV-neurophysins*

As is the case for mammalian VLDV-neurophysins, avian VLDV-neurophysins have 93 amino acids (21) (Fig. 2). There are 28 substitutions and one insertion in goose VLDV-neurophysin when compared with bovine VLDV-neurophysin. In the N-terminal segment (1-9), there is only a substitution in position 5 whereas the four marker amino acids, V, L, D, V in positions 2, 3, 6 and 7 are conserved. In the central part (amino acids 10-76) which is nearly invariant in mammalian neurophysins, there are 18 substitutions in goose protein compared to the bovine protein. Interestingly, most of the positions substituted in goose are also substituted in ostrich (Fig. 2). In the C-terminal segment (77-93) which is hypervariable in mammalian neurophysins, there are also many substitutions in the avian neurophysins (Fig. 2).

It is of interest to note that Phe-35 of mammalian VLDV-neurophysins is replaced by Tyr-35 in the avian VLDV-neurophysins and conversely Tyr-49 of mammalian neurophysins is replaced by Phe-49. Identical substitutions are found in goose and ostrich VLDV-neurophysins in positions 16, 17, 30, 35, 41, 49, 52, 55, 58 and 77 of the central region compared to the bovine counterpart. Furthermore an identical insertion (Ser) in position 66 is found in both avian neurophysins.

```
          1   2   3       5   6   7           10                  15                  20                  25
Bovine    Ala.Val.Leu.Asp.Leu.Asp.Val.Arg.Thr.Cys.Leu.Pro.Cys.Gly.Pro.Gly.Gly.Lys.Gly.Arg.Cys.Phe.Gly.Pro.Ser.
Porcine   --------------------------------Lys-----------------------------------------------------------------
Equine    ----Ala-------------------------Lys-----------------------------------------------------------------
Rat       ----Ala-----------------Met-----Lys-----------------------------------------------------------------
Human     ----Ala.Pro---------------------Lys-------------------------------------------------------------Asn

Goose     ----------------Gly-------------Lys-------------------------Arg.Asn.Arg-------------------------Arg
Ostrich   ----------------Met-----Ile-----Lys-----Met-----------------Arg.Asn---------His-----------------Asn

          26              30                  35                  40                  45                  50
Bovine    Ile.Cys.Cys.Gly.Asp.Glu.Leu.Gly.Cys.Phe.Val.Gly.Thr.Ala.Glu.Ala.Leu.Arg.Cys.Gln.Glu.Glu.Asn.Tyr.Leu
Porcine   ---------------------------------------------------------------------------------------------------
Equine    ---------------------------------------------------------------------------------------------------
Rat       ------------Ala------------------------------------------------------------------------------------
Human     ------------Ala.Glu--------------------------------------------------------------------------------

Goose     --------------------Glu-------------Tyr.Leu---------Pro-----Thr-------------------------Ser.Phe-----
Ostrich   --------------------Glu-------------Tyr.Phe---------Ser-----Thr-----------------------------Phe-----

          51              55                  60                  65     66'          70                  75
Bovine    Pro.Ser.Pro.Cys.Gln.Ser.Gly.Gln.Lys.Pro.Cys.Gly.Ser.Gly.Gly.Arg[   ]Cys.Ala.Ala.Ala.Gly.Ile.Cys.Cys.Ser
Porcine   ----------------------------------------------------Glu--------[   ]--------------------------------Asn
Equine    ---------------------------------------------------------------[   ]-----------------------------------
Rat       ---------------------------------------------------------------[   ]--------Thr------------------------
Homme     ------------------------------------Ala------------------------[   ]--------Val.Leu-----Leu------------

Goose     ----Thr---------Glu---------Arg-----------------Gly.Asp-----Ala.Ser-------------Pro--------------------
Ostrich   ----Thr---------Glu---------Arg-----------------Asn.Asn.Glu.Gly.Ser-------------Ser--------------------

          76              80                  85                  90      92  93
Bovine    Pro.Asp.Gly.Cys.His.Glu.Asp.Pro.Ala.Cys.Asp.Pro.Glu.Ala.Ala.Phe.Ser.Gln
Porcine   ----------------Arg.Phe---------------------------------Thr------------
Equine    ----------------Leu.Ala---------Ser-----His.Asp-------------------------
Rat       ----------------Arg.Thr-----------------------------Ser-----------------
Human     --------------------Ala-------------------Ala---------Thr--------------

Goose     Ser.Glu---------Val.Ala-----------------Glu.Arg---------Leu-----Ala-----
Ostrich   Asn.Glu---------Met.Val-----Ser.Ser---------Gln-----Val.Met-----Pro-----
```

Fig. 2 A comparison of mammalian, goose and ostrich VLDV-neurophysins.

*Structure of MSEL-neurophysins and processing of pro-vasotocin*

Avian MSEL-neurophysins differ from their mammalian counterparts in that an amino acid sequence 'copeptin', which is cleaved from mammalian MSEL neurophysin during post translational processing, is not cleaved from avian MSEL neurophysin. As a result, 'big' MSEL-neurophysins with 131/132 residues instead of 93/95 residues as in mammals, are found in the avian neurohypophysis (22).

Similarly, the MSEL-neurophysin from the frog is a 'big' neurophysin with 129 residues (26). It seems that, during processing, pro-vasotocin in non-mammalian tetrapods is only cleaved once whereas mammalian pro-vasopressin is cleaved twice. A comparison of ox, goose and frog MSEL-neurophysins is shown in Figure 3. When the sequences of goose big MSEL-neurophysin and ox MSEL-neurophysin-copeptin are compared, there are four substitutions in region 1-9, 14 substitutions and one insertion in region 10-75, and 37 substitutions and five deletions or shortening in region 76-131/134. The C-terminal part displays a higher variability than the N-terminal and central parts.

It is puzzling to note that in the N-terminal and the central parts of the protein, most of the positions that have been substituted in the goose when compared to the ox, are also substituted in the frog. The preserved residues might be important in maintaining the conformation of neurophysin, which is crucial for interactions with processing enzymes. On the other hand, it has been shown by enzymatic dissection of human MSEL-neurophysin and hormone-complex precipitation that the hormone binding-site is located in the N-terminal subdomain (residues 8-59) (27).

```
        1                                   10                                  20
Ox      Ala.Met.Ser.Asp.Leu.Glu.Leu.Arg.Gln.Cys.Leu.Pro.Cys.Gly.Pro.Gly.Gly.Lys.Gly.Arg.
Goose   ---Leu.Ala ---Thr.Ala -----------------------------------------Asn.Arg ------
Frog    Ser.Tyr.Pro ---Thr --- Val---------- Ile ---------------------Asn.Arg ---Asn

        21                                  30          33    33'  34                     40
Ox      Cys.Phe.Gly.Pro.Ser.Ile.Cys.Cys.Gly.Asp.Glu.Leu.Gly[        ]Cys.Phe.Val.Gly.Thr.Ala.Glu
Goose   --------------- Gly---------------- Val-------------Gly ---- Tyr ----------------------
Frog    ---------------Asn----------------- Glu.Asp ------- [        ]----Tyr.Ile --------Pro ----

        41                                  50                                  60
Ox      Ala.Leu.Arg.Cys.Gln.Glu.Glu.Asn.Tyr.Leu.Pro.Ser.Pro.Cys.Gln.Ser.Gly.Gln.Lys.Pro
Goose   Thr.Arg-------- Ala---------Asp ---------------------------- Pro ----Gly.Gln ---
Frog    Thr -----------Val---------------------------------------Glu.Ala ----Gly -------

        61                                  70                                  80
Ox      Cys.Gly.Ser.Gly.Gly.Arg.Cys.Ala.Ala.Ala.Gly.Ile.Cys.Cys.Asn.Asp.Glu.Ser.Cys.Val
Goose   ----------------------------------- Asp ----Val---------Ser.Ala.Asp.Thr-----Ala
Frog    -------- Ala ---------------------- Pro ----Val ------------------Gln--------Thr

                                                       Neurophysin<-┐   ┌-> Copeptin
        81                              Ile 90                      |96 |              100
Ox      Thr.Glu.Pro.Glu.Cys.Arg.Glu.Gly.Val.Gly.Phe.Pro.Arg.Arg.Val.Arg.Ala.Asn.Asp.Arg
Goose   Ala.Asp.Ala.Ser -----Leu -----Glu.Gly.Ser.Glu[           ]-----Ala.Glu.Glu.Ala.Ala.Glu
Frog    Met.Asp.Ser.Ser -----Leu.Asp.Glu.Asp.Ser.Glu[           ]-----Gln -----Val.Ser.Pro.Asp

        101                                 110                                 120
Ox      Ser.Asn.Ala.Thr.Leu.Leu.Asp.Gly.Pro.Ser.Gly.Ala.Leu.Leu.Leu.Arg.Leu.Val.Gln.Leu
Goose   Lys ---- Leu---- Val------------ Ser.Ala---- Asp ----------------------- Met-----------
Frog    Gln ---- Met---- Gln.Met.Asn ---- Ser.Ala.Ser.Asp ------------------------ Met.His.Met

        121                                 130 131 132         135
Ox      Ala.Gly.Ala.Pro.Glu.Pro.Ala.Glu.Pro.Ala.Gln.Pro.Gly.Val.Tyr
Goose   -------- Arg.Gln.Gln.Gly.Arg.Gln ---- Gly.Leu.Leu
Frog    ----Asn.Arg.Gln.Gln.Gln.Gln.Thr.Lys.His.Tyr
```

Fig. 3 A comparison of ox, goose and frog MSEL-neurophysins. In ox pro-vasopressin there is a cleavage site at the level of Arg-96 giving rise to 'short' MSEL-neurophysin and copeptin but this cleavage does not occur in goose and frog pro-vasotocins resulting in the formation of 'big' MSEL-neurophysins.

## Evolution of neurohypophysial precursors

### *Organization of the genes*

Genes encoding mammalian oxytocin and vasopressin precursors have been identified in rat (6), ox (28), and man (29). In mammals there is a single gene for each prohormone and their general organization involves 3 exons separated by 2 introns. The first exon encodes the signal sequence, the hormone, the signal processing sequence and the first 9 residues of the associated neurophysin. The second exon encodes residues 10 to 76/77 of neurophysin; the third exon encodes for residues 77 to 93/95 of neurophysin and, in vasopressin precursors, for the C-terminal glycopeptide extension, copeptin. Precursors may have evolved as a composite system since each exon could have its own autonomous mutations.

The chicken vasotocin gene also comprises three exons with two short intervening sequences (30). The positions within the mRNA sequence where splicing of the pre-mRNA occurs are homologous to those of other characterized neurohypophysial hormone genes (30). Chicken 'big' MSEL-neurophysin has 130 amino acids compared with 131 for the goose (22) and 129 for the frog (26).

In bony fish, a different gene organization has been observed. In the white sucker (*Catostomus commersoni*) the two genes for pro-vasotocin (31) are constructed in the same way as the mammalian pro-vasopressin gene, but the two genes for pro-isotocin are intronless (32).

*Gene duplication*

Duplication either of the oxytocin-like hormone or vasopressin-like hormone or both has been observed in dogfish and in marsupials. In the spiny dogfish, aspargtocin ([$Asn^4$]-oxytocin) and valitocin ([$Val^8$]-oxytocin) have been identified (9) and in the spotted dogfish, asvatocin ([$Asn^4$,$Val^8$]-oxytocin) and phasvatocin ([$Phe^3$,$Asn^4$,$Val^8$]-oxytocin) have been characterized (10). Duplicated hormones suggest gene duplication but this remains to be shown since the corresponding neurophysins have not been yet isolated. In Australian marsupials belonging to the *Macropodidae* family, two vasopressin-like peptides, lysipressin and phenypressin, have been found, while in the Northern bandicoot (*Isoodon macrourus*) belonging to the *Peramelidae* family, lysipressin and vasopressin have been identified (5, 8). Furthermore, in the latter species, both oxytocin and mesotocin have been isolated. All four hormones have also been isolated from the neurohypophysis of the North-American opossum *Didelphis virginiana* (5, 8). Since marsupials are known not to be tetraploid, any putative gene duplications giving rise to these hormones are likely to be of a tandem rather than a polyploid type. Duplication of neurohypophysial hormone precursor genes has been observed in tetraploid fish such as the white sucker (31, 32) and the chum salmon (33, 34). Although the hormone moiety is identical in the duplicated genes, so that only isotocin and vasotocin are found as in the other bony fish, differences in VLDV- and MSEL-neurophysins reveal the gene duplication. Curiously, in the white sucker, although there is only a 10% difference between the two isotocin precursors, there is a 45% difference between the two vasotocin precursors. It may therefore be questioned whether they derive from the same duplication event (31, 32).

As far as avian neurohypophysial hormones genes are concerned, there is currently no evidence of gene duplication of the type found in marsupials or fish.

*Gene conversion*

The striking identity in the central parts of the MSEL- and VLDV-neurophysins corresponding to amino acids 10-76 from mammalian species suggests a link between the two encoding genes (35). Direct sequencing of rat (6), ox (28) and human (29) genes reveals nearly perfect identity in the central exons encoding the central parts of both neurophysins of a given species. This can be explained by gene conversion involving non-reciprocal exchange between the two genes that are close together on the same chromosome (29).

Although gene conversion seems to have occurred in mammalian species, the occurrence of such an event in birds remains to be proved. A comparison of the amino acid sequences of the goose MSEL- and VLDV-neurophysins shows that several positions are substituted by the same amino acid in both neurophysins when compared to the bovine counterparts (Fig. 4). This identical poly-substitution in the two proteins is consistent with a recent gene conversion.

```
                10                               17  18       20      22
MSEL Ox         Cys.Leu.Pro.Cys.Gly.Pro.Gly.Gly.Lys.Gly.Arg.Cys.Phe
VLDV Ox         Cys.Leu.Pro.Cys.Gly.Pro.Gly.Gly.Lys.Gly.Arg.Cys.Phe

MSEL Goose      Cys.Leu.Pro.Cys.Gly.Pro.Gly.Asn.Arg.Gly.Arg.Cys.Phe
VLDV Goose      Cys.Leu.Pro.Cys.Gly.Pro.Arg.Asn.Arg.Gly.Arg.Cys.Phe

                30                  33'  34  35                 40  41
MSEL Ox         Asp.Glu.Leu.Gly[        ]Cys.Phe.Val.Gly.Thr.Ala.Glu.Ala
VLDV Ox         Asp.Glu.Leu.Gly[        ]Cys.Phe.Val.Gly.Thr.Ala.Glu.Ala

MSEL Goose      Val.Glu.Leu.Gly. Gly. Cys.Tyr.Val.Gly.Thr.Ala.Glu.Thr
VLDV Goose      Glu.Glu.Leu.Gly[        ]Cys.Tyr.Leu.Gly.Thr.Pro.Glu.Thr
```

Fig. 4 Indirect evidence for gene conversion of goose neurohypophysial pro-hormone genes. The identical substitutions in positions 17, 18, 35 and 41 in the two goose neurophysins suggest conversion between the two genes.

*The two evolutionary lineages*

When neurohypophysial hormones and neurophysins of mammals, birds and reptiles are compared, two precursor lineages are apparent, the pro-oxytocin-like lineage and the pro-vasopressin-like lineage (Fig. 5). Because of the similarity between the amino acid sequences, it is assumed that there are two gene lineages deriving from a common ancestral gene through a primordial duplication. This primordial duplication is supposed to have occurred before the emergence of bony fish because both types of precursors occur in this vertebrate class (1, 5, 8).

The great evolutionary stability of mesotocin and vasotocin which is present in all non-mammalian tetrapods (including lungfishes) is striking. The principal function of vasotocin is anti-diuretic, but the major function of mesotocin is uncertain. Mammalian oxytocin and vasopressin receptors have recently been identified (36-39) and those of mesotocin and vasotocin will be soon discovered. Identification of the cells bearing these receptors will throw new light on the physiological roles of mesotocin and vasotocin in all tetrapods, including birds.

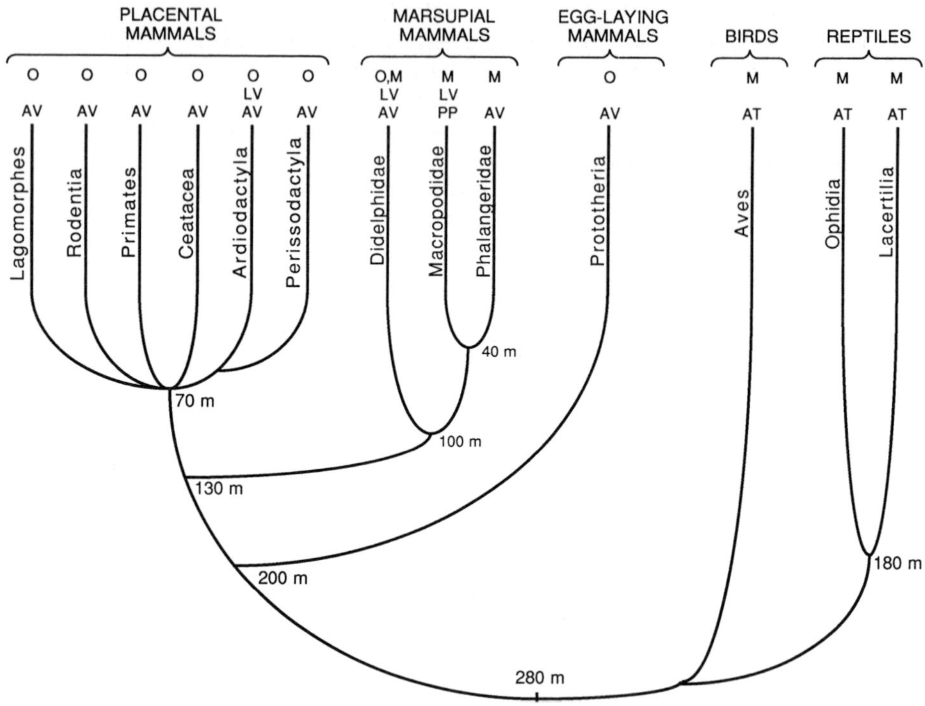


Fig. 5 Neurohypophysial hormones and tetrapod evolution. The letters indicate hormones identified in modern representatives of the groups: O, oxytocin; AV, arginine vasopressin; LV, lysine vasopressin; M, mesotocin; PP, phenypressin; AT, arginine vasotocin. The numbers give the time in millions of years (m) since the divergence.

## Acknowledgements

The authors thank Mrs Danielle Thevenet and Mrs Christine Jeanney for their skilled technical assistance. This investigation has been supported in part by CNRS (SDI 5662), Ministère de la Recherche et de l'Enseignement Supérieur (Action 'Biologie') and the Fondation pour la Recherche Médicale.

## References

1 Heller H (1974) History of neurohypophysial research. *Handbook of Physiology* Section 7 'Endocrinology' vol. 4, Part 1, pp 103-117. Eds E Knobil & W H Sawyer. Washington, DC: American Physiological Society

2 Stürmer E (1968) Bioassay procedures for neurohypophysial hormones and similar polypeptides. *Handbook of Experimental Pharmacology* New Series. Vol. XXIII, pp 130-189. Ed. B Berde. Berlin: Springer-Verlag

3 Bentley P J (1974) Actions of neurohypophysial peptides in amphibians, reptiles, and birds. *Handbook of Physiology* Section 7 'Endocrinology' Vol. 4, Part 1, pp 545-563. Eds E Knobil & W H Sawyer. Washington DC: American Physiological Society

4 Acher R, Chauvet J & Chauvet M T (1970) Phylogeny of the neurohypophysial hormones. The avian active peptides. *European Journal of Biochemistry* **17** 509-513

5 Acher R (1988) The neurohypophysial hormone-neurophysin precursors: composite biosynthesis and composite evolution. *Progress in Endocrinology 1988* Proceedings of the 8th International Congress of Endocrinology Vol. 2, pp 1505-1526. Eds H Imura, K Shizume & S Yoshida. Amsterdam: Elsevier, Excerpta Medica

6 Richter D & Ivell R (1985) Gene organization, biosynthesis, and chemistry of neurohypophysial hormones. *The Pituitary Gland* pp 127-148. Ed. H Imura. New York: Raven Press

7 Chauvet M T, Chauvet J & Acher R (1975) Phylogeny of neurophysins: partial amino acid sequence of a sheep neurophysin. *FEBS Letters* **52** 212-215

8 Acher R (1990) Structure, evolution and processing adaptation of neurohypophysial hormone-neurophysin precursors. *Progress in Comparative Endocrinology* pp 1-9. Eds A Epple, C G Scanes & M H Stetson. New York: Wiley-Liss

9 Acher R, Chauvet J & Chauvet M T (1972) Phylogeny of the neurohypophysial hormones. Two new active peptides isolated from a cartilaginous fish *Squalus acanthias*. *European Journal of Biochemistry* **29** 12-19

10 Acher R, Chauvet J, Chauvet M T & Rouillé Y (1992) Two new neurohypophysial hormones, asvatocin and phasvatocin: evolutionary duplication of the oxytocin-like peptide in dogfish. *Peptides: Chemistry and Biology* pp 937-938. Eds J A Smith & J Rivier. Leiden: ESCOM Science Publishers BV

11 Rouillé Y, Chauvet M T, Chauvet J & Acher R (1986) Phylogénie des hormones neurohypophysaires des oiseaux: microidentification de la mésotocine et de la vasotocine chez l'Autruche (*Struthio camelus*). *Comptes Rendus Société de Biologie* **180** 35-42

12 Berde B & Boissonnas R A (1968) Basic pharmacological properties of synthetic analogues and homologues of the neurohypophysial hormones. *Handbook of Experimental Pharmacology* New Series Vol. XXIII, pp 802-870. Ed. B Berde. Berlin: Springer-Verlag

13 Braun E J (1982) Glomerular filtration in birds - its control. *Federation Proceedings* **41** 2377-2381

14 Heller H & Pickering B T (1961) Neurohypophysial hormones of non-mammalian vertebrates. *Journal of Physiology London* **155** 98-114

15 Koike T I, Shimada K & Cornett L E (1988) Plasma levels of immunoractive mesotocin and vasotocin during oviposition in chickens: relationships to oxytocic action of the peptides *in vitro* and peptide interaction with myometrial membrane binding sites. *General and Comparative Endocrinology* **70** 119-126

16 Chauvet M T, Hurpet D, Chauvet J & Acher R (1981) A reptilian neurohypophysial hormone, mesotocin ([Ile8]-oxytocin), in Australian marsupials. *FEBS Letters* **129** 120-122

17 Robinzon B, Koike T I, Neldon H L & Kinzler S L (1988) Distribution of immunoreactive mesotocin and vasotocin in the brain and pituitary of chickens. *Peptides* **9** 829-834

18 Rouillé Y, Spang, A, Chauvet J & Acher R (1992) Evidence for distinct dibasic processing endopeptidases with Lys-Arg and Arg-Arg specificities in neurohypophysial secretory granules. *Biochemical Biophysical Research Communications* **183** 128-137

19 Rouillé Y, Chauvet J & Acher R (1992) Partial conversion of vasopressinyl-Gly-Lys-Arg into pharmacologically active vasopressin through secretory granule carboxypeptidase E and α-amidating processing enzymes. *Biochemistry International* **26** 739-746

20 Chauvet M T, Lévy B, Michel G, Chauvet J & Acher R (1986) Precursors of mesotocin and vasotocin in birds: identification of VLDV- and MSEL-neurophysins in chicken, goose and ostrich. *Bioscience Reports* **6** 381-395

21 Michel G, Lévy B, Chauvet M T, Chauvet J & Acher R (1990) Complete amino acid sequence of goose VLDV-neurophysin. Traces of a putative gene conversion between promesotocin and provasotocin genes. *International Journal of Peptide and Protein Research* **36** 457-464

22 Michel G, Lévy B, Chauvet M T, Chauvet J & Acher R (1990) Non-mammalian 'big' neurophysins - complete amino acid sequence of a two-domain MSEL-neurophysin from goose. *International Journal of Peptide and Protein Research* **36** 302-307

23 Rouillé Y, Lévy B, Chauvet M T, Chauvet J & Acher R (1989) Isolation of neurosecretory granules containing vasotocin, mesotocin, MSEL- and VLDV-neurophysins from goose neurohypophysis. *Neuropeptides* **13** 187-190

24 Lazure C, Saayman H S, Naude R J, Oelofsen W & Chretien M (1987) Complete amino acid sequence of a VLDV-type neurophysin from ostrich differs markedly from known mammalian neurophysins. *International Journal of Peptide Protein and Research* **30** 634-645

25 Lazure C, Saayman H S, Naude R J, Oelofsen W & Chretien M (1989) Ostrich MSEL-neurophysin belongs to the class of two-domain 'big' neurophysin as indicated by complete amino acid sequence of neurophysin/copeptin. *International Journal of Peptide and Protein Research* **33** 46-58

26 Michel G, Chauvet J, Chauvet M T & Acher R (1987) One-step processing of the amphibian vasotocin precursor: structure of a frog (*Rana esculenta*) 'big' neurophysin. *Biochemical Biophysical Research Communications* **149** 538-544

27 Michel G, Chauvet J & Acher R (1991) The hormone-binding site of neurophysins: binding of vasopressin to the N-terminal sub-domain dissected from human MSEL-neurophysin through endopeptidase Lys-C. *Neuropeptides* **20** 211-216

28 Ruppert S, Scherer G & Schütz G (1984) Recent gene conversion involving bovine vasopressin and oxytocin precursor genes suggested by nucleotide sequence. *Nature* **308** 554-557

29 Sausville E, Carney D & Battey J (1985) The human vasopressin gene is linked to the oxytocin gene and is selectively expressed in a cultured lung cancer. *Journal of Biological Chemistry* **260** 10236-10241

30 Hamann D, Hunt N & Ivell R (1992) The chicken vasotocin gene. *Journal of Neuroendocrinology* **4** 505-513

31 Morley S D, Schönrock C, Heierhorst J, Figueroa J, Lederis K & Richter D (1990) Vasotocin genes of the teleost fish *Catostomus commersoni*: gene structure, exon-intron boundary, and hormone precursor organization. *Biochemistry* **29** 2506-2511

32 Figueroa J, Morley S D, Heierhorst J, Krentler C, Lederis K & Richter D (1989) Two isotocin genes are present in the white sucker *Catostomus commersoni* both lacking introns in their protein coding regions. *The EMBO Journal* **8** 2873-2877

33 Heierhorst J, Mahlmann S, Morley S D, Coe I R, Scherwood N M & Richter D (1990) Molecular cloning of two distinct vasotocin precursor cDNAs from chum salmon (*Oncorhyncus keta*) suggests an ancient gene duplication. *FEBS Letters* **260** 301-304

34 Hyodo S, Kato Y, Ono M & Urano A (1991) Cloning and sequence analyses of cDNAs encoding vasotocin and isotocin precursors of chum salmon, *Oncorhyncus keta*: evolutionary relationships of neurohypophysial hormone precursors. *Journal of Comparative Biochemistry* **160** 601-608

35 Chauvet M T, Hurpet D, Chauvet J & Acher R (1983) Identification of human neurophysins: Complete amino acid sequences of MSEL- and VLDV-neurophysins. *Proceedings of the National Academy of Sciences of the USA* **80** 2839-2843

36 Kimura T, Tanizawa O, Mori K, Brownstein M J & Okayama H (1992) Structure and expression of a human oxytocin receptor. *Nature* **356** 526-529

37 Morel A, O'Carroll A M, Brownstein M J & Lolait S J (1992) Molecular cloning and expression of a rat Vla arginine vasopressin receptor. *Nature* **356** 523-526

38 Birnbaumer M, Seibold A, Gilbert S, Ishido M, Barberis C, Antaramian A, Brabet P & Rosenthal W (1992) Molecular cloning of the receptor for human antidiuretic hormone. *Nature* **357** 333-335

39 Lolait S J, O'Carroll A M, McBride O W, Konig M, Morel A & Brownstein M J (1992) Cloning and characterization of a V2 receptor and possible link to nephrogenic diabetes insipidus. *Nature* **357** 336-339

Avian Endocrinology
Ed. P J Sharp, pp 161-176
Journal of Endocrinology Ltd, Bristol (1993)

# Molecular biology of the chicken growth hormone receptor

J Burnside and L A Cogburn
Delaware Agricultural Experiment Station, Department of Animal Science and Agricultural Biochemistry, College of Agricultural Sciences, University of Delaware, Newark, DE 19717, USA

## Introduction

The growth promoting effects of growth hormone (GH) are mediated in part by the interaction of GH with its hepatic receptor and the subsequent higher production of insulin-like growth factor-I (IGF-I). Although much is known about the factors that regulate synthesis and secretion of GH, a great deal is yet to be learned about the role of the GH receptor (GHR) in regulation of the somatotrophic axis. The presence of high affinity GH-binding proteins on the surface of cells and in the circulation has been known for some time, although their function is still unknown (1, 2). Recently, the structural relationship between the cell-bound GHR and the soluble GH binding protein (GHBP) has been demonstrated. Within the last five years, cDNAs for GHR and GHBP have been cloned for a number of mammalian species (3-9) and the chicken (10). These clones have provided the tools for detailed study of regulation and function of the GHR and the GHBP. The purpose of this review is to provide insight into structure and regulation of expression of the chicken GHR (cGHR) since it was first cloned (10).

## Isolation and characterization of the chicken growth hormone receptor cDNA

A cDNA encoding cGHR was isolated from a chicken liver cDNA library, using polymerase chain reaction (PCR) with primers derived from highly conserved sequences of the mammalian GHR (10). The cGHR resembles the mammalian GHR with the exception of low overall sequence homology and the absence of the exon 3 homologue of the human GHR. The nucleotide sequence of the cGHR predicts a mature protein of 592 amino acids (MW = 67,000) that is only 50-58% homologous to the rabbit, rat and human GHR (Fig. 1). The cGHR and mammalian GHR have an extracellular GH-binding domain, a single transmembrane domain and an intracellular domain, which is presumably involved in signal transduction. Structural features of the GHR which are conserved in chickens include 7 cysteine residues in the extracellular domain and 5 in the intracellular region and 4 potential glycosylation sites in the mature peptide. Regions in the extracellular domain critical for GH binding to the human GHR (amino acids 101-106 and 125-132 of the human GHR) (11) are also conserved in the cGHR. The absence of exon 3 at first appeared to be a unique feature of the cGHR; however, an isoform of the human GHR lacking exon 3 has been recently identified in placenta (12). Apparently, the deletion of exon 3 in the human GHR does not affect binding of GH to this GHR isoform (11). The sequential binding of one GHR molecule to site 1 on human GH (hGH), followed by binding of a second GHR to site 2 on hGH causes receptor dimerization which is required for signal transduction (13). The overall similarity among the mammalian and chicken GHRs suggests that they function in an analogous manner. However, additional studies are needed to determine the intracellular signalling mechanism utilized by the GHR.

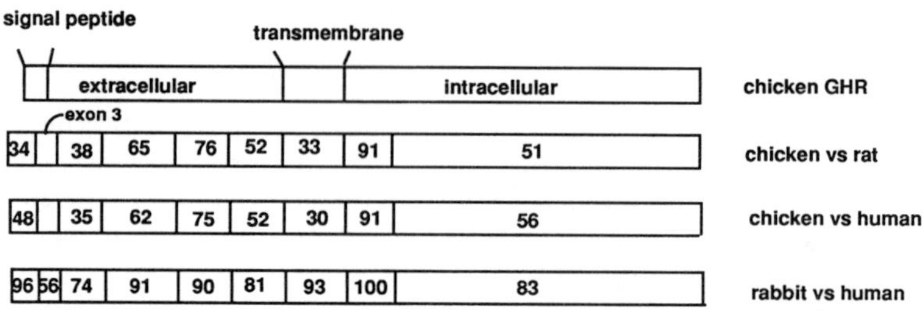


Fig. 1 Schematic comparison of the structure of the chicken growth hormone receptor (GHR) with that of mammalian GHR. The signal peptide and transmembrane domain are indicated. The chicken cDNA is lacking sequences corresponding to exon 3 of the human GHR. Regions are delineated according to the exon boundaries. Numbers inside the boxes indicate the percent identity of that region with the chicken GHR.

## Identification of cDNAs encoding other growth hormone receptor isoforms

The presence of soluble growth hormone binding proteins in the circulation of mammals is well documented (1, 2). These GHBPs could play a significant role in the delivery and action of GH in target tissues by dampening oscillations of GH secretion or serving as a reservoir of GH by prolonging its half-life in blood. In rats (6) and mice (7), the GHBP is encoded by a discrete mRNA which arises from alternative splicing of the GHR gene. Splicing takes place at the junction of exon 7/8 where the transmembrane and intracellular domains (exons 8-10) are replaced by a short alternative exon encoding a unique tail of hydrophilic amino acids. Although alternatively spliced forms of the receptor have been found in rabbits (14), it is thought that the GHBP in both rabbits and humans is derived from proteolytic cleavage of the membrane-anchored GHR.

Some preliminary reports suggest the presence of GHBP in chicken serum (15, 16). The identification of alternatively spliced, truncated forms of the GHR in the rat and mouse prompted us to screen a chicken liver cDNA library for putative GHBP cDNAs. The library was initially screened with a restriction fragment corresponding to the extracellular domain of the cGHR. The resultant GHR-positive clones were then rescreened with either the extracellular probe or a probe corresponding to the intracellular region. Several unique clones which contained only the extracellular GH-binding domain were isolated (Fig. 2). Three clones were identified and partially sequenced in our laboratory. In addition, two truncated cGHR clones have been identified by Baumbach and colleagues (17, 18). One clone, identified in both laboratories, terminates at nucleotide 360 and contains a poly A tail of 12-40 bases (numbering is based on our original published sequence) (10). These small, truncated-cGHR clones (designated cGHBP-1) have been studied in detail by Baumbach's group, and could be involved in a unique regulatory mechanism related to expression of the cGHR (17). A second receptor-related cDNA (cGHBP-2) diverges at nucleotide 387, which corresponds to the exon 5/6 junction of the human GHR gene. The open reading frame extends for an additional 8 unique amino acids before encountering a stop codon. A third type of cDNA, cGHBP-3, has been described by Baumbach (18). This cDNA appears to result from the use of an alternative splice donor site within intron 6. This transcript contains a 17-base insertion which shifts the reading frame of exon 7 such that a stop codon is encountered, eliminating translation of the transmembrane and intracellular domains. A fourth clone (cGHBP-4) has been identified which also diverges at the exon 6/7 junction of the human GHR gene and uses either an alternative exon or a retained intron to extend the open reading frame for an additional 28 amino acids (19).

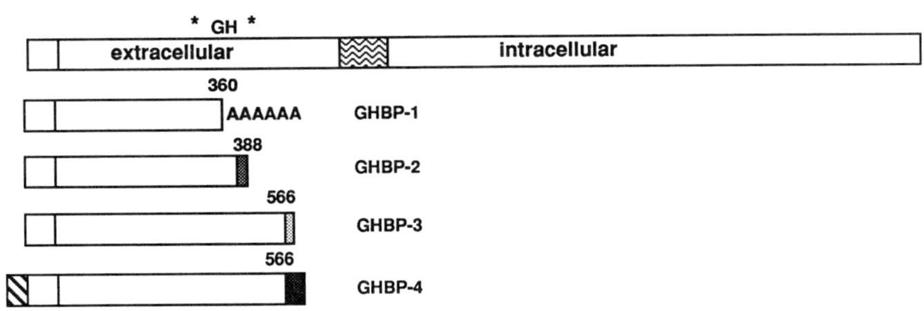


Fig. 2 Linear representation of chicken growth hormone receptor (cGHR) isoforms isolated from a chicken liver cDNA library. Approximate location of the GH-binding domain is indicated by *GH*. These truncated forms of the GHR are candidates for a soluble chicken growth hormone binding protein (cGHBP) since they lack the transmembrane domain (wavy pattern). Structures of the cGHBP are described in text. Stippled regions of GHBP-2,3,4 indicate unique 3′ extensions of the open reading frame. Hatched region indicates a unique 5′-untranslated sequence in GHBP-4.

The cDNA for the putative cGHBP should contain regions, identified by mutational analysis of the human GHR (11), that are critical for GH binding. The shorter cDNA (cGHBP-1) is missing one of the GH-binding sites. The remaining isoforms, however, are potential candidates for a functional cGHBP. It is interesting that cGHBP-4 contains a novel noncoding exon at the 5′ end, where the sequence diverges from the cGHR 10 bp upstream of the initiator methionine. Sequence heterogeneity has also been found in the 5′-untranslated region of the GHR cDNA of rats, mice, rabbits and humans (3, 5-7). The presence of divergent 5′ sequences in the cGHR and cGHBP-4 suggests that the regulation of expression of these two transcripts differs. Whether any of these transcripts are translated *in vivo* remains to be determined. Nonetheless, the identification of the various receptor isoforms indicates that alternatively-spliced transcripts exist in the cGHR and that specific isoforms could be responsible for the generation of the cGHBP.

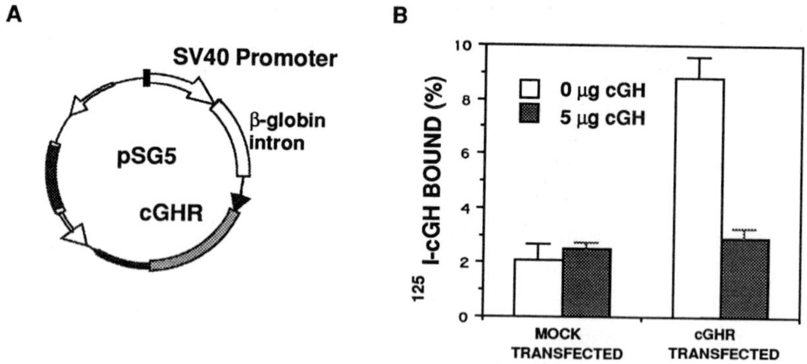


Fig. 3 Expression of chicken growth hormone binding activity in transfected COS-7 cells. COS-7 cells at 50% confluency were transfected with chicken growth hormone receptor (cGHR) expression vector (Panel A) using $CaPO_4$. Total binding of $^{125}$I-cGH to membranes prepared from mock transfected or transfected cells was determined in the presence or absence of 5µg unlabelled cGH (Panel B). Values are means ± s.e. of triplicate determinations.

## Binding properties of the chicken growth hormone receptor expressed in mammalian cells

The full-length chicken growth hormone receptor cDNA has been inserted into a mammalian expression vector and transiently expressed in COS-7 cells (Fig. 3A). The COS-7 cell line has been used to demonstrate expression of a functional GHR for the rabbit, rat and human (3, 6). Transfection of COS-7 cells with the cGHR cDNA results in expression of a high-affinity membrane-bound GHR (Fig. 3B). This expression system has been used to produce recombinant cGHR for characterization of the binding properties of a homogenous cGHR.

Binding studies show that the expressed cGHR has a high affinity for cGH ($K_a$ = 2.3 x $10^9$ L/M) which is similar to the binding affinity reported for the native cGHR prepared from liver membranes (20). Different mammalian GH and prolactin (PRL) preparations have been used to demonstrate displacement of radioligand ($^{125}$I-cGH) from the cGHR expressed in COS-7 cells (Fig. 4). The recombinant cGHR displays similar relative binding affinities for different mammalian GH (ovine, o; porcine, p; bovine, b) and PRL preparations as does the native hepatic cGHR (cGH>oGH>pGH>-oPRL>bPRL) (20). It is of particular interest that the cGHR has a 100-fold less affinity for turkey PRL (tPRL) than that of the mammalian PRL preparations, whereas chicken PRL shows virtually no binding to the recombinant-derived cGHR. These observations suggest that mammalian PRL binds to the cGHR because of sequence homology to regions in the ligand that are critical for receptor discrimination between GH and PRL.

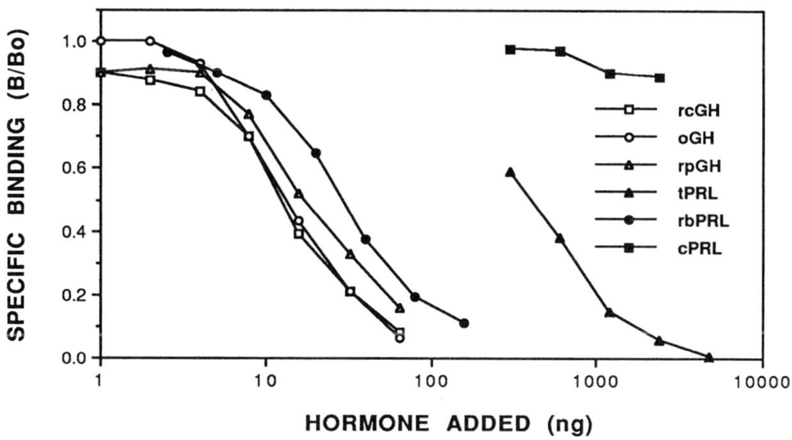


Fig. 4 Displacement of $^{125}$I-chicken growth hormone (cGH) from membranes of COS-7 cells transfected with the chicken growth hormone receptor (cGHR) by various avian and mammalian hormones. The following abbreviations were used for different hormone preparations: GH, growth hormone; PRL, prolactin; o, ovine; rp, recombinant-derived porcine; rb, recombinant-derived bovine; t, turkey; c, chicken; rc, recombinant-derived chicken. Values are means ± s.e. of triplicate determinations.

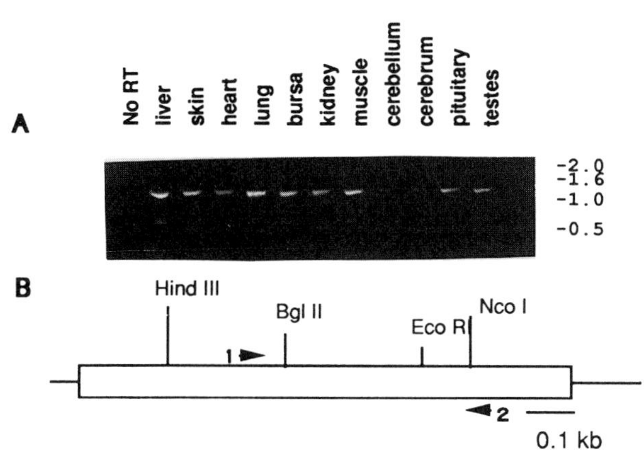


Fig. 5 Expression of chicken growth hormone receptor (cGHR) mRNA in different tissues of the chicken. An ethidium bromide gel (Panel A) shows a 1043 base pair fragment amplified from RNA using reverse transcriptase coupled to PCR. No fragment was detected in the absence of reverse transcriptase using liver RNA (No RT). A DNA ladder (1 kbp) was used for size determination. Restriction map of the cGHR cDNA indicating the location of the primers used in the PCR reaction (Panel B).

Although the liver is the most abundant source of GHR, we have used reverse transcriptase coupled to PCR to demonstrate expression of the GHR in a number of other tissues in the chicken (Fig. 5). These tissues include the heart, skeletal muscle, brain, adrenal and the pituitary gland where it may function in an autoregulatory loop (21, 22). The detection of the GHR in many different tissues of mammals and birds supports the diverse actions of GH.

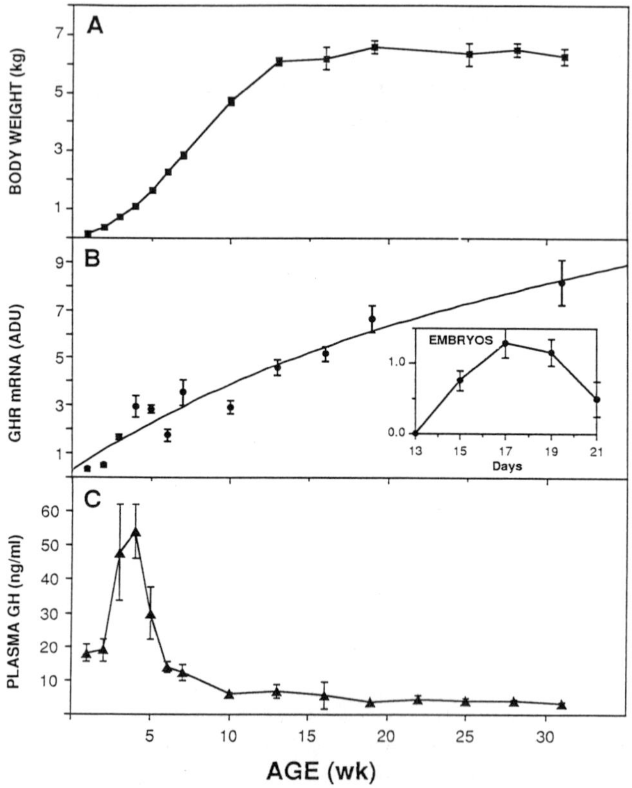


Fig. 6 Ontogeny of (A) body weight, (B) hepatic growth hormone receptor (GHR) mRNA and (C) concentrations of plasma GH in broiler chickens. GHR mRNA were determined by Northern blot hybridization analysis of total RNA using the cGHR cDNA as a probe. Amounts of GHR mRNA were quantitated after standardization with β-actin mRNA and are expressed in arbitrary densitometric units (ADU). Values are means ± s.e. for four different birds. (Reprinted from 23 with permission, Elsevier Publishers Ireland Ltd.)

## Regulation of growth hormone receptor expression

The ontogeny of cGHR has been examined in the broiler chicken and its expression in the liver has been compared to that of IGF-I (23). The relationship between body weight, GHR mRNA and plasma GH levels is shown in Figure 6. In the embryo, hepatic cGHR mRNA is not detectable until Day 15, is highest on Days 17 and 19, and then declines at hatching (Day 21) (Fig. 6B inset). A similar pattern of hepatic GH-binding activity has been reported in the chicken embryo (24). Following an initial two week delay after hatching, there is a progressive increase in hepatic GHR mRNA until the chickens reach a mature body weight. Expression of hepatic GHR mRNA increases about 20-fold between hatching and 31 weeks of age. On the other hand, concentrations of plasma GH reach a peak at 3 to 4 weeks of age and then fall sharply until maintenance of a low basal concentration between 10 and 31 weeks of age. Down-regulation of the GHR may account for the initial lag in expression of GHR mRNA seen after hatching due to increased concentrations of plasma GH (25-27). The rapid fall in plasma GH seen after 4 weeks of age may lift the GH inhibition of GHR gene expression.

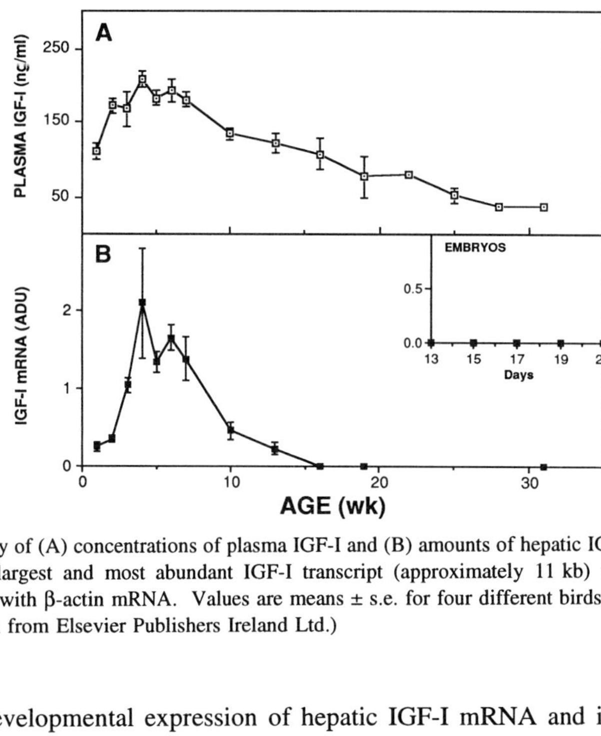


Fig. 7 Ontogeny of (A) concentrations of plasma IGF-I and (B) amounts of hepatic IGF-I mRNA in broiler chickens. The largest and most abundant IGF-I transcript (approximately 11 kb) was quantitated after standardization with β-actin mRNA. Values are means ± s.e. for four different birds. (Reprinted from 23 with permission from Elsevier Publishers Ireland Ltd.)

The developmental expression of hepatic IGF-I mRNA and its relationship to plasma IGF-I is shown in Fig 7. Hepatic IGF-I mRNA is undetectable in the chicken embryo between 13 and 21 days of incubation (Fig. 7A inset). Although both GH and IGF-I are important for embryogenesis in the chicken (28), expression of the GHR (23) and IGF-I mRNA (27, 28) are not coordinate in embryonic liver. The highest amounts of hepatic IGF-I mRNA are found at 4 weeks of age and coincide with the peak in concentrations of plasma IGF-I. It appears that the decline in plasma IGF-I lags considerably behind the sharp fall in concentrations of plasma GH and expression of hepatic IGF-I mRNA. The slower decay in expression of IGF-I could be due to increased abundance of GHR which compensates for the reduced availability of circulating ligand (GH).

In mammals, GH appears to up-regulate expression of its receptor (29-31), while the GHR in chickens appears to be down-regulated by its ligand (32-35). The pattern of GH administration (pulsatile versus continuous) appears to effect GHR expression in mammals (36) and chickens (35). The expression of GHR gene has been examined in 6-week-old broiler chickens continuously infused with 80 or 160μg cGH/kg/day via osmotic minipumps for 6 days (Fig. 8). Both doses of cGH depress hepatic GH-binding activity by about 25%. Northern blot analysis shows that the low dose (80μg cGH/kg/day) increases cGHR mRNA levels by 2-fold, whereas the higher dose (160μg cGH/kg/day) depress cGHR expression by 60% (Fig. 8). Three different species of GHR-related transcripts (3.2, 4.3 and 6.4 kilobases) have been found that appear to be coordinately regulated (Fig. 8). Despite down-regulation of the cGHR by a 3-fold elevation of plasma GH levels, plasma IGF-I levels are increased by 35% after 6 days of continuous infusion of 160μg cGH/kg/day. Thus, there appears to be a strong inverse relationship between concentrations of plasma high GH and expression of the cGHR mRNA in the liver, which is observed during normal development and with GH administration in young broiler chickens.

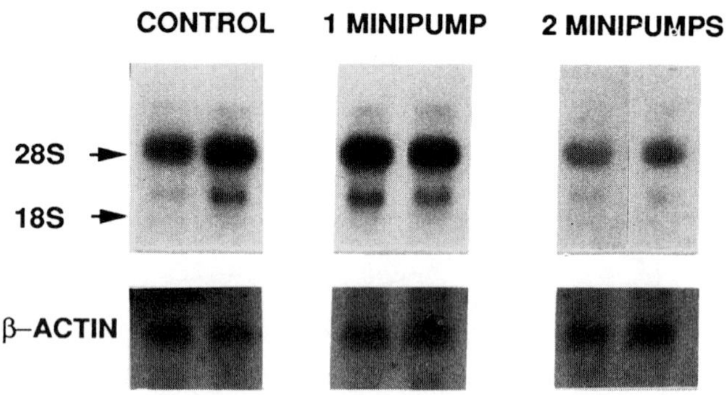


Fig. 8 Northern blot analysis of hepatic growth hormone receptor (GHR) mRNA in 6-week-old broiler chickens continuously infused with chicken growth hormone (cGH) from osmotic minipumps for 6 days. Minipumps delivered 0 (control), 80 (1 minipump) or 160 (2 minipumps) μg cGH/kg-day. Total mRNA was prepared from the livers and GHR mRNA measured by Northern blot analysis after standardization with β-actin mRNA.

## Analysis of growth hormone receptor in sex-linked dwarf chickens

A line of dwarf chickens has provided a unique opportunity to study the role of the GHR in growth. The sex-linked dwarf (SLD) mutation of chickens was first described by Hutt (37) and is characterized by a mature body weight that is 30-40% lower than that of normal chickens (38, 39). The dwarfism in chickens is due to a recessive gene (gene symbol *dw*) present on the Z chromosome. The endocrine abnormalities and growth characteristics of the SLD chicken have been extensively studied in several laboratories (see 40 and 41 for recent reviews). In general, concentrations of plasma IGF-I are low despite elevated plasma GH (42-45). The *dwdw* chickens are also hypothyroid, since concentrations of plasma triiodothyronine ($T_3$) and hepatic 5′-monodeiodinase activity are low, while concentrations of plasma thyroxine ($T_4$) are normal or elevated in comparison to *DwDw* chickens (44, 46). However, supplementation with either GH, $T_3$ or $T_4$ does not correct the growth retardation of SLD chickens (47-50). The absence of hepatic GH-binding activity in SLD chickens (51, 52) suggests that the primary lesion is in the cGHR gene. This type of dwarfism represents a form of GH resistance similar to Laron syndrome described in humans (53). The structure and expression of the cGHR has been examined in two different lines of SLD (54). One of these lines (CT) was obtained from Dr Bill Rishell of Arbor Acres Farm in Glastonbury, CT. The second line (GA) is maintained by Dr Henry Marks at the University of Georgia.

*Hepatic growth hormone-binding activity*

Specific binding of $^{125}$I-cGH to liver membranes from the two different lines of male dwarf chickens and their normal brothers is shown in Figure 9. Specific GH binding activity [expressed as (bound - nonspecific binding)/total radioactivity x 100] of normal CT males was about 20% at 35 weeks of age, whereas GH-binding activity was undetectable in *dwdw* chickens from this genetic background. Specific GH-binding activity was about 7% in the normal chickens obtained from the GA line. In contrast, specific GH binding in the liver membranes of *dwdw* males from the GA line was just at the limit of detection (0.34% binding). The difference in specific binding seen in the *DwDw* birds from the two different lines was due to the difference in the

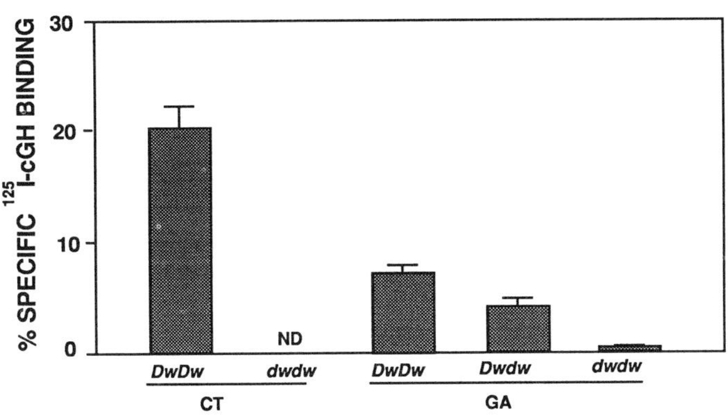


Fig. 9 Specific binding of $^{125}$I-chicken growth hormone (cGH) to hepatic membranes isolated from two different lines (CT and GA) of male sex-linked dwarf (*dwdw*) chickens. Chickens from the CT line were 35 weeks of age and the GA line were 11 weeks of age. Hepatic GH-binding activity was non-detectable (ND) in the *dwdw* chicken from the CT line. Values are means ± s.e. for four different birds. (Reprinted from 54 with permission, Academic Press, Inc.)

age (35 weeks for the CT and 11 weeks for the GA line) since GH binding activity (51) and GHR gene expression (23) progressively increase with age. It is interesting that the level of GH-binding activity of heterozygous males (*Dwdw*) which have one copy of the dwarfing gene and a single normal gene was about one-half that observed for the homozygous normal. This level of binding activity is similar to that seen in normal, hemizygous females (*Dw-*), which also have only a single copy of the normal gene (data not shown). Both normal females and heterozygous males have reduced body weights compared to normal males. It is tempting to speculate that amplification of the GHR gene, either by natural selection or direct genetic manipulation would enhance the growth of normal chickens.

*Southern blot analysis*

An analysis of restriction fragment length polymorphism of the DNA of the two different lines of chickens has been used to determine if the GHR gene in the *dwdw* chicken has undergone any gross structural changes (Fig. 10). The restriction fragment pattern of DNA from the CT line of *dwdw* chickens differs from that of *DwDw* chickens of the same genetic background. In DNA cut with the restriction enzyme *Hind*III, the predominant hybridizing band is approximately 2.2 kilobase pairs (kbp) in normal birds and about 3 kbp in *dwdw* chickens. This suggests the presence of a mutation in the GHR gene itself, or a closely-linked gene, in the CT line of SLD chickens. The *DwDw* chickens of GA and CT lines show identical DNA restriction patterns with *dwdw* chickens of the GA line, which indicates an absence of a polymorphism with the restriction enzymes tested.

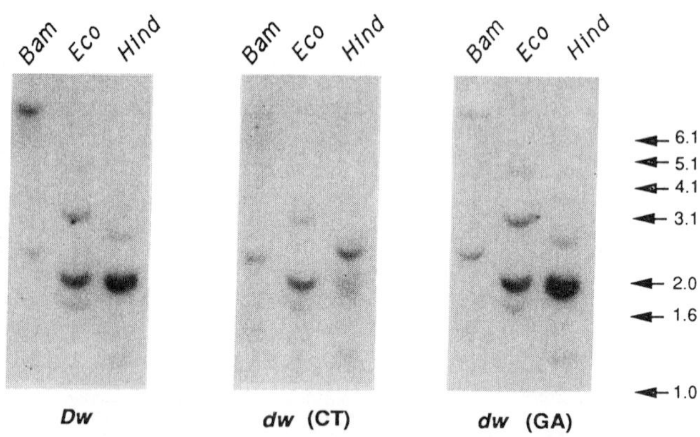


Fig. 10 Autoradiograph of Southern blot analysis of genomic DNA using chicken growth hormone receptor (cGHR) or cDNA prepared from a normal (*Dw*) chicken and two different lines of sex-linked dwarf (*dwdw)* chickens (CT and GA). DNA was digested with the indicated restriction enzymes (*Bam, Bam* HI; *Eco, Eco* RI; *Hind, Hind* III), electrophoresed, transferred to a nylon membrane, and hybridized to the cGHR cDNA. A 1 kbp DNA ladder was used for size determination. (Reprinted from 54 with permission, Academic Press, Inc.)

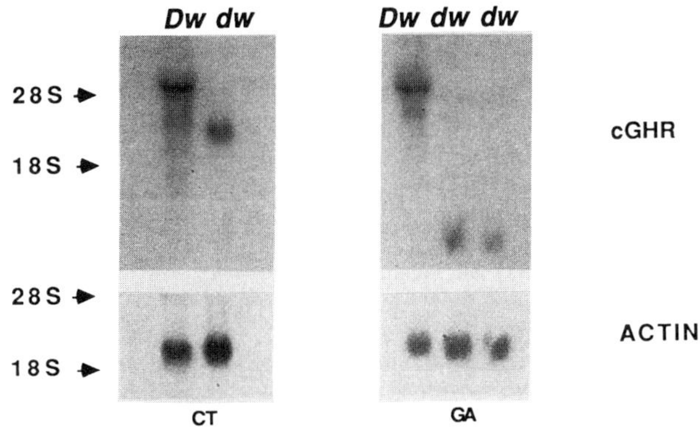


Fig. 11 Northern blot analysis of liver RNA using chicken growth hormone receptor (cGHR) cDNA prepared from normal (*Dw*) and sex-linked dwarf (*dwdw*) chickens (CT and GA). RNA was electrophoresed in a formaldehyde/agarose gel, transferred to nylon and hybridized to the cGHR cDNA (upper panel). The blot was also hybridized to β-actin to confirm the presence of intact RNA in all lanes. The position of 28S and 18S ribosomal RNAs are marked. (Reprinted from 54 with permission, Academic Press, Inc.)

*Expression of hepatic growth hormone receptor mRNA*

The expression of the GHR in the two lines of SLD chickens has been examined by Northern blot analysis (Fig. 11). In *DwDw* chickens of both lines, a major transcript of approximately 4.3 kilobase (kb) occurs along with a less abundant transcript of 3.2 kb. A single, aberrantly-sized transcript (2.5 kb) is detected in the RNA prepared from livers of the CT line of *dwdw* chickens. The aberrantly-sized transcript could be due to either a large deletion or premature termination of transcription of the GHR gene. Evidence has been found for a different type of mutation in *dwdw* birds from the GA line in which no large hybridizing species corresponding to the GHR mRNA is detected. Although a smaller band (0.7 kb) is apparent on these blots, the small transcript does not contain sufficient coding region for a functional GHR protein. In SLD of the Leghorn lineage, no alteration in expression of GHR mRNA is observed (55), despite an absence of hepatic GH-binding activity (51). Genomic DNA in the SLD Leghorn shows no evidence for a polymorphic site linked to the GHR gene (data not shown). A point mutation leading to premature termination of the protein could lead to a non-functional protein, despite the presence of a normal-sized transcript. Further analysis of the GHR gene sequence in both normal and SLD chickens will be necessary to determine the exact molecular nature of these defects.

It is clear that the SLD trait arose independently in different populations of chickens (56) and has been transferred to other lines by out-breeding. This would account for the heterogeneity in the mutations that have been described. Regardless of the nature of the mutation, the GHR gene must be found on the Z chromosome if it is the dwarfing gene found in SLD chickens. Our cGHR cDNA probe has recently been used to localize the GHR gene to the Z chromosome of chickens (57). These data and the fact that at least two different types of receptor mutations produce the same phenotype provide direct evidence that a mutation in the GHR gene is the primary cause of sex-linked dwarfism in chickens.

Table 1 Stimulation of IGF-I secretion in primary culture, hepatocytes from chickens with the sex-linked dwarfing gene (*dwdw*) after transfection with the gene encoding chicken growth hormone receptor.

| | **IGF-I in culture media (ng/ml)** | |
| --- | --- | --- |
| | **No growth hormone added** | **Growth hormone** |
| *DwDw* hepatocytes | 0.42 ± 0.05 | 1.87 ± 0.22* |
| *dwdw* hepatocyte mock transfected | 0.19 ± 0.01 | 0.23 ± 0.02[NS] |
| *dwdw* hepatocytes cGHR transfected | 0.19 ± 0.02 | 0.51 ± 0.03* |

Hepatocytes from *DwDw* and *dwdw* chickens (GA line) were isolated and cultured as described by O'Neill *et al* (58). The *dwdw* hepatocytes were transfected with the cGHR expression vector, using Lipofectin (25 μg) to mediate DNA transfer. Cells were incubated with or without cGH (100 ng/ml) in the presence of insulin (25 ng/ml) for 48 h. IGF-I was concentrated from media by reverse phase chromatography (Sep-Pak cartridges) and assayed by radioimmunoassay (26). Data are presented as the mean ± s.e. IGF-I concentration (ng/ml) in media collected from three different plates of cells. IGF-I concentrations were compared between GH-treated and non-treated hepatocytes. * indicates a significant difference ($P \leq 0.05$) and NS indicates no significant difference.

*Hepatocyte culture and transfection*

Primary cultures of hepatocytes represent a unique system for reconstitution of the defective GHR gene in the *dwdw* chicken. The chicken hepatocyte culture system described by O'Neill *et al* (58) has been used for primary culture of hepatocytes from the *dwdw* chicken (GA line). Primary cultures of hepatocytes from *DwDw* birds increase IGF-I secretion by 5-fold in response to cGH added to media; however, hepatocytes from the *dwdw* chicken are unresponsive to cGH (Table 1). Transfection of *dwdw* hepatocytes with an expression vector containing the cGHR cDNA (Fig. 3A) restores the GH-GHR-IGF-I pathway as indicated by a 2.6-fold increase in IGF-I secretion into media in response to GH stimulation. The reconstitution of the GH-GHR-IGF-I pathway by transfection with the cloned cGHR proves the deficiency of the GHR in the *dwdw* chicken. This culture system of chicken hepatocytes lacking a functional GHR will be useful for structural and functional analysis of the cloned cGHR.

## Conclusion

Broiler chickens which have been selected for growth for many generations represent an interesting model for studying the endocrine factors that control growth.

There are several aspects of the endocrine system that are unique to birds, which limit direct extrapolation of information obtained from similar studies in mammals. One dramatic difference is that chickens appear to be GH-resistant, since exogenous GH does not enhance growth rate or improve body composition (33). This GH resistance could occur at the level of the GHR, where GHR expression could be a rate-limiting step in further enhancement of growth in poultry. An understanding of the molecular biology of the cGHR will provide new information on regulation of somatic growth in chickens.

## References


1 Baumann G (1991) Growth hormone heterogeneity: genes, isohormones, variants, and binding proteins. *Endocrine Reviews* **12** 424-449

2 Herington A C, Tiong T S & Ymer S I (1991) Serum binding proteins for growth hormone: origins, regulation of gene expression and possible roles. *Acta Paediatrica Scandinavica Supplement* **379** 61-69

3 Leung D W, Spencer S A, Cachianes G, Hammonds R G, Collins C, Henzel W J, Barnard R, Waters M J & Wood W I (1987) Growth hormone receptor and serum binding protein: purification, cloning and expression. *Nature* **330** 537-543

4 Godowski P J, Leung D W, Meacham L R, Galgani J P, Hellmiss R, Keret R, Rotwein P S, Parks J S, Laron Z & Wood W I (1989) Characterization of the human growth hormone receptor gene and demonstration of a partial gene deletion in two patients with Laron-type dwarfism. *Proceedings of the National Academy of Sciences of the USA* **86** 8083-8087

5 Mathews L S, Enberg B & Norstedt G (1989) Regulation of rat growth hormone receptor gene expression. *Journal of Biological Chemistry* **264** 9905-9910

6 Baumbach W R, Horner D L & Logan J S (1989) The growth hormone-binding protein in rat serum is an alternatively spliced form of the rat growth hormone receptor. *Genes & Development* **3** 1199-1205

7 Smith W C, Kuniyoshi J & Talamantes F (1989) Mouse serum growth hormone (GH) binding protein has GH receptor extracellular and substituted transmembrane domains. *Molecular Endocrinology* **3** 984-990

8 Hauser S D, McGrath M F, Collier R J & Krivi G G (1990) Cloning and *in vivo* expression of bovine growth hormone receptor mRNA. *Molecular and Cellular Endocrinology* **72** 187-200

9 Adams T E, Baker L, Fiddes R J & Brandon M R (1990) The sheep growth hormone receptor: molecular cloning and ontogeny of mRNA expression in the liver. *Molecular and Cellular Endocrinology* **73** 135-145

10 Burnside J, Liou S S & Cogburn L A (1991) Molecular cloning of the chicken growth hormone receptor complementary DNA: mutation of the gene in sex-linked dwarf chickens. *Endocrinology* **128** 3183-3192

11 Bass S H, Mulkerrin M G & Wells J A (1992) A systematic mutational analysis of hormone binding determinants in the human growth hormone receptor. *Proceedings of the National Academy of Sciences of the USA* **88** 4498-4502

12 Urbanek M, Macleod J N, Cooke N E & Liebhaber S A (1992) Expression of a human growth hormone (hGH) receptor isoform is predicted by tissue-specific alternative splicing of exon-3 of the hGH receptor gene transcript. *Molecular Endocrinology* **6** 279-287

13 Fuh G, Cunningham B C, Fukunaga R, Nagata S, Goeddel D V & Wells J A (1992) Rational design of potent antagonists to the human growth hormone receptor. *Science* **256** 1677-1680

14 Ymer S I & Herington A C (1992) Developmental expression of the growth hormone receptor gene in rabbit tissues. *Molecular and Cellular Endocrinology* **83** 39-49

15 Vasilatos-Younken R, Anderson B J, Rosebrough R W, McMurtry J P & Bacon W L (1991) Identification of circulating growth hormone-binding proteins in domestic poultry: an initial characterization. *Journal of Endocrinology* **130** 115-122

16 Davis S L, Graf M, Morrison C A, Hall T R & Swift P J (1992) Identification and partial purification of serum growth hormone binding protein in domestic animal species. *Journal of Animal Science* **70** 773-780

17 Oldham E R, Bingham B & Baumbach W R (1992) Regulation of avian GH receptor RNA by alternate poly (A) addition. Abstract 701. *Abstracts of Endocrine Society*

18 Baumbach W R (1991) *Personal communication*

19 Burnside J, Liou S S & Cogburn L A (1992) Molecular cloning of the chicken serum growth hormone binding protein cDNA. *Poultry Science* **71** (Supplement 1) 75

20 Krishnan K A, Proudman J A & Bahr J M (1989) Avian growth hormone receptor assay: use of chicken and turkey liver membranes. *Molecular and Cellular Endocrinology* **66** 125-134

21 Fraser R A, Siminoski K & Harvey S (1991) Growth hormone receptor gene: novel expression in pituitary tissue. *Journal of Endocrinology* **128** R9-R11

22 Fraser R A & Harvey S (1992) Ubiquitous distribution of growth hormone receptors and/or binding proteins in adenohypophyseal tissue. *Endocrinology* **130** 3593-3600

23 Burnside J & Cogburn L A (1992) Developmental expression of hepatic growth hormone receptor and insulin-like growth factor-I mRNA in the chicken. *Molecular and Cellular Endocrinology* **89** 91-96

24 Vanderpooten A, Huybrechts L M, Decuypere E & Kühn E R (1991) Differences in hepatic growth hormone receptor binding during development of normal and dwarf chickens. *Reproduction Nutrition Development* **31** 47-55

25 Harvey S, Davison T F & Chadwick A (1979) Ontogeny of growth hormone and prolactin secretion in the domestic fowl (*Gallus domesticus*). *General and Comparative Endocrinology* **39** 270-273

26 McGuinness M C & Cogburn L A (1990) Measurement of developmental changes in plasma insulin-like growth factor-I levels of broiler chickens by radioreceptor assay and radioimmunoassay. *General and Comparative Endocrinology* **79** 446-458

27 Kikuchi K, Buonomo F C, Kajimoto Y & Rotwein P (1991) Expression of insulin-like growth factor-I during chicken development. *Endocrinology* **128** 1323-1328

28 Serrano J, Shuldiner A R, Roberts C T, LeRoith D & De Pablo F (1990) The insulin-like growth factor I (IGF-I) gene is expressed in chick embryos during early organogenesis. *Endocrinology* **127** 1547-1549

29 Grant A L, Helferich W G, Kramer S A, Merkel R A & Bergen W G (1991) Administration of growth hormone to pigs alters the relative amount of insulin-like growth factor-I mRNA in liver and skeletal muscle. *Journal of Endocrinology* **130** 331-338

30 Sauerwein H, Breier B H, Bass J J & Gluckman P D (1991) Chronic treatment with bovine growth hormone upregulates high-affinity hepatic somatotropic receptors in sheep. *Acta Endocrinologica* **124** 307-313

31 Ambler G R, Breier B H, Surus A, Blair H T, McCutcheon S N, Silbergeld A & Gluckman P D (1992) The interrelationship between and the regulation of hepatic growth hormone receptors and circulating GH binding protein in the pig. *Acta Endocrinologica* **126** 155-161
32 Leung F C, Taylor J E, Wien S & Van Iderstine A (1986) Purified chicken growth hormone (GH) and a human pancreatic GH-releasing hormone increase weight gain in chickens. *Endocrinology* **118** 1961-1965
33 Cogburn L A (1991) Endocrine manipulation of body composition in broiler chickens. *Critical Reviews in Poultry Biology* **3** 283-305
34 Vanderpooten A, Darras V M, Huybrechts L M, Rudas P, Decuypere E & Kühn E R (1991) Effect of hypophysectomy and acute administration of growth hormone (GH) on GH-receptor binding in chick liver membranes. *Journal of Endocrinology* **129** 275-281
35 Zhong C, Cogburn L A, Moellers R F & Burnside J (1991) Effects of pulsatile and continuous chicken growth hormone administration on expression of growth hormone receptor and insulin-like growth factor-I in liver and muscle tissue. *Poultry Science* **70** (Supplement 1) 137
36 Maiter D, Walker J L, Adam E, Moatsstaats B, Mulumba N, Ketelslegers J M & Underwood L E (1992) Differential regulation by growth hormone (GH) of insulin-like growth factor-I and GH receptor binding protein gene expression in rat liver. *Endocrinology* **130** 3257-3264
37 Hutt F B (1959) Sex-linked dwarfism in the fowl. *Journal of Heredity* **50** 209-221
38 Marks H L (1982) Selection for body weight at eight weeks of age in dwarf and normal meat-type chickens originating from a common control population background. *Poultry Science* **62** 227-234
39 Stewart P A, Washburn K W & Marks H L (1984) Effect of the *dw* gene on growth, plasma hormone concentrations and hepatic enzyme activity in a randombred population of chickens. *Growth* **48** 59-73
40 Decuypere E, Huybrechts L M, Kühn E R, Tixier-Biochard M & Mérat P (1992) Physiological alterations associated with the chicken sex-linked dwarfing gene. *Critical Reviews in Poultry Biology* **3** 191-221
41 Tixier-Boichard M, Huybrechts L M, Kühn E R, Decuypere E, Charrier J & Mongin P (1989) Physiological studies on the sex-linked dwarfism of the fowl: a review on the search for the gene's primary effect. *Génétique Selection Evolution* **21** 217-234
42 Hoshino S & Yamamoto K (1977) Synthesis and release of growth hormone, prolactin, and other proteins from the anterior pituitary of normal and dwarf chickens. *General and Comparative Endocrinology* **32** 7-16
43 Hoshino S, Wakita M, Suzuki M & Yamamoto K (1982) Changes in a somatomedin-like factor and immunoassayable growth hormone during growth of normal and dwarf pullets and cockerels. *Poultry Science* **61** 777-784
44 Scanes C G, Marsh J, Decuypere E & Rudas P (1983) Abnormalities in the plasma concentrations of thyroxine, triiodothyronine and growth hormone in sex-linked dwarf and autosomal dwarf White Leghorn domestic fowl (*Gallus domesticus*). *Journal of Endocrinology* **97** 127-135
45 Huybrechts L M, King D B, Lauterio T J, Marsh J & Scanes C G (1985) Plasma concentrations of somatomedin-C in hypophysectomized, dwarf and intact growing

domestic fowl as determined by heterologous radioimmunoassay. *Journal of Endocrinology* **104** 233-239

46 May J D & Marks H L (1983) Thyroid activity of selected, nonselected, and dwarf broiler lines. *Poultry Science* **62** 1721-1724

47 Leung F C, Taylor J E & Van Iderstine A (1984) Effects of dietary thyroid hormones on growth and serum $T_3$, $T_4$ and growth hormone in sex-linked dwarf chickens. *Proceedings of the Society for Experimental Biology and Medicine* **177** 77-81

48 Bowen S J, Huybrechts L M, Marsh J A & Scanes C G (1987) Influence of triiodothyronine and growth hormone on growth of dwarf and normal chickens: interaction of hormones and genotype. *Comparative Biochemistry and Physiology* **86A** 137-142

49 Huybrechts L M, Michielsen R, Darras V M, Berghman L R, Decuypere E & Kühn E R (1988) Effect of methimazole-induced hypothyroidism on a growth hormone induced insulin-like growth factor I response in normal and sex-linked dwarf domestic fowl. *Medical Science Research* **16** 85-86

50 Tixier-Boichard M, Decuypere E, Huybrechts L, Kühn E R & Mérat P (1990) Effects of dietary $T_3$ on growth parameters and hormone levels in normal and sex-linked dwarf chickens. *Domestic Animal Endocrinology* **7** 573-586

51 Leung F C, Styles W J, Rosenblum C I, Lilburn M S & Marsh J A (1987) Diminished hepatic growth hormone receptor binding in sex-linked dwarf broiler and Leghorn chickens. *Proceedings of the Society for Experimental Biology and Medicine* **184** 234-238

52 Kühn E R, Huybrechts L M, Vanderpooten A & Berghman L (1989) A decreased capacity of hepatic growth hormone (GH) receptors and failure of thyrotrophin-releasing hormone to stimulate the peripheral conversion of thyroxine into triiodothyronine in sex-linked dwarf broiler hens. *Reproduction Nutrition Development* **29** 461-467

53 Laron Z, Pertzelan A, Karp M, Kowadlo-Silbergeld A & Daughaday W H (1971) Administration of growth hormone to patients with familiar dwarfism with high plasma immunoreactive growth hormone: measurement of sulfation factor, metabolic and linear growth responses. *Journal of Clinical Endocrinology and Metabolism* **33** 333-342

54 Burnside J, Liou S S, Zhong C & Cogburn L A (1992) Abnormal growth hormone receptor gene expression in the sex-linked dwarf chicken. *General and Comparative Endocrinology* **88** 20-28

55 Harvey S, Hull K L & Marsh J A (1992) Sex-linked dwarfism: evidence against a growth hormone-receptor gene deletion in Leghorn chickens. Abstract 703. *Abstracts of the Endocrine Society*

56 Guillaume J (1976) The dwarfing gene *dw*: its effects on anatomy, physiology, nutrition and management. *World's Poultry Science Journal* **32** 285-304

57 Crittenden L, Provencher L, Santangelo L, Levin I, Abplanalp H, Briles W E & Dodgson J B (1993) Characterization of a red jungle fowl by White Leghorn backcross reference population for molecular mapping of the chicken genome. *Poultry Science* **72** 334-348

58 O'Neill I E, Houston B & Goddard C (1990) Stimulation of insulin-like growth factor-I production in primary cultures of chicken hepatocytes by chicken growth hormone. *Molecular and Cellular Endocrinology* **70** 41-47

# Part Four
# Endocrine Control of Metabolism and Water Balance

Avian Endocrinology
Ed. P J Sharp, pp 177-188
Journal of Endocrinology Ltd, Bristol (1993)

# The insulin receptor in chicken tissues

J Simon and M Taouis
Station de Recherches Avicoles, Endocrinologie de la Croissance et du Métabolisme, INRA, 37380 Nouzilly, France

## Introduction

Insulin, in common with other peptidic hormones, binds to a specific membrane receptor to exert its biological actions. Studies on the insulin receptor began in 1971 when a suitable insulin tracer became available (1, 2). This review summarizes current knowledge of insulin receptors in avian tissues with an emphasis on the chicken. Binding characteristics, ontogeny, and tissue distribution of insulin receptors will be considered first, followed by an analysis of the structure of the receptor and its interaction with insulin. The regulation of insulin release, its biological actions, and insulin sensitivity in birds are not considered since they have been reviewed recently by Rideau (3) and Simon (4).

## Binding characteristics

The interaction of insulin with its receptor has not been fully analyzed in birds, but the main characteristics of this interaction in chicken and turkey tissues are similar to those described in mammalian tissues. This also applies to solubilized receptors. For example, the binding of insulin to receptors in turkey erythrocytes is dependent on time, temperature, and pH, with an optimal pH for binding around 7.8. At lower temperatures binding increases, while the time course required for binding to occur is slower (5, 6). Insulin binding to turkey erythrocytes is also affected by ionic environment (5, 6). Insulin binding is low in the absence of cations, except Tris-$H^+$; it is decreased by $Mn^{++}$ and increased by $K^+$ and $Na^+$ and, most markedly, by $Ca^{++}$ or $Mg^{++}$.

The receptor is specific in that unrelated peptides do not bind to it although related peptides and insulin analogues bind with different affinities, according to their biological potencies. The rank order of potencies in bird tissues is chicken insulin > porcine insulin >> guinea pig insulin or pro-insulin. The higher affinity of chicken insulin for its receptor is due to the histidine residue at position 8 of the A chain in the chicken insulin molecule (4). The insulin-like growth factors I and II (IGF-I and IGF-II) which possess some structural analogy with pro-insulin, also bind with a low affinity to the insulin receptor in tissues including liver, muscles, brain and cultured neurones from embryos and growing chickens.

The binding of insulin is reversible so that after reaching equilibrium, addition of excess of insulin, or dilution, causes the insulin-receptor complex to dissociate. In kinetic experiments where dissociation is measured after dilution or dilution with the addition of insulin, the presence of insulin accelerates and extends the rate of dissociation of the hormone-receptor complex. This is interpreted as an example of negative co-operativity in which inhibitory interactions between binding sites lead to a gradual decrease in the affinity of the receptors as their occupancy increases. This does not seem to be a universal phenomenon because no evidence has been found for negative cooperativity between insulin receptors in chicken embryo fibroblasts (7).

In membranes or cells, the average affinity of insulin receptor, as estimated by the insulin concentration which inhibits tracer binding by 50%, is in the nanomolar range. When results from competitive binding experiments are expressed in Scatchard co-ordinates, the plot is curvilinear upward. This is accounted for by either the existence of negative cooperativity between binding sites, or the existence of two types of binding sites exhibiting a high and a low affinity for insulin.

## Ontogeny of insulin receptors

Insulin receptors occur during embryogenesis and after hatching in many tissues from chickens and turkeys, including brain, liver, cardiac and skeletal muscles, erythrocytes, thymocytes, fibroblasts, chondrocytes and myoblasts. They are present in membranes prepared from whole embryos as early as 2-3 days of embryogenesis and develop at different rates in different tissues during differentiation and development (8, 10, 11).

Insulin receptors localized by autoradiography are seen in whole chicken embryos as early as stage 4 of development (12). In 6-day embryos, insulin binding is highest in the retina and lateral wall of the growing encephalic vesicles (9). In older embryos, and in the adult brain, insulin binding is high in the paleostriatum augmentatum, lobus parolfactorius and molecular layer of the cerebellum (9). This localization of insulin receptor does not fully coincide with that of IGF-I receptor suggesting that insulin may have a specific role in brain development and physiology (13). In chickens, insulin stimulates neuronal growth (14, 15) and choline acetyltransferase activity in retinal neurones (16). Evidence that these insulin receptors are functional early during chicken embryogenesis is suggested by the observation that insulin and insulin analogues injected on day 2 of development stimulate metabolism, growth and differentiation two days later (17). In contrast, injections of insulin antibody or insulin receptor antibody have depressive effects on most of these parameters (18).

## Insulin receptor concentration

In the liver and thymus of growing chickens, the concentration of insulin receptor is 2-5 times lower than in the corresponding tissues of the rat or calf (19-22). This may contribute to the relative state of insulin resistance which is observed in chickens compared to mammals (4). The low number of insulin receptors in chicken tissues may result from a 'down-regulation' induced by the hyperactivity of endogenous insulin. However, observations in ducks suggest other mechanisms are involved. In ducks, Pekin duck (*Anas platyrhyncos*) insulin is less active, while chicken insulin is more active, than mammalian insulin. This difference might be expected to result in high concentrations of insulin receptor in duck tissues. In fact, the insulin receptor number is similar in Pekin duck and chicken liver membranes but is lower in liver membranes from the Muscovy duck (*Cairina moschata*) (23).

Alteration of nutritional state in chickens modifies insulin receptor numbers in the liver without affecting affinity. Prolonged fasting for 48 h increases the number of insulin receptors per mg of membrane protein, while refeeding following a fast decreases it (24). The same nutritional conditions do not modify brain membrane insulin receptor number or affinity (24). In muscles, the low level of insulin binding and high non-specific binding do not permit such quantification, irrespective of the nutritional state (25).

There is evidence that the concentration of insulin receptors in chicken liver membranes is related to insulin sensitivity. Daily injections of corticosterone, which reduce the hypoglycemic response to insulin, also reduce insulin receptor number (26). Conversely, dwarf chickens (*dw*, a recessive sex-linked gene) which are more sensitive to insulin, have higher concentrations of insulin receptor in the liver compared to normal (*Dw*) chickens (27). This evidence for a relationship between insulin sensitivity and insulin receptor number is not supported by studies in genetically fat and lean chickens. Genetically fat chickens are more sensitive to exogenous insulin (28) than genetically lean chickens despite the fact that there are no differences in liver insulin receptor number or affinity between genotypes (29).

In mammals, the cell surface insulin receptor number is mainly regulated by insulin itself. In response to high concentrations of insulin, membrane insulin receptor number decreases as a consequence of 'down-regulation'. This mechanism has been extensively studied in chicken hepatocytes (30-32). Following exposure to 50 nM insulin for 18 h, cell surface insulin receptor number decreases from 10,000 to 4,000 sites/cell, i.e. by 60% with no apparent alteration in the affinity of the receptor for insulin (31). In both normal or down-regulated states, the total cellular insulin receptor number estimated following cell solubilization remains constant: about 17,000 sites/cell. The synthesis and degradation rates of insulin receptor have also been estimated, using heavy isotope-labelled amino-acids (30-32). The half life of insulin receptor in chicken hepatocytes is 10 h and is not modified by chloroquine or other inhibitors of lysosomal proteases. Therefore, following down-regulation and internalization, most of the insulin receptors remain intact, suggesting that in the chicken liver, as in mammalian tissues, internalized insulin receptor is recycled back to the cell surface. In contrast, internalized insulin is degraded principally by a chloroquine-sensitive pathway. Therefore, in chicken liver, following internalization, insulin receptors and insulin are processed by different routes which are similar to those described in mammalian cells (for a review see 33). Down-regulation of insulin receptors has also been demonstrated in quail oviduct cells (34), in a chicken hepatoma cell line (LMH cells, 35), in chicken neurones (36) and in chicken embryo heart cells (37, 38). In heart cells, down-regulation decreases cell responsiveness to insulin (37, 38). In contrast, in cultured chicken retina neurones, insulin binding changes during culture independently of insulin concentration (16). Similarly, in turkey erythrocytes maintained in culture in the presence of a high concentration of insulin, insulin receptors are not down-regulated but surprisingly, their affinity is increased (39). Therefore, in birds, insulin receptor down-regulation may not occur in all the tissues, or alternatively may be related to other factors in addition to insulin concentration.

**Insulin receptor structure**

The structure of the avian insulin receptor has been characterised in several tissues in studies involving cross-linking $^{125}$I insulin or auto-phosphorylation. These tissues include turkey erythrocytes (40, 41), chicken embryo fibroblasts (7) and heart cells (42), quail oviduct cells (34) and membranes from chicken brain, liver and skeletal muscle (10, 25, 43, 44). As in mammals, two subunits have been identified: the β subunit is about 95KDa while the α subunit is about 135KDa. The α subunit binds insulin, causing a conformational change in the insulin receptor, autophosphorylation of the β subunit, and activation of an intrinsic tyrosine kinase (45, 46). These properties are shared with several other growth factor receptors. A full length gene encoding the chicken insulin receptor has not yet been cloned but sequence information is available for the autophosphorylation domain which is homologous with the corresponding domain of the human insulin receptor (47). Results from structural studies suggest that, as in mammals, the avian insulin receptor α subunit does not contain a hydrophobic fragment and is therefore extracellular. The β subunit contains one hydrophobic fragment, representing a trans-membrane domain which extends on both sides of the membrane. The cytosolic portion includes the kinase domain and both subunits are linked by disulfide bonds.

The biosynthesis of insulin receptor has not been studied in chicken tissues, but in mammals the biosynthetic pathway involves the synthesis of a glycosylated (high mannose content) proreceptor precursor in the endoplasmic reticulum which is cleaved to produce the mature α- and β-subunits (48, 49). These are further glycosylated in the Golgi apparatus (48, 49). Evidence that chicken insulin receptor is also glycosylated is suggested by the finding that Triton-solubilized receptors adsorb to, and can be eluted from, various lectins (24). Further, desialylation of chicken liver insulin receptors by neuraminidase digestion accelerates their migration in polyacrylamide gel electrophoresis (44) and concanavalin A lectin inhibits insulin binding to chicken myoblasts (50).

Insulin receptors are present on the membrane in an heterotetrameric form with the 2α-2β subunits being linked by disulfide bonds (40-42). However, heterodimeric forms (α-β) may also occur (40-42, 51). The kinase of the (α-β) form has been found to be inactive in some studies (see references in 51), suggesting that the presence of this form in the membrane does not necessarily represent a functional receptor. Only the heterotetrameric form of the receptor (2α-2β) is observed in turkey erythrocytes (52), chicken liver and LMH cells (35) after polyacrylamide gel electrophoresis, provided it is protected from 'mild' reduction by iodoacetamide or N-ethylmaleimide, and provided the samples are not boiled.

## Properties of the α-insulin receptor subunit

In chickens and mammals, differences between brain and peripheral tissue insulin receptors have been described (13). The insulin receptor in chicken brain (44), and neurones (13), has a higher affinity than the insulin receptor in peripheral tissue, although other binding characteristics are the same. The molecular size of the chicken brain insulin receptor α-subunit is about 10 kDa less than the liver insulin receptor and is insensitive to neuraminidase digestion (44), suggesting a tissue-specific glycosylation for the insulin receptor in the brain. This difference between brain and liver insulin receptors is present during embryonic development although at day 2 of embryogenesis only 'brain type' receptors are expressed (43). In mammals, brain-type insulin receptors are expressed exclusively in neurones of the central nervous system and not in peripheral neurones (13). The physiological significance of a different type of insulin receptor in the nervous system is still uncertain because the brain is generally considered to be insensitive to insulin. However, insulin stimulates growth, differentiation and protein synthesis in chicken neurones (14, 15, 36). A specific type of insulin receptor has also been described in chicken muscles before or after hatching (25, 43): the α-subunit of the muscle insulin receptor is intermediate in size between those of brain and liver. The possibility that variations in the glycosylation of muscle insulin receptor accounts for the intermediate size of the muscle α-subunit has not been examined. The tissue specificities of the size of the α-insulin receptor subunit are not modified by prolonged fasting for 48 h, or after refeeding following prolonged fasting (24). It is not known whether the heterogeneity which is observed for insulin receptor α-subunit is only due to differences in glycosylation between tissues. In man there is evidence for differences in the amino acid core of insulin receptors. Although there is only one gene encoding the insulin receptor, two distinct insulin receptors can be generated by alternative splicing of one exon (53). These two forms of insulin receptors differ by the presence or the absence of 12 amino acids at the C-terminal end of the α-subunit. The truncated form of insulin receptor exhibits increased affinity for insulin and is exclusively expressed in normal human muscle whereas the full length insulin receptor is present in diabetic human muscles (53).

## Properties of the β-insulin receptor subunit

In mammals, insulin-stimulated autophosphorylation of the insulin receptor β-subunit and the subsequent activation of the constitutive tyrosine kinase are the first events following insulin binding and they initiate the pathways for insulin signalling (54). Those properties have been demonstrated for turkey (51, 52), duck (23) and chicken (12, 24, 25, 44) receptors prepared from various tissues of growing birds or embryos. Chicken brain and liver insulin receptor β-subunits can be immunoprecipitated by a human anti-insulin receptor antibody (24). Furthermore, insulin receptor kinase is active as early as day 2 of chicken embryonic development (12), suggesting that at this stage of development, biological effects of insulin are mediated through insulin receptor kinase.

In chicken and duck liver, an unidentified ATPase, which co-purifies with the insulin receptor during purification using wheat germ agglutinin lectin affinity chromatography, masks the kinase activity of the receptor when measured in standard conditions (23, 24, 44). Modification of the phosphorylation medium by the addition of 20 mM $Mg^{++}$, 3 mM $Mn^{++}$ and 1 mM vanadate overcomes this problem, making it possible to demonstrate liver insulin receptor kinase activity (44). This ATPase activity is absent or inactive in preparations of insulin receptors from mammalian livers, chicken muscle and chicken brain (24, 25, 44). It is not known whether this ATPase activity affects insulin sensitivity in chicken and duck liver *in vivo*. It is not inhibited by ATP competitors (CTP or GTP), indicating its specificity, or by the usual ATPase inhibitors (23, 24) and is not immunoprecipitated by a chicken $Na^+/K^+$ ATPase antibody (23). This unusual ATPase is not intrinsic to the receptor since a fraction of insulin receptor free of ATPase can be prepared from chicken or duck liver using lentil or pea lectin affinity chromatography (23, 44).

The insulin-dependent autophosphorylation of the chicken insulin receptor β-subunit, as observed in chicken liver and a chicken liver hepatoma (LMH) cell line, is manganese dependent and very rapid with maximum autophosphorylation being almost achieved within 30 s at room temperature (35). Phosphorylation is tyrosine specific since phosphorylated receptors from chicken liver and muscle can be immunoprecipitated using anti-phosphotyrosine antibodies (35, J Simon, unpublished observations). The specific residues which are phosphorylated in response to insulin have not been identified. However, since the fragment of the chicken insulin receptor gene so far identified is highly conserved in the kinase domain, it is very likely that the phosphorylation involves at least Tyr 1146, Tyr 1150 and Tyr 1151. In mammals, those sites are important for full activation of the insulin receptor. The possibility that the receptor is further phosphorylated on other Tyr residues and also on Ser/Thr residues is also likely. In rat, it has been shown that there is a non-constitutive serine kinase tightly attached to insulin receptor which is activated by insulin (55). The significance of this Sr/Thr kinase and the phosphorylation of serine and threonine residues in the insulin receptor has not yet been elucidated.

The tyrosine kinase activity of insulin receptors from chicken brain and liver has the same specificity as in other species (44). When insulin receptor tyrosine kinase activity is measured using (Glu, Tyr) polymer in a ratio 4 to 1, as a substrate, it is high both in the absence and presence of insulin. Using this substrate, insulin receptor kinase activity has been measured in various nutritional or physiological states associated with changes in insulin sensitivity. Prolonged fasting, which is likely to evoke insulin resistance (24), impairs basal and insulin-stimulated phosphorylation of the artificial substrate by liver insulin receptor but does not impair the kinase activity of brain or muscle insulin receptors (24, 25). Prolonged fasting may also affect the autophosphorylation of the liver insulin receptor, but this is difficult to quantify due to the low incorporation of radioactivity (24). Treatment with corticosterone which depresses growth (56) and sensitivity to exogenous insulin (26) also impairs the basal and insulin-stimulated tyrosine kinase activity of liver insulin receptor (26). The basal activity of muscle insulin receptor is also transiently altered following corticosterone treatment. The physiological significance of basal kinase activity is unknown.

Two situations have been described where insulin sensitivity is increased. In dwarf chickens (27), liver insulin receptor number is higher than in normal chickens, which may improve insulin sensitivity, whereas the insulin-stimulated kinase activity of the liver insulin receptor is the same in both types of bird. However, the basal activity of the liver insulin receptor is decreased in dwarf chickens. Genetically fat chickens are slightly more sensitive to insulin than genetically lean chickens in the fasting state (28). In contrast, obese mammals are resistant to insulin compared to their lean counterparts. Furthermore, genetically fat chickens exhibit enhanced liver lipogenesis and a faster restoration of glycogen when refed following a fast, than genetically lean chickens (see references listed in 57). However, liver insulin receptor number and kinase activity are not different in genetically lean and fat chickens (29). From those studies, it appears therefore that the functional coupling between insulin binding to the α-subunit of its receptor and the activation of the β-subunit kinase is not always the same. Furthermore, changes in functional coupling can vary independently between tissues. In mammals, several mechanisms have been suggested to account for changes in insulin receptor tyrosine kinase activity and insulin sensitivity. These include the state of phosphorylation of serine residues which is likely to be controlled by several serine kinases including protein kinase C, protein kinase A (see the references listed in 4) and an unidentified serine kinase (55). An alternative mechanism may involve the state of phosphorylation of tyrosine residues which appear to be under the control of different phosphatases (58). Whether the same mechanisms are present and active to the same extent in chicken tissues is at present unknown.

In mammals and chickens, alterations of insulin sensitivity are not always coupled to an alteration of insulin receptor number or kinase activity (59). Several steps in the insulin signal transduction pathway subsequent to the receptor are therefore predicted to play a modulatory role in the control of insulin responsiveness.

**Insulin signalling pathways**

Insulin signal transduction is mediated principally, if not exclusively, by the insulin receptor tyrosine kinase (54). However, the activities of many insulin-sensitive cellular enzymes are controlled through phosphorylation and dephosphorylation reaction cascades on serine and threonine residues. The link or the 'switch' between the insulin receptor tyrosine kinase and these serine kinases is not yet understood (60). In mammals, several cellular substrates have been proposed for the insulin receptor tyrosine kinase which are phosphorylated on tyrosine residues in response to insulin (54). These include insulin receptor substrate 1 (IRS-1, 185 kDa, a cytosolic protein), a 120 kDa protein which appears to be an ecto-ATPase, a 42 kDa protein (MAP kinase, a serine/threonine kinase) and in 3T3-L1 adipocytes, a 15 kDa fatty acid binding protein. The precise role of these proteins is not yet elucidated. In parallel, or subsequent to, the phosphorylation of these proteins, other insulin signalling pathways have been proposed including activation of phosphatidyl inositol-3 kinase and protein kinase C and interactions with unidentified G-proteins (54).

The existence of only one insulin receptor substrate, a 70 kDa protein, has been suggested in the chicken (61). This protein, which is phosphorylated on tyrosine residues in response to insulin and IGF-I, is transiently expressed in cultured chicken embryo neurones (61). Further characterization is required before this protein is fully accepted as an insulin receptor substrate. In chicken LMH cells (35), a 72 kDa phosphoprotein also fulfils the criteria of an insulin receptor substrate. It has the appropriate time course for phosphorylation, phosphorylation on tyrosine residues, and it responds to insulin in a dose-dependent manner. However the $Mn^{++}$ dependency of the phosphorylation of this protein, its immunoprecipitation by two insulin receptor antibodies, and the fact that it disappears on polyacrylamide gel electrophoresis in non-reducing conditions, strongly suggest that this protein is a truncated insulin receptor β-subunit or a degradation product of insulin receptor β-subunit (35). Other possible signal transduction pathways in chicken neurones are suggested by the observation that insulin stimulates the activity of one of the multiple forms of protein kinase C, protein kinase C-ε (62). Finally, indirect evidence suggests in the turkey erythrocyte that the binding of insulin receptor is regulated by an unidentified G-protein (63).

## Conclusion

The properties of the insulin receptor in chicken and bird tissues are similar to those described in mammals. However, in the chicken, as in mammals, the mechanisms by which insulin acts are not yet fully elucidated. The elucidation of those mechanisms in chicken is of practical interest because of the role of insulin in the control of metabolism and growth (57) and because they would contribute to a better understanding of 'insulin resistance' which is a natural condition in the chicken (4).

## Acknowledgments

The excellent typing of the manuscript by Roseline Gauthier and help in the preparation of the bibliography by Michèle Plouzeau and Bernadette Chevalier are gratefully acknowledged.

## References

1 Cuatrecasas P (1971) Insulin-receptor interactions in adipose tissue cells: direct measurement and properties. *Proceedings of the National Academy of Sciences of the USA* **68** 1264-1268

2 Freychet P, Roth J & Neville D M Jr (1971) Insulin receptors in the liver: specific binding of [$^{125}$I] insulin to the plasma membrane and its relation to insulin bioactivity. *Proceedings of the National Academy of Sciences of the USA* **68** 1833-1837

3 Rideau N (1988) Insulin secretion in birds. *Leanness in domestic birds* pp 269-294. Eds B Leclercq & C Whitehead. London: Butterworths

4 Simon J (1989) Chicken as a useful species for the comprehension of insulin action. *Critical Reviews in Poultry Biology* **2** 121-148

5 Ginsberg B H, Cohen R M, Kahn C R & Roth J (1978) Properties and partial purification of the detergent-solubilized insulin receptor. A demonstration of negative cooperativity in micellar solution. *Biochimica et Biophysica Acta* **542** 88-100

6 Ginsberg B H, Kahn C R & Roth J (1977) The insulin receptor of the turkey erythrocyte: similarity to mammalian insulin receptors. *Endocrinology* **100** 82-90

7 Kasuga M, Van Obberghen E, Nissley S P & Rechler M (1982) Structure of insulin-like growth factor in chicken embryo fibroblasts. *Proceedings of the National Academy of Sciences of the USA* **79** 1864-1868

8 Bassas L, de Pablo F, Lesniak M A & Roth J (1985) Ontogeny of receptors for insulin-like peptides in chick embryo tissues: early dominance of insulin-like growth over insulin receptors in brain. *Endocrinology* **117** 2321-2329

9 Bassas L, Girbau M, Lesniak M A, Roth J & de Pablo F (1989) Development of receptors for insulin and insulin-like growth factor-I in head and brain of chick embryos: autoradiographic localization. *Endocrinology* **125** 2320-2327

10 Bassas L, Lesniak M A, Serrano J, Roth J & de Pablo F (1988) Developmental regulation of insulin and type I insulin-like growth factor receptors and absence of type II receptors in chicken embryo tissues. *Diabetes* **37** 637-644

11 Bassas L, Zelenka P S, Serrano J & de Pablo F (1987) Insulin and IGF receptors are developmentally regulated in the chick embryo eye lens. *Experimental Cell Research* **168** 561-566

12 Girbau M, Bassas L, Alemany J & de Pablo F (1989) *In situ* autoradiography and ligand-dependent tyrosine kinase activity reveal insulin receptors and insulin-like growth factor I receptors in prepancreatic chicken embryos. *Proceedings of the National Academy of Sciences of the USA* **86** 5868-5872

13 Heidenreich K A (1991) Insulin in the brain. What is its role? *Trends in Endocrinology and Metabolism* **2** 9-12

14 Heidenreich K A & Toledo S P (1989) Insulin receptors mediate growth effects in cultured fetal neurons. II. Activation of a protein kinase that phosphorylates ribosomal protein S6. *Endocrinology* **125** 1458-1463

15 Heidenreich K A & Toledo S P (1989) Insulin receptors mediate growth effects in cultured fetal neurons. I. Rapid stimulation of protein synthesis. *Endocrinology* **125** 1451-1457

16 Kyriakis J M, Hausman R E & Peterson S W (1987) Insulin stimulates choline acetyltransferase activity in cultured embryonic chicken retina neurons. *Proceedings of the National Academy of Sciences of the USA* **84** 7463-7467

17 Girbau M, Gomez J A, Lesniak M A & de Pablo F (1987) Insulin and insulin-like growth factor I both stimulate metabolism, growth, and differentiation in the postneurula chick embryo. *Endocrinology* **121** 1477-1482

18 Girbau M, Lesniak M A, Gomez J A & de Pablo F (1988) Insulin action in early embryonic life: anti-insulin receptor antibodies retard chicken embryo growth but not muscle differentiation *in vivo*. *Biochemical and Biophysical Research Communications* **153** 142-148

19 Cramb G, Langslow D R & Philipps J H (1982) The binding of pancreatic hormones to isolated chicken hepatocytes. *General and Comparative Endocrinology* **46** 297-309

20 Kemmler W, Renner R, Zynamon A & Hepp K D (1978) Interactions between insulins and liver membrane receptors of guinea pig, calf and chicken. Exclusion of a species-specific insulin receptor. *Biochimica et Biophysica Acta* **543** 349-356

21 Simon J, Freychet P & Rosselin G (1977) A study of insulin binding sites in the chicken tissues. *Diabetologia* **13** 219-228

22 Simon J (1979) Effect of insulin on glucose oxidation and amino isobutyric acid transport and binding of insulin of in chicken thymocytes. *Biochimica et Biophysica Acta* **585** 563-574

23 Constans T, Chevalier B, Lerouet M & Simon J (1991) Insulin sensitivity and liver insulin receptor structure in ducks from two genera. *American Journal of Physiology* **261** R882-R890

24 Simon J, Rosebrough R W, McMurtry J P, Steele N C, Roth J, Adamo M & LeRoith D (1986) Fasting and refeeding alter the insulin receptor tyrosine kinase in the chicken liver but fail to affect brain insulin receptors. *Journal of Biological Chemistry* **261** 17081-17088

25 Adamo M, Simon J, Rosebrough R W, McMurtry J P, Steele N C & LeRoith D (1987) Characterization of the chicken muscle insulin receptor. *General and Comparative Endocrinology* **68** 456-465

26 Taouis M, Derouet M, Chevalier B & Simon J (1993) Corticosterone effect on insulin receptor number and kinase activity in chicken muscle and liver. *General and Comparative Endocrinology* **89** 167-175

27 Gueritault I, Simon J, Chevalier B, Derouet M, Tixier-Boichard M & Mérat P (1990) Increased in-vivo insulin sensitivity but normal liver insulin receptor kinase activity in dwarf chickens. *Journal of Endocrinology* **126** 67-74

28 Saadoun A, Simon J, Williams J & Leclercq B (1988) Levels of insulin, corticosterone, $T_3$, $T_4$ and insulin sensitivity in fat and lean chickens. *Diabetes and Metabolism* **14** 97-103

29 Simon J, Chevalier B, Derouet M & Leclercq B (1991) Normal number and kinase activity of insulin receptors in liver of genetically fat chickens. *Journal of Nutrition* **121** 379-385

30 Krupp M N & Lane D M (1982) Evidence for different pathways for the degradation of insulin and insulin receptor in the chick liver cell. *Journal of Biological Chemistry* **257** 1372-1377

31 Krupp M & Lane D M (1981) On the mechanism of ligand-induced down-regulation of insulin receptor level in the liver cell. *Journal of Biological Chemistry* **256** 1689-1694

32 Lane D M (1981) The regulation of insulin receptor level and activity. *Nutrition Reviews* **39** 417-425

33 Carpentier J-L (1989) The cell biology of the insulin receptor. *Diabetologia* **32** 627-635

34 Kato S, Itoh S, Miura Y, Naito H & Noguchi T (1990) Characterization of insulin receptors in primary cultures of quail (*Coturnix coturnix japonica*) oviduct cells. The level of insulin receptor is regulated by steroid and peptide hormones. *Comparative Biochemistry and Physiology* **97B** 783-791

35 Taouis M, Derouet M, Caffin J P, Chavanieu A & Simon J (1993) Insulin receptor and insulin sensitivity in a chicken hepatoma cell line. *Molecular and Cellular Endocrinology* (in press)

36 Heidenreich K A, de Vellis G & Gilmore R (1988) Functional properties of the insulin receptor found on neurons. *Journal of Neurochemistry* **51** 878-887

37 Serravezza J C, Endo F, DeHaan R L & Elsas L J (1983) Insulin-induced receptor loss reduces responsiveness of chick heart cells to insulin. *Endocrinology* **113** 497-507

38 Serravezza J C, Wheeler F B, DeHaan R L & Elsas L J (1981) Characterization and regulation of insulin binding by embryonic heart cells. *Developmental Biology* **84** 417-424

39 Kosmakos F C & Roth J (1980) Insulin-induced loss of the insulin receptor in IN-9 lymphocytes. A biological process mediated through the insulin receptor. *Journal of Biochemistry* **255** 9860-9869

40 Aiyer R A (1983) Structural characterization of insulin receptors. I. Hydrodynamic properties of receptors from turkey erythrocytes. *Journal of Biological Chemistry* **258** 14992-14999

41 Aiyer R A (1983) Structural characterization of insulin receptors. II. Subunit composition of receptors from turkey erythrocytes. *Journal of Biological Chemistry* **258** 15000-15003

42 Endo F & Elsas L J II (1984) Structural analysis and subunit interaction of insulin receptor from membranes of cultured embryonic chick heart cells. *Endocrinology* **115** 1828-1837

43 Bassas L, de Pablo F, Lesniak M A & Roth J (1987) The insulin receptors of chick embryo show tissue-specific structural differences which parallel those of the insulin-like growth factor I receptors. *Endocrinology* **121** 1468-1476

44 Simon J & LeRoith D (1986) Insulin receptors of chicken liver and brain. Characterization of α and β subunit properties. *European Journal of Biochemistry* **158** 125-132

45 Ballotti R, Baron V, Gautier N, Hainaut P, Scimeca J C, Dolais-Kitabgi J, Lammers R, Schlessinger J, Ullrich A & Van Obberghen E (1992) Activation and regulation of the insulin receptor kinase. *Diabetes and Metabolism* **18** 98-103

46 Perlman R, Bottaro D P, White M F & Kahn C R (1989) Conformational changes in the α- and β-subunits of the insulin receptor identified by anti-peptide antibodies. *Journal of Biological Chemistry* **264** 8946-8950

47 Scavo L M, Serrano J, Roth J & de Pablo F (1991) Genes for the insulin receptor and the insulin-like growth factor-I receptor are expressed in the chicken embryo blastoderm and throughout organogenesis. *Biochemical and Biophysical Research Communications* **176** 1393-1401

48 Hedo J A & Gorden P (1985) Biosynthesis of the insulin receptor. *Hormone and Metabolic Research* **17** 487-490

49 Olson T S, Bamberger M J & Lane M D (1988) Post-translational changes in tertiary and quaternary structure of the insulin proreceptor. *Journal of Biological Chemistry* **263** 7342-7351

50 Sandra A, Leon M A & Przybylski R J (1979) Reversal by insulin of concanavalin A inhibition of myotube formation and evidence for common binding sites. *Endocrinology* **105** 391-401

51 Mortensen E R, Drachman J G & Guidotti G (1991) The α-β monomer of the insulin receptor has hormone-responsive tyrosine kinase activity. *Biochemical Journal* **273** 49-56

52 Treadway J L, Morrison B D, Wemmie J A, Frias I, O'Hare T, Pilch P F & Pessin J E (1990) The endogenous functional turkey erythrocyte and rat liver insulin receptor is an α2-β2 heterotetrameric complex. *Biochemical Journal* **271** 99-105

53 Mosthaf L, Vogt B, Hering H U & Ullrich A (1991) Altered expression of insulin receptor types A and B in the skeletal muscle of non-insulin-dependent diabetes mellitus patients. *Proceedings of the National Academy of Sciences of the USA* **88** 4728-4730

54 Roth R A, Zhang B, Chin J E & Kovacina K (1992) Substrates and signalling complexes: the tortured path to insulin action. *Journal of Cellular Biochemistry* **48** 12-18

55 Lewis R E, Wu G P, MacDonald R G & Czech M P (1990) Insulin-sensitive phosphorylation of serine 1293/1294 on the human insulin receptor by a tightly associated serine kinase. *Journal of Biological Chemistry* **265** 947-954

56 Simon J (1984) Effects of daily corticosterone injections upon plasma glucose, insulin, uric acid and electrolytes and food intake pattern in the chicken. *Diabetes and Metabolism* **10** 211-217

57 Simon J (1988) Insulin in birds: metabolic effects and possible implications in genetically fat and lean chickens. *Leanness in domestic birds* pp 253-268. Eds B Leclercq & C Whitehead. London: Butterworths

58 King M J & Sale G J (1990) Dephosphorylation of insulin-receptor autophosphorylation sites by particulate and soluble phosphotyrosyl-protein phosphatases. *Biochemical Journal* **266** 251-259

59 Adamo M, LeRoith D, Simon J & Roth J (1988) Effect of altered nutritional states on insulin receptors. *Annual Review of Nutrition* **8** 149-166

60 Czech M P, Klarlund J K, Yagaloff K A, Bradford A P & Lewis R E (1988) Insulin receptor signalling. Activation of multiple serine kinases. *Journal of Biological Chemistry* **263** 11017-11020

61 Kenner K A & Heidenreich K A (1991) Insulin and insulin-like growth factors stimulate *in vivo* receptor autophosphorylation and tyrosine phosphorylation of a 70k substrate in cultured fetal chick neurons. *Endocrinology* **129** 301-311

62 Heidenreich K A, Toledo S P, Brunton L L, Watson M J, Daniel-Issakani S & Strulovici B (1990) Insulin stimulates the activity of a novel protein kinase C, PKC-ε, in cultured fetal chick neurons. *Journal of Biological Chemistry* **265** 15076-15082

63 Mortensen E R, Drachman J & Guidotti G (1992) Guanosine nucleotides regulate hormone binding of insulin receptors. *Biochemical Journal* **281** 735-743

Avian Endocrinology
Ed. P J Sharp, pp 189-200
Journal of Endocrinology Ltd, Bristol (1993)

# From avian pancreatic polypeptide to mammalian neuropeptides: carbohydrate metabolism implications

R L Hazelwood
Department of Biology-Program in Physiology, University of Houston, Houston, TX 77204-5513, USA

## Introduction

The discovery of the first of the pancreatic polypeptide (PP) family in chickens in 1968, followed by the isolation and characterization of two other family members (namely, polypeptide-YY or PYY, and neuropeptide-Y or NPY) in mammals in 1980-82, ushered in an era in which families of structurally-related peptides were first recognised to exist in nature. Examples of such families include the gastrin family, the secretin family, the growth factors, and the insulin family. The pancreatic polypeptide family, sometimes referred to as the PP-fold family, is characterized by a 36 amino acid double helix structure, tightly folded, with a highly conserved carboxyl terminal hexapeptide. Structural homology of mammalian PPs with PYY is approximately 50-55%, and that of PYY with NPY is 70-74%. The folded molecule of each peptide brings both N- and C-termini in close approximation, a feature that is important to receptor recognition (1).

The vast majority of PP emanates from the F-cell of the pancreatic islets of birds and mammals, or the islet equivalent in lower vertebrate forms. A small (~5%) contribution of PP to plasma apparently comes from the antrum or duodenal area in a few mammals such as the dog and monkey. Gut distention, gut peptide release, and direct F-cell stimulation by absorbed nutrients play significant secretogogic roles.

The majority of F-cells in higher vertebrates are located in the pancreatic endocrine mass, mainly situated at the islet periphery in juxtaposition with the A-cell. Additionally, F-cells can frequently be found in clusters outside the islet, embedded in the acinar mass. Once released from the pancreas, PP circulates as a dimer (>80%), with a half-life of 6-7 minutes, and at concentrations of approximately 40-80 pM/L in normal humans, or 30-50 times higher in birds. This concentration of plasma PP may triple or even quintuple within minutes after eating a standard meal. Also, it is abnormally high in the pancreas of those with severe obesity and in the plasma of those diseases wherein the pancreas is stressed, including diabetes mellitus (2). In those animals lacking a discrete pancreas, and/or an islet mass, PP-like material is located only in the central nervous system (CNS).

The various actions that have been ascribed to PP in both mammals and birds do not give any indication that a singular mechanistic process is common to all. Of the more prominent effects, those reported to inhibit exocrine pancreatic secretion, intestinal motility and secretion, gall bladder contraction and excretion appear to be dominant, though some may be indirect actions of PP (3). Metabolic effects of PP include antilipolytic at the fat cell (birds), suppressive of glucose-induced insulin release (rodents), and inductive of satiety in certain rodents such as those used as animal models for studying Type II diabetes mellitus (2). Of particular interest have been the observations that PP binds avidly and specifically to brain membranes (4) and that when introduced to lateral brain ventricles, the polypeptide induces alterations in the feeding behaviour of mammals (1).

Polypeptide-YY (PYY) mainly emanates from the lower intestinal tract, the colon, and the rectum where it is co-localized with glicentin in the so-called L-cell. Like PP, meal eating is the main trigger of PYY release, although the time coordinates of release differ markedly from those of the former. While PP is released within 60 seconds of commencement of eating, PYY is released slowly over the first 30-60 minutes, only to peak in the circulation some two or three hours later. Obviously, vagal efferents are not involved in PYY secretion. The quality of the ingesta also dictates the stimulus-force for PYY secretion. High lipid meals are much more effective than isocaloric carbohydrate meals. Dinner is a better secretogogue than breakfast, also indicating a role for a volume-stimulus. Cholecystokinin is a powerful trigger of PYY release also, thus a foregut-lower gut axis exists.

Polypeptide-YY also may be released from some pancreatic islet cells, specifically the peripherally-located A-cells wherein the polypeptide is co-localized with glucagon. Once PYY reaches the circulation it achieves a normal plasma concentration of 6-10 pM/L in humans. Polypeptide-YY decreases gastric motility and secretion, gall bladder activity, exocrine pancreatic secretion, and intestinal motility and secretion. This polypeptide possesses all the attributes of the long sought-after putative hormone, enterogastrone, described some 60 years ago. The action of PYY on the gut is not mediated by a release of somatostatin but rather by an inhibition of both intestinal and pancreatic blood flow (1, 2, 5), as well as by direct effects on the intestinal crypt region where intestinal water and electrolyte secretion are decreased (6, 7). Thus, PYY may be considered the perfect anti-diarrhoeal agent in that it slows passage of the digesta, it decreases fluid and electrolyte loss via secretion, and it promotes absorption processes to take place.

Additionally, PYY receptors have been identified in the brain where the peptide has been shown to be one of the most powerful endogenous orexigenic agents known. In contrast to its satiety effects when administered peripherally, PYY is a strong stimulator of robust feeding (via an action on hypothalamic nuclei) in bird and mammal alike (8). As in other parts of the body, PYY and NPY share a common receptor in the brain (9). PYY receptors not only are to be found in the hypothalamic and arcuate areas but also in the pancreatic islet mass, more specifically the centrally located B-cell mass. Thus, in addition to its effect on feeding behaviour, PYY may play a central role in the modulation of insulin, the hormone which plays a key role in the metabolic distribution of ingested nutrients.

The discovery of neuropeptide-Y (NPY) and its localization throughout the vertebrate brain and peripheral autonomic nervous and vascular systems has resulted in many studies of its pressor and orexigenic actions. Neuropeptide-Y is a 36 amino acid polypeptide found in large, dense-core vesicles of cell bodies and terminals throughout the brain, most particularly the hypothalamic-arcuate areas. Additionally, it is found in terminal neural elements either alone, or co-localized with nor-epinephrine. Co-localization in the brain most frequently involves arteriolar innervation, while most peripherally located NPY is in elements of the sympathetic nervous system where it is frequently co-localized with the catecholamines. Such elements reach the endocrine pancreas. In humans, NPY circulates at concentrations approximating 20-30 pM/L plasma. The polypeptide has been described as the most widely distributed of all neuropeptides in the vertebrate brain, with the hypothalamic regions that are associated with feeding and drinking behaviour containing the highest concentrations. Many studies have been carried out relative to its effects on those hypothalamic centres, studies which led to the conclusion that like PYY, NPY is an extremely powerful orexigenic agent as well as a powerful pressor agent (10, 11).

On an equimolar basis, NPY is as effective as PYY and PP in suppressing exocrine pancreatic and biliary secretion in mammals. Further, its pressor activity is exerted by a PYY-shared receptor on vascular smooth muscle (especially in the intestinal region). Also, inhibition of gut chloride secretion - and thus water reabsorption - by NPY contributes to the 'quieting' action that the polypeptide has as a modulator of intestinal activity. Lowered hydrostatic pressure in the gut favours water reabsorption. Finally, the role, if any, NPY plays in regulating endocrine secretion from the pancreas has been investigated, since neural elements containing NPY (or NPY with nor-epinephrine) have been identified in the islets of Langerhans.

In conclusion, a family of pancreatic polypeptides has been identified that possesses strong structural homology and overlapping biological actions. Of considerable interest are the effects of these polypeptides on feeding behaviour, including selectivity of nutrients to be ingested, and as well on regulating those pancreatic hormones that are known to affect the final distribution of nutrients.

## Carbohydrate metabolism and the pancreatic polypeptide-family

### *Pancreatic polypeptide (PP)*

Avian pancreatic polypeptide (APP) was the first family member to be isolated and characterized when Kimmel and colleagues were studying avian insulin and carbohydrate metabolism in chickens. Biological studies initially carried out with this polypeptide were restricted to bolus injections (of what now appears to be pharmacological amounts) into chicks to study the effects on gastrointestinal function. However, since the only other (then) recognized pancreatic islet polypeptides were known to alter carbohydrate metabolism, APP also was evaluated as to its effects on plasma glucose, liver glycogen, and plasma lipid components. To the investigators' surprise, APP was found to be glycogenolytic but not hyperglycaemic (12). Also, it was effective in lowering plasma glycerol, cholesterol, fatty acids, and apolipoprotein while increasing triglyceride levels. Thus, the impression gained was that APP was effective in converting glucose to lipid substrate, and that hepatic glycogen was the primary (initial) source for such a conversion. Further, APP was found to be antilipolytic in isolated adipocytes, a finding that has been confirmed and which pointed investigators toward investigating the role PP may have in obesity or obese-related diseases such as Type II diabetes mellitus in animal models and humans alike.

Considerable interest in APP (or PPs in general) was generated by the report (13) that chronic injections of APP (or bovine PP, BPP) into New Zealand obese mice corrected their glucose intolerance, hyperglycaemia, hyperinsulinaemia, and polyphagia (with associated weight gain). While Floyd and colleagues had established that most human diabetics had grossly elevated concentrations of human pancreatic polypeptide, a case for cause or effect between PP and the disease could not be firmly established. Further, the early observations that PP may be involved in normal lipid metabolism were extended to human and rodent models of obesity where it was documented that in such cases where concentrations of plasma are low, pancreatic PP concentrations are high, and meal eating fails to release PP in normal amounts, suggesting a defective release mechanism. Should PP be renamed 'satietin'? The possibility that it should was indicated also by the interesting observation that APP specifically binds to avian brain membrane preparations more avidly than to other tissues (4). Does a functional pancreatic-brain axis exist in vertebrates?

The localization of pancreatic islet F-cells (PP-cells) in juxtaposition with other hormone-secreting cells suggested that PP may be a modulator of islet hormone secretion. To date no evidence exists that indicates PP alters either glucagon or somatostatin (A-cell and D-cell, respectively) synthesis or secretion. Neither is there evidence that PP alters basal B-cell release of insulin, though there are a few reports that the polypeptide inhibits glucose-induced insulin release. Possibly, PP acts as an intraislet modulator of insulin release. Insulin-induced hypoglycaemia in birds and in mammals releases PP to the circulation, a response that is mediated mainly by muscarinic mechanisms (14). This response is not related either to the rate of decrease or the extent of depression of concentrations of plasma glucose. In chickens, glucose infusion depresses APP levels via a β-adrenergic mechanism.

*Polypeptide-YY (PYY)*

The distribution of PYY along the lower gastrointestinal tract (co-localized with glicentin in the L-cell) is complemented by the finding that in mammals the polypeptide also emanates from peripheral islet cells where it is normally co-localized with glucagon. No information is available indicating whether or not the hormone is released with glucagon in response to a given stimulus, as opposed to possible singular release events (avian PYY is yet to be characterized as to its origin, its chemistry, and its function). When PYY is infused in the mouse at modest to moderate levels, no effects are observed on basal concentrations of plasma glucose, insulin or glucagon. In contrast, however, cholinergic and/or glucose-induced insulin release is inhibited at moderate PYY concentrations (15). Cholinergic secretion of glucagon also is inhibited. Thus, at least in the mouse, intraislet regulation of stimulated hormone release may include that of PYY along with other modulators, PYY being an inhibitory substance. In this sense, PYY appears to act much like PP (see above), but is in contrast to documented effects of other gut peptides such as cholecystokinin (CCK), secretin and gastric inhibitory peptide (GIP), which stimulate insulin (all) and/or glucagon (secretin) release. Again, as with PP, a striking feature of PYY binding is its predilection for plasma membranes prepared from brain tissue, especially in hypothalamic regions that are intimately concerned with feeding behaviour.

*Neuropeptide-Y (NPY)*

The third PP-family component, NPY, is almost exclusively found in neural tissue (synaptosomes, axons, dendrites, etc.), both centrally and peripherally located. Those elements serving the pancreas include NPY only, as well as NPY co-localized with nor-epinephrine, terminations resting on B-cells and/or smooth muscle vascular elements. The stimuli controlling the release of this neuropeptide *in vivo* are yet to be determined, but experimental evidence implicates NPY as a significant factor in regulating insulin release. Thus, when isolated rat islets are exposed to increasing concentrations of NPY, no effect is observed on basal insulin release (at 5.5 mM glucose) but profound inhibition (over 40%) of glucose-stimulated (11.0 mM glucose) insulin release is seen (16). Similar observations have been observed at the organ level when rat pancreas is perfused with NPY with basal or stimulatory amounts of glucose (5.5 and 11.0 mM). Thus, peripherally, the possibility that NPY is involved in the neural control of insulin release appears reasonable, especially from the *in vitro* data cited (16). Further, the role of NPY (as well as PP and PYY) in effecting a pressor activity within sympathetic neural fibres may well contribute to the observed inhibition of insulin secretion. Thus, while co-release of NPY and nor-epinephrine in the perfused - but not the isolated - islet preparation may be responsible for the inhibition of insulin secretion observed, it is possible that NPY effects release of the catecholamine. Whether or not the NPY-effected inhibition is mediated solely by a regulation of sympathetic neural activity, or by a direct action on the B-cell, or both, is yet to be defined clearly.

Just as the effects of insulin on feeding behaviour differ (opposite effects) depending on whether the hormone is administered peripherally (subcutaneously) or centrally (via intracerebral ventricular (ICV) microinjection), the effects of NPY on insulin release *in vivo* also differ depending on the route of administration. Unstressed, unanaesthetized rats injected with small amounts of NPY via ICV respond with quick increases in insulin secretion (16). Such injections, made into the third cerebral ventricle, allow the neuropeptide to perfuse hypothalamic centres that regulate both insulin release (suprachiasmatic region) and feeding behaviour (paraventricular (PVN) and arcuate) areas. These areas are densely innervated by NPY neural elements and presumably contain NPY-receptors. Thus, it is probable that NPY acts as a central modulator of both feeding and drinking behaviour, and is involved in the release of a pancreatic hormone intricately involved in regulating carbohydrate metabolism. Such a suggestion is reinforced by the many observations linking insulin (both peripheral and central) with feeding patterns in mammals. The orexigenic effect seen following adrenergic stimulation of the rodent PVN is probably due to vagal stimulation of the B-cell.

In summary, all three members of the pancreatic polypeptide family have been implicated experimentally with regulation of insulin secretion, feeding and drinking behaviour, and possibly satiety (depending on the route of administration).

## Feeding behaviour and the pancreatic polypeptide-family

*Satiety - peripheral and central*

Satiety has two major components, peripheral and central, and is achieved by neural, endocrine and nutrient substrate mechanisms. The urge to eat is nullified by volume distention of the upper gastrointestinal tract and thus over neural afferent pathways, by hormone release such as that of CCK and GIP, (which have been shown to interact with specific receptors in the higher CNS), and by the quality of the nutrients, as shown by the antiphagic effects of a high carbohydrate snack taken immediately before a mixed meal.

Generally, central hunger mechanisms override peripheral satiety mechanisms in the control of feeding behaviour. In this regard, the PVN appears to be far more important than associated hypothalamic nuclei, including the glucostat area of the ventral medial hypothalamus (VMH) in initiating the satiety phase of feeding behaviour. The observations (described above) that chronic APP or BPP injections into obese mice curtail the hyperphagia and hyperglycaemia characteristic of such animals, argues for a peripheral role that this family member may play in achieving satiety. Further, the prolonged release of PYY during and after a meal, also, would appear to be a component of a battery of gut peptides that have specific receptors in various CNS areas in addition to receptors in the foregut and exocrine pancreas (17). PYY is known to have specific receptors in the hypothalamus, distinct from those for NPY.

*Orexigenic effects - central*

Direct application of NPY or PYY to the PVN induces increased feeding behaviour in rodents, squirrels, pigs and chickens. Drinking behaviour is also increased, even in the absence of food, indicating that the neuropeptides have two distinct effects in this respect (18, 19). Polypeptide-YY and NPY, when microinjected or topically applied to the region of the PVN, are extremely effective orexigenic stimuli. This hypothalamic area is inhibited by corticotrophin (CRF) application, which in turn is inhibited by norepinephrine (NE) application. Thus, the orexigenic effects of NE at the PVN are in fact due to inhibition of inhibition. Neural pathways running to and from the PVN, VMH and arcuate tracts are densely populated with NPY, and NPY co-localized with NE fibres. However, the orexigenic drive observed with NPY applied centrally cannot be attributed to co-release of NE because the latter's effect is blocked by alpha adrenergic blockade, adrenalectomy, double vagotomy, and pre-treatment with CCK, while that of NPY is not. Further, NPY causes a selective, preferential ingestion of high carbohydrate meals as well as increased water intake, while NE depresses water intake and exerts equal preference for fatty foods as well as carbohydrate substrates (18, 19). The ingestive drive observed with NPY is relatively long lasting, at least four hours, after application to the PVN. Of considerable interest is the fact that the ingestive effects of NPY can be observed either during day time or night time, in fasted or satiated rats, and even in the hibernating squirrel during its pre-hibernation anoretic period.

On an equimolar basis, both PYY and NPY are far more potent (>1000x) orexigenic agents than is NE. In fact, these two PP-family members appear to be the most powerful endogenous orexigenic agents known. Could variations in their concentrations at the hypothalamus play a vital role in hunger-satiety phenomena? Could the waxing and waning of the concentrations at the PVN be a factor in selective nutrient preference, such as carbohydrate in diabetes mellitus?

## Pancreatic polypeptide family and pathophysiologic considerations

*Diabetes mellitus*

Collectively, observations that APP (and other PPs) bind avidly to brain plasma membrane preparations, that frequently human diabetic subjects have very high circulating HPP levels, that NPY and PYY 'drive' feeding behaviour when applied centrally, and that rodents receiving NPY at the PVN level preferentially seek high carbohydrate substrates, have led investigators to examine what role, if any, the PP-family plays in diabetes mellitus. There is no evidence whatsoever that the family plays a causative role in this disease. It is relevant to note that concentrations of plasma APP in normal birds are 40-80 times higher than those observed in most mammals, including man. Normal avian blood glucose concentrations are two to three times those of normal humans and are similar to those in humans with uncontrolled diabetes. Evidence gathered in the last few years implicates NPY, in particular, in the behavioural expression of diabetes mellitus as related to food and water intake. Thus, rats treated with the B-cell cytotoxin, streptozotocin, are found to have significantly higher amounts of hypothalamic amounts of NPY than normal rats as early as two-three weeks after being made diabetic. The increase in hypothalamic NPY content is closely related, therefore, to the induced hypoinsulinaemia, not to the attending hyperglycaemia which is immediate (20). Further, there is an increase in the population of F-cells in the periphery of the diabetic pancreatic islet. Also, studies on pre-pro-NPY mRNA in the diabetic rat hypothalamus indicate elevations of up to 500% over control values in NPY mRNA several weeks after the induction of a Type I diabetic state (21). Elevated amounts of NPY are readily released from diabetic hypothalami when subjected to depolarization (22). All of these alterations in NPY physiology are corrected with insulin replacement.

Diabetes mellitus is a condition characterized by hyperglycaemia, glycosuria, osmotic diuresis, polydipsia and hyperphagia. Neuropeptide-Y is a proven hyperphagic agent as well as a stimulant to water ingestion. Further, the peptide favours carbohydrate ingestion over that of other nutrients. Considering the marked increase in the amount of NPY in diabetic (rodent) hypothalami, and the known pressor effect of NPY, focal reduction in the delivery of plasma glucose to certain hypothalamic nuclei may result from the attending vasoconstriction. 'Starvation for glucose' may result. This condition may be aggravated by the concomitant decreased delivery of insulin which is necessary for the intracellular transport of glucose into certain hypothalamic cells. Thus, a reasonable homeostatic response to nutrient deprivation would be to increase the volume of food intake as well as to increase carbohydrate intake. Neuropeptide-Y could serve this purpose in diabetes mellitus. Finally, findings of increased NPY-immunostaining in the area of the supraoptic nucleus (SON) of diabetic rats may indicate an important water homeostatic role for the polypeptide in this disease. In face of polyuria due to the attendant osmotic diuresis, increased anti-diuretic hormone (ADH) release by the SON would compensate for the water loss. So a possible corrective role can be suggested for hypothalamic NPY in diabetes, one that favours intake of nutrients (especially carbohydrates) to replenish 'starved' cells (hyperphagia), conservation of water (polydipsia) to offset loss due to diuresis, and one that attempts to achieve normoglycaemia (B-cytotropic insulin release).

*Obesity*

The antilipolytic effect of avian pancreatic polypeptides in isolated chicken adipocytes has already been noted. This may have important implications in that, unlike mammals, insulin is not antilipolytic in birds. Possibly, APP substitutes in this role for insulin in the ultimate distribution of lipids in birds. Also cited above is the fact that obese human subjects fail to respond to meal eating with a rapid release of HPP from the pancreas. Prader-Willi subjects, who frequently exhibit obesity, hyperglycaemia and glucose intolerance, fail to respond to the meal-eating stimulus with a release of HPP. Collectively, these observations lead us to focus on a faulty trigger-release of PP from the pancreatic islets. Experimentally, congenitally obese (ob/ob) mice do not respond to the meal-eating stimulus as do their lean littermates. Yet, the ob/ob pancreas contains at least twice as much PP as does the pancreata of lean control mice. If such an obese mouse is parabiosed with a lean littermate, or receives transplanted normal or alloxanized pancreatic islet tissue, the usual characteristics attending this form of obesity (hyperphagia, hyperglycaemia, glycosuria, obesity, and glucose intolerance) disappear (13). Collectively, the parabiotic studies and the pancreatic islet transplant studies indicate that the obese animal possesses a functional satiety centre, one that can be stimulated by a blood borne factor that emanates from non-B-cell elements.

Fundamental differences in the regulation of hypothalamic NPY have been reported in Zucker (fa/fa) obese rats (23). Such obese rats have increased amounts of NPY in hypothalamic areas that are considered important to feeding behaviour. When injected into the PVN, NPY causes hyperphagia, obesity and increased insulin and glucagon secretion. The amounts of neuropeptide-Y in the PVN increase during periods of fasting, only to return to normal, control levels upon feeding. The fa/fa data strongly suggest that NPY acts on certain hypothalamic nuclei as an orexigenic agent. Could the polypeptide play a significant role in the obese condition so prevalent in America today?

*Bulimia and anorexia nervosa*

Consider the following:

1 The PP-family has been identified as an important modulator of gastrointestinal activity;

2 At least two of the family are very potent pressor neuropeptides, including cerebral arteriolar elements and/or focal cerebral vasoconstriction;
3 Neuropeptide-Y and PYY are important endogenous orexigenic peptides acting on specific hypothalamic nuclei.

It may then follow that waxing and waning of amounts of the PP-fold family members in the central nervous system may be the basis for aberrant eating patterns observed in certain pathophysiological states. For instance, could highly variable peaks and valleys of NPY at the hypothalamic level be responsible for the 'binge eating' associated with bulimia? This condition is found most frequently in young adult women, previously normal, who develop an abnormal fear of becoming fat. The amount of food ingested can be enormous, up to 50,000 kcal/eating period. The eating pattern has been described as 'dietary chaos', but an underlying characteristic worth mentioning is the fact that bulimic subjects have a strong preference for carbohydrate-rich foods! Dental caries are common in these subjects. Again, could an accumulation of NPY (or PYY) at the site of certain hypothalamic nuclei be responsible for the binge eating syndrome, and with vascular removal, the hiatus that occurs between eating periods? Studies in this area are called for and certainly should be rewarding.

Anorexia nervosa is considered by some workers as a form of bulimia, while other workers consider it a separate disease because of a number of distinct differences that have been observed between the two. This disease, too, is found mostly in young women of middle- to upper-class backgrounds, and like bulimia is characterized by a paralyzing fear of obesity. A radical restriction of caloric intake results, leading to a morbid emaciation. Primary hypothalamic disorders have been suggested as the basis of this condition, but whether or not they are truly causal or secondary to other factors, is yet to be established. The question may be asked "Is a valley of plasma NPY (or PYY) at the level of the hypothalamus a contributing factor to the anorectic period?" If so, then the episodic binge eating that spells the prolonged anoretic period in the anorexic patient may be attributed to a central accumulation of the polypeptide(s). Definitive answers to these challenging questions are not yet available but the questions indeed encourage investigation.

In summary, this field has come a long, interesting way from the observations of robust feeding in birds treated with APP, to the possibility that sister polypeptides play a causative role in certain human pathophysiological disorders, such as diabetes mellitus, obesity, bulimia, and anorexia nervosa.

**Future work on the pancreatic polypeptide family**

The preceding paragraphs outline some of the fascinating areas of study that lie ahead for investigators. In addition to the vascular importance of the PP-family, particularly in cerebral and coronary regions, the molecular mechanisms of intestinal inhibition, and the possibility of additional family members, are being identified and characterized. There is an obvious need for specific agonists and antagonists. Such developments would make it possible to establish (experimentally) a truly deficient state involving each of the family members, and thereby establish the real physiological significance of each polypeptide. Further, such work should lead to the 'perfect' anti-diarrhoeal agent, one that would have considerable use in clinical situations. Finally, such antagonists may be just what is needed in treating Type II diabetics where, most frequently, failure of dietary compliance hinders successful treatment. Also, a similar antagonist may be useful in countering peaks and valleys of NPY levels at the hypothalamus, thereby alleviating symptomatology associated with 'binge eating' and/or anorexia nervosa, if not obesity itself (1).

From APP in 1968 to NPY in 1992, avian pancreatic polypeptide, discovered accidentally, remains a somewhat enigmatic relative to other PP-family members. Less is known about this (first) member of the family after 25 years than about the most recently discovered member, NPY. Yet, it was the discovery of APP and the subsequent (early) observations that high, specific binding of the polypeptide to brain membranes occurs that led investigators to the current fascinating arena of PP family-CNS interaction, including the study of some important pathophysiological problems.

**Acknowledgements**

Sincere appreciation is extended to Leticia Ramirez for her role in manuscript preparation and to NSF award DCB-87-02206 for partial support.

## References

1 Hazelwood R L (1993) The pancreatic polypeptide (PP-fold) family: gastrointestinal, vascular and feeding behavioural implications. *Proceedings of the Society for Experimental Biology and Medicine* **202** 44-63

2 Hazelwood R L (1990) Pancreatic polypeptide (PP) and its relevant relatives. *Progress in Comparative Endocrinology* **342** 250-256

3 Lin T M, Evans D C, Chance R E & Spray G F (1977) Bovine pancreatic peptide: action on gastric and pancreatic secretion in dogs. *American Journal of Physiology* **232** E311-E315

4 Adamo M L & Hazelwood R L (1990) Characterization of liver and cerebellar binding sites for avian pancreatic polypeptide. *Endocrinology* **126** 434-440

5 Sheikh S P (1991) Neuropeptide Y and peptide YY: major modulators of gastrointestinal blood flow and function. *American Journal of Physiology* **261** G701-G715

6 Cox H M & Cuthbert A W (1988) Neuropeptide Y antagonizes secretagogue evoked chloride transport in rat jejunal epithelium. *Pflügers Archives* **413** 38-42

7 Laburthe M (1990) Peptide YY and neuropeptide Y in the gut: availability, biological actions, and receptors. *Trends in Endocrinology and Metabolism* **1** 168-174

8 Kuenzel W J, Douglass L W & Davison B A (1987) Robust feeding following central administration of neuropeptide Y or peptide YY in chicks. *Peptides* **8** 823-828

9 Inui A, Okita M, Inoue T, Sakatani N, Oya M, Morioka H, Shii K, Yokono K, Mizuno N & Baba S (1989) Characterization of peptide YY receptors in the brain. *Endocrinology* **124** 402-409

10 Gray T S & Morley J E (1986) Neuropeptide Y: anatomical distribution and possible function in the mammalian nervous system. *Life Sciences* **38** 389-401

11 Stanley B G & Leibowitz S F (1984) Neuropeptide Y: stimulation of feeding and drinking by injection into paraventricular nucleus. *Life Sciences* **35** 2635-2642

12 Hazelwood R L, Turner S D, Kimmel J R & Pollock H G (1973) Spectrum effects of a new polypeptide (third hormone?) isolated from the chicken pancreas. *Endocrinology* **21** 485-497

13 Gates R J & Lazarus N R (1977) The ability of pancreatic polypeptides (APP and BPP) to return to normal the hyperglycaemia, hyperinsulinaemia and weight gain of New Zealand obese mice. *Hormone Research* **8** 189-202

14 Havel P J, Parry S J, Curry D L, Stern J S, Akpan J O & Gingerich R L (1992) Autonomic nervous system mediation of the pancreatic polypeptide response to insulin-induced hypoglycaemia in conscious rats. *Endocrinology* **130** 2225-2229

15 Böttcher G, Ahrén B, Lundquist I & Sundler F (1989) Peptide YY: intrapancreatic localization and effects on insulin and glucagon secretion in the mouse. *Pancreas* **4** 282-288

16 Moltz J H & McDonald J K (1985) Neuropeptide Y: direct and indirect action on insulin secretion in the rat. *Peptides* **6** 1155-1159

17 Pappas T N, Debas H T, Goto Y & Taylor I L (1985) Peptide YY inhibits meal-stimulated pancreatic and gastric secretion. *American Journal of Physiology* **248** G118-G123

18 Morley J E, Levien A S, Gosnell B A, Kneip J & Grace M (1987) Effect of neuropeptide Y on ingestive behaviours in the rat. *American Journal of Physiology* **252** R599-R609

19 Clark J J, Kalra P S, Crawley W R & Kalra S P (1984) Neuropeptide Y and human pancreatic polypeptide stimulate feeding behaviour in rats. *Endocrinology* **115** 427-429

20 Williams G, Sted J H, Cardoso H, Ghatei M A, Lee Y C, Gill J S, Burrin J M, Polak J M & Bloom S R (1988) Increased hypothalamic neuropeptide Y concentration in diabetic rats. *Diabetes* **37** 763-772

21 White J D, Olchovsky D, Kershaw M & Berelowitz M (1990) Increased hypothalamic content of preproneuropeptide-Y messenger ribonucleic acid in streptozotocin-diabetic rats. *Endocrinology* **126** 765-772

22 Sahu A, Sninshy C A, Kalra P S & Kalra S P (1990) Neuropeptide-Y concentration in microdissected hypothalamic regions *in vitro* release from the medial basal hypothalamic-preoptic area of streptozotocin-diabetic rats with and without insulin substitution therapy. *Endocrinology* **126** 193-198

23 McKibbin P E, Cotton S J, McMillan S, Holloway B, Mayers R, McCarthy D & Williams G (1991) Altered neuropeptide Y concentrations in specific hypothalamic regions of obese (fa/fa) Zucker rats: possible relationship to obesity and neuroendocrine disturbances. *Diabetes* **40** 1423-1429

Avian Endocrinology
Ed. P J Sharp, pp 201-216
Journal of Endocrinology Ltd, Bristol (1993)

# Some actions of angiotensin II in gallinaceous and anseriform birds

W N Holmes and J Cronshaw
Biological Sciences, University of California, Santa Barbara, CA 93106, USA

## Introduction

Studies that were started in the late nineteenth century ultimately led to the elucidation of the renin-angiotensin II system as an important blood pressure-regulating mechanism in mammals (1). Angiotensin II (AII) is generated in a two-stage reaction that occurs in the plasma. First the enzyme renin, which is secreted into the circulation by the kidney, acts on angiotensinogen, an $\alpha_2$ globulin synthesized in the liver, to form the decapeptide angiotensin I (AI). The latter is then cleaved by a circulating converting enzyme of pulmonary origin to yield the octapeptide AII. This peptide was first defined in terms of its extremely potent vasoconstrictor properties, and in the horse, the pig, the rat, and the human it has the following amino acid sequence: $Asp^1$-$Arg^2$-$Val^3$-$Tyr^4$-$Ile^5$-$His^6$-$Pro^7$-$Phe^8$.

Angiotensin II is rapidly metabolized, by enzymatic removal of the $Asp^1$ residue at the N-terminal, to form angiotensin III (AIII). Alternatively, the same enzymatic action may convert AI to (des-$Asp^1$)-AI, and from this compound AIII may be formed directly through the action of converting enzyme. In mammals, AIII has approximately 40% of the pressor activity of AII.

Following the identification of aldosterone as the principle mineralocorticoid secreted by the mammalian adrenal cortex, and the demonstration that this hormone is synthesized primarily in the cells of the zona glomerulosa (2, 3), it was shown that AII also functions as an important regulator of aldosterone synthesis (4-7). Although AIII may retain much of the aldosterone-stimulating activity of AII, ligand-receptor binding studies suggest that it may not function as an important steroidogenic regulator.

## Angiotensin II-mediated aldosterone synthesis: the mammalian paradigm

Most of the work designed to elucidate the molecular and cellular basis of AII action has been conducted on just a few mammalian species using tissue samples and dispersed cells derived from the zona glomerulosa. This action involves the binding of AII to specific receptors located in the plasma membranes of the glomerulosal cells. High circulating concentrations of AII resulting from the changes in electrolyte intake or the intravenous administration of AII are each accompanied by increases in the number of receptors (8-10). Increases and decreases in the rates of synthesis and degradation of receptor protein, and the accelerated recycling of internalized receptor-ligand complexes, have been suggested as possible mechanisms whereby changes in the abundance of receptors may be evoked.

Several lines of evidence suggest that the diverse actions of AII can be accounted for by the existence of AII receptor sub-types. The existence of two distinct sub-populations of receptors has been unequivocally demonstrated through ligand-binding studies using both peptide (11, 12) and non-peptide (13-15) antagonists. The terms $AT_1$ and $AT_2$ receptors have been proposed as standard nomenclature for these two receptors. The $AT_1$ receptor is defined as having a greater affinity for AII than AIII, having DuP753 as its prototypical biphenylimidazole antagonist, and to have effector pathways mediated through the stimulation and inhibition of the inositol 1,4,5-triphosphate ($IP_3$), diacylyglycerol (DAG) and the cyclic adenosine 3′,5′-monophosphate (cAMP) second messenger systems respectively (16). On the other hand, the $AT_2$ receptor is defined as one with a similar affinity for AII and AIII, and to have PD123177 and PD123319 as prototypical tetrahydroimidazopyridine antagonists, and CGP 42112A as a modified peptide antagonist (16).

Both $AT_1$ and $AT_2$ receptors are present in the rat adrenal gland, but only the $AT_1$ receptor is present in the bovine and rabbit adrenal cortex (12, 17, 18). All of the known effects of AII on glomerulosal cells are mediated through the $AT_1$ receptor subtype (12). The $AT_2$ receptors in the rat adrenal gland are more abundant in the medulla than in the cortex, and their function in both tissues is not known. Recently, screening cDNA libraries by the polymerase chain reaction combined with restriction enzyme analysis has revealed a third AII receptor subtype which has been designated $AT_3$ (19). Compared with the $AT_1$ receptor, the $AT_3$ receptor has different pharmacological binding properties and two additional potential phosphorylation sites for protein kinase C. It is becoming clear that the different responses evoked by AII in the various target cell-types is at least in part a reflection of the receptor population.

The high- and low-affinity binding sites described in glomerulosal cell membranes (20, 21), which may or may not be associated with the same receptor, have been shown to fall roughly into two categories, namely those that act through a putative G protein which is *insensitive* to pertussis toxin, and those that act through a pertussis *toxin-sensitive* G protein which inhibits the activation of adenylate cyclase (22, 23).

The onset of aldosterone synthesis is preceded by the binding of AII to the high-affinity class of receptors, and the response appears to be mediated via a series of second messengers produced following the activation of phosphodiesterase molecules resident in the plasma membrane. It has been proposed that the overall response consists of two distinct temporal phases, each defined by the distribution and turnover of $Ca^{++}$, but each distinguished according to the spatial domain and the molecular targets of the $Ca^{++}$. Data supporting this hypothesis have been derived primarily from studies on the responses *in vitro* of rat and bovine zona glomerulosa cells. These studies have been extensively reviewed by Barrett *et al* (24) and a model has been proposed to explain the overall biphasic response of the glomerulosal cell to AII (Fig. 1).

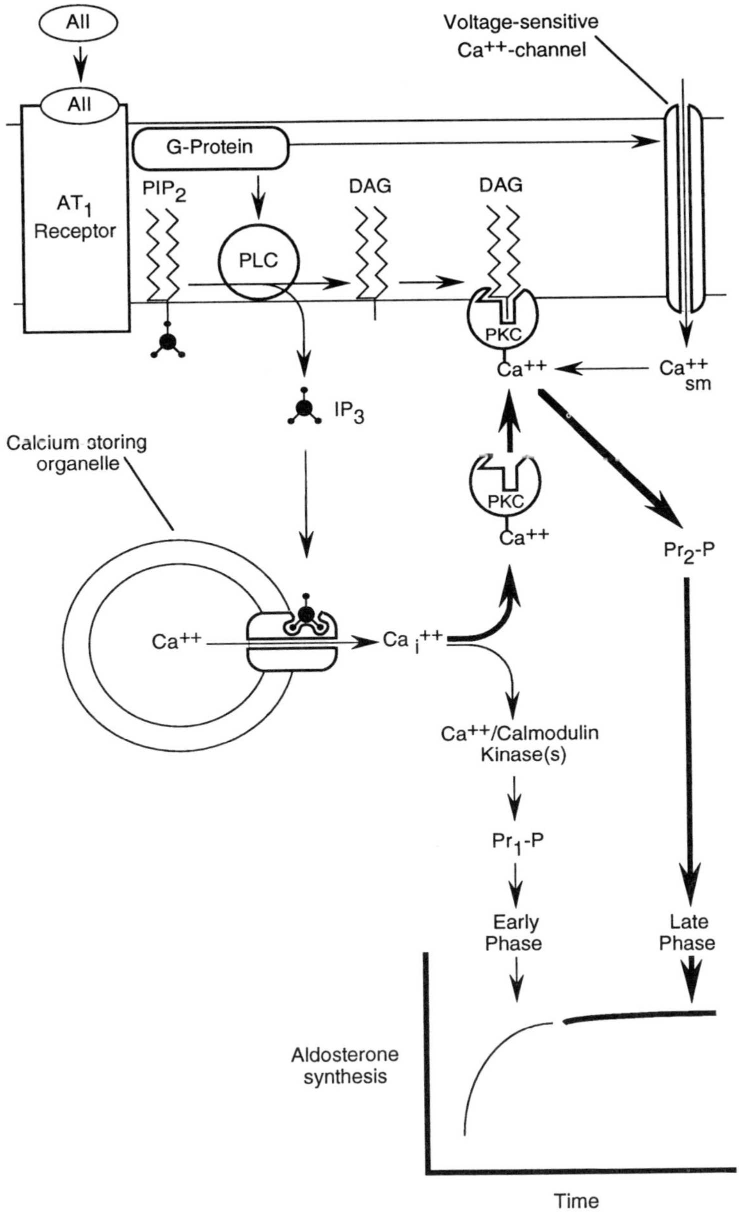


Fig. 1 A model describing the biphasic steroidogenic response of the mammalian zona glomerulosa cell to angiotensin II (adapted from 24). AII, angiostensin-II; $AT_1$, angiotensin-II type 1 receptor; $Ca^{++}_{sm}$, calcium ions in sub-membrane domain; DAG, diacylyglycerol; $IP_3$, inositol 1,4,5-triphosphate; PKC, protein kinase C; PLC, phospholipase C; $PIP_2$, phosphatidylino-sitol-4,5-biphosphate; $Pr_1$.P, $Pr_2$.P, phosphoproteins.

The following stepwise sequence of the responses portrayed in the model is as follows:

1 The interaction of AII with the $AT_1$ receptor promotes the binding of GTP to a G-protein located on the inner surface of the plasma membrane and in this state the G-protein activates a nearby phospholipase C molecule (PLC);
2 The hormonal effect of AII is now amplified through the enzymatic hydrolysis of phosphatidylinositol-4,5-biphosphate ($PIP_2$), a phospholipid component of the plasma membrane, to yield two second messengers, namely $IP_3$ and DAG;
3 During the early transient phase of the response to AII, $IP_3$ stimulates a liberation of $Ca^{++}$ from intracellular storage pools to generate a rise in cytosolic calcium ion concentration ($[Ca^{++}{}_i]$); it has been suggested that these pools are not associated with the endoplasmic reticulum (25). This $IP_3$-dependent increase in $[Ca^{++}{}_i]$ induces a short-lived activation of a group of $Ca^{++}$/calmodulin-dependent protein kinases and as a result there is transient increase in the phosphorylation of 'early phase' phosphoproteins ($Pr_1$.P) that results in a similarly transient increase in aldosterone synthesis;
4 A secondary sustained phase of aldosterone synthesis is induced primarily through the action of DAG, although it is linked to events occurring previously during the $IP_3$-dependent early phase of the response. Thus, the $IP_3$-induced rise in $[Ca^{++}{}_i]$ not only activates calmodulin (CaM) but also activates the cytosolic protein kinase C (PKC) which is then translocated to the plasma membrane where it binds to the DAG. The DAG-PKC complex then acts to stimulate the formation of a discrete set of 'late phase' phosphoproteins ($Pr_2$.P), and this in turn leads to period of sustained aldosterone synthesis.
5 In addition to stimulating the synthesis of $IP_3$ and DAG, the association of AII with its receptor also leads to a G-protein mediated activation of voltage-dependent $Ca^{++}$ channels in the plasma membrane. It has been proposed that the resulting net influx of $Ca^{++}$ leads to a critical increase in the concentration of calcium ions in a sub-membrane domain $[Ca^{++}]_{sm,}$ and that this in turn facilitates the formation of the DAG-PKC complex that initiates the sustained late phase of the response.

The adrenal glomerulosal cell is also extremely sensitive to changes in extracellular $K^+$ concentration and, even in the absence of AII, only a small increase induces significant increases in aldosterone synthesis. This agonistic effect of the $K^+$ per se may reflect the opening of voltage-dependent $Ca^{++}$ channels following depolarization of the plasma membrane (26). The agonistic effect of $K^+$, however, may also complement the receptor-mediated responses of glomerulosal cells to AII (27).

The stages in the steroidogenic pathway at which the early- and late-phase phosphoproteins ($Pr_1$.P and $Pr_2$.P) act to stimulate increases in aldosterone synthesis are not well understood. It has been shown, however, that when AII is intravenously infused to elevate the plasma AII concentration in rats to a level similar to that seen when the animals are fed a low-sodium diet, the 21-hydroxylase and 11-hydroxylase activities of the adrenocortical tissue remain unchanged; but the characteristically lowered activities of these enzymes that occur following hypophysectomy are partially restored following the prolonged administration of AII (9).

**Angiotensin II in birds**

Interactions between renin derived from the kidney tissue and the substrate present in homologous plasma have revealed differences between the primary sequences of both the angiotensin I (AI) and the AII peptides found in different mammalian species. For example, the AI isolated from the horse, the pig, the rat, and the ox all have histidine and leucine residues at the positions 9 and 10 respectively, but bovine AII has a valine instead of an isoleucine residue at the 5-position. In the chicken the AI peptide differs from these mammalian peptides by having a serine residue at the 9-position, but the AII peptide is identical to bovine AII (28, 29).

Although the vasoactive effects of AII have been studied extensively, especially in the chicken (30-32), there is not yet a universal consensus regarding the role of AII as a steroidogenic regulator in birds. This failure to develop a consensus is probably due to the fact that almost all of the early investigations were conducted on the domestic chicken. Until recently there has been a tendency to presume that responses observed in the chicken were representative of birds in general. However, the chicken is not a good model for studies on the effects of AII on steroidogenesis, and several earlier generalizations must now be modified.

**Adrenal steroidogenesis and angiotensin II in the domestic hen**

Zones of structurally distinct steroidogenic cell-types are not readily discernible in the adrenal gland of the adult intact chicken. Following adenohypophysectomy, however, the cells in the inner region of the gland in mature birds tend to degenerate to a greater extent than those at the periphery (33). Moreover, rod-shaped mitochondria like those found in the glomerulosal cells in rats, mice, and cattle have been reported to be characteristic of the peripheral cells in the chicken (33). The zone of cells in this region of the gland has been shown to become wider in $Na^+$-depleted cockerels and narrower in $Na^+$-repleted birds (33-35) and increases in juxtaglomerular granulation and renal pressor activity accompany the development of a wider peripheral zone of tissue in the $Na^+$-depleted state (35). These observations imply that the inner region of the gland may consist principally of corticotrophin-dependent cells while those located in the subcapsular region may represent the avian counterpart of the mammalian zona glomerulosa.

As far as the subcapsular region of the gland is concerned, functional studies have not confirmed this hypothesis. Thus, although the plasma concentration of aldosterone varies inversely with the amount of $Na^+$ in the diet, the increase that occurs when the birds are fed a low $Na^+$ diet is not correlated with increases in either renal-pressor activity or plasma AII concentration (35-38). These observations are consistent with the results of an earlier study *in vitro* in which AII was shown not to induce steroidogenesis in incubated adrenal gland slices (39).

Some aspects of these earlier studies have recently been re-examined. In a study on the ontogeny of adrenal function it has been reported that adrenal steroidogenic cells in the chicken at the end of embryogenesis are not absolutely refractory to AII (40). At this time the cells in the subcapsular region of the gland are small and undifferentiated, and they contain fewer lipid droplets than the cells in the inner region of the gland. When dispersed cells derived from the whole gland are incubated in medium containing angiotensin II, a small but significant increase aldosterone release is induced; this response is not dose-related and is only one-twentieth to one-fiftieth of the maximum inducible response seen when cells from neonatal mallard, Pekin and Muscovy ducklings are similarly incubated (40). The mechanism whereby this small response is induced and the location of the responsive cells in the gland is not known.

In another study on cell suspensions prepared from the adrenal glands of adult chickens, however, neither [$Val^5$]-AII nor [$Ile^5$]-AII had any significant effect on the rate of aldosterone synthesis (41). Since high-affinity binding sites (Kd=1.11 nM, concentration=60,000 sites per cell) have been identified in chicken adrenal steroidogenic cells, this failure to respond was probably not attributable to a complete absence of AII receptors (R Carsia, personal communication). The addition of the $Ca^{++}$ ionophore (A23187) and the phorbol ester, 12-*O*-tetradecanoyl phorbol 13-acetate (TPA) to the incubation medium, however, also failed to stimulate aldosterone synthesis in these cells. In glomerulosal cells these compounds act synergistically to induce a full and sustained release of aldosterone by changing the intracellular distribution of $Ca^{++}$, and activating PLC, respectively. Thus, the $Ca^{++}$/ $IP_3$/ PLC second messenger system probably does not contribute to the enhanced rate of aldosterone synthesis seen when the chickens are fed a low-$Na^+$ diet (41).

**Adrenal steroidogenesis and angiotensin II in the domestic turkey**

Maintenance of 6-week-old turkeys on a low $Na^+$ diet stimulates significant increases in absolute and relative adrenal gland weight (42, 43). At the same time, while the concentration of plasma ACTH remains unchanged, the concentration of plasma AII and aldosterone increase three-fold and six-fold respectively (42, 44). Studies *in vitro* have shown that both $[Val]^5$- and $[Ile]^5$-AII are potent and specific concentration-dependent stimulants of aldosterone synthesis in steroidogenic cells isolated from the adrenal glands of domestic turkeys (J F Kocsis *et al.*, unpublished observation). Indeed, the potency of $[Val]^5$-II is 3.6 times that of ACTH although it has only one-third of the effectiveness of ACTH. This discrepancy may be due to the fact that AII, unlike ACTH, does not stimulate cholesterol side-chain cleavage (cyanoketone-isolated pregnenolone synthesis). Thus, these investigators have suggested that AII, per se, stimulates only the later steps in the steroidogenic pathway, and so, in the absence of ACTH, its effectiveness is dependent upon the prevailing pool of precursors (45)

Histological studies have failed to show any zonation of the steroidogenic cells, but centrifugation of collagenase-dispersed cells in a Percoll gradient has revealed the presence of high- and low-density subpopulations of steroidogenic cell (43). Although aldosterone synthesis increases in both subpopulations following maintenance of the birds on a low-$Na^+$ diet, the AII- and $K^+$-inducible rates of synthesis increase to a greater degree in the high-density than in the low-density cells (43).

The binding characteristics of the presumptive native peptide, $[Val]^5$-AII, suggest the presence of only one class of AII receptor in a mixed population of low- and high-density steroidogenic cells isolated from the turkey adrenal (46). In this regard the cells appear to resemble cultured zona fasciculata cells (47) rather than glomerulosal cells which possess two distinct classes of AII receptors (48-53). The $K_d$ value for this single class of receptors ($K_d$ = 1.0-2.0 nM), however, is within the range of the values recorded for high-affinity binding sites in zona glomerulosa cells (8, 20, 49, 50, 55), cultured zona fasciculata cells (47) and particulate fractions from homogenates of mammalian adrenal capsular tissue (11, 21, 51-54, 56). The abundance of the AII receptor in turkey steroidogenic cells (63,500 sites per cell) is less than one-third the number of $AT_1$ receptors found in the glomerulosal cells (150,000-250,000 sites per cell) of several mammalian species (8, 50, 51). It is estimated that, at equilibrium, approximately 70% of the turkey receptors are in the membrane while about 30% are internalized; this estimate is similar to the estimated distribution reported for glomerulosal cells (55, 57). An increase in receptor occupancy is accompanied by an increase in aldosterone synthesis, suggesting that the AII-induced synthesis of aldosterone in the turkey cells occurs via a non-'spare' receptor mode (45). Several properties of the AII receptor in the turkey steroidogenic cells are similar to those of the $AT_1$ receptor found in glomerulosal cells (11-13, 15). For example, upon binding to the receptor, AII induces concentration-dependent increases in cytosolic $Ca^{++}$ concentration (58), and the receptor, like the $AT_1$ receptor in glomerulosal cells (11, 13, 59, 60), is sensitive to dithiothreitol (DTT) (45). On the other hand, the AII receptor in turkey steroidogenic cells is pharmacologically distinct from each of the *two* receptors found in glomerulosal cells; for example, the potencies of the nonpeptidic antagonists of $AT_1$ and $AT_2$, DuP753 and PD123177 respectively, are much lower than those reported for the receptors in glomerulosal cells (14, 59, 60). Indeed, in terms of the potency and binding-inhibition properties of peptide analogues and nonpeptide antagonists, the AII receptor in the turkey steroidogenic cell has some characteristics in common with the myocardial AII receptor in *Xenopus laevis* (61).

Studies on the binding characteristics of non-native AII analogues in mixed populations of cells suggest that at least two isomorphs of the AII receptor may occur in some of the turkey steroidogenic cells, and that these isomorphs may activate distinct pathways (62). One of the isomorphs is coupled to the $Ca^{++}$-$IP_3$-PKC second messenger system and when activated the resultant increases in $[Ca^{++}]_i$ and PKC are postulated to lead to the formation of a domain of phosphorylated proteins which in turn activate immediate-early growth response genes. It is also suggested that the gene-products of this activation mechanism primarily regulate cell differentiation and growth, and probably do not regulate acute AII-induced aldosterone synthesis. The other isomorph is thought to activate some part of the steroidogenic pathway leading to aldosterone synthesis. Although this appears to require some background $Ca^{++}$-calmodulin activity it does not appear to require an elevation in $[Ca^{++}]_i$ per se. This signal transduction pathway is yet to be defined, but there is some evidence to suggest that the arachidonic acid cascade may play a role.

The notion that the AII receptors in turkey steroidogenic cells have a conformation distinct from that of mammalian AII receptors is supported by recent molecular analyses (63). A partial cDNA has been generated by reverse transcription-polymerase chain reaction of turkey steroidogenic cell mRNA using primers specific for the rat aortic smooth muscle $AT_1$ cDNA (60) and bovine adrenal $AT_1$ receptor cDNA (59). This cDNA fragment lacks the coding regions for the amino- and carboxy-terminals of the receptor, the first transmembrane domain, and a portion of the seventh transmembrane domain. Nevertheless, sequence alignment of this cDNA fragment with the rat aortic smooth muscle cDNA and subsequent deduced amino acid sequence show 78% identity. There are, however, some noteworthy differences; the peptide has two additional cysteines for potential disulphide bridging in the first extracellular loop, an amino acid is deleted in the second extracellular loop, and two additional potential protein kinase C phosphorylation sites are present in that portion of the third intracellular loop which interacts with the G-protein(s) (63).

**Adrenal steroidogenesis and angiotensin II in Japanese quail**

During the course of a study on the dypsogenic effect of AII, it was noted that in addition to stimulating dose-dependent increases in water intake, a single large intraperitoneal injection of AII (10 mg per 100 g body weight) also stimulated a significant increase in plasma aldosterone concentration (64).

**Adrenal steroidogenesis and angiotensin II in mallard, Pekin and Muscovy ducks**

In Pekin ducklings, subcapsular zone (SCZ) of steroidogenic cells that contain a greater abundance of lipid droplets than cells located in the inner zone (IZ) of the adrenal gland is readily discernible at the end of embryogenesis (65). Incubated cell suspensions and superfused tissue samples prepared from adrenal glands of these neonates respond to both ACTH and AII in a concentration-dependent manner (40, 65). These patterns of responsiveness *in vitro* are undoubtedly related to the adrenal steroidogenic responses *in vivo* seen in newly hatched ducklings.

On the first day of post-natal life both the plasma corticosterone and aldosterone concentrations increase in response to mild stress; this advanced state of adrenal development is in marked contrast to the relatively low level of responsiveness seen in neonatal chickens (66, 67). Moreover, as soon as the newly-hatched mallard ducklings start to forage, the circulating concentrations of plasma aldosterone and corticosterone may change independently in response to different dietary electrolyte regimes (38). Thus, an unstressed bird given a diet that is low in $Na^+$ shows a significant increase in the resting concentration of aldosterone, but no associated rise in the concentration of plasma corticosterone. Conversely, when similarly unstressed birds are fed a high $Na^+$ diet, the concentration of plasma corticosterone becomes elevated and the aldosterone concentration remains unchanged. Furthermore, although exposure to stressors is always accompanied by increases in the concentrations of both hormones, the stress-induced increases in aldosterone concentration are greatest when the birds are consuming a low-$Na^+$ diet.

In both the Pekin and the mallard duck, the mechanisms leading corticosterone and aldosterone synthesis develop early during embryogenesis (68, 69). Portions of the nephrotome containing the differentiating adrenal anlage are able to synthesize both hormones on about the 10th day of incubation, and by the 20th day a concentration-dependent response to 1-24 ACTH has developed; and during the balance of the incubation period the gradient of response-curve becomes greater and the range of effective concentrations shifts to the left (K E Rohde, W N Holmes and J Cronshaw, unpublished observations). The responsiveness of the developing steroidogenic tissue to AII, however, develops more slowly; on the 20th day of incubation the cells are still refractory, and a full concentration-dependent response does not become established until just before hatching on the 27th day of incubation (65; W N Holmes, B I Nave and J Cronshaw, unpublished observations). Although the steroidogenic cells in the neonate Pekin duckling are less sensitive to AII than those in mallard duckling, the net effect of AII in both instances is to decrease the characteristically high ratio of corticosterone to aldosterone secreted when the cells are solely under the influence of corticotrophin (40, 65). Thus, the maximal AII-inducible rate of corticosterone synthesis is only one-hundredth of the corresponding rate induced by a similarly effective concentration of 1-24 ACTH.

Hyponatremia in the adult Pekin duck is associated with increases in the circulating concentrations of both AII and aldosterone (70). Moreover, the increase in the concentration of plasma AII following intravenous infusion is accompanied by dose-dependent increases in plasma aldosterone concentration (70). The pattern of this dose-response, however, is influenced by the sodium balance of the bird, the response to a given dose of AII being greater in the hyponatremic than in the hypernatremic individual. This phenomenon may reflect the prevailing balance between the inhibitory effect of atrial natriuretic factor and the stimulatory effect of AII on aldosterone synthesis.

Morphometric analysis has shown that in the adult Pekin duck the adrenal steroidogenic tissue consists of two zones of structurally distinct cells (71). A subcapsular zone (SCZ), located immediately inside the connective tissue capsule, consists of cells that are significantly larger than those located in the inner zone (IZ) of the gland. The volume densities (volume per unit volume of cytoplasm) of the mitochondria and lipid droplets, and the surface density (area per unit volume of cytoplasm) of the smooth endoplasmic reticulum, are each significantly greater than they are in the IZ cells. In contrast the surface density of the outer mitochondrial membranes in the SCZ cells is significantly less than it is in the IZ cells. Also the SCZ cells have pleomorphic rather than the rounded nuclei that characterize IZ cells.

In studies designed to determine whether these two morphologically distinct cell-types have discrete physiological functions, tissue slices derived from each zone have been superfused separately with buffer containing either ACTH or AII. Thus, while ACTH is a potent stimulant of both corticosterone and aldosterone synthesis in each zone of tissue (72, 73), $[\mathrm{Val}]^5$-AII and $[\mathrm{Ile}]^5$-AII act to stimulate only aldosterone synthesis in the SCZ cells and have little steroidogenic effect on the IZ cells (65). This specific stimulatory action of AII on the SCZ cells, which is concentration-dependent between $10^{-12}$ and $10^{-9}$ M (65), is consistent with the observation that the SCZ of steroidogenic tissue in the Pekin duck adrenal gland is the principal locus of AII receptors (70). Moreover, it is of interest to note that the ligand-binding characteristics of these receptors have much in common with those in the turkey cells which are tentatively presumed to reside predominantly in the high-density steroidogenic cells (cf 43, 46, 70).

In the domestic Muscovy duck (*Cairina moschata*), incubated cell suspensions prepared from the adrenal glands of embryos immediately prior to their emergence are more responsive to AII than similar preparations from either mallard or Pekin duck embryos (40). Although the minimally- and maximally-effective concentrations of AII are same, the maximum AII-inducible rate of aldosterone synthesis in the Muscovy embryo is approximately twice that which is inducible in the emergent mallard and Pekin embryos. The longer period of embryogenesis in the Muscovy duck (36 days) compared with the mallard and Pekin ducks (27 days) may account for this difference. No studies have been conducted on adrenal steroidogenic function in juvenile and adult individuals of this species.

## Conclusion

The effects of AII on specific aldosterone synthesis in mammals have been studied far more extensively and in greater detail than they have in birds. There is now strong evidence, however, that the renin-angiotensin system described in several species of birds (70, 74, 75) may, with the possible exception of the domestic chicken, be directly involved in the regulation of aldosterone synthesis in several species of birds. The consensus emerging from studies on the domestic turkey and the domestic Pekin duck is that discrete steroidogenic cell-types are present in the adrenal gland, and that one of these cell-types may possess most of the specific AII receptors responsible for the signal transduction that ultimately results in preferential aldosterone synthesis. In the duck adrenal gland these cells form a discrete subcapsular zone of tissue, and they can be distinguished structurally from those present in the inner region of the gland. Thus, the SCZ has many functional characteristics that are reminiscent of the zona glomerulosa in mammals and the cells of the inner zone, which are more sensitive and more responsive to corticotrophin, probably represent the function counterparts of the zona fasciculata cells.

Much more work must be done, both in greater depth and on more species, before a fully comprehensive account of the role angiotensin as a steroidogenic regulator in birds can be presented.

## Acknowledgements

The work of the authors was supported by grants from the National Science Foundation, Washington DC (DCR91-04560) and the Committee on Research, University of California.

## References


1 Page I H & Bumpus F M (1961) Angiotensin. *Physiological Reviews* **41** 331-390

2 Ayers P J, Gould R P, Simpson S A S & Tait J F (1956) The *in vitro* demonstration of different corticosteroid production within the ox adrenal gland. *Biochemical Journal* **63** 19P

3 Giroud C J P, Stachenko J & Piletta P (1958) *In vitro* studies on the functional zonation of the adrenal cortex and of the production of aldosterone. *International Symposium on Aldosterone* pp 56-72. Eds A F Muller & C M O'Conner. Boston: Little Brown

4 Laragh J H, Angers M, Kelly W G & Lieberman S (1960) Hypotensive agents and pressor substances. *Journal of the American Medical Association* **194** 234-240

5 Genest J, Nowaczynski W, Koiw E, Sandor T & Biron P (1960) Adrenocortical function in essential hypertension. *Essential Hypertension, an International Symposium* pp 126-146. Eds K D Bock & P T Cottier. Berlin: Springer Verlag

6 Blair-West J R, Coghlan J P, Denton D A, Goding J R, Monroe J A, Peterson R E & Wintour M (1962) Humoral stimulation of adrenal cortical secretion. *Journal of Clinical Investigation* **41** 1606-1627

7 Ames R P, Borkowski A J, Sicinski A M & Laragh J H (1965) Prolonged infusions of angiotensin II and norepinephrine and blood pressure, electrolyte balance and aldosterone and cortisol secretion in normal man and in cirrhosis with ascites. *Journal of Clinical Investigation* **44** 1171-1186

8 Douglas J, Saltman S, Fredlund P, Kondo T & Catt K J (1976) Receptor binding of angiotensin II and antagonists. Correlation with aldosterone production by isolated canine adrenal glomerulosa cells. *Circulation Research* **38** (Supplement II) 108-112

9 Aguilera G, Menard R H & Catt K J (1980) Regulatory actions of angiotensin II on receptors and steroidogenic enzymes in adrenal glomerulosa cells. *Endocrinology* **107** 55-60

10 Aguilera G, Schirar A, Baukal A & Catt K J (1980) Angiotensin II receptors: properties and regulation in adrenal glomerulosa cells. *Circulation Research* **46** (Supplement 1) I118-I127

11 Whitebread S M, Mele B, Kamber B & de Gasparo M (1989) Preliminary biochemical characterization of two angiotensin II receptor subtypes. *Biochemical & Biophysical Research Communications* **163** 284-291

12 Balla T, Baukal A J, Eng S & Catt K J (1991) Angiotensin II receptor sub-types and biological responses in the adrenal cortex and medulla. *Molecular Pharmacology* **40** 401-406

13 Chiu A T, McCall D E, Nguyen T T, Carini D J, Duncia J V, Herblin W F, Uyeda R T, Wong P C, Wexler R R, Johnson A L & Timmermans P B M W M (1989) Discrimination of angiotensin II receptor subtypes by dithiothreitol. *European Journal of Pharmacology* **170** 117-118

14 Chiu A T, Herblin W F, McCall D E, Ardecky R J, Carini D J, Duncia J V, Pease L J, Wong P C, Wexler R R, Johnson A L & Timmermans P B M W M (1989) Identification of Angiotensin II receptor subtypes. *Biochemical & Biophysical Research Communications* **165** 196-203

15 Chang R S L & Lotti V J (1989) Two distinct angiotensin II receptor binding sites in rat adrenal revealed by new selective nonpeptide ligands. *Molecular Pharmacology* **29** 347-351

16 Bumpus F M, Catt K J, Chiu A T, DeGasparo M, Goodfriend T, Husain A, Peach M J, Taylor D G & Timmermans P B M W M (1992) Nomenclature for angiotensin receptors. *Hypertension* **17** 720-721

17 Dudley D T, Panek R L, Major T C, Lu G H, Bruns R F, Klinkefus B A, Hodges J C & Weishaar R E (1990) Subclasses of angiotensin II binding sites and their functional significance. *Molecular Pharmacology* **38** 370-377

18 Hajnoczky G, Csordas G, Bago A, Chiu A T & Spät A (1992) Angiotensin II exerts its effect on aldosterone production and potassium permeability through receptor subtype $AT_1$ in rat adrenal glomerulosa cells. *Biochemical Pharmacology* **43** 1009-1012

19 Sandberg K, Ji H, Clark A J L, Shapira H & Catt K J (1992) Cloning and expression of a novel angiotensin II receptor subtype. *Journal of Biological Chemistry* **267** 9455-9458

20 Glossman H, Baukal A J & Catt K J (1974) Properties of angiotensin II receptors in the bovine and rat adrenal cortex. *Journal of Biological Chemistry* **249** 825-834

21 Brown G J, Douglas & Bravo E (1980) Angiotensin II receptors and *in vitro* aldosterone responses of aldosterone producing adenomas, adjacent nontumorous tissue, and normal human adrenal glomerulosa. *Journal of Clinical Endocrinology and Metabolism* **51** 718-723

22 Exton J H (1988) Mechanism of action of calcium-mobilizing agonists: some variations on a young theme. *FASEB Journal* **2** 2670-2676

23 Marie J & Jard S (1983) Angiotensin II inhibits adenylate cyclase from adrenal cortex glomerulosa zone. *FEBS Letters* **159** 97

24 Barrett P Q, Bollag W B, Isales C M, McCarthy R T & Rasmussen H (1989) Role of calcium in angiotensin II-mediated aldosterone secretion. *Endocrine Reviews* **10** 496-518

25 Rossier M, Capponi A & Vallotton M (1989) The inositol 1,4,5-triphosphate-binding site in adrenal cortical cells is distinct from the endoplasmic reticulum. *Journal of Biological Chemistry* **264** 14078-14084

26 Kojima I, Kojima K & Rasmussen H (1985) Intracellular calcium and adenosine 3′, 5′-cyclic monophosphate as mediators of potassium-induced aldosterone secretion. *Biochemical Journal* **228** 69-76

27 Fredlund P, Saltman S, Kondo T, Douglas J & Catt K J (1977) Aldosterone production by isolated glomerulosa cells: modulation of sensitivity to angiotensin II and ACTH by extracellular potassium concentration. *Endocrinology* **100** 481-486

28 Khosla M C, Smeby R R & Bumpus F M (1974) Structure-activity relationship in angiotensin II analogs. *Handbook of Experimental Pharmacology: Angiotensin.* Vol 37 pp 126-161. Eds I H Page & F M Bumpus. Heidelberg: Springer Verlag

29 Nakayama T, Nakajima T & Sokabe H (1973) Comparative studies on angiotensins. III. Structure of fowl angiotensin and its identification by DNS-method. *Chemical Pharmacology Bulletin* **21** 2085-2087

30 Nishimura H, Nakamura Y, Sumner R P & Khosla M C (1982) Vasopressor and depressor actions of angiotensin in the anesthetized fowl. *American Journal of Physiology* **242** H314-H324

31 Stallone J N, Nishimura H & Khosla M C (1989) Angiotensin II vascular receptors in fowl aorta: binding specificity and modulation by divalent cations and guanine nucleotides. *The Journal of Pharmacology & Experimental Therapeutics* **251** 1076-1082

32 Yamaguchi K & Nishimura H (1988) Angiotensin II-induced relaxation of fowl aorta. *American Journal of Physiology* **255** R591-R599

33 Kondics L & Kjaerheim A (1966) The zonation of interrenal cells of the fowl. *Zeitschrift Zellforschung Mikroskopie Anatomie* **70** 81-90

34 Kondics L (1963) Adrenal zone controlling mineral metabolism in pigeon. *Acta Biologica Academica Scienica Hungarica* **13** 233-240

35 Taylor A A, Davis J O, Breitenbach R P & Hartroft P M (1970) Adrenal steroid secretion and a renal-pressor system in the chicken (*Gallus domesticus*). *General and Comparative Endocrinology* **14** 321-333

36 Radke W J, Albasi C M & Harvey S (1984) Dietary sodium and adrenocortical activity in ducks (*Anas platyrhynchos*) and chickens (*Gallus domesticus*). *General and Comparative Endocrinology* **56** 121-129

37 Rosenberg J & Hurwitz S (1987) Concentration of adrenocortical hormones in relation to cation homeostasis in birds. *American Journal of Physiology* **253** R20-R24

38 Redondo J L, Holmes W N & Cronshaw J (1988) Effects of short-term changes in electrolyte intake of the adrenal steroidogenic responses of juvenile mallard ducks (*Anas platyrhynchos*). *Journal of Comparative Biochemistry & Physiology* **91A** 513-518

39 De Roos R & De Roos C C (1963) Angiotensin II: Its effects on corticoid production by chicken adrenals *in vitro*. *Science* **141** 1284

40 Collie M A, Holmes W N & Cronshaw J (1992) A comparison of the responses of dispersed steroidogenic cells derived from the embryonic adrenal tissue from the domestic chicken (*Gallus domesticus*), the domestic Pekin duck and the wild mallard duck (*Anas platyrhynchos*), and the domestic Muscovy duck (*Cairina moschata*). *General and Comparative Endocrinology* **88** 375-387

41 Rosenberg J, Pines M & Hurwitz S (1988) Regulation of aldosterone secretion by avian adrenocortical cells. *Journal of Endocrinology* **118** 447-453

42 Kocsis J F & Carsia R V (1989) Steroidogenic properties of isolated turkey adrenocortical cells. *Domestic Animal Endocrinology* **6** 121-131

43 Kocsis J F, Lam E T & Carsia R V (1989) Dietary sodium restriction enhances turkey adrenocortical cell function. *Poultry Science* **68** (Supplement 1) 80

44 Kocsis J F, Lam E T & Carsia R V (1989) Angiotensin II stimulates aldosterone secretion by isolated turkey adrenocortical cells. *Poultry Science* **68** (Supplement 1) 80

45 Kocsis J F, Carsia R V & McIlroy P J (1991) Modulation of Angiotensin II-induced aldosterone secretion by avian adrenocortical cells: lack of correlation with changes in AII receptor parameters. *Endocrinology* 126 Supplement A923

46 Kocsis J F, Carsia R V & McIlroy P J (1990) Partial characterization of angiotensin II receptors in avian adrenocortical cells. *Endocrinology* **124** Supplement A1031

47 Viard I, Rainey W E, Capponi A M, Begeot M & Saez J M (1990) Ovine adrenal fasciculata cells contain angiotensin II receptors coupled to intracellular effectors but are resistant to the steroidogenic effects of this hormone. *Endocrinology* **127** 2071-2078

48 Carroll J E, Landry A S & Elliott M E (1983) Cholesteryl hemisuccinate alters membrane fluidity, angiotensin receptors, and responses in adrenal glomerulosa cells. *Life Sciences* **32** 1573-1581

49 Carroll J E & Goodfriend T L (1984) Androgen modulation of adrenal angiotensin receptors. *Science* **224** 1009-1011
50 Douglas J G (1980) Effect of high potassium diet on angiotensin II receptors and angiotensin-induced aldosterone production in rat adrenal glomerulosa cells. *Endocrinology* **106** 983-990
51 Douglas J, Aguilera G, Kondo T & Catt K J (1978) Angiotensin II receptors and aldosterone production in rat adrenal gomerulosa cells. *Endocrinology* **102** 685-696
52 Douglas J G, Brown G P & White C (1984) Angiotensin II receptors of human and primate adrenal fasciculata and glomerulosa: correlations of binding and steroidogenesis. *Metabolism* **33** 685-688
53 Douglas J G, Khosla M C & Bumpus F M (1985) Efficacy of octa- and heptapeptide antagonists of angiotensin II as inhibitors of angiotensin III binding in the rat adrenal glomerulosa. *Endocrinology* **116** 1598-1602
54 Douglas G M, Michailov M, Khosla M C & Bumpus F M (1980) Comparative receptor-binding properties of heptapeptide and octapeptide antagonists of angiotensin II in rat adrenal glomerulosa and uterine smooth muscle. *Endocrinology* **106** 120-124
55 Crozat A, Penhoat A & Saez J M (1986) Processing of angiotensin II (A-II) and ($Sar^1Ala^8$)A-II by cultured bovine adrenocortical cells. *Endocrinology* **118** 2312-2318
56 Capponi A M & Catt K J (1979) Angiotensin II receptors in adrenal cortex and uterus. *Journal of Biological Chemistry* **254** 5120-5127
57 Bianchi C, Gutkowska J, De Lean A, Ballak M M B, Anand-Srivastava M B, Genest J & Cantin M (1986) Fate of [$^{125}$I]angiotensin II in adrenal zona glomerulosa cells. *Endocrinology* **118** 2605-2607
58 Shimmel R J, Kocsis J F & Carsia R V (1990) Angiotensin II induces sustained elevations of cytosolic calcium in avian adrenocortical cells. *Endocrinology* **124** Supplement A991
59 Sasaki K, Yamano Y, Bardhan S, Iwai N, Murray J J, Hasegawa M, Matsuda Y & Inagami T (1991) Cloning and expression of a complementary DNA encoding a bovine angiotensin II type-1 receptor. *Nature* **351** 230-233
60 Murphy T J, Alexander R W, Griendling K K, Runge M S & Bernstein K E (1991) Isolation of a cDNA encoding the vascular type-1 angiotensin II receptor. *Nature* **351** 233-236
61 Sandberg K, Ji H, Millan M A & Catt K J (1991) Amphibian myocardial angiotensin II receptors are distinct from mammalian $AT_1$ and $AT_2$ receptor subtypes. *FEBS* **284** 281-284
62 Kocsis J F, Schimmel R J, Carsia R V & McIlroy P J (1992) Turkey adrenocortical cells possess aldosteronogenic angiotensin II (AII) receptors that are not coupled to elevations in intracellular calcium. *Endocrinology* **128** Supplement A62
63 Carsia R V, McIlroy P J, Kowalski K I & Tilly J L (1993) Isolation of turkey adrenocortical cell angiotensin II (AII) receptor partial cDNA: evidence for a single copy gene expressed predominantly in the adrenal gland. *Biochemical and Biophysical Research Communications* **191** 1073-1080
64 Kobayashi H & Takei Y (1982) Mechanisms for induction of drinking with special reference to angiotensin II. *Comparative Biochemistry and Physiology* **71A** 485-494
65 Holmes W N, Al-Ghawas S C, Cronshaw J & Rohde K E (1991) The structural organization and the steroidogenic responsiveness *in vitro* of adrenal gland tissue from the neonatal mallard duck (*Anas platyrhynchos*). *Cell and Tissue Research* **263** 557-566

66 Holmes W N, Cronshaw J & Redondo R L (1989) Changes in the adrenal steroidogenic responsiveness of the mallard duck (*Anas platyrhynchos*) during early postnatal development. *Comparative Biochemistry and Physiology* **92A** 403-408

67 Holmes W N, Cronshaw J & Redondo R L (1990) Stress-induced adrenal steroidogenesis in neonate mallard ducklings and domestic chickens. *Zoological Science* **7** 723-730

68 Holmes W N, Cronshaw J & Redondo R L (1990) The ontogeny of adrenal steroidogenic function in the mallard duck (*Anas platyrhynchos*). *Endocrinology of Birds* pp 143-158. Eds M Wada, S Ishii & C G Scanes. Tokyo: Japanese Science Society Press. Berlin: Springer Verlag

69 Cronshaw J, Holmes W N, Ely J A & Redondo J L (1989) Pre-natal development of the adrenal steroidogenic tissue in the mallard duck (*Anas platyrhynchos*). *Cell and Tissue Research* **258** 593-601

70 Gray D A, Gerstberger R & Simon E (1989) Role of angiotensin II in aldosterone regulation in the Pekin duck. *Journal of Endocrinology* **123** 445-452

71 Pearce R B, Cronshaw J & Holmes W N (1978) Evidence for the zonation of interrenal tissue in the adrenal gland of the duck (*Anas platyrhynchos*). *Cell and Tissue Research* **196** 363-379

72 Cronshaw J, Ely J A & Holmes W N (1985) Functional differences between two structurally distinct types of steroidogenic cell in the avian adrenal gland: the corticotropic responsiveness *in vitro* of tissue samples from the subcapsular and inner zones of the duck adrenal. *Cell and Tissue Research* **240** 561-567

73 Holmes W N & Cronshaw J (1984) Adrenal gland: some evidence for the structural and functional zonation of the steroidogenic tissues. *Journal of Experimental Zoology* **232** 627-631

74 Nishimura H (1980) Comparative endocrinology of renin and angiotensin. *The Renin-angiotensin System* pp 29-77. Eds J A Johnson & R R Anderson. New York: Plenum Press

75 Chan M Y & Holmes W N (1971) Studies on a 'renin-angiotensin' system in the normal and hypophysectomized pigeon (*Columbia livia*). *General and Comparative Endocrinology* **16** 304-311

Avian Endocrinology
Ed. P J Sharp, pp 217-226
Journal of Endocrinology Ltd, Bristol (1993)

# Dietary and aldosterone effects on the morphology and electrophysiology of the chicken coprodeum

V S Elbrønd[1], V Dantzer[1], T M Mayhew[2] and E Skadhauge[1]
[1]Department of Anatomy and Physiology, The Royal Veterinary and Agricultural University, DK-1870 Frederiksberg C, Denmark; [2]Department of Human Morphology, Queen's Medical Centre, Nottingham NG7 2UH, UK

## Introduction

The chicken cloaca is divided into three compartments: copro-, uro- and proctodeum, of which the coprodeum is located orally. The organ has various functions, including the storage of uretheral urine and faeces, osmoregulation and ion transport. The coprodeal epithelium provides an excellent model for the study of ion transport mechanisms, and in particular $Na^+$-transport. A close correlation has been observed between dietary NaCl load, the plasma concentration of the mineralocorticoid, aldosterone, and amiloride-inhibitable transepithelial $Na^+$-transport (1-5). Long-term adaptation to a high NaCl diet depresses the concentration of plasma aldosterone and reduces transepithelial $Na^+$-transport across the coprodeum, whereas adaptation to a low NaCl diet results in a hundred-fold higher net $Na^+$-transport and nine-fold higher concentration of plasma aldosterone. The transepithelial transport of $Na^+$ across the coprodeal epithelium is best described by the classic Kofoed-Johnsen-Ussing model, as studied on frog skin, showing a passive 'downhill' flow of $Na^+$ through apical $Na^+$-channels into the cell, and an active 'uphill' basolateral exchange of $Na^+$ and $K^+$ in ATP-ase-dependent, oxygen-consuming and ouabain-inhibitable $Na^+/K^+$-pumps. Absorption of $Na^+$ and net $Na^+$-flux across the coprodeal epithelium is nearly equal to the short-circuit current (1). Histologically, the coprodeal wall is very similar to that of the caudal part of the intestinal tract, with a folded mucosa covered by a simple columnar epithelium, a lamina propria, a submucosa, a tunica muscularis and a peripheral adventitia. Studies of transepithelial $Na^+$-transport using the vibrating microprobe technique on stretched mucosal membranes demonstrate a flow of positive ions from the mucosal to the serosal side in cells located on top of the mucosal folds (6). Morphological investigations carried out in conjunction with functional studies have therefore focused on these $Na^+$-transporting cells (7-12).

## Physiological aspects of transepithelial $Na^+$-transport

In 1979 Bindslev (13) demonstrated a correlation between the net $Na^+$-flux and the apical permeability of epithelial cells, which was more thoroughly tested by Clauss *et al.* (14), using rapid current-voltage measurements. After depolarization of the basolateral membrane with a high concentration of $K^+$, specific $Na^+$-permeability was determined as the difference in values measured before and after mucosal exposure to amiloride. In hens fed a low NaCl diet, an apical permeability of (7.3±0.6) x $10^{-2}$ cm/h was measured at an amiloride-inhibitable $Na^+$-current of 182±18 μA/$cm^2$ h, whereas in hens fed a high NaCl diet, in which the $Na^+$-transport is close to zero, no measurements of the permeability were obtainable. Transport of $Na^+$ through the apical $Na^+$-channels is passive, and no effects are observed after addition of glucose or amino acids to the mucosa (15). Enzyme-histochemical methods demonstrate that ouabain-inhibitable $Na^+/K^+$ pumps are located at the baso-lateral plasma membranes (12, 16). A two-fold higher activity of ouabain-inhibitable $K^+$-para-nitrophenyl-phosphatase is found in cell homogenates of coprodeum from hens fed a low NaCl diet compared with hens fed a high NaCl diet. This moderate difference in enzyme activity is in contrast to the large differences in the apical $Na^+$-permeability. It is therefore concluded that the physiological adaptation of the transepithelial $Na^+$-transport in the coprodeal epithelial cells is mainly regulated *apically*.

**Aldosterone and transepithelial $Na^+$-transport**

A correlation between the transepithelial $Na^+$-transport across the coprodeal epithelium and the concentration of plasma aldosterone has been observed in several studies (2-4, 17-21). The concentration of plasma aldosterone is hardly detectable in the hens fed a high NaCl diet, whereas in the hens fed a low NaCl diet it increases up to 300-400 pg/ml. The plasma concentrations of other osmoregulatory hormones, such as arginine vasotocin, corticosterone and prolactin show negligible variations between hens fed a high and low NaCl diet and only minor responses to osmotic stimuli are seen in hens fed a high NaCl diet (17, 22, 23). Acute induction of $Na^+$-transport across the coprodeal epithelium is measured *in vivo* after aldosterone perfusion (24) and *in vitro* after aldosterone injection (14). However, coprodeal epithelial cells sampled from hens 24 h after addition of NaCl to a low NaCl diet show no specific responses to aldosterone *in vitro* after 8 hours of incubation with the steroid (25). Chronic effects of steady-state concentrations of plasma aldosterone are induced after long-term adaptation to a low NaCl diet which results in coprodeal cells being physiologically and morphologically adjusted to retain high amounts of $Na^+$. *In vivo* injections of aldosterone three times daily in hens on a high NaCl diet induce an increase in amiloride-inhibitable $Na^+$-transport within 24 hours to 2.1 μmol/$cm^2$h, and within 6 days to a third of the $Na^+$-transport measured in hens adapted to a low NaCl diet (18). This indicates that other factors such as cellular morphology are involved in the regulation of $Na^+$-transport and might explain the delay in adaptation. It is suggested that aldosterone is the main factor involved in the transepithelial $Na^+$-transport, but other time-limiting factors, particularly cellular development, modify the rate of adaptation.

**Effects of dietary NaCl on plasma aldosterone and transepithelial $Na^+$-transport**

As previously described, differences in the amount of NaCl in the diet result in different concentrations of plasma aldosterone and transepithelial $Na^+$-transport through the coprodeal epithelium. The mean intake of NaCl in hens fed a high NaCl diet is

10-13.5 mmol NaCl/kg body weight/day and 0.2-0.3 mmol NaCl/kg body weight/day in hens fed a low NaCl diet (1, 2, 20). These diets represent the extremes with respect to $Na^+$-transport and concentrations of plasma aldosterone (22). The coprodeal epithelium becomes fully electrophysiologically adapted 8 days after dietary $Na^+$-depletion of hens fed a high NaCl diet, and 24 hours after resalination of hens, by changing from a low to a high NaCl diet with an initial oral load of NaCl (2, 5, 18, 19). Acute resalination of hens on a low NaCl diet induces, within 24 hours, a drastic decline in $Na^+$-transport and in plasma aldosterone. A high rate of $Na^+$-transport in tissue from hens recently resalinated can only be maintained by aldosterone stimulation *in vivo* for a few days (3). This indicates that other factors than aldosterone are essential to maintain a high rate of transepithelial $Na^+$-transport.

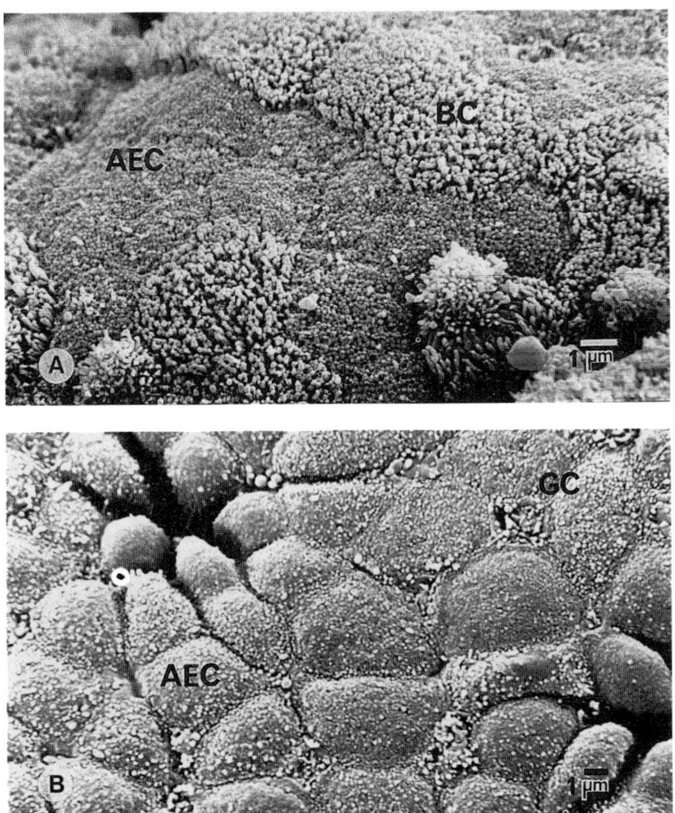


Fig. 1 Scanning electron microscopy of the mucosal surface of the coprodeum from hens fed A) a low NaCl or B) a high NaCl diet. The epithelial surface from the hen fed a low NaCl diet is dominated by long and densely packed microvilli, whereas the epithelial surface from the hens fed a high NaCl diet is covered with short and sparsely distributed microvilli. Other cell types shown are absorptive intestinal epithelial cells (AEC), goblet cells (GC), and brush cells (BC).

**Effect of dietary NaCl on the morphology of the coprodeal epithelium**

The morphology of the coprodeal epithelium has been studied in relation to $Na^+$-transport (8, 12, 26, 27). The epithelium is composed of three different cell types: 'normal' absorptive intestinal epithelial cells, goblet cells, and 'dark cells' (12, 26, 27, Fig. 1). The later cell type, distinguished by its dark appearance in toluidine blue-stained light microscopical sections and by the electron-dense cytoplasm in electron microscope preparations, has been identified as a brush cell (8, 9). This cell type is of endodermal origin and until recently, has only been described in laboratory animals (28-34) and some reptiles (32, 35). In the coprodeum the brush cells are localized in clusters in the upper part of the transitional zone between the top of mucosal folds and the openings to the crypts. Light- and electron microscopical preparations show that the brush cells have long microvilli that protrude over the apical surface level of neighbouring cells, and their distribution is particularly well demonstrated in scanning electron microscope preparations (Fig. 1). An increased number of rod-shaped particles ($Na^+$-channels?) in the apical plasma membrane, as identified by freeze fracture techniques in hens fed a low NaCl diet suggests that the brush cell is a 'supercell', involved in $Na^+$-transport (26, 27). Recent electrophysiological results using the vibrating microprobe technique do not support this view because $Na^+$-transport is evenly distributed among the epithelial cells (6). Some (8, 26, 27) but not all (12) studies show that the frequency of the three main epithelial cell types differ between hens fed high or low NaCl diets, the main difference being the frequency of the dark cell/brush cell ratio, which on biased material is found to be highest in hens on a low NaCl diet. The discrepancies between studies on the ratios of coprodeal epithelial cell types in hens fed low or high NaCl salt diets probably involve several factors such as biased and unbiased material, an uneven distribution of the brush cells and ease of identification (8, 9).

At the ultrastructural level the main morphological difference between coprodeal epithelia from hens fed low or high NaCl diets is the length and packing density of the apical microvilli on the surface of the absorptive intestinal epithelial cells (7, 8, 10, 11, 20, 26, 27). In tissue from hens fed a low NaCl diet, the longest and most densely packed microvilli are present, whereas they are short and sparsely distributed in hens fed a high NaCl diet (Fig. 1). Quantitative analyses demonstrate that the cell number and the length and packing density of the microvilli are significantly greater in tissue from hens fed a low NaCl diet, than from hens fed on high NaCl diet (7, 11, 12, 16). A morphometric analysis of the surface area of different domains of the plasma membrane of absorptive epithelial cells shows a significant increase in the apical surface of cells from hens fed a low NaCl diet (Table 1). This increase in surface area might be associated with insertion of apical $Na^+$-channels. Whether the regulation of cellular $Na^+$-transport is due to insertion of new $Na^+$-channels in the membrane or to the opening of quiescent channels is still an unsolved question (36-40).

Kidney cells from the distal tubules are, like coprodeal epithelial cells, aldosterone-sensitive and show adaptations of the plasma membranes to changes in the $Na^+$-absorption. The two cell types show similarities in their general morphology, such as apical surface amplification with microvilli and a relative large amount of mitochondria, promoting energy for the baso-lateral $Na^+/K^+$ pumps. During adaptation from low to high $Na^+$-transport the apical surface area of the average coprodeal epithelial cell increases x 2.3 in hens fed a low NaCl diet (Table 1), whereas no changes are seen in the apical surface of kidney cells (41, 42). The lateral cell membranes in both cell types are folded to increase the area for insertion of the baso-lateral $Na^+/K^+$-pumps. Lateral plasma membrane foldings in the coprodeal cells create complex spiral-like structures in the cytoplasm of neighbouring cells. The lateral surface area per cell is the same in hens fed a low and a high NaCl diet, but due to a higher cell number in hens fed a low NaCl diet the total lateral surface area increases x 1.5-2. The total surface area of the basal plasma membrane of coprodeal cells is also greatest in hens fed a low NaCl diet, but this area is negligible compared to that of the lateral membrane area (12). In kidney cells a demand for increased $Na^+$-absorption induces an increase in the surface gain of the basal plasma membrane (41, 42). A comparison of membrane adaptations between the coprodeal epithelial cells and the kidney cells in response to regulation of $Na^+$-absorption show discrepancies in the membrane domains recruited. Kidney cells adapt basally, while coprodeal cells mainly adjust to higher $Na^+$-absorption by increasing the apical surface area of the single cell and increase the total cell number in the whole organ (see Table 1) (8, 10-12).

Table 1 A comparison of the rate of $Na^+$-transport and morphological features of absorptive intestinal epithelial cells from the coprodeum of hens fed on low or high NaCl diets (data from 7, 10, 11, 12).

| | **High NaCl diet** | **Low NaCl diet** |
|---|---|---|
| $Na^+$-transport | 0-1 μmol/cm²h | 7-12 μmol/cm²h |
| Cell number | 270 x $10^6$ | 420 x $10^6$ |
| Membrane area | 2660 cm² | 4310 cm² |
| Apical surface area | | |
| Total | 86 cm² | 202 cm² |
| per cell | 32 μm² | 49 μm² |
| Lateral surface area | | |
| Total | 2532 cm² | 4036 cm² |
| per cell | 932 μm² | 932 μm² |
| Basal surface area | | |
| Total | 45 cm² | 68 cm² |
| per cell | 17 μm² | 17 μm² |
| Number of microvilli | | |
| Total | 35 x $10^9$ | 71 x $10^9$ |
| per cell | 131 | 171 |
| Length of microvilli | | |
| Total | 7280 m | 19738 m |
| per cell | 0.21 μm | 0.28 μm |

### $Na^+$-channels in the coprodeal epithelium

The apical $Na^+$-channels in the coprodeal epithelium have been studied in detail. The diuretic, amiloride, blocks apical $Na^+$-channels and abolishes $Na^+$-transport, thereby reducing the short-circuit current to a value close to zero (43). Immunocytochemical studies of $Na^+$-channels in the chicken coprodeal epithelium show that they are located in the apical plasma membrane and are immunologically related to bovine renal $Na^+$-channels (39). Polypeptide fractions with MWs of 235 and 150 kDa from the apical plasma membrane of the coprodeal epithelial cells cross-react with purified bovine renal $Na^+$-channel antibodies, and their molecular masses correspond to two subunits of the bovine renal $Na^+$-channel, of which one binds amiloride (39). The number of open and conducting $Na^+$-channels changes during adaptation to alterations in dietary salt, but the mechanism controlling the number of conducting channels is not fully understood. The number of $Na^+$-channels inserted in the apical plasma membrane might be constant and the ratio of open or closed channels may depend on demand for $Na^+$-transport (36, 40). Alternatively, the number of $Na^+$-channels inserted in the membrane may change in relation to the rate of $Na^+$-transport (38, 44). The latter possibility is suggested by the finding that a transfer of mRNA from coprodeal epithelial cells from hens fed a low, but not a high, NaCl diet into *Xenopus* oocytes induces an increase in the amiloride-inhibitable $Na^+$-transport (37). This suggests that the transferred mRNA mediates a *de novo* synthesis and subsequently an insertion of $Na^+$-channels into the oocyte membrane.

In the coprodeal cells a positive immunoreaction is seen in the upper apical cytoplasma (39), where a variable amount of vesicles with changing morphology are present (8, 20, 26, 27). This might indicate that the vesicles are involved in the transport of $Na^+$-channel components. The recent findings do not lead to any final conclusion but suggest a regulation of coprodeal $Na^+$-transport by changing the number of $Na^+$-channels inserted in the apical membrane.

### Recent morphological observations on coprodeal epithelial cells

Recent morphological studies provide further evidence that changes in $Na^+$-transport over coprodeal epithelial cells are mediated through morphological changes in the apical membrane (14). An increase in the packing density and height of the apical microvilli is observed in association with increased $Na^+$-transport induced by the administration of aldosterone for 6 days. Conversely, a decrease in these parameters, associated with decreased $Na^+$-transport occur within 72 h after resalination of hens fed low NaCl diets. Morphological observations also show that the number of brush cells increases during aldosterone treatment and decrease during resalination. The general morphological adaptations of the coprodeal cells prompt questions about the origin and nature of the cells. Are adapting cells newly recruited from the crypts or are they remodelled from existing cells? Recent immunocytochemical studies of cell turnover using bromo-deoxy-uridine show that the cell turnover of the coprodeal epithelium exceeds 8 days (personal observation). This period is greater than the period over which aldosterone or resalination effects occur (6 days and 72 hours respectively). It is therefore concluded that initial morphological adaptations to changes in the demand for $Na^+$-transport are due to a remodelling of the apical membrane of pre-existing coprodeal epithelial cells.

**Conclusion**

Amiloride inhibitable $Na^+$-transport through the chicken coprodeal epithelium shows the greatest variation seen in any vertebrae. Dietary NaCl and concentration of plasma aldosterone regulate the amount of transepithelial $Na^+$-transport in this organ. $Na^+$-ions enter the epithelial cells passively through apical intramembraneous $Na^+$-channels containing proteins with molecular masses and immunoreactivity which are similar to those comprising bovine renal $Na^+$-channels. The intracellular $Na^+$-ions are removed actively from the cell by $Na^+$-/$K^+$- pumps located at the baso-lateral plasma membranes. The coprodeal organ adjusts morphologically to changes in $Na^+$-absorption by adapting the apical surface area of individual cells, and by changing the total cell number and the proportion between the cell types in the organ. Sudden changes in the demand for $Na^+$-absorption are physiologically regulated within within a short period (1 to 8 days) by opening/closing or inserting/degrading $Na^+$-channels in the apical membrane. Although the single epithelial cell demonstrates a certain capacity for remodelling, gross morphological adaptation is first fully completed after 8-14 days. In general, the relationship between the structure and function of the coprodeal epithelium is demonstrated here by the correlation between transepithelial $Na^+$-transport and morphological adaptations.

**Acknowledgements**

We thank Inge Bjerring Nielsen for excellent technical assistance. Studies were supported by Grant ST2J-0392-C from the European Community and SJVF 13-4224 from The Danish Agricultural and Veterinarian Research Council. Equipment was donated by The Velux (1981) Foundation and The NOVO Foundation.

**References**

1 Choshniak I, Munck B G & Skadhauge E (1977) Sodium chloride transport in the chicken coprodeum: basic characteristics and dependence on the sodium chloride intake. *Journal of Physiology* (London) **271** 489-504

2 Skadhauge E, Thomas D H, Chadwick A & Jallageas M (1983) Time course of adaptation to low and high NaCl diets in the domestic fowl: effects on electrolyte excretion and on plasma hormone levels (aldosterone, corticosterone and prolactin). *Pflügers Archiv* **396** 301-307

3 Skadhauge E (1989) Hormonal regulation of sodium absorption and chloride secretion across the lower intestine of birds. *Zoological Science* **6** 437-444

4 Skadhauge E (1989) Regulation of NaCl transport across avian hindgut. *Adaptation and Development of Gastrointestinal Function* pp 41-48. Eds M W Smith & F V Sepulveda. Manchester: Manchester University Press

5 Thomas D H & Skadhauge E (1982) Time course of adaptation to low and high NaCl diets in the domestic fowl. Effects on electrical behaviour of isolated epithelia from the lower intestine. *Pflügers Archiv* **395** 301-307

6 Holtug K, Shipley A, Dantzer V, Sten-Knudsen O & Skadhauge E (1991) Localization of sodium absorption and chloride secretion in an intestinal epithelium. *Journal of Membrane Biology* **122** 215-229

7 Elbrønd V S, Dantzer V, Mayhew T M & Skadhauge E (1991) Avian lower intestine adapts to dietary salt (NaCl) depletion by increasing transepithelial sodium transport and microvillous membrane surface area. *Experimental Physiology* **76** 733-744

8 Elbrønd V S (1992) Morphology in relation to Na-transport in chicken coprodeal epithelium. Lightmicroscopical, ultrastructural and electrophysiological studies pp 1-90. Ph.D. Thesis, The Royal Veterinary and Agricultural University, Copenhagen

9 Elbrønd V S & Dantzer V (1993) Brush cells in the chicken lower intestine (coprodeum). *Journal of Anatomy* **182** 132-133

10 Mayhew T M, Dantzer V, Sødring-Elbrønd V & Skadhauge E (1990) A sampling scheme for studying the functional morphology of the coprodeum in hens on high- and low-NaCl diets. *Journal of Anatomy* **170** 246-248

11 Mayhew T M, Elbrønd V S, Dantzer V & Skadhauge E (1992) Quantitative analysis of factors contributing to expansion of microvillous surface in the coprodeum of hens transferred to a low-NaCl diet. *Journal of Anatomy* **81** 73-77

12 Mayhew T M, Elbrønd V S, Dantzer V & Skadhauge E (1992) Structural and enzymatic studies on plasma membrane domains and sodium pump enzymes of absorptive epithelial cells in the avian lower intestine. *Cell and Tissue Research* **270** 577-585

13 Bindslev N (1979) Sodium transport in the lower intestine. Induction of sodium sites in the brush border by a low sodium diet. *Journal of Physiology* (London) **288** 449-446

14 Clauss W, Dürr J E, Guth D & Skadhauge E (1987) Effects of adrenal steroids on $Na^+$-transport in the lower intestine (coprodeum) of the hen. *Journal of Membrane Biology* **96** 141-152

15 Thomas D J, Jallaegas M, Munck B G & Skadhauge E (1980) Aldosterone effects on electrolyte transport of the lower intestine (coprodeum and colon) of the fowl (*Gallus domesticus*) *in vitro*. *General and Comparative Endocrinology* **40** 165-170

16 Dantzer V, Møller O & Skadhauge E (1988) Morphological and enzymic adaptation to aldosterone of the epithelium of the caudal hindgut of the hen. *Journal of Physiology* (London) **396** 30P

17 Arnason S S, Rice G, Chadwick A & Skadhauge E (1986) Plasma levels of arginine vasotocin, aldosterone and corticosterone during prolonged dehydration in the domestic fowl: effect of dietary NaCl. *Journal of Comparative Physiology B* **156** 386-397

18 Clauss W, Arnason S S, Munck B G & Skadhauge E (1984) Aldosterone-induced sodium transport in the lower intestine. Effects of varying NaCl intake. *Pflügers Archiv* **401** 354-360

19 Clauss W & Skadhauge E (1988) Modulation of Na and Cl transport by mineralocorticoids. *Comparative Biochemistry and Physiology* **90A** 583-589

20 Clauss W, Dantzer V & Skadhauge E (1988) A low salt diet facilitates Cl secretion in hen lower intestine. *Journal of Membrane Biology* **102** 83-96

21 Skadhauge E, Clauss W, Arnason S S & Thomas D H (1985) Mineralocorticoid regulation of lower intestinal ion transport. *Transport Processes. Iono- and Osmoregulation* pp 118-133. Eds R Gilles & M Gilles-Baillien. Berlin-Heidelberg: Springer Verlag

22 Arnason S S & Skadhauge E (1991) Steady-state sodium absorption and chloride secretion of colon and coprodeum, and plasma levels of osmoregulatory hormones in hens in relation to sodium intake. *Journal of Comparative Physiology B* **161** 1-14

23 Rice G E, Arnason S S, Arad Z & Skadhauge E (1985) Plasma concentrations of arginine vasotocin, prolactin, aldosterone and corticosterone in relation to oviposition and dietary NaCl in the domestic fowl. *Comparative Biochemistry and Physiology* **81A** 769-777

24 Thomas D H, Skadhauge E & Read M W (1979) Acute effects of aldosterone on water and electrolyte transport in the colon and coprodeum of the domestic fowl (*Gallus domesticus*) *in vivo*. *Journal of Endocrinology* **83** 229-237

25 Elbrønd V S & Skadhauge E (1992) $Na^+$-transport during long-term incubation of hen lower intestine: no aldosterone effect. *Comparative Biochemistry and Physiology* **101A** 203-208

26 Eldrup E, Møllgård K & Bindslev N (1979) Possible epithelial sodium channels in the luminal membrane of the hen lower intestine visualized by freeze-fracture. *Hormonal Control of Epithelial Transport* pp 253-262. Eds J Bourguet, J Chevalier, M Parisi & P Ripoche. Paris: INSERM

27 Eldrup E, Møllgård K & Bindslev N (1980) Possible epithelial sodium channels visualized by freeze-fracture. *Biochimica et Biophysica Acta* **596** 152-157

28 Iseki S & Kondo H (1989) Specific localisation of hepatic fatty acid binding protein in the gastric brush cells of rats. *Cell and Tissue Research* **257** 545-548

29 Iseki S (1991) Postnatal development of the brush cells in the common bile duct of the rat. *Cell and Tissue Research* **266** 507-510.

30 Luciano L, Wermbter G & Reale R (1973) Die Feinstruktur der Gallenblase und der Gallengange. III. Beobachtungen an gefriergeatzten Präparaten der Gallenblase der Maus inter besonderer Berucksichtigung der dichten Korper und des Verbindungskomplexes. *Cytobiologie* **1** 76-88

31 Luciano L, Castellucci M & Reale E (1977) Die Feinstruktur der Gallenblase und der Gallengange VI. Rasterelektronenmikroskopische Beobachtungen an den Bürstenzellen des Choledochus der Ratte. *Beiträge zur Elektronenmikroskopichen Direktabbildung von Oberflächen* **10** 469-476

32 Luciano L & Reale E (1979) A new morphological aspect of the brush cells of the mouse gallbladder epithelium. *Cell and Tissue Research* **210** 36-44

33 Luciano L, Castellucci M & Reale E (1981) The brush cells of the common bile duct of the rat. Thin section, freeze-fracture and scanning electron microscopy. *Cell and Tissue Research* **218** 403-420

34 Luciano L & Reale E (1990) Brush cells of the mouse gallbladder. A correlative light- and electron-microscopical study. *Cell and Tissue Research* **262** 339-349

35 Trier J S, Allan C H, Marcial M A & Madara J L (1987) Structural feature of the apical and tubulovesicular membranes of rodent small intestinal tuft cells. *The Anatomical Record* **219** 69-77

36 Benos D J (1991) Purification of an epithelial sodium channel: mechanism of aldosterone action. *Aldosterone: Fundamental Aspects* pp 261-271. Eds J P Bonvalet, N Farman, M Lombès & M E Rafestin-Oblin. Colloque INSERM/John Libbey Eurotext Ltd

37 Garty H & Asher C (1991) Does aldosterone induce *de novo* synthesis of $Na^+$ channels? *Aldosterone: Fundamental Aspects* pp 273-283. Eds J P Bonvalet, N

Farman, M Lombès & M E Rafestin-Oblin. Colloque INSERM/John Libbey Eurotext Ltd

38 Palmer L G, Pácha J & Frindt G (1991) Regulation of apical sodium channels in the rat cortical collecting tubule by aldosterone. *Aldosterone: Fundamental Aspects* pp 285-293. Eds J P Bonvalet, N Farman, M Lombès & M E Rafestin-Oblin. Colloque INSERM/John Libbey Eurotext Ltd

39 Smith P R, Saccomani G, Bradford A L, Dantzer V, Benos D J & Skadhauge E (1993) Immunochemical identification of amiloride-sensitive sodium channels from an intestinal epithelium. *Cell and Tissue Research* **272** 129-136

40 Tousson A, Alley C D, Sorcher E J, Brinkley B R & Benos D J (1989) Immunochemical localization of amiloride-sensitive sodium channels in sodium-transporting epithelia. *Journal of Cell Science* **93** 349-362

41 Stanton B A (1991) Aldosterone regulation of sodium and potassium transport by the distal tubule and collecting duct: structure-function relationships. *Aldosterone: Fundamental Aspects* pp 187-197. Eds J P Bonvalet, N Farman, M Lombès & M E Rafestin-Oblin. Colloque INSERM/John Libbey Eurotext Ltd

42 Wade J B, Stanton B A, Field M J, Kashgarian M & Giebisch G (1990) Morphological and physiological responses to aldosterone: time course and sodium dependence. *American Journal of Physiology* **259** (Renal Fluid Electrolyte Physiology) F88-F94

43 Bindslev N, Cuthbert A W, Edwardson J M & Skadhauge E (1982) Kinetics of amiloride action in the hen coprodeum *in vitro*. *Pflügers Archiv* **392** 340-346

44 Asher C, Singer D, Eren R, Yeger O, Dascal N & Garty H (1992) NaCl dependent expression of an amiloride-blockable $Na^+$ channel in *Xenopus* oocytes. *American Journal of Physiology (Gastro-intestinal Liver Physiology)* **262** G244-G248

# Part Five
# Endocrine Control of Bone and Soft Tissue Growth

Avian Endocrinology
Ed. P J Sharp, pp 227-238
Journal of Endocrinology Ltd, Bristol (1993)

# Parathyroid hormone regulation of avian osteoclasts

L G May[1], V R Gilman[1] and C V Gay[1,2]
[1]Department of Molecular and Cell Biology, [2]Department of Poultry Science, The Pennsylvania State University, University Park, PA 16802, USA

## Introduction

Metabolic demands on the avian skeleton are substantially greater than in other vertebrates because of extremely rapid initial growth rates and because of the process of egg shell formation. Consequently the avian skeleton is frequently used as a model for skeletal studies. During the past decade avian bone has been exploited as a source of the cells which resorb bone, osteoclasts. Osteoclast isolation methods have been reviewed recently (1, 2). There are three phases of avian life when osteoclasts are particularly abundant: late-stage embryos (15-19 day), post-hatch chicks (up to about six weeks of age), and adult laying females. In the first two cases long-bone growth is notably rapid; in the latter case medullary bone forms within the marrow cavity of long bones, vertebrae and ribs, and serves as a source of rapidly mobilizable calcium for eggshell formation. Avian osteoclasts are even more aggressive bone resorbing cells than their mammalian counterparts (3). This suggests that very precise control mechanisms may have evolved to regulate osteoclast activity. In avian cells the mechanisms may be amplified and so may serve as useful model systems.

Parathyroid hormone (PTH) has long been attributed with stimulating bone resorption, a process carried out by osteoclasts. It was both surprising and intriguing when the discovery was made, in the mid 1970s, that osteoblast-like cells, rather than osteoclasts, responded to PTH in a substantial manner (4). Subsequently it has been shown that osteoblasts specifically bind PTH in amounts detectable by immmunocyto-chemistry and autoradiography (5-7). PTH-stimulated osteoblasts are believed to dominate the regulation of osteoclasts and considerable research has been focused on characterizing the indirect regulatory pathway (8).

PTH is also a direct stimulator of osteoclasts. One observation that implicated this was that PTH-dependent increases in cAMP in cell populations enriched in osteoclast-like or osteoblast-like cells occurred at different rates (4). This suggested that two types of cells may have been responding, one directly and one indirectly.

The focus of this paper will be on the binding of PTH directly to osteoclasts and on the signalling mechanisms that may be involved.

## Parathyroid hormone and osteoclasts

In 1983, Rao and colleagues reported specific immunocytochemical staining of PTH binding to both osteoclasts and osteoblasts in sections of rat fore limb (9).

Although that investigation did not reveal the presence of receptors for PTH nor address the specificity of binding, it nevertheless suggested direct binding of PTH to osteoclasts. Three recent studies revealed specific binding of radiolabelled or biotinylated PTH, which was competitive with unlabelled PTH, to isolated chicken (10-12) and rat (11) osteoclasts. Duong and colleagues (11) further reported that the plasma membrane protein to which the PTH was bound had a molecular mass of ~70,000 daltons. Agarwala and Gay (12) found that biotinylated PTH adhered to cultured chicken osteoclasts in a manner indicative of receptor binding. The process was saturable, competitive with unlabelled PTH, not affected by other peptides of similar molecular weight, and had relatively high binding affinity (half-maximal binding ~$10^{-7}$M). The time course of the binding was rapid, with the osteoclast surface being cleared of hormone in about 20 minutes. Osteoblasts, on the other hand, retained bound PTH throughout the experiment. Experiments repeated with time extensions to 80 minutes, revealed that osteoblasts continued to retain bound PTH. The rapid clearance of PTH from osteoclast surfaces and the relatively low sensitivity of autoradiography may account for the failure of earlier studies to identify specific binding of PTH to osteoclasts (5-7).

Osteoclasts from all species are required to differentiate between, and accurately respond to, a variety of hormones and other chemical signals. These hormones, which either enhance or diminish the process of bone resorption, are expected to be mediated along different pathways within the cell. Several aspects of the resorption process have been identified. There is substantial evidence that hydrogen ions are secreted by an ATP-driven proton pump located at the ruffled border (13-17). The protons for this pump are believed to be provided by carbonic anhydrase, an enzyme that is responsible for the reversible hydration of carbon dioxide (1, 18, 19). Carbonic anhydrase has also been localized at the ruffled border (20). Osteoclast acidity is increased under the influence of PTH in organ culture (21) and in isolated cell preparations (18, 22, 23). These studies support the view that PTH directly stimulates osteoclast acidification. However, the molecular mechanisms utilized by PTH, which stimulates the osteoclast to resorb bone, have only begun to be investigated.

PTH alters osteoclast activity in several other ways. The spreading of osteoclasts onto bone surfaces was found to increase when PTH was administered to cultured bone slices (24). However, in a study of isolated osteoclasts which were at approximately 95% purity, PTH had no effect on the distribution of microtubules and microfilaments, whereas the distribution of the cytoskeletal elements changed in response to calcitonin (25). These studies suggest that osteoclast spreading is an indirect response to PTH.

PTH stimulates the release of lysosomal enzymes, including cysteine proteinases and procollagenase, from cultured bone (19). PTH stimulates osteoclastic bone resorption *in vivo* by increasing the development of new osteoclasts and by increasing the metabolic activity of osteoclasts (26, 27). Mears (28) has shown that PTH alters membrane potential of osteoclasts. It would be timely to repeat the latter study now that technology for isolating osteoclasts has emerged.

## G-proteins as transducers for calcium regulatory hormones

A signalling pathway exists between a hormone receptor and its effector for peptide hormones and for some lipid soluble hormones as well. Signal transduction is itself a complex array of a transducer interacting with one of several possible effectors which, in turn, produces a diverse group of second messengers. Guanine nucleotide binding proteins (G-proteins) are the only class of transducers yet known (for reviews see 29-31). Discovered in the 1970s, the G-proteins that function as intermediaries in transmembrane signalling are heterotrimers with subunits designated $\alpha$, $\beta$, and $\gamma$, in order of decreasing mass. There have been at least 16 $\alpha$, 3 $\beta$, and 4 $\gamma$ subunits identified so far (32). G-proteins are classified according to function; for example, $G_s$ stimulates adenylate cyclase and $G_i$ inhibits adenylate cyclase. Effector systems include adenylate cyclase, the phosphoinositide pathway, potassium and calcium channels, and cyclic GMP phosphodiesterase. Unique combinations of the components of the signalling system confer the potential for enormous diversity for regulatory pathways.

A variety of hormones, mainly peptides which cannot pass through the plasma membrane, have been studied in a wide variety of systems in an attempt to elucidate the precise mechanisms by which hormones mediate their effects on cells. For example, receptors for calcitonin, a calcium sparing hormone, were found to couple with a distinct G-protein in a pig kidney cell line (LLC-PK1) in a cell cycle dependent manner (33). Zaidi and colleagues (34) showed two effects of calcitonin on rat osteoclasts mediated by two distinct G-proteins. These were cessation of cell motility and cellular retraction. The ability of PTH, 1,25 dihydroxyvitamin $D_3$ and prostaglandin $E_2$ ($PGE_2$) to stimulate bone resorption was blocked when fetal rat limb bones were exposed to pertussis toxin (35), a G-protein inhibitor. Whether or not the effect is indirectly mediated through osteoblasts has not been determined. Tokuda and colleagues (36) also found $PGE_2$ to be coupled to a pertussis toxin sensitive G-protein in the osteoblast cell line, MC3T3-E1. A pertussis toxin sensitive G-protein has been shown to link prostaglandin $F_2\alpha$ to phospholipase C in cloned osteoblast-like MC3T3-E1 cells (37). In the osteoblast-like osteosarcoma cell line, UMR-106, pertussis toxin enhanced PTH-sensitive adenylate cyclase activity and diminished the incremental increase in the PTH response produced by activation of protein kinase C (38). In contrast, under conditions favouring the down-regulation of PTH receptors in UMR-106 cells, PTH treatment was associated with decreased cholera toxin sensitive G-protein mediated adenylate cyclase activity (39). Most of the studies on G-proteins in bone cells have focused on osteoblasts; little has been reported for osteoclasts.

## G-proteins and parathyroid hormone regulation of osteoclasts

One effect of PTH on osteoclasts, enhanced acid production, is transduced across the plasma membrane through a stimulatory G-protein which is linked to adenylate cyclase (40). Activation of adenylate cyclase and subsequent formation of cAMP results in the production of acid by osteoclasts *in vitro*, presumably through the action of carbonic anhydrase. In order to trace the activation pathway, agents have been used which allow the G-protein-adenylate cyclase pathway to be analysed, namely dibutyryl cyclic AMP as a second messenger; alloxan, which inhibits adenylate cyclase; lithium chloride, which uncouples G-proteins from their receptors; and cholera toxin, which permanently activates adenylate cyclase.

Osteoclasts were isolated from 2-3 week old chick long bones by sequential filtration, as previously described (22, 41) and were used for experiments between 4 and 6 days of culture. Image analysis of TRAP-positive cells showed approximately 90% purity. Acid production in isolated osteoclasts was analysed using the pH-sensitive dye, acridine orange. Acridine orange behaves as a weak base that diffuses through intact biological membranes but is sequestered when protonated in acidic compartments. The fluorescence intensity of the trapped dye is proportional to the hydrogen ion concentration. The average fluorescence intensity of acidic compartments in individual osteoclasts was estimated using a Leitz microscope MPV photometer. Analysis of the osteoclasts began with neutralization in 20mM ammonium chloride, followed by rinsing, and subsequent recovery in the presence of various stimulators or inhibitors for 15 minutes to 2 hours, as described earlier (41). The cells were then stained for 10 minutes with acridine orange to detect acidic compartments. The stimulatory agents employed were parathyroid hormone ($10^{-8}$M, bovine), dibutyryl cyclic AMP ($10^{-4}$M, Sigma, St Louis, MO, USA), and cholera toxin (0.01 μg/ml, Sigma). Inhibitory agents included alloxan ($10^{-8}$M, Sigma) and lithium chloride ($10^{-4}$M).

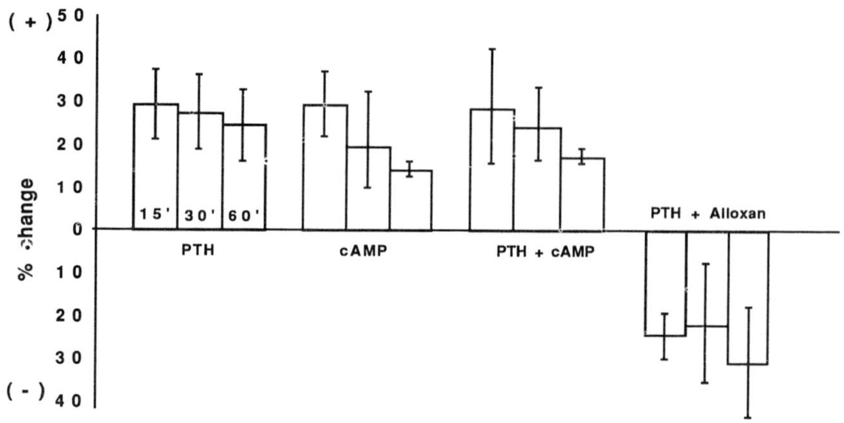


Fig. 1 The effects of parathyroid hormone (PTH), dibutyryl cyclic AMP (cAMP) and alloxan on acid production by cultured osteoclasts. The cells were stained with the pH-sensitive fluorescent dye, acridine orange. An increase fluorescence in individual cells indicates an increase in acidity. Shown is the percentage change of fluorescence relative to untreated control cells measured at 15, 30 and 60 min after addition of the test substance to the culture medium. Each bar is the mean value of fluorescence intensity in 40-60 osteoclasts ± se. The data show that PTH stimulates the osteoclast through the effector adenylate cyclase.

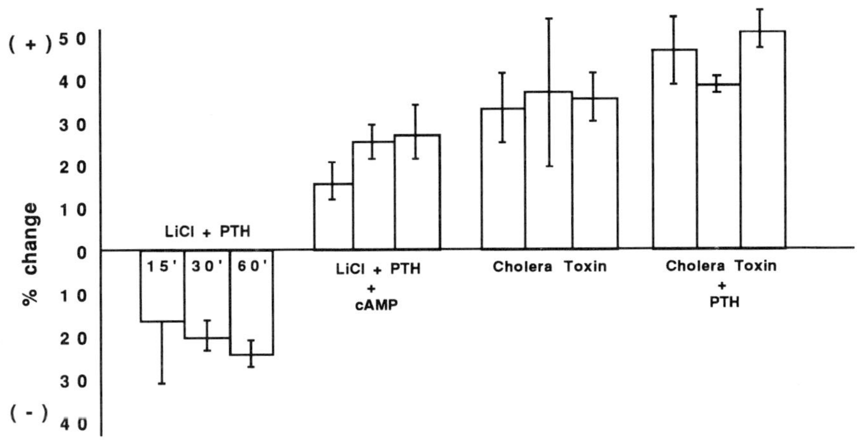


Fig. 2 Evidence showing that a G-protein of the $G_s$ type links the activated PTH-receptor to its effector. The experimental design is the same as in Figure 1. Treatments were PTH, LiCl, dibutyryl cyclic AMP and cholera toxin, in the combinations shown.

Figure 1 shows that the PTH effect is cAMP dependent. Osteoclasts, in the presence of PTH, were stimulated to produce approximately 30% more acid as compared to untreated osteoclasts. The PTH-stimulated acid production was generally maximal at 15 minutes and did not increase with time of treatment. This effect was mimicked by dibutyryl cAMP; however, the effects of PTH and cAMP together were not additive. Acid production stimulated by PTH could be blocked by the adenylate cyclase inhibitor, alloxan. Lithium chloride which uncouples G-proteins from receptors was used as a probe for a possible G-protein (Figure 2) linking the PTH receptor to the adenylate cyclase enzyme (42). Lithium chloride treatment caused a dramatic decrease in acidity of PTH-stimulated osteoclasts. The reduction in acidity was prevented by adding cAMP to the culture wells. These findings suggest that a transducing G-protein molecule is involved in the stimulation of osteoclast acidification by PTH. Zaidi *et al.* (43) also found evidence of G-protein involvement in calcitonin-treated osteoclasts utilizing lithium chloride. Caution must be used in interpretation of the lithium effect since lithium could cause effects on osteoclasts other than that on guanine nucleotide binding (44). However, since the addition of cAMP bypassed the lithium effect and stimulated osteoclasts to produce acid, the view that a G-protein links the PTH receptor to adenylate cyclase is reinforced and suggests that the major, and perhaps the only, effect of lithium on osteoclasts is prevention of guanine nucleotide binding. In addition, our data supports the observation of Zaidi and colleagues (43) that lithium is fully effective in whole cell systems in uncoupling hormone receptors from G-protein-modulated hormone effects.

Cholera toxin, which permanently activates many G-proteins of the $G_s$ type, was used to further characterize the identity of the G-protein implied by the lithium effect. This permanent activation is a consequence of ADP-ribosylation on $Arg^{201}$ of the $\alpha$ subunit of $G_s$ which results in decreased GTPase activity and decreased affinity for $\beta\gamma$ subunits (45, 46). Figure 2 shows that cholera toxin mimicked the effects of PTH and cAMP on acid production and the PTH and cholera toxin effects were not additive. The results shown in Figures 1 and 2 demonstrate that PTH-stimulated acid production by osteoclasts is regulated by a stimulatory G-protein coupled to adenylate cyclase. Since image analysis revealed an osteoclast purity of approximately 90% and osteoclasts have previously been shown to specifically bind PTH (10-12), the responses can be attributed to direct stimulation of osteoclasts.

Insight into the mechanism that follows PTH stimulation of osteoclasts is provided by Silverton *et al.* (47) and Dietsch (48). PTH treatment of osteoclasts has been shown to activate carbonic anhydrase (47, 48). The activation appears to be due to a cAMP-dependent phosphorylation resulting from increased activity of a protein kinase (48). This implies protein kinase A involvement. Figure 3 is a diagrammatic representation of the probable signal pathway.

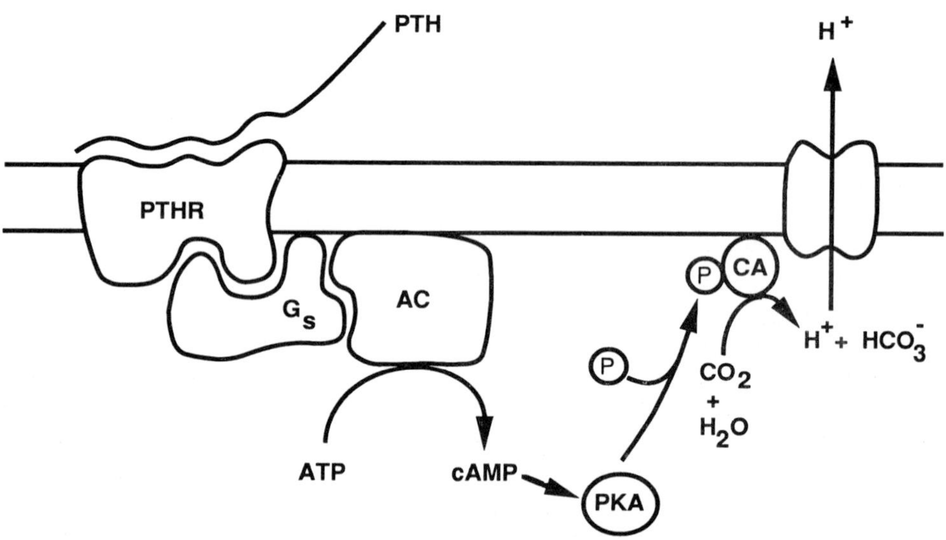


Fig. 3 Diagram of the likely signal pathway for PTH-stimulation of increased osteoclast acidity: PTH binds to its receptor, PTHR; $G_s$ is activated and, in turn, activates adenylate cyclase (AC) which produces cyclic AMP (cAMP); cAMP stimulates protein kinase A (PKA); carbonic anhydrase (CA) activation occurs by PKA-mediated phosphorylation; the resulting acidification is dependent on carbonic anhydrase activity (21, 22, 41, 49).

**Effector systems for parathyroid hormone signal transduction**

In osteoclasts, the only PTH regulating pathway described so far utilizes adenylate cyclase as its effector, as described in the previous section. Progress of other laboratories in elucidating signalling mechanisms of PTH in other systems include a study by Freyaldenhaven and colleagues (38) who found that activation of protein kinase C in osteoblast-like cells increased PTH-stimulated adenylate cyclase activity by either action on the inhibitory G-protein or the catalytic subunit. Pertussis toxin, which impairs the ability of $G_i$ to interact with a receptor and thus impairs the inhibition of adenylate cyclase, caused PTH sensitive adenylate cyclase activity to increase (38). Rat osteoblast-like cells also exhibited a pertussis toxin sensitive $G_i$ that modulated adenylate cyclase activity (50). PTH stimulation of adenylate cyclase activity has been reported in the human osteosarcoma cell for two osteoblast cell lines, SaOS-2 (51) and UMR-106 (39).

Protein kinase A, which is dependent upon cAMP for activation, has been shown to be important in PTH-stimulated bone resorption in mouse bone marrow cells (52) and in PTH action on renal cells (53, 54). Reshkin and Murer (53) found involvement of both protein kinase A and protein kinase C in the regulation of PTH in opossum kidney cells. Further, two receptor subtypes were identified. The high affinity receptor subtype was coupled to phospholipase C, whereas the low affinity receptor was linked to adenylate cyclase (54). Whether these effector systems are also utilized by osteoclasts remains to be determined.

Across a spectrum of tissues and cells, several different adenylate cyclase isotypes have been identified, including membrane-bound and soluble forms. In the studies discussed above, the adenylate cyclase subtype is not yet known. As specific probes become available and antibodies are produced for the different subtypes, it should be possible to locate adenylate cyclases within cell compartments. Histochemical methods for localizing adenylate cyclase have been applied to bone sections (55-57). Initially, enzyme activity was found in osteoblast plasma membranes but was not detectable in osteoclasts (55, 56). Subsequently, a report by Eliam *et al.* (58) revealed that osteoclasts from calcium deficient chicks responded to calcitonin by a doubling of cAMP production and an increase in the number of calcitonin receptors; in osteoclasts from calcium sufficient chicks no cAMP response was detected. This suggested that adenylate cyclase activity or concentration was increased during calcium insufficiency. When the adenylate cyclase localization study was repeated on both calcium sufficient and deficient chicks, the reaction product was also found on osteoclast plasma membrane under the calcium deficient condition (57). While this evidence was not readily quantifiable, it suggests another type of regulatory mechanism, namely modulation of the concentration and/or activity of the effector molecule, adenylate cyclase.

**Conclusion**

It is concluded that, in osteoclasts, 1) the effector molecule for PTH stimulation of acid production is adenylate cyclase, 2) cAMP is the second messenger, and 3) at least one G-protein, of the $G_s$ type, acts as a transducer.

## Acknowledgements

This research was funded by NIH grant DE04345. The parathyroid hormone was provided by the National Hormone and Pituitary Program.

## References

1 Gay C V (1988) Avian bone resorption at the cellular level. *CRC Critical Reviews in Poultry Biology* **1** 197-210

2 Osdoby P, Krukowski M & Collin-Osdoby P (1992) Experimental systems for studying osteoclast biology. *Biology and Physiology of the Osteoclast* pp 1-32. Eds B R Rifkin & C V Gay. Boca Raton: CRC Press

3 Jones S J, Ali N N & Boyde A (1986) Survival and resorptive activity of chick osteoclasts in culture. *Anatomy and Embryology* **174** 265-275

4 Wong G L & Cohn D V (1975) Target cells in bone for parathyroid hormone and calcitonin are different: enrichment for each cell type by sequential digestion of mouse calvaria and selective adhesion to polymeric surfaces. *Proceedings of the National Academy of Sciences of the USA* **72** 3167-3171

5 Silve C M, Hradek G T, Jones A L & Arnaud C D (1982) Parathyroid hormone receptor in intact embryonic chick bone: characterization and cellular localization. *Journal of Cell Biology* **94** 379-386

6 Barling P M & Bibby N J (1985) Study of the localization of [$^3$H] bovine parathyroid hormone in bone by light microscope autoradiography. *Calcified Tissue International* **37** 441-446

7 Rouleau M F, Warshawsky H & Goltzman D (1986) Parathyroid hormone binding *in vivo* to renal, hepatic, and skeletal tissues of the rat using a radioautographic approach. *Endocrinology* **118** 919-931

8 Simmons D J & Grynpas M D (1990) Mechanisms of bone formation *in vivo*. *The Osteoblast and Osteocyte* pp 193-302. Ed. B K Hall. New Jersey: Telford Press

9 Rao L G, Murray T M & Heersche J N M (1983) Immunohistochemical demonstration of parathyroid hormone binding to specific cell types in fixed rat bone tissue. *Endocrinology* **113** 805-810

10 Teti A, Rizzoli R & Zambonin Zallone A (1991) Parathyroid hormone binding to cultured avian osteoclasts. *Biochemica Biophysica Research Communications* **174** 1217-1222

11 Duong L T, Grasser W, DeHaven P A & Sato M (1990) Parathyroid hormone receptors identified on avian and rat osteoclasts. *Journal of Bone and Mineral Research* **5** S203

12 Agarwala N & Gay C V (1992) Specific binding of parathyroid hormone to living osteoclasts. *Journal of Bone and Mineral Research* **7** 531-539

13 Baron R, Neff L, Louvard D & Courtoy P J (1985) Cell-mediated extracellular acidification and bone resorption: evidence for a low pH in resorbing lacunae and localization of a 100 kD lysosomal membrane protein at the osteoclast ruffled border. *Journal of Cell Biology* **101** 2210-2222

14 Akisaki T & Gay C V (1986) Ultracytochemical evidence for a proton-pump adenosine triphosphatase in chick osteoclasts. *Cell and Tissue Research* **245** 507-512

15 Blair H C, Teitelbaum S L, Ghiselli R & Gluck S (1989) Osteoclastic bone resorption by a polarized vacuolar proton pump. *Science* **245** 855-857

16 Bekker P J & Gay C V (1990) Biochemical characterization of an electrogenic vacuolar proton pump in purified chicken osteoclast plasma membrane vesicles. *Journal of Bone and Mineral Research* **5** 569-579

17 Väänänen H K, Karhukorpi E K, Sundquist K, Wallmark B, Roininen I, Hentunen T, Tuukkanen J & Lakkakorpi P (1990) Evidence for the presence of a proton pump of the vacuolar proton ATPase type in the ruffled border of osteoclasts. *Journal of Cell Biology* **111** 1305-1312

18 Gay C V (1991) The role of carbonic anhydrase in bone resorbing cells. *Carbonic Anhydrase from Biochemistry and Genetics to Physiology and Clinical Medicine* pp 419-422. Eds F Botrè, G Gross & B T Storey. Weinheim: VCH Verlagsgesellschaft

19 Delaissé J-M & Vaes G (1992) Mechanism of mineral solubilization and matrix degradation in osteoclastic bone resorption. *Biology and Physiology of the Osteoclast* pp 289-314. Eds B R Rifkin & C V Gay. Boca Raton: CRC Press

20 Anderson R E, Schraer H & Gay C V (1982) Ultrastructural immunocytochemical localization of carbonic anhydrase in normal and calcitonin treated chick osteoclasts. *Anatomical Record* **204** 9-20

21 Anderson R E, Jee W S S & Woodbury D M (1985) Stimulation of carbonic anhydrase in osteoclasts by parathyroid hormone. *Calcified Tissue International* **37** 646-650

22 Hunter S J, Schraer H & Gay C V (1988) Characterization of isolated and cultured chick osteoclasts: the effects of acetazolamide, calcitonin, and parathyroid hormone on acid production. *Journal of Bone and Mineral Research* **3** 297-303

23 Gay C V & Kief N (1990) Effects of estrogen, prostaglandin $E_2$ and parathyroid hormone on isolated osteoclasts. *Journal of Bone and Mineral Research* **5** S96

24 Pandalai S & Gay C V (1990) Effects of parathyroid hormone, calcitonin, and dibutyryl cyclic-AMP on osteoclast area in cultured chick tibia. *Journal of Bone and Mineral Research* **5** 701-705

25 Hunter S J, Schraer H & Gay C V (1989) Characterization of the cytoskeleton of isolated chick osteoclasts: effects of calcitonin. *Journal of Histochemistry and Cytochemistry* **37** 1529-1537

26 Stern P H & Lakatos P (1992) Effects of pharmacologic agents on osteoclasts. *Biology and Physiology of the Osteoclast* p 367. Eds B R Rifkin & C V Gay. Boca Raton: CRC Press

27 Heersche J N M (1992) Systemic factors regulating osteoclast function. *Biology and Physiology of the Osteoclast* p 159. Eds B R Rifkin and C V Gay. Boca Raton: CRC Press

28 Mears D C (1971) Effects of parathyroid hormone and thyrocalcitonin on the membrane potential of osteoclasts. *Endocrinology* **88** 1021-1028

29 Gilman A G (1987) G-proteins: transducers of receptor-generated signals. *Annual Reviews in Biochemistry* **56** 615-649

30 Birnbaumer L (1990) Transduction of receptor signal into modulation of effector activity by G-proteins: the first 20 years or so... *FASEB Journal* **4** 3068-3078

31 Naccache P H (1990) *G Proteins and Calcium Signalling*. Boca Raton: CRC Press

32 Tamir H, Fawzi A B, Tamir A, Evans T & Northup J K (1991) G-protein $\beta\gamma$ forms: identity of $\beta$ and diversity of $\gamma$ subunits. *Biochemistry* **30** 3929-3936

33 Chakraborty M, Chatterjee D, Kellokumpu S, Rasmussen H & Baron R (1991) Cell cycle-dependent coupling of the calcitonin receptor to different G-proteins. *Science* **251** 1078-1082

34 Zaidi M, Datta H K, Moonga B J & MacIntyre I (1990) Evidence that the action of calcitonin on rat osteoclasts is mediated by two G-proteins acting via separate post-receptor pathways. *Journal of Endocrinology* **126** 473-481

35 Klein R F, Nissenson R A & Strewler G J (1991) Pertussis toxin inhibits hormonal stimulation of bone resorption in fetal rat limb bones. *Journal of Pharmacology and Experimental Therapeutics* **258** 877-881

36 Tokuda H, Kozawa O, Yoneda M, Oiso Y, Takatsuki K, Asano T & Kato K (1991) Possible coupling of prostaglandin $E_2$ receptor with pertussis toxin sensitive guanine nucleotide-binding protein in osteoblast like cells. *Journal of Biochemistry* **109** 229-233

37 Miwa M, Tokuda H, Tsushita K, Kotoyori J, Takahashi Y, Ozaki N, Kozawa O & Oiso Y (1990) Involvement of pertussis toxin-sensitive GTP-binding protein in prostaglandin $F_2$ alpha-induced phosphoinositide hydrolysis in osteoblast like cells. *Biochemica Biophysica Research Communications* **171** 1229-1235

38 Freyaldenhoven A M, Gutierrez G E, Lifschitz M D & Katz M S (1992) Protein kinase C differentially modulates PTH- and $PGE_2$-sensitive adenylate cyclase in osteoblast-like cells. *American Journal of Physiology* **262** E87-E95

39 Mitchell J & Goltzman D (1990) Mechanisms of homologous and heterologous regulation of parathyroid hormone receptors in the rat osteosarcoma cell line UMR-106. *Endocrinology* **126** 2650-2660

40 May L G & Gay C V (1992) Acid production by osteoclasts is regulated by parathyroid hormone through G-protein-linked adenylate cyclase. *Journal of Bone and Mineral Research* **7** (Supplement 2) 856

41 Hunter S J, Rosen C J & Gay C V (1991) *In vitro* resorptive activity of isolated chick osteoclasts: effects of carbonic anhydrase inhibition. *Journal of Bone and Mineral Research* **6** 61-66

42 Avissar S, Schreiber G, Danon A & Belmaker R H (1988) Lithium inhibits adrenergic and cholinergic increases in GTP binding in rat cortex. *Nature* **331** 440-442

43 Zaidi M, Patchell A, Datta H K & MacIntyre I (1989) Uncoupling of receptor mediated cellular responses by ionic lithium. *Journal of Endocrinology* **123** R5-R7

44 Berridge M J, Downes C P & Hanley M R (1982) Lithium amplifies agonist-dependent phosphatidylinositol responses in brain and salivary glands. *Biochemical Journal* **206** 587-595

45 Neer E J & Clapham D E (1988) Roles of G-protein subunits in transmembrane signalling. *Nature* **333** 129-134

46 Kaziro Y, Itoh H & Nakafuku M (1990) Organization of genes coding for G-protein $\alpha$ subunits in higher and lower eukaryote. *G-proteins* pp 63-80. Eds R Iyengar & L Birnbaumer. San Diego: Academic Press

47 Silverton S F, Dodgson S F, Fallon M D & Forster R E (1987) Carbonic anhydrase activity of chick osteoclasts is increased by parathyroid hormone. *American Journal of Physiology* **253** E670-E674

48 Dietsch P (1987) Parathyroid hormone, carbonic anhydrase and calcium homeostasis. *Generalized Bone Diseases* pp 279-289. Eds F Kuhlencordt, P Dietsch, E Keck & H P Kruse. Berlin: Springer Verlag

49 Raisz L G, Simmons H A, Thompson W J, Shepard K L, Anderson P S & Rodan G A (1988) Effects of a potent carbonic anhydrase inhibitor on bone resorption in organ culture. *Endocrinology* **122** 1083-1086

50 Pines M, Santori A, Gierschik P, Menczel J & Spiegel A (1986) The inhibitory guanine nucleotide regulatory protein modulates agonist-stimulated cyclic AMP production in rat osteosarcoma cells. *Bone and Mineral* **1** 15-26

51 Ikeda K, Imai Y, Fukase M & Fujita T (1990) The effect of 1,25-dihydroxyvitamin $D_3$ on human osteoblast-like osteosarcoma cell; modification of response to PTH. *Biochemica Biophysica Research Communications* **168** 889-897

52 Kaji H, Sugimoto T, Kanatani M & Fukase M (1992) The activation of cAMP-dependent protein kinase is directly linked to the stimulation of bone resorption by parathyroid hormone. *Biochemica Biophysica Research Communications* **182** 1356-1361

53 Reshkin S J & Murer H (1991) Parathyroid hormone regulation in opossum kidney cells *in situ* protein phosphorylation reactions involving protein kinase A, protein kinase C, and GTP-binding protein. *Cellular Physiology and Biochemistry* **1** 143-159

54 Reshkin S J, Forgo J & Murer H (1991) Apical and basolateral effects of parathyroid hormone in OK cells: transport inhibition, messenger production, effects of pertussis toxin, and interaction with a PTH analog. *Journal of Membrane Biology* **124** 227-237

55 Walzer C (1980) An attempt at localizing adenylate cyclase in rat calvaria. Influence of sodium fluoride and parathyroid hormone. *Histochemistry* **68** 281-296

56 Yamamoto T & Gay C V (1989) Ultrastructural localization of adenylate cyclase in chicken bone cells. *Journal of Histochemistry and Cytochemistry* **37** 1705-1709

57 Fukushima O, Yamamoto T & Gay C V (1991) Ultrastructural localization of adenylate cyclase activity in chicken osteoclasts. *Journal of Histochemistry and Cytochemistry* **39** 1207-1213

58 Eliam M C, Baslé M, Bouizar Z, Bielakoff J, Moukhtar M & de Vernejoul M C (1988) Influence of blood calcium on calcitonin receptors in isolated chick osteoclasts. *Journal of Endocrinology* **119** 243-248

Avian Endocrinology
Ed. P J Sharp, pp 239-248
Journal of Endocrinology Ltd, Bristol (1993)

# Calcitrophic peptides and rapid calcium fluxes into chick bone *in vivo*

C G Dacke[1], A K Ancill[2], G Whitaker[1] and Z A Bascal[1]
[1]Division of Pharmacology, School of Pharmacy and Biomedical Sciences, University of Portsmouth, Portsmouth, PO1 2DT, UK; [2]Biomedical Research Department, AEA Environment and Energy, Harwell Laboratory, Oxfordshire, OX11 0RA, UK

## Introduction

Calcium (Ca) is one of the most efficiently regulated plasma constituents in birds. All the classical Ca- regulating hormones, parathyroid hormone (PTH), calcitonin (CT) and 1,25-dihydroxy vitamin $D_3$, are recognised in this class, although their actions and/or sensitivities can be quite different from those in mammalian species (1). In recent years it has become clear that other putative Ca- and bone-regulating factors, such as prostaglandins (PGs) (2) or calcitonin gene-related peptide (CGRP) (3), can affect avian Ca metabolism in ways which are profoundly different from those in mammals. This may, in part, be related to enhanced requirements for Ca mobilization during eggshell calcification; however it should be noted that much of the work reviewed here was carried out in juvenile, growing birds in which the requirements of Ca metabolism are similar to those in mammals.

Mammals given either an acute intravenous Ca load, or EDTA, which lowers plasma Ca levels, respond to this challenge within tens of minutes to a few hours (4). Growing chickens aged about 1 week will correct such challenges within a few minutes. An intensive study has been made of factors which bring about rapid changes in levels of plasma Ca in chicks. Using acute labelling with bone-seeking isotopes, combined with a microwave fixation technique to stabilise subsequent isotope levels in bone, it has been demonstrated that humoral and paracrine Ca-regulating factors rapidly modulate exchanges of Ca occurring between blood and exchangeable bone compartments. An intravenous dose of $^{45}Ca$ in a week-old chick is rapidly cleared from the plasma pool and within 15 minutes approximately 40% of the original dose is located within the skeleton (5), compared with growing rats, where 50% of an injected dose of $^{47}Ca$ is estimated to be present in the skeleton within 10 minutes (6). By calculating the unidirectional plasma-bone clearance constant ($K_{pb}{}^{45}Ca$) and hence net $Ca^{++}$ influx, and by estimating total rate of Ca accretion into the skeleton, total skeletal outflux of $Ca^{++}$ is estimated to be approximately 80% of influx. From this it can be deduced that the plasma pool of $Ca^{++}$ is cleared into the skeleton every few minutes. Typically, net $Ca^{++}$ accumulation into the chick femur would be in the order of 0.28 mole $min^{-1}g^{-1}$ wet weight, although this varies according to prevailing rates of dietary Ca absorption and other factors (5). Clearly any factor which modifies either the rapid influx or outflux of $Ca^{++}$ in this system is likely to have a profound effect on minute to minute plasma Ca modulation in the rapidly growing animal. Bronner and Stein (7) calculated from our data (5) that the $t_{½}$ of $^{45}Ca$ uptake by the chick femur is less than 10 minutes compared with around 30 minutes in rabbit, dog and rat.

**Parathyroid hormone**

Parathyroid hormone (PTH) causes a slow hypercalcaemic response in mammals which involves recruitment and activation of osteoclasts in bone, mediated via receptors located in osteoblasts (8). It has been recognised for more than 35 years (9) that birds are exquisitely sensitive to PTH. Furthermore the maximum hypercalcaemic response to this hormone occurs in egg-laying hens as early as 8 minutes after administration (10), a time scale too brief for significant osteoclastic resorption (8) and unlikely to be accounted for by the relatively slow changes in intestinal (11) or renal (12) transport mechanisms. Similarly, rapid hypercalcaemic responses to PTH are detected in immature chickens and/or Japanese quail (13), providing the basis for the development of simple *in vivo* bioassays for this hormone (14, 15). These assays employ a technique by which calcium salts incorporated into the hormone vehicle transiently raise concentrations of plasma Ca following intravenous injection. This greatly enhances the sensitivity and precision of the bioassay. Using a technique of acute $^{45}Ca$ labelling, the initial (0-30 minute) response to PTH has been shown to involve inhibition of plasma $Ca^{++}$ clearance (16).

The basis for this decreased clearance remained obscure for more than a decade until a method was developed for temporal microwave fixation of injected radioisotopes (5). This made it possible to demonstrate that the decreased plasma clearance of $^{45}Ca$ in chicks can be accounted for, in part, by an inhibition of net isotope uptake into the skeleton. Dose-dependent responses are observed using the active bPTH(1-34) fragment, and following intravenous injection of a prostaglandin $E_2$ analogue, 16,16-dimethyl $PGE_2$ (17). These responses are rapid (3 min) and most apparent in the femur and, to a lesser extent, calvaria (18). It had previously been claimed (6) that the initial effect of PTH(1-84) in young adult rats is to cause a transient $^{47}Ca$ movement into bone resulting in the hypocalcaemic 'blip' which characterises the mammalian response. This Ca influx was considered to trigger the classical bone resorption mechanisms which accounted for the hypercalcaemia of the chick PTH assay. This type of response to PTH is only seen in rats in the absence of microwave fixation; when microwave fixation is applied the response to bPTH(1-34) or intact PTH(1-84) is similar to that seen in chicks (18). Using the microwave fixation technique, 16,16-dimethyl $PGE_2$ induced a profound inhibition of net $^{45}Ca$ uptake into both femur and calvaria in chicks, but paradoxically, in rats, only the response in femur was similar to that in chicks (18). The prostaglandin tended to enhance the uptake of the isotope into rat calvaria (18). These results show that microwave fixation is an essential technique for studying short term *in vivo* uptake of bone-seeking radio nuclides. Thus $^{45}Ca$ uptake continues for 45 min or more in unfixed animals *post mortem* (6), a similar situation to that in rats in which post-mortem uptake of bone-seeking isotopes was prevented by heat fixation (19).

The effects of PTH on bone $^{45}Ca$ uptake are mimicked by injecting phosphodiesterase inhibitors such as caffeine or 3-isobutyl-1-methylxanthine intravenously into chicks, suggesting a possible role for cAMP in this response (20). Oxidation of bPTH(1-34) with hydrogen peroxide abolishes the $^{45}Ca$ uptake response and reduces the concurrent cAMP activation, but does not reduce the hypercalcaemic response to this molecule. The analogue [$Nle^8$, $Nle^{18}$, $Tyr^{34}$]-bPTH(1-34) gives a much reduced hypercalcaemic response and slightly reduces effects on plasma $^{45}Ca$ clearance and bone uptake. This suggests that the initial hypercalcaemic response to PTH is not merely a reflection of its acute effects on net Ca uptake into bone (21).

Parathyroid hormone receptors are located on the surfaces of osteoblasts and are considered to be absent in osteoclasts (8), although this is disputed (22). There is considerable evidence suggesting that PTH induces rapid changes in Ca transfer by osteoblasts and osteocytes. Thus PTH-stimulated increases in Ca uptake by these cells have been observed, while others have reported either no response in intracellular Ca concentration or a net Ca efflux from bone cells, at least in embryonic chick bone *in vitro* (see 8).

Parathyroid hormone also affects cell spread area in osteoblasts (23) and osteoclasts (24). In osteoblasts from juvenile rats *in vitro*, PTH but not $PGE_2$ causes elongation and inhibition of border ruffling, although the minimum time course of these responses was not reported (23). Electron microscopy has been used to determine the responses of medullary bone osteoblasts to PTH in laying Japanese quail during the inactive phase of their eggshell calcification cycle (24). Parathyroid hormone induces the osteoclasts to form ruffled borders bounded by filamentous-rich clear zones within 20 minutes of hormone treatment, these changes being characteristic of 'active' bone-resorbing cells found during shell calcification. In tibiae from Ca-deficient chicks, osteoclasts *in situ* increase their cell spread area by 40% within 2-4 minutes of PTH challenge (25). Irrespective of whether or not this response represents direct or indirect effects of the hormone on these cells, it is remarkably fast compared with classical responses to this hormone (1, 8).

Parathyroid hormone has potent hypotensive activity in a variety of vertebrates including birds (26) and stimulates their atrial activity *in vitro* (27). In the domestic hen a PTH bolus causes a drop in bone blood flow by 3 minutes associated with transient hypocalcaemia, followed by hyperaemia by 30 minutes associated with an increased venous/arterial Ca gradient and hypercalcaemia (28). These changes in blood flow are probably due to systemic hypotension rather than local factors, as PTH does not affect vascular smooth muscle of bone (29).

**Parathyroid hormone-related peptide**

Clinical research in bone and Ca metabolism has focused on parathyroid hormone-related peptide (PTHrp), which is found in the circulation of certain patients with malignancy-associated hypercalcaemia and also as the predominant peptide in the circulation of fetal mammals. Parathyroid hormone and PTHrp genes are located on different chromosomes but share common organisational features, and the peptides show much structural homology, suggesting a common ancestral gene. Parathyroid hormone-related peptide exists in three known forms ranging in size from 139 to 173 amino acids, as the result of alternative gene splicing (30). It has a wide spectrum of actions in mammalian bone and Ca metabolism, many in common with PTH; these range from stimulation of osteoclastic bone resorption to enhancement of placental mineral transport (30). A recent report indicates that PTHrp is expressed in a variety of tissues in chick embryos, this molecule having a highly conservative structural homology with the human sequence, the first 21 residues being identical (31). Both PTH and PTHrp enhance cAMP and inhibit collagen synthesis in avian epiphysial cartilage cells, an effect which is blocked by the PTH antagonist PTH(3-34) (32). Parathyroid hormone-related peptide (1-34) when tested in the chick hypercalcaemic assay has only slight PTH agonist activity with respect to either plasma Ca levels or to $^{45}$Ca clearance from plasma (A K Ancill & C G Dacke, unpublished data). In femur, the peptide causes a substantial decrease in $^{45}$Ca uptake while in calvarium the opposite effect may occur. This phenomenon is currently undergoing further investigation.

## Calcitonin

Despite the discovery of calcitonin (CT) more than three decades ago, its role in bone and Ca metabolism remains an enigma. Only in mammalian species has CT been shown to regulate plasma Ca levels, the basis of its hypocalcaemic action lying in an ability to inhibit osteoclastic bone resorption (33). High concentrations of circulating CT are present in sub-mammalian species, including birds, and are detectable in the plasma of Japanese quail, using bioassays (1). Sub-mammalian vertebrates appear to be refractory to dosage with CT (1). It is not clear whether or not this is due to the high circulating concentrations of biologically active hormone causing down-regulation of receptors. Initial studies in avian osteoclasts showed no response to CT in terms of post receptor events (34, 35); however more recently cAMP responses have been demonstrated in osteoclasts from chicks maintained on low Ca or rachitogenic diets (36, 37). Moreover, osteoclasts from Ca deficient chicks respond to CT challenge *in vitro* within 4 minutes by a 58% reduction in cell spread area (25). These findings support the view that, in the chick, CT receptors are down-regulated under normal physiological conditions. The administration of salmon CT to 22h-fasted chicks causes a rapid (10 minutes) but variable inhibition of net $^{45}$Ca uptake into the skeleton, the femur being most affected (38). This effect is similar to that of PTH and $PGE_2$, but its physiological significance is obscure.

## Calcitonin gene-related peptide and amylin

Calcitonin gene-related peptide is a 37-amino acid neuropeptide derived from the same gene as CT, both belonging to the amylin super family (39). However, unlike CT, which is expressed mainly within the thyroid parafollicular cells in mammals, or ultimobranchial body in sub-mammals, it is found within the central and peripheral nervous system, including bone neurones (40), where its distribution corresponds with that of substance P (30). It is expressed within the dense distributions of peptidergic neurones found within the chick ultimobranchial gland (41). The structure of CGRP, unlike that of CT, is well conserved between vertebrate species with around 90% structural homology between chicken and human CGRPs, compared with only 50% between the respective CT molecules (42). The presence of CGRP in bone neurones, coupled with the interaction of this peptide with osteoclastic CT receptors (43, 44), suggests a paracrine role, modulating bone turnover and hence Ca homeostasis.

Calcitonin gene-related peptide has several putative physiological functions. It is a potent vasodilator and is implicated in central and peripheral neurotransmission and modulation (42). Only relatively recently has its role in bone and Ca metabolism been recognised. Calcitonin gene-related peptide shares the acute hypocalcaemic effects of CT, albeit at around 1000-fold less potency in rodents, as well as inhibiting bone resorption (45), stimulating cAMP production in mouse calvaria (46) and inhibiting neonatal rat osteoclastic spreading (45). In the rabbit *in vivo*, CGRP causes transient hypocalcaemia followed by a more sustained hypercalcaemia, while in chicks the only response is hypercalcaemia (3). In chicks, CGRP and PTH are approximately equipotent on a molar basis in inducing hypercalcaemia (47). Intravenous injection of rat CGRP in chicks induces a rapid hypercalcaemic response measured by the clearance of simultaneously injected $^{45}$Ca label from the plasma, which is sustained for at least 1 hour (48). This was most evident in chicks which are fully fed till the time of the experiment. Fasted chicks show a hypophosphataemic response and an increased plasma clearance of the $^{45}$Ca label. The effect of CGRP on $^{45}$Ca uptake into the chick skeleton has also been investigated using the microwave fixation technique (38). Both rat and chicken CGRP sequences cause transient (10 minutes) increases in $^{45}$Ca uptake into a variety of bone types. These responses were well developed in 22 h fasted chicks, but absent in fully fed birds. They are variable between bone types with the greatest responses in calvaria and vertebrae. Furthermore, with low doses of CGRP (70 pmole/100g body wt) reversal of the response is noticed in calvaria but not other bone types, while in fed animals this is the only response seen, again in calvaria. Interpretation of these findings is difficult but they suggest that in the chick, CGRP may have a variety of effects on bone and Ca metabolism. These may involve acute effects on net movement of Ca into and out of the skeleton. However, while consistent with reported changes in plasma $^{45}$Ca clearance, they seem too transient to account for them. Furthermore, they do not seem to account for the hypercalcaemic responses found, although some alternative target site such as the kidney (49) may be responsible for these.

Another member of the CT/CGRP family is amylin, a CGRP-like peptide from the pancreatic islet cells. This is the most potent non-calcitonin peptide thus far discovered, at least in mammalian assay systems (42). Amylin has been found to have no effect on plasma Ca and $^{45}$Ca concentrations when tested in the chick model (A K Ancill, Z A Bascal & C G Dacke, unpublished result).

Table 1 Acute effects of calcitrophic agents on chick Ca metabolism

| **Agent** | **Plasma Ca concentration** | **Plasma $^{45}$Ca concentration**[1] | **Uptake of $^{45}$Ca into bone** |
|---|---|---|---|
| Bovine parathyroid hormone (1-34) | ↑↑ | ↑↑ | ↓↓ |
| Parathyroid hormone-related peptide (1-34) | ↑ | ↑ | ↓↓↑ |
| 16,16-dimethyl prostaglandin $E_2$ | ↑↑↑ | ↑↑↑ | ↓↓↓ |
| Calcitonin | 0 | - | ↓↓ |
| Calcitonin gene-related peptide | ↑↑ | ↓↓ | ↓↑ |
| Amylin | 0 | 0 | - |
| IBMX (phosphodiesterase inhibitor) | ↑↑ | ↑↑ | ↓↓ |

Key: ↑ = increase, ↓ = decrease, 0 = no change from control value
[1] $^{45}$Ca administered simultaneously with the calcitrophic agent

## Conclusion

Pharmacological evidence is presented here which indicates that several factors, including PTH, $PGE_2$, CGRP, and possibly CT, can acutely influence exchange of Ca between avian blood and bone (Table 1). There are at least three putative mechanisms which could account for these effects.

A simple model has been proposed in which the rapid effects of PTH and $PGE_2$ on skeletal $^{45}$Ca uptake could be explained in terms of cAMP-mediated inhibitions of outwardly directed $Ca^{++}$ pumps located in the cell membranes of bone-lining cells (21). A criticism of this model is that it is metabolically wasteful, since the intracellular free Ca level would have to remain undisturbed in the face of a high rate of bi-directional transcellular $Ca^{++}$ flux. Moreover it is suggested that the extreme rapidity of Ca pumping which would be necessary is inconsistent with any known epithelial transport system (7).

An alternative model would involve rapid changes in blood circulation to the bone. Parathyroid hormone, CGRP and $PGE_2$ are all vasoactive and PTH and $PGE_2$, at least, can directly affect skeletal blood flow (28, 29). Our data (18) do not fully support this model, in that co-injected $^{125}$I-albumen or $^{14}$C-mannitol markers do not show changes in bone consistent with a change in blood or diffusible space, following PTH or 16,16-dimethyl $PGE_2$ treatment. PTH and $PGE_2$ also caused decreased clearance of $^{85}Sr$ into rat bone which was inconsistent with observed changes in bone blood flow (50).

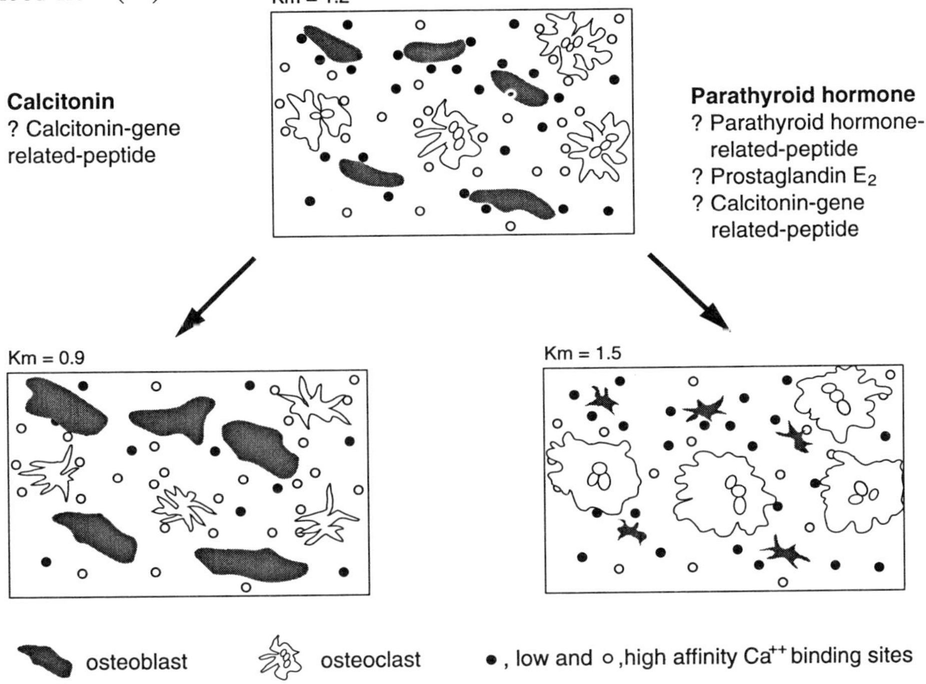


Fig. 1 Effect of calcitrophic agents on spatial relationships of osteoclasts and osteoblasts on a bone surface. Km=1.2, normocalcaemic situation with an equal number of high and low affinity $Ca^{++}$ binding sites; Km=0.9, hypocalcaemic situation where agents cause exposure of high-affinity binding sites due to osteoclast shrinkage; Km=1.5, hypercalcaemic situation where exposure to agents causes exposure of low-affinity binding sites due to shrinkage of osteoblasts. Apparent Km values are given in terms of extra cellular free Ca concentration (m moles $l^{-1}$) for convenience. (Modified after 7).

Recently Bronner and Stein (7) proposed that Ca influx into and out of the exchangeable bone surface is acutely controlled by osteoclastic and osteoblastic shape changes in response to Ca-regulating factors (Fig. 1). They propose that at least two classes of $Ca^{++}$-binding sites exist on the bone surface, i.e. those with high and those with low affinity. They suggest that osteoblasts may normally cover, and then shrink away from, the low-affinity sites in response to PTH, thus allowing Ca effluxes to occur. In contrast, high-affinity sites are normally occluded by osteoclasts which retract and move away under the influence of CT, thereby enabling an influx of Ca.

Currently, this model provides the most attractive explanation for the actions of calcitrophic agents on chicken Ca metabolism.

## References


1 Dacke C G (1979) *Calcium Regulation in Sub-Mammalian Vertebrates.* London: Academic Press

2 Dacke C G (1989) Eicosanoids, steroids and miscellaneous hormones. *Vertebrate Endocrinology: Fundamentals and Biomedical Implications* vol 3, pp 171-210. Eds P K T Pang & M P Schreibman. New York: Academic Press

3 Tippins J R, Morris H R, Panico M, Etienne T, Bevis P, Girgis S, MacIntyre I, Azria M & Attinger M (1984) The myotropic and plasma-calcium modulating effects of calcitonin gene-related peptide (CGRP). *Neuropeptides* **4** 425-434

4 Bronner F (1982) Calcium homeostasis. *Disorders of Mineral Metabolism* vol 2, pp 43-97. Eds F Bronner & J W Coburn. New York: Academic Press

5 Shaw A J, Whitaker G & Dacke C G (1989) Kinetics of rapid $^{45}$Ca uptake into chick skeleton *in vivo*: effects of microwave fixation. *Quarterly Journal of Experimental Physiology* **74** 907-915

6 Parsons J A & Robinson C J (1971) Calcium shift into bone causing transient hypocalcaemia after injection of parathyroid hormone. *Nature* **230** 581-582

7 Bronner F & Stein W D (1992) Modulation of bone calcium binding sites regulates plasma calcium: an hypothesis. *Calcified Tissue International* **50** 483-489

8 Hurwitz S (1989) Parathyroid hormone. *Vertebrate Endocrinology: Fundamentals and Biomedical Implications* vol 3, pp 45-77. Eds P K T Pang & M P Schreibman. New York: Academic Press

9 Polin D, Sturkie P D & Hunsaker W (1957) The blood calcium response of the chicken to parathyroid extracts. *Endocrinology* **60** 1-5

10 Candlish J K & Taylor T G (1970) The response-time to the parathyroid hormone in the laying fowl. *Journal of Endocrinology* **48** 143-144

11 Bronner F, Pansu D & Stein W D (1986) An analysis of intestinal calcium transport across the rat intestine. *American Journal of Physiology* **250** G561-G569

12 Laverty G & Clark N B (1989) The kidney. *Vertebrate Endocrinology: Fundamentals and Biomedical Implications* vol 3, pp 277-317. Eds P K T Pang & M P Schreibman. New York: Academic Press

13 Dacke C G & Kenny A D (1972) Avian bioassay method for parathyroid hormone. *Endocrinology* **92** 463-470

14 Parsons J A, Reit B & Robinson C J (1973) Bioassay for parathyroid hormone using chicks. *Endocrinology* **92** 454-462

15 Kenny A D & Dacke C G (1974) The hypercalcaemic response to parathyroid hormone in Japanese quail. *Endocrinology* **62** 15-23

16 Shaw A J, Whitaker G & Dacke C G (1989) Kinetics of rapid $^{45}$Ca uptake into chick skeleton *in vivo*. *Quarterly Journal of Experimental Physiology* **74** 907-915

17 Shaw A J & Dacke C G (1985) Evidence for a novel inhibition of calcium uptake into chick bone in response to bovine parathyroid hormone (1-34) or 16,16-dimethyl prostaglandin $E_2$ *in vivo*. *Journal of Endocrinology* **105** R5-R8

18 Dacke C G & Shaw A J (1987) Studies of the rapid effects of parathyroid hormone and prostaglandins on $^{45}$Ca uptake into chick and rat bone *in vivo*. *Journal of Endocrinology* **115** 369-377

19 Tothill P & MacPherson J N (1978) *Post mortem* migration of bone seeking radio nuclides in the rat and rabbit and its effects on estimates of bone uptake. *Clinical Science and Molecular Medicine* **55** 221-223

20 Shaw A J & Dacke C G (1989) Cyclic nucleotides and the rapid inhibitions of bone $^{45}Ca$ uptake in response to bovine parathyroid hormone and 16,16-dimethyl prostaglandin $E_2$ in chicks. *Calcified Tissue International* **44** 209-213

21 Dacke C G & Shaw A J (1988) Effects of synthetic bovine parathyroid hormone (1-34) and its analogues on $^{45}Ca$ uptake and adenylate cyclase activation in bone and plasma calcium levels in the chick. *Quarterly Journal of Experimental Physiology* **73** 573-584

22 Teti A, Rizzoi R & Zambonin-Zallone A (1991) Parathyroid hormone binding to cultured avian osteoclasts. *Biochemical and Biophysical Research Communications* **174** 1217-1222

23 Ali N N, Melhuishe P B, Boyde A, Bennet A & Jones S (1990) Parathyroid hormone, but not prostaglandin $E_2$, changes the shape of osteoblasts maintained on bone *in vitro*. *Journal of Bone and Mineral Research* **5** 115-121

24 Miller S C (1978) Rapid activation of the medullary bone osteoclast cell surface by parathyroid hormone. *The Journal of Cell Biology* **76** 615-618

25 Pandala S & Gay C V (1990) Effects of parathyroid hormone, calcitonin, and dibutyryl-cyclic AMP on osteoclast area in cultured chick tibiae. *Journal of Bone and Mineral Research* **5** 701-705

26 Pang P K T, Zhang R H & Yang M C M (1984) Hypotensive action of parathyroid hormone in chicken. *Journal of Experimental Zoology* **232** 691-696

27 Sham J K S & Pang P K T (1986) Calcium-dependent chronatropic action of parathyroid hormone (1-34) in isolated atria of Japanese quail. *General and Comparative Endocrinology* **64** 250-253

28 Boelkins J N, Mazurkiewicz M, Mazur P E & Mueller W J (1976) Changes in blood flow to bone during hypocalcemic and hypercalcemic phases of the response to parathyroid hormone. *Endocrinology* **98** 403-412

29 Driessens M & Vanhoutte P M (1981) Effect of calcitonin, hydrocortisone and parathyroid hormone on canine blood vessels. *American Journal of Physiology* **241** H92-H94

30 Mallette L E (1991) The parathyroid polyhormones: new concepts in the spectrum of peptide hormone action. *Endocrine Reviews* **12** 110-117

31 Schermer D T, Chan S D H, Bruce R, Nissenson R A, Wood W I & Strewler G J (1991) Chicken parathyroid hormone-related protein and its expression during embryologic development. *Journal of Bone and Mineral Research* **6** 149-155

32 Pines M, Granot I & Hurwitz S (1990) Cyclic AMP-dependent inhibition of collagen synthesis in avian epiphysial cartilage cells: effect of chicken and human parathyroid hormone and parathyroid hormone-related peptide. *Bone Mineralisation* **9** 23-34

33 Copp D H & Kline L W (1989) Calcitonin *Vertebrate Endocrinology: Fundamentals and Biomedical Implications* vol 3, pp 79-103. Eds P K T Pang & M P Schreibman. New York: Academic Press

34 Miyaura C, Nagata N & Suda T (1981) Failure to demonstrate the stimulatory effect of calcitonin on cyclic AMP accumulation in avian bone *in vitro*. *Endocrinologica Japonica* **28** 403-408

35 Ito M B, Schraer H & Gay C V (1985) The effects of calcitonin, parathyroid hormone and prostaglandin $E_2$ on cyclic AMP levels of isolated osteoclasts. *Comparative Biochemistry and Physiology* **81A** 653-657

36 Eliam M C, Baslé M, Bouziar B, Bielakoff J, Moukhtar M & deVernejoul M C (1988) Influence of blood calcium on calcitonin receptors in isolated chick osteoclasts. *Journal of Endocrinology* **119** 243-248

37 Rifkin B R, Auszmann J M, Kleckner A P, Vernillo A T & Fine A S (1988) Calcitonin stimulates cAMP accumulation in chick osteoclasts. *Life Sciences* **42** 799-804

38 Ancill A K, Bascal Z A, Whitaker G & Dacke C G (1991) Calcitonin gene-related peptide promotes transient radio calcium uptake into chick bone *in vivo*. *Experimental Physiology* **76** 143-146

39 Rosenfield M G, Mermod J J, Amara S G, Swanson L W, Sawchenko P E, Rivier J, Vale W W & Evans R M (1983) Production of a novel peptide encoded by the calcitonin gene via tissue specific RNA processing *Nature* **304** 129-135

40 Bjurholm A, Kreicbergs A, Brodin E & Schultzberg M (1985) Substance P- and CGRP-immunoreactive nerves in bone. *Peptides* **9** 165-171

41 Kameda Y (1991) Immunocytochemical localization and development of multiple kinds of neuropeptides and neuroendocrine proteins in the chick ultimobranchial gland. *Journal of Comparative Neurology* **304** 373-386

42 Zaidi M, Moonga B, Bevis J R, Bascal Z A & Breimer M D (1990) The calcitonin gene peptides: biology and clinical relevance. *Critical Reviews in Clinical Laboratory Science* **28** 109-174

43 Goltzman D & Mitchell J (1985) Interaction of calcitonin and calcitonin gene-related peptide at receptor sites in target tissues. *Science* **227** 1343-1345

44 Zaidi M, Chambers T J, Gaines Das R E, Morris H R & MacIntyre I (1987) A direct action of human calcitonin gene-related peptide on isolated human osteoclasts. *Journal of Endocrinology* **155** 511-518

45 Roos B A, Fischer J A, Pignat W, Allander C B & Raisz L J (1986) Evaluation of *in vivo* and *in vitro* calcium regulating actions of non calcitonin peptides produced via calcitonin gene expression. *Endocrinology* **118** 46-51

46 Zaidi M, Chambers T J, Bevis P J R, Beacham J L, Gaines Das R E & MacIntyre (1988) Effect of the peptides from the calcitonin genes on bone and bone cells. *Quarterly Journal of Experimental Physiology* **73** 471-486

47 Bevis P J R, Zaidi M & MacIntyre I (1990) A dual effect of calcitonin gene-related peptide on plasma calcium levels in the chick. *Biochemical and Biophysical Research Communications* **169** 846-850

48 Ancill A K, Bascal Z A, Whitaker G & Dacke C G (1990) Effects of rat and chicken calcitonin gene-related peptides (CGRP) on calcium metabolism in chicks. *Regulatory Peptides* **30** 231-238

49 Zaidi M, Datta H K & Bevis P J R (1990) Kidney: a target organ for calcitonin gene-related peptide. *Experimental Physiology* **75** 27-32

50 Cochrane E & McCarthy I D (1991) Rapid effects of parathyroid hormone (1-34) and prostaglandin $E_2$ on bone blood flow and strontium clearance in the rat *in vivo*. *Journal of Endocrinology* **131** 359-365

Avian Endocrinology
Ed. P J Sharp, pp 249-260
Journal of Endocrinology Ltd, Bristol (1993)

# Effect of hormones and growth factors on cell proliferation and extracellular matrix protein synthesis by avian epiphyseal growth plate chondrocytes

M Pines[1], E Monsonego[1], T Shalom-Barak[1], O Halevy[2] and S Hurwitz[1]
[1]Institute of Animal Science, Agricultural Research Organization, The Volcani Center, Bet Dagan, Israel; [2]Faculty of Agriculture, The Hebrew University of Jerusalem, Rehovot, Israel

## Introduction

Chondrocyte proliferation within the cartilaginous growth plate is the first step in bone elongation. This process of cell replication and the synthetic activities of the chondrocytes are regulated by a variety of hormones and growth factors. The function of the former is to transduce systemic information about the metabolic state of the animal to the growing cell, while the latter may be responsible for local modulation of cellular activities important in the development of proper bone architecture. Several peptide hormones and growth factors, such as parathyroid hormone (PTH) and insulin-like growth factor (IGF-I), stimulate cell proliferation and may actually cause de-differentiation. Others, such as retinoic acid and steroid hormones, facilitate differentiation and inhibit cell proliferation. This review is concerned with the effects of growth hormone (GH), epidermal growth factor (EGF), and their interactions, on cell proliferation and synthesis of some matrix proteins.

## Morphology of the growth plate

The growth plate is populated by chondrocytes, which are arranged in columns parallel to the long axis of the bone. Passing from the proximal to distal border of the plate, several zones are found: (a) the resting or reserve zone, containing stem cells; (b) the proliferative zone, with stacks of flattened cells (cell division occurs mostly in the longitudinal direction and leads to cell column formation); (c) the hypertrophic zone, with hypertrophic chondrocytes; and (d) the degenerative zone, with a partially calcified matrix and invading capillaries (1, 2). Each chondrocyte, once formed, remains in a spatially fixed location throughout its cellular life cycle.

Chondrocytes located in the different regions differ in their morphology, secretion of extracellular matrix components and activities of various enzymes (3, 4). Synthesis of collagen type II is characteristic of chondrocytes in their proliferative stage (5), whereas that of collagen type X is restricted to the hypertrophic state (6-8). Alkaline phosphatase activity is of particular interest, since its appearance marks the onset of hypertrophy and calcification (9, 10). Chondrocyte maturation is accompanied also by changes in the rate of proteoglycan synthesis and in the size of proteoglycan monomers (11-13).

One of the non-collagenous matrix proteins secreted by hypertrophic chondrocytes is osteopontin. This protein is a glycosylated phosphoprotein with a large number of acidic amino acid residues containing a cell-binding site defined by the RGD sequence (14). Osteopontin is secreted into the extracellular matrix concomitantly with mineral deposition, which is consistent with the hypothesis that it may serve as a bridge between the cell and apatitic mineral (15). In addition, chondrocytes from different maturational stages possess a variable number of receptors for vitamin D hormones (16), parathyroid hormone (17), TGF-β (4), basic fibroblast growth factor (18) and other hormones and growth factors.

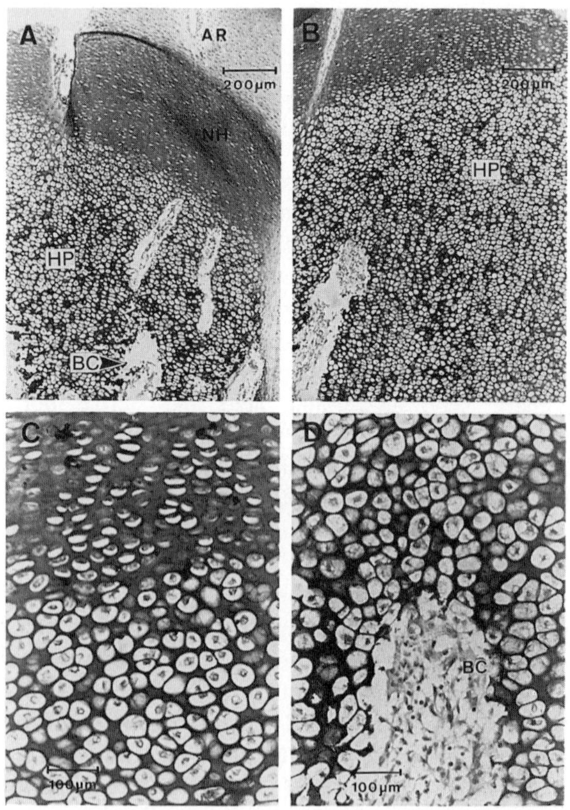


Fig. 1 Cross sections of the growth plate from a 2-week-old chicken through A) the proximal section showing articular cartilage (AR), non-hypertrophic cartilage (NH), hypertrophic cartilage (HP) and invading blood capillaries (BC) and B) the distal section showing the large number of cells in the hypertrophic zone. At a higher magnification details can be seen of C) the border between the hypertrophic and the non-hypertrophic zones and D) of the invasion of a blood capillary with the associated destruction of cartilage.

## The avian growth plate

Compared with the mammalian growth plate, the avian growth plate contains longer columns of cells which become randomly orientated in the hypertrophic and calcified zones (Fig. 1). Furthermore, more cells are found in each zone and the metaphyseal blood vessels penetrate deeper into the growth plate (19, 20). The growth plate of a 4-7 week-old chicken contains approximately 200 cells per column (21), compared with 25 cells in the rat (22). In Leghorns, the transition between the zones is more orderly than in broilers (23). A comparison of turkey and rabbit growth plates shows that the individual columns are less distinguishable and are not widely separated (24). Further, 50% more columns are found per unit of lateral measurement in the turkey than in the rabbit growth plate, and 60% more cells per unit of vertical measurement (24). A wider proliferative zone has been found in a heavy- than in a light-bodied turkey strain (25).

## Longitudinal bone growth

The rate of longitudinal bone growth is modulated by the rate of production of new chondrocytes in the proliferative zone, by the diameter of chondrocytes in the most distal portion of the hypertrophic zone, and by matrix production in the transversa septa. The process begins with the division of the stem cells at the top of each column to produce the cells of the proliferative zone. These proliferative chondrocytes divide, with the highest rate of division occurring in the mid region of the proliferative zone. In chickens and turkeys, age-dependent growth rates were correlated with the width of the hypertrophic zone rather than with that of the proliferative zone (26), suggesting that, as in mammals (27), cartilage cell hypertrophy is the main determinant of longitudinal growth.

The importance of a normal regulation of chondrocyte metabolism is suggested by the gross abnormalities caused by defective regulation of growth and differentiation. Dwarfism, rickets and dyschondroplasia are examples of such disorders (19, 28).

## Avian chondrocytes in culture

Investigation of chondrocyte metabolism is essential for the identification and manipulation of the pathways of normal and abnormal development of bone. Chondrocytes, which are part of the complex spatial organization of the growth plate, are influenced by chemical signals, cytokines, systemic hormones, growth factors and physical forces emanating from the surrounding tissues and the cartilage itself. *In vitro* cultures of isolated chondrocytes are ideally suited for studying *de novo* synthesis and deposition of extracellular matrix compounds and cell proliferation pathways. The existence of chondrocyte subpopulations in the growth plate and the possibility of culturing cells at various densities to test the influence of cell-cell interactions and cell-derived factors is additional justification for using cultured chondrocytes. Further, the absence of an organized matrix allows evaluation of defined substances for their individual effects. Various methods have been established for preparing and culturing epiphyseal growth plate chondrocytes (29).

Chondrocytes grown in culture may deviate from their original state of differentiation. For example, in contrast to chondrocytes *in situ*, chondrocytes in culture express the gene coding for collagen type I. This must be considered when the *in vitro* preparation is used as a model system. If the experimental objective is to determine the physiological responses of chondrocytes, the state of differentiation of the cells must be defined using criteria including cell morphology, specific matrix protein synthesis and enzyme activities.

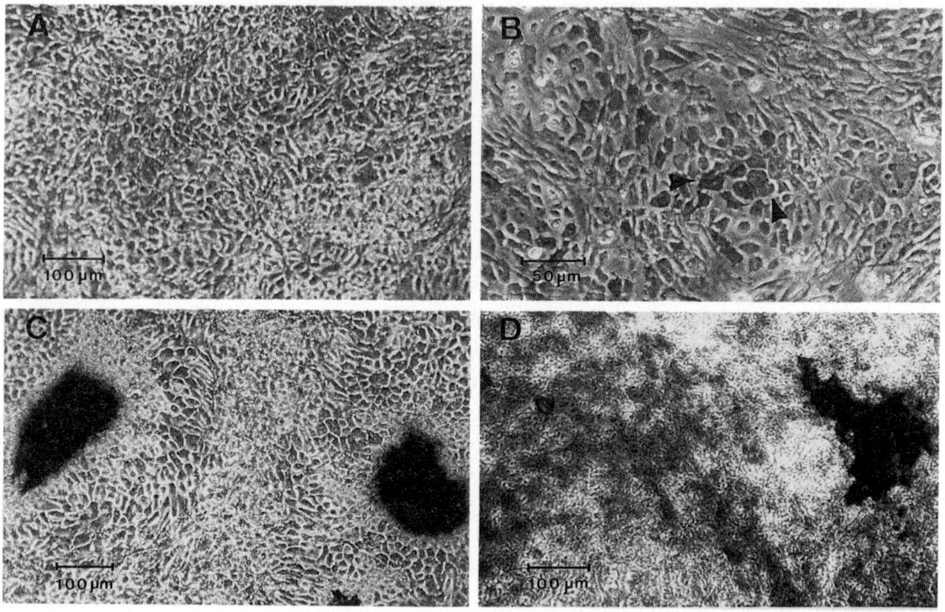


Fig. 2 Avian growth plate chondrocytes in culture showing A) a confluent culture without any areas of specialization, B) the appearance of differentiated individual and groups of cells identified by positive alkaline phosphatase (arrows) and C, D) the progressive formation of nodules containing alkaline phosphatase-positive cells.

Avian epiphyseal growth-plate chondrocytes in culture are polygonal in shape and grow as a monolayer (20, 30). When cultured with 5% fetal calf serum, the cells proliferate with a doubling time of approximately 40-48 h (30), express the collagen type II gene (31), and show no alkaline phosphatase activity (Fig. 2A), suggesting that they are in the proliferative state. Addition of ascorbic acid (5mg/ml) daily for a week, without replacing the medium, facilitates differentiation, leading to the expression of an alkaline phosphatase activity (Fig. 2 B-D).

**Epidermal growth factor**

Epidermal growth factor (EGF) is a 53-amino acid polypeptide characterized by three disulphide bonds, which is derived from a large precursor protein (32). The peptide binds to a cell membrane receptor, a 170kDa transmembrane glycoprotein (33) which, like other growth factor receptors, has intrinsic tyrosine kinase activity important in signal transduction. Epidermal growth factor binding to the receptor stabilizes an oligomeric configuration of the receptor, activates tyrosine kinase and induces receptor autophosphorylation (33, 34). In addition, EGF binding evokes phosphorylation of a variety of cellular substrates such as phospholipases and protein kinases (35, 36), which may stimulate the phosphoinositide signalling pathway by activating protein kinase C (37). Thus, EGF binding to its receptor may stimulate a variety of tyrosine as well as serine and threonine phosphorylations. It has been postulated that the complex of EGF and EGF receptor is internalized and that some EGF is transported to the nucleus and binds to the nuclear receptor. This EGF-nuclear receptor complex may bind to DNA to modulate gene transcription (38). Both *in vivo* and in cell culture systems, EGF enhances proliferation and inhibits differentiation of target cells (39).

**Epidermal growth factor and chondrocytes**

In chondrocytes, EGF stimulates DNA synthesis and inhibits proteoglycan synthesis (40, 41). Epidermal growth factor receptors have been identified in avian and rabbit chondrocytes (41, 42). In the former, the number of EGF receptors is inversely correlated with chondrocyte differentiation. Although EGF synthesis has not been demonstrated in bone or cartilage, it has been suggested that the multiple EGF-like domains of extracellular matrix proteins are signals for cellular growth and differentiation (43, 44).

Incubation of avian chondrocytes with EGF results in the appearance of phosphorylated osteopontin in the extracellular milieu. No EGF-dependent osteopontin phosphorylation has been observed within the cells, suggesting that osteopontin is secreted immediately after phosphorylation. Fibroblast growth factor and TGF-β, but not IGF-I or IGF-II, also cause phosphorylation of osteopontin. Whereas EGF affects osteopontin phosphorylation, activation of protein kinase C by phorbol esters augments osteopontin gene expression and osteopontin synthesis. It should be noted that although osteopontin synthesis is associated with hypertrophic chondrocytes, proliferating avian chondrocytes also synthesize osteopontin.

**Growth hormone and bone growth**

Early work showed that hypophysectomy in mammalian species results in a time-dependent decrease in the rate of longitudinal bone growth. Local administration of GH into the proximal tibial epiphyseal plate stimulates longitudinal growth on the injected side (45). GH receptors are found throughout the growth plate in the germinative, proliferative and hypertrophic cell layers (46). Furthermore, incubation of chondrocytes with GH stimulates IGF-I gene expression and hence IGF-I production. This locally produced IGF-I interacts with receptors on the proliferative chondrocytes by autocrine or paracrine mechanisms (47). Thus GH appears to act on longitudinal bone growth in mammals by direct stimulation of the differentiated epiphyseal growth plate precursor cells, by stimulation of IGF-I production, and enhancement of the cell responsiveness to IGF-I.

The mechanism by which GH binds to its receptors has recently been elucidated, and the hGH epitopes which interact with the extracellular part of the hGH receptor, also known as hGH binding protein (hGHBP), have been identified. Two separate binding sites in the hGH molecule interact simultaneously with two molecules of hGHBP (48, 49), to form a 1:2 hGH:hGHBP complex (50).

The function of GH in growth regulation in avian species is not clear. Lines of birds with the slow rates of growth have the higher concentrations of circulating GH than fast growing lines (51, 52). Restriction of growth (53) or protein intake (54) increases circulating levels of GH. Finally, injection of GH in young broilers does not alter growth performance (55, 56) or longitudinal bone growth and growth-plate width (57), although in some cases a transient increase in body weight is observed (58). Thus, GH cannot be regarded as a simple growth-promoting hormone in birds.

## Growth hormone and avian chondrocytes

In contrast to mammalian chondrocytes, GH alone does not affect proliferation or metabolic activity of avian growth-plate chondrocytes (59). Cloning of the gene encoding the chicken growth hormone receptor (60) has facilitated the demonstration of the presence of GH receptors. Using this clone as a probe, the expression of the GH receptor (GH-R) gene in avian growth-plate cartilage has been investigated by Northern blot analysis (61). A single transcript of ~5.4 kb was found in cultured growth-plate chondrocytes as well as in growth-plate extracts. Chondrocytes bear ~1300 GH receptors/cell with a dissociation constant of approximately 0.1 nM as estimated by binding studies using radiolabelled human GH (hGH). Chicken GH (cGH) has a lower affinity for avian chondrocyte receptors than hGH, is unable to bind to the mammalian GH-R, and does not form a complex with purified liver hGH-binding-protein. In avian chondrocytes growth hormone increases the expression of the gene encoding the proto-oncogene c-*myc* and increases the synthesis of a yet unidentified 45 kDa matrix protein. In all cases, cGH is more potent than hGH despite its lower binding affinity for the GH receptor. Chicken GH also stimulates the expression of the genes encoding for collagen type II and IGF-I, both associated with proliferative chondrocytes. The disparity between the affinity of chicken and human GH for the receptors on the one hand, and their potency in eliciting physiological responses on the other, can be exploited in studies of the structure-function relationship of the hormone.

## Conclusion

Chondrocyte proliferation and production of extracellular matrix proteins are complex processes controlled through the actions of several hormones, growth factors and proto-oncogenes (Fig. 3). Interactions between the regulatory components must exist to allow precise control of these pathways when cells are exposed to multiple signals. The effect of one factor may depend on the presence of other local growth factors: for example, TGF-β and FGF exert a synergistic effect on proliferation of avian chondrocytes (62).

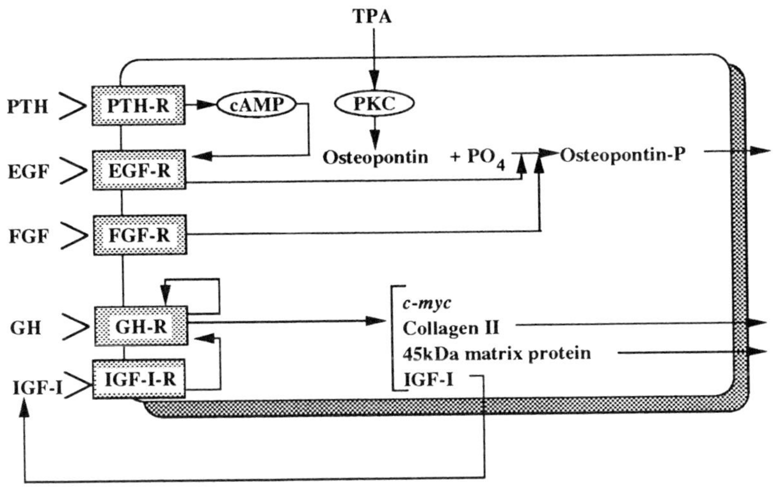


Fig. 3 Summary of interactions among hormones and growth factors involved in the growth and differentiation of avian growth-plate chondrocytes. PTH, parathyroid hormone; EGF, epidermal growth factor; FGF, fibroblast growth factor; GH, growth hormone; IGF-I, insulin-like growth factor; TPA, phorbol ester; PKC, protein kinase C; cAMP, adenosine 3′,5′ cyclic phosphoric acid; R, receptor.

One mechanism by which an agent can affect the response of one cell to another, is to modulate the expression of the genes encoding their respective receptors. In avian chondrocytes, parathyroid hormone, which inhibits collagen synthesis (63) and acts as a mitogen (30), also attenuates EGF receptor gene expression (41). This effect of PTH results in a decrease in the EGF-dependent osteopontin phosphorylation. In rabbit chondrocytes, EGF reduces the number of PTH receptors (64). Atrial natriuretic peptide, a hormone whose action is mediated by cGMP (65) and which inhibits avian chondrocyte proliferation (30), attenuates EGF receptor gene expression. Chicken GH receptor is attenuated by GH itself, whereas in mammalian chondrocytes GH up-regulates it own receptors (66), while IGF-I up-regulates chicken growth hormone receptor.

Although GH increases the expression of genes encoding c-*myc* and IGF-I, both of which are associated with the initiation of cell proliferation, GH acting alone does not affect avian cell proliferation. This may be because other growth factors are necessary in order to obtain any effect of GH on cell proliferation. In support of this view a synergistic effect of GH and EGF has been observed on avian growth plate chondrocytes (61). This may be due to the effect of GH on EGF receptor binding and mRNA levels, as was observed in other systems (67, 68). Alternatively, a GH-induced increase in IGF production may be associated with the secretion of an IGF-I binding protein which may bind the IGF-I and prevent the growth factor from stimulating cell proliferation (69). This view is supported by preliminary experiments indicating the presence of a specific IGF-I binding protein in the medium of cultured chondrocytes.

## References

1 Hunziker E B, Schenk R K & Cruz-Orive L M (1987) Quantitation of chondrocyte performance in growth-plate cartilage during longitudinal bone growth. *Journal of Bone Joint Surgery* **69A** 162-173

2 Poole A R, Matsui Y, Hinek A & Lee R E (1989) Cartilage macromolecules and the calcification of cartilage matrix. *Anatomical Record* **224** 167-179

3 Reinholt F P, Engfelt B, Heinegard D & Hjerpe A (1985) Proteoglycans and glycosaminoglycans of epiphyseal chondrocytes in florid and healing low phosphate, vitamin D deficiency rickets. *Collagen and Related Research* **5** 55-64

4 O'Keefe R J, Rosier R N & Puzas J E (1990) Differential expression of biological effects in maturationally distinct subpopulations of growth plate chondrocytes. *Connective Tissue Research* **24** 53-66

5 Sandberg M & Vuorio E (1987) Localization of type I, II and III collagen mRNAs in developing human skeletal tissues by *in situ* hybridization. *Journal of Cell Biology* **104** 1077- 1084

6 Kwan A P L, Freemont A J & Grant M E (1986) Immunoperoxidase localization of type X collagen in chick tibiae. *Bioscience Reports* **6** 155-162

7 Kwan A P L, Dickson I R, Freemont A J & Grant M E (1989) Comparative studies of type X collagen expression in normal and rachitic chicken epiphyseal cartilage. *Journal of Cell Biology* **109** 1849-1856

8 Haynes S J & Goff J (1991) Distribution of type X collagen in tibiotarsi of broiler chickens with vitamin D deficiency. *Calcified Tissue International* **49** 188-192

9 Vaananen H K (1980) Immunohistochemical localization of alkaline phosphatase in the chicken epiphyseal growth plate. *Histochemistry* **65** 143-148

10 Morris D C, Randall J C, Stechschulte D J, Zeiger S, Mansur D & Anderson H C (1990) Enzyme cytochemical localization of alkaline phosphatase in cultures of chondrocytes derived from normal and rachitic rats. *Bone* **11** 345-352

11 Campo R D & Romano J E (1986) Changes in cartilage proteoglycans associated with calcification. *Calcified Tissue International* **39** 175-184

12 Makower A M, Wroblewski J & Pawlowski A (1988) Effects of IGF-I, EGF and FGF on proteoglycans synthesized by fractionated chondrocytes of rat rib growth plate. *Experimental Cell Research* **179** 498-506

13 Pawlowski A, Makower A M, Madsen K, Wroblewski J & Friberg U (1986) Cell fractions from rat rib growth cartilage: biochemical characterization of matrix molecules. *Experimental Cell Research* **164** 211-222

14 Castagnola P, Bet P, Quarto R, Gennari M & Cancedda R (1991) cDNA cloning and gene expression of chicken osteopontin. Expression of osteopontin mRNA in chondrocytes is enhanced by trypsin treatment of cells. *Journal of Biological Chemistry* **266** 9944-9949

15 Gerstenfeld L C, Gotoh Y, McKee M D, Nanci A, Landis W J & Glimcher M J (1990) Expression and ultrastructural immunolocalization of a major 66kDa

phosphoprotein synthesized by chicken osteoblasts during mineralization *in vitro*. *Anatomical Record* **228** 93-103
16 Suda S, Takahashi N, Shinki T, Horiuchi N, Yamaguchi A, Yoshiki S, Enomoto S & Suda T (1985) 1α25-dihydroxyvitamin D receptors and their action in embryonic chick chondrocytes. *Calcified Tissue International* **37** 82-90
17 Barling P M & Bibby N J (1985) Study of the localization of [3H] bovine parathyroid hormone in bone light microscope autoradiography. *Calcified Tissue International* **37** 441-446
18 Iwamoto M, Shimazu A, Nakashima K, Suzuki F & Kato Y (1991) Reduction in basic fibroblast growth factor receptor is coupled with terminal differentiation of chondrocytes. *Journal of Biological Chemistry* **266** 461-467
19 Leach R M & Gay C V (1987) Role of epiphyseal cartilage in endochondral bone formation. *Journal of Nutrition* **117** 784-790
20 Pines M & Hurwitz S (1991) The role of the growth plate in longitudinal bone growth. *Poultry Science* **70** 1806-1814
21 Howlett C R (1979) The fine structure of the proximal growth plate of avian tibia. *Journal of Anatomy* **128** 377-399
22 Kember N F (1960) Cell division in endochondral ossification. A study of cell proliferation in rat bones by the method of tritiated thymidine autoradiography. *Journal of Bone Joint Surgery* **42B** 824-839
23 Reiland S, Olsson S E, Poulos P W & Elwinger K (1978) Normal and pathological skeletal development in broiler and Leghorn chickens. A comparative investigation. *Acta Radiologia Supplement* **358** 277-298
24 Wise D R & Jennings A R (1973) The development and morphology of the growth plates of two long bones of the turkey. *Research in Veterinary Science* **14** 161-166
25 LeBlanc B, Wyers M, Cohn-Bendit F, Legall J M, Thibault E & Florent J M (1986) Histology and histomorphometry of the tibia growth in two turkey strains. *Poultry Science* **65** 1787-1795
26 Hurwitz S, Livne E, Plavnik I, Pines M & Silberman M (1992) Tibial development in turkeys and chickens as affected by early age feed restriction. *Growth, Development and Aging* **59** 191-203
27 Breur G J, VanEnkevort B A, Farnum C E & Wilsman N J (1991) Linear relationship between the volume of hypertrophic chondrocytes and the rate of longitudinal bone growth in growth plates. *Journal of Orthopedic Research* **9** 348-359
28 Bashey R I, Leach R M, Gay C V & Jimenez S A (1989) Type X collagen in avian tibial dyschondroplasia. *Laboratory Investigation* **60** 106-112
29 Leach R M & Rosselot G E (1992) The use of avian epiphyseal chondrocytes for *in vitro* studies of skeletal metabolism. *Journal of Nutrition* **122** 802-805
30 Pines M & Hurwitz S (1988) The effect of parathyroid hormone and atrial natriuretic peptide on cyclic-nucleotide production and proliferation of avian epiphyseal growth plate chondroprogenitor cells. *Endocrinology* **123** 360-365
31 Granot I, Halevy O, Hurwitz S & Pines M (1993) Halofuginone: an inhibitor of collagen type I synthesis. *Biochimica et Biophysica Acta* **1156** 107-112
32 Fisher D A & Lakshmanan J (1990) Metabolism and effects of epidermal growth factor and related growth factors in mammals. *Endocrine Reviews* **11** 418-442

33 Schlessinger J, Ullrich A, Honegger A M & Moolenaar W H (1988) Signal transduction by epidermal growth factor receptor. *Cold Spring Harbor Laboratory NY* **LIII** 515-519

34 Carpenter G (1987) Receptors for epidermal growth factor and other polypeptide mitogens. *Annual Review Biochemistry* **56** 881-914

35 Margolis B, Rhee S G, Felder S, Mervic M, Lyall R, Levitzki A, Ullrich A, Zilberstein A & Schlessinger J (1989) EGF induces phosphorylation of Phospholipase C-II: a potential mechanism for EGF receptor signalling. *Cell* **57** 1101-1107

36 Ackerman P, Glover C V C & Osheroff N (1990) Stimulation of casein kinase II by epidermal growth factor: relationship between the physiological activity of the kinase and phosphorylation site of its subunit. *Proceedings of the National Academy of Sciences of the USA* **87** 821-825

37 Susa M, Olivier A R, Fabbro D & Thomas G (1989) EGF induces biphasic S6 kinase activation: late phase is protein kinase C-dependent and contributes to mitogenicity. *Cell* **57** 817-824

38 Rakowicz-Szulczynska E M, Otwiąska D, Rodeck U & Koprowski H (1989) Epidermal growth factor (EGF) and monoclonal antibody to cell surface EGF receptor bind to the same chromatin receptor. *Archives of Biochemistry and Biophysics* **268** 456-464

39 Knecht M & Catt K J (1983) Modulation of cAMP-mediated differentiation in ovarian granulosa cells by epidermal growth factor and platelet-derived growth factor. *Journal of Biological Chemistry* **258** 2789-2794

40 Hiraki Y, Inoue H, Kato Y, Fukuya M & Suzuki F (1987) Combined effects of somatomedin like growth factor with fibroblast growth factor or epidermal growth factor in DNA synthesis in rabbit chondrocytes. *Molecular and Cellular Biochemistry* **76** 185-193

41 Halevy O, Schindler D, Hurwitz S & Pines M (1991) Epidermal growth factor receptor gene expression in avian epiphyseal growth-plate cartilage cells: effect of serum, parathyroid hormone and atrial natriuretic peptide. *Molecular and Cellular Endocrinology* **75** 229-235

42 Kinoshita A, Takigawa M & Suzuki F (1992) Demonstration of receptors for epidermal growth factor on cultured rabbit chondrocytes and regulation of their expression by various growth and differentiation factors. *Biochemical and Biophysical Research Communications* **183** 14-20

43 Baldwin C T, Reginato A M & Prockop D J (1989) A new epidermal growth factor-like domain in the human core protein for the large cartilage-specific proteoglycan. *Journal of Biological Chemistry* **264** 15747-15750

44 Stanescu V, Chaminade F & Do Pham T (1991) Immunological detection of the EGF-like domain of the core proteins of large proteoglycans from human and baboon cartilage. *Connective Tissue Research* **26** 283-293

45 Isaksson O P G, Jansson J O & Gause I A M (1982) Growth hormone stimulates longitudinal bone growth directly. *Science* **216** 1237-1239

46 Bernard R, Haynes K M, Werther G A & Waters M J (1988) The ontogeny of growth hormone receptors in the rabbit tibia. *Endocrinology* **122** 2562-2569

47 Isaksson O P G, Lindahl A, Nilsson A & Isgaard J (1987) Mechanism of the stimulatory effect of growth hormone on longitudinal bone growth. *Endocrine Reviews* **8** 426-438

48 Cunningham B C & Wells J A (1991) Rational design of receptor-specific variants of human growth-hormone. *Proceedings of the National Academy of Sciences of the USA* **88** 3407-3411
49 Cunningham B C, Ultsch M, De-Vos A M, Mulkerrin M G, Clauser K R & Wells J A (1991) Dimerization of the extracellular domain of the human growth hormone receptor by a single hormone molecule. *Science* **254** 821-825
50 De-Vos A M, Ultsch M & Kossiakoff A A (1992) Human growth hormone and extracellular domain of its receptor: crystal structure of the complex. *Science* **255** 306-312
51 Goddard C, Wilkie R S & Dunn I C (1988) The relationship between insulin-like growth factor-1, growth hormone, thyroid hormones and insulin in chickens selected for growth. *Domestic Animal Endocrinology* **5** 165-176
52 Bacon W L, Vasilatos-Younken R, Nestor K E, Andersen B J & Long D W (1989) Pulsatile patterns of plasma growth hormone in turkeys: effect of growth rate, age and sex. *General and Comparative Endocrinology* **75** 417-426
53 Proudman J A & Opel H (1981) Effect of feed and water restriction on basal and TRH-stimulated growth hormone secretion in the growing turkey poult. *Poultry Science* **60** 659- 667
54 Scanes C G, Griminger P & Buonomo F C (1981) Effect of dietary protein restriction on circulating concentrations of growth hormone in growing domestic fowl *Gallus domesticus*. *Proceedings of the Society for Experimental Biology and Medicine* **168** 334-337
55 Scanes C G, Harvey S, Marsh J A & King D B (1984) Hormones and growth in poultry. *Poultry Science* **63** 2062-2074
56 Burke W H, Moore J A, Ogez J R & Builder S E (1987) The properties of recombinant chicken growth hormone and its effects on growth, body composition, feed efficiency, and other factors in broiler chickens. *Endocrinology* **120** 651-658
57 Cravener T L, Vasilatos-Younken R & Wellenreiter R H (1989) Effect of subcutaneous infusion of pituitary-derived chicken growth hormone on growth performance of broiler pullets. *Poultry Science* **68** 1133-1140
58 Leung F C, Taylor J E, Wien S & Van Iderstine A (1986) Purified chicken growth hormone (GH) and human pancreatic GH-releasing hormone increase body weight gain in chickens. *Endocrinology* **118** 1961-1965
59 Rosselot G E, Leach R M & Vasilatos-Younken R (1990) Effect of growth hormone and insulin-like growth factor I on cell proliferation and metabolic activity of cultured chondrocytes. Abstract 202. *72nd Annual Meeting of the Endocrine Society*
60 Burnside J, Liou S S & Cogburn L A (1991) Molecular cloning of the chicken growth hormone receptor complementary deoxyribonucleic acid: mutation of the gene in sex-linked dwarf chickens. *Endocrinology* **128** 3183-3192
61 Monsonego E, Halevy O, Hurwitz S & Pines M (1993) Effect of growth hormone on cell proliferation and matrix protein synthesis by avian epiphyseal growth plate chondrocytes. *Calcified Tissues International* **52** 526
62 Crabb I D, O'Keefe R J, Puzas J E & Rosier R N (1990) Synergistic effect of transforming growth factor β and fibroblast growth factor on DNA synthesis in chick growth plate chondrocytes. *Journal of Bone and Mineral Research* **5** 1105-1112

63 Pines M, Granot I & Hurwitz S (1990) Cyclic AMP-dependent inhibition of collagen synthesis in avian epiphyseal cartilage cells: effect of chicken and human parathyroid hormone and parathyroid hormone-related peptide. *Bone and Mineral* **9** 23-33

64 Takigawa M, Kinoshita A, Enomoto M, Asada A & Suzuki F (1991) Effects of various growth and differentiation factors on expression of parathyroid hormone receptors on rabbit costal chondrocytes in culture. *Endocrinology* **129** 868-876

65 Pines M & Hurwitz S (1989) Atrial natriuretic peptide and sodium nitroprusside stimulate cyclic GMP accumulation by avian skin fibroblasts and epiphyseal growth-plate chondroprogenitor cells. *Journal of Endocrinology* **120** 319-324

66 Nilsson A, Carlsson B, Mathews L & Isaksson O P G (1991) Growth hormone regulation of the growth hormone receptor mRNA in cultured rat epiphyseal chondrocytes. *Molecular and Cellular Endocrinology* **70** 237-246

67 Johansson S, Husman B, Norstedt G & Andersson G (1989) Growth hormone regulates the rodent hepatic epidermal growth factor receptor at a pretranslational level. *Journal of Molecular Endocrinology* **3** 113-120

68 Ekberg S, Carlsson L, Carlsson B, Billing H & Jansson J O (1989) Plasma growth hormone pattern regulates epidermal growth factor receptor messenger ribonucleic acid level and EGF binding in the rat liver. *Endocrinology* **125** 2158-2166

69 Liu L, Brinkman A, Blat C & Harel L (1991) IGFBP-1, an insulin like growth factor binding protein, is a cell growth inhibitor. *Biochemical and Biophysical Research Communication* **174** 673-679

Avian Endocrinology
Ed. P J Sharp, pp 261-274
Journal of Endocrinology Ltd, Bristol (1993)

# Chemistry and physiology of poultry growth hormone

C G Scanes[1], C Aramburo[2], R M Campbell[3], J J Kopchick[4] and S V Radecki[1]
[1]Department of Animal Sciences, Rutgers The State University, PO Box 231, New Brunswick, NJ 08903-0231, USA; [2]Department of Physiology, Instituto de Investigaciones Biomedicas, Apdo Postal 70-228, UNAM 04510 Mexico DF, Mexico; [3]Department of Animal Science Research, Hoffmann-La Roche Building 86/Room 650, 340 Kingsland Street, Nutley, NJ 07110, USA; [4]Irvine Hall, Ohio University, Athens, OH 45701-4712, USA

## Introduction

Our knowledge of the physiology of growth hormone (GH) in poultry and other birds has accumulated rapidly in the last twenty years. The present paper is not a comprehensive review of the physiology and chemistry of GH. Rather, it is intended to consider aspects of recent work with chicken GH (cGH) related to a series of specific questions. These will be considered under three sub-headings: post-translational variants of cGH; GH and chicken adipose tissue and structural-activity relationship of GH. The following questions will be addressed. Do variants of cGH exist and, if so, what is their relevance? Is mammalian GH active in the chicken? Is the lipolytic effect of GH artifactual, real or physiologically relevant? Does GH exert its varied effects on growth, lipolysis and insulin-like effects via one GH receptor (GH-R) and one signal transduction mechanism?

## Variants of chicken growth hormone

Analysis by non-denaturing polyacrylamide gel electrophoresis (PAGE) shows the presence of at least three charge variants of cGH (Rf=0.23, 0.30 and 0.35) (1, 2). When the resolution is increased by use of isoelectric focusing (IEF) and Western blotting, it is possible to identify eight to ten charge variants (3, 4), some of which appear to be developmentally regulated (Fig. 1).

The apparent lack of more than one GH mRNA in chicken pituitary tissue (5) suggests that most cGH variants come from post-translational modifications. One cGH variant has been isolated (6, 7). This isoform is glycosylated (G-cGH), binds to concanavalin A, and has an apparent molecular size, under reducing conditions, of approximately 29 kDa. Under non-reducing conditions, higher molecular size bands corresponding to dimeric forms have been described. Analysis by IEF shows that G-cGH consists of at least three charge isoforms (pI=6.52, 6.7 and 7.17) (7). Chicken GH contains at least one consensus sequence for N-glycosylation in position 188-190 (Asn-Cys-Thr) of its primary structure.

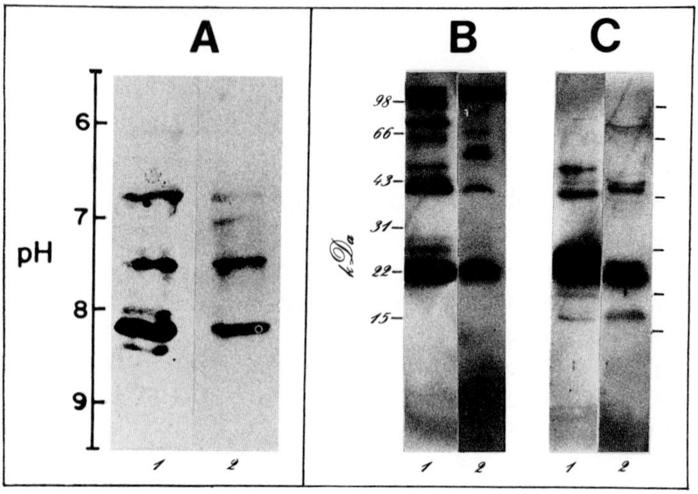


Fig. 1 Western immuno-blots showing charge and size heterogeneity of GH immunoreactivity in chicken anterior pituitary tissue(1) and plasma(2). A: Analysis by isoelectrofocusing (IEF); B: SDS-PAGE under non-reducing conditions; C: SDS-PAGE under reducing conditions.

A series of size variants of cGH are found in the pituitary and plasma (Fig. 1). Analysis by SDS-PAGE under non-reducing conditions shows that a monomeric form (22 kDa) is the most prominent in the gland. However, higher molecular size forms of GH, corresponding probably to disulphide-linked oligomers (dimer, trimer, etc.), occur simultaneously in the pituitary (3, 4) (Fig. 1). When analyzed under reducing conditions, most of the high molecular size forms disappear, except for a small amount of dimer, increasing the proportion of the monomeric band (Fig. 1). Additionally, a proteolytically cleaved 'two-chain' form (25 kDa) is observed in the pituitary when resolved by SDS-PAGE under non-reducing conditions and this seems to give rise to an immunoreactive 16 kDa fragment upon reduction (8).

There is evidence for other types of post-translational modification of GH (2, 9). Chicken pituitary cells in culture incorporate $^{32}$P-phosphate into a cGH-like immunoreactive protein, as analyzed by SDS-PAGE, Western blotting and autoradiography. Furthermore, synthesis and secretion of a phosphorylated variant of cGH is stimulated by growth hormone releasing factor (GHRF) (10) and phorbol ester (9). *In vitro*, protein kinase A (PKA) phosphorylates native and recombinant cGH but not glycosylated cGH (9, 10).

Evidence for a cleaved 'two-chain' form of cGH (25 kDa) comes from Western blotting/PAGE analyses of chicken anterior pituitary extracts and purified cGH under non-reducing conditions. After reduction, this form disappears and is replaced by an immunoreactive fragment of 16 kDa (8). In an analogous manner, thrombin and collagenase cleave cGH *in vitro* into a 25 kDa 'two-chain' form, which upon reduction produces the 16 kDa fragment. The putative site of cleavage corresponds to residues

132-135. This can be deduced from the size of the fragment and the existence of specific recognition sites for both enzymes at this location. Similar proteolytic cleavage of cGH can be achieved by an acid protease present in chicken liver membrane preparations (8).

*Differential control of the synthesis and release of GH variants*

At present there is little evidence in support of the independent/ differential control of synthesis and/or release of GH variants. However, the relative contents of GH variants in the pituitary gland and plasma vary during growth and development (11).

*The biological effects of GH variants*

There is little data demonstrating that different GH variants have unique biological activities. Glycosylated cGH is active in stimulating $T_4$ metabolism, as is the non-glycosylated form (L Berghman, J Buyse, L Huybrechts, E Decuypere & C Scanes, unpublished observations). The most abundant charge isomer of cGH contains all the biological activities ascribed to GH (12). However, some studies suggest that the cGH variants may have different biological activities (1). Two monomeric charge variants of cGH (Rf = 0.23 and 0.30) show growth-stimulating activity in a rodent system (13), but have a differential effect on chicken adipose tissue explants, with one variant being mainly lipolytic (Rf = 0.23), the other (Rf = 0.30) being non-lipolytic (2).

To have physiological significance, the GH isoforms should be released into the circulation in order to reach their target tissues and elicit their effects. It has been shown that, when isolated by monoclonal immunoaffinity chromatography, size and charge variants of cGH are present in chicken serum and co-migrate with those present in fresh pituitary extract (14) (Fig. 1). This is indicative of the variants being synthesized in the pituitary gland and released into the blood.

**Growth hormone and adipose tissue metabolism**

*In vitro*, GH exerts three effects on chicken adipose tissues; it stimulates basal lipolysis (15), inhibits glucagon-stimulated lipolysis in an insulin-like manner (16) and stimulates glucose uptake (17). The effects on lipolysis appear to be contradictory from a physiological viewpoint. *In vivo*, the acute administration of GH to conscious (18) or anaesthetized chickens (19) is followed by an increase in concentrations of circulating free fatty acid (FFA) indicating stimulation of lipolysis. Circulating concentrations of FFA are increased after the administration of cGH or bovine GH (18, 19). This supports the view that the lipolytic effect of GH in chickens is not an artifact of the *in vitro* system.

*In vitro* lipolytic and insulin-like (anti-lipolytic in the presence of glucagon) assays for GH using chicken adipose explants have been previously described (15, 16). Several aspects critical to these assays have not been discussed elsewhere in detail. The abdominal adipose tissue is obtained without delay following sacrifice and immediately placed in oxygenated (by bubbling 95% $0_2$, 5% $C0_2$) Krebs-Hepes buffer (with no sodium bicarbonate added) containing fatty acid-depleted bovine serum albumin (Armor/Intergen, Purchase, NY, USA) and 15mM glucose at 37 °C. Tissue is sliced in a petri-dish by gentle 'shearing' with two scalpels. Care is taken to ensure that the tissue is in media at 37 °C through this process which may take between 15 and 30 minutes. Only the adipose tissue explants (~4-6 mg each), which have been separated from connective tissue and which float, are used in the assays. The tissue explants (8 per vial) are subsequently pre-incubated (in 20 ml scintillation vials) for 1 hour under 95% $0_2$, 5% $C0_2$. The media are then removed and the explants are incubated with treatments (again under 95% $0_2$/5% $C0_2$) for a subsequent hour in a shaker water bath (60 oscillations per minute). It is essential that tissue is maintained in oxygenated media and at 37 °C throughout the incubations. The media are freshly made prior to each study to ensure calcium is in solution, as the lipolytic effect of GH is not observed at low calcium concentrations (20). Moreover, the media must contain the optimum concentrations of glucose and bovine serum albumin (BSA) for the lipolytic effect of GH to be manifest. The consequences of inadequate glucose or BSA concentrations on the lipolytic effects of GH are shown in Table 1.

Table 1 Characterization of the lipolytic response/assay for bovine growth hormone (bGH) using chicken adipose tissue *in vitro*.

| | **Glycerol release (nmoles/g/hour ± s.e.m.)** | |
|---|---|---|
| **Treatment (n=3)** | **no bGH** | **bGH (1mg/1)** |
| Effect of glucose concentration | | |
| 0 mM glucose | 168±12.5 | 174± 5.1 |
| 1.7 mM glucose | 186±12.1 | 186± 8.3 |
| 5 mM glucose | 223± 9.2 | 244± 4.8 |
| 15 mM glucose | 229± 9.4 | 342±12.2$^{**}$ |
| Effect of bovine serum albumin concentration | | |
| 0 g/l BSA | 154± 5.5 | 187±10.9$^{*}$ |
| 0.01 g/l BSA | 309± 7.4 | 416± 9.3$^{**}$ |
| 0.1 g/l BSA | 312±12.6 | 473±10.4$^{**}$ |
| 1 g/l BSA | 307± 9.3 | 450±15.0$^{**}$ |

Compared with values in the absence of GH, $^{*}p < 0.05$; $^{**}p < 0.001$.

*Comparison of the lipolytic and insulin-like responses of chicken and rat adipose tissue to GH in vitro*

There is still controversy as to whether GH *per se* stimulates lipolysis by rat adipose tissue *in vitro* (21-24). Two systems have been used extensively by Goodman and co-workers (24-26, 28) which are referred to as lipolytic; these require the presence of either theophylline or glucocorticoids for GH to have the optimal effect. The first rat system involves a preincubation period during which adipose tissue segments from hypophysectomized rats are incubated in the presence of GH for 3 hours. Lipolysis, measured as glycerol release, is then determined in a subsequent fourth hour of incubation in fresh media containing theophylline (25, 26). In the second rat system, adipose tissue segments from intact rats are pre-incubated for 3 hours with GH and the glucocorticoid, dexamethasone. Lipolysis measured as glycerol release is measured in the fourth hour of incubation in fresh media containing both GH and dexamethasone (27, 28). It is not clear whether these responses are both indices of the same biological response to GH. They involve a time lag before a response is observed and in both, the effect of GH can be blocked by the protein kinase C inhibitors, staurosporine and sphingosine (28). However, inhibitors of protein and RNA synthesis only prevent the lipolytic effect of GH in the presence of dexamethasone (25).

The lipolytic assay for GH using chicken adipose tissue *in vitro* is more straightforward. The assay employs adipose tissue from intact animals, the presence of glucocorticoids is not required and GH stimulates lipolysis in the first hour of incubation (15). While an unambiguous stimulation of lipolysis by GH is observed during the first hour of incubation, the magnitude of the response increases with the duration of incubation (15). The chicken lipolytic GH assay is somewhat less sensitive to GH than more widely used GH/theophylline rat assay (Table 2). Like the rat GH glucocorticoid assay, inhibitors of RNA and protein synthesis block the lipolytic effect of GH on chicken adipose tissue *in vitro* (20).

Table 2 Comparison of the sensitivities of biological assays for GH using chicken and rat adipose tissue *in vitro*.

| | $ED_{50}$ (ng/ml)[a] | |
|---|---|---|
| **Biological response** | **Chicken adipose tissue** | **Rat adipose tissue** |
| Lipolytic | 27.5[b] | 3.8[d] |
| Insulin-like | | |
| $^{14}C$-glucose → $CO_2$ | - | 53.2[d] |
| $^{14}C$-Leucine → $CO_2$ | - | 20[e] |
| Anti-lipolytic | 9.4[c] | 42.5[e] |

[a]The amount of GH required to induce a half maximal response ($ED_{50}$). Recombinant bovine GH was used in the chicken systems and human GH in the rat systems. [b]Mean of data from 29 and R M Campbell and C G Scanes, unpublished. [c]Mean of data from 29, 30 and R M Campbell and C G Scanes, unpublished. [d]Mean of data from 26, 31. [e]from 26.

Three insulin-like responses of rat adipose tissue to GH have been observed *in vitro* and these have been used as systems to characterize the actions of GH (Table 2). These responses have similar sensitivities and hence may be manifestations of the same overall effect. One insulin-like response, the anti-lipolytic effect of GH in the presence of norepinephrine, is observed using both chicken and rat adipose tissue (Table 2). The anti-lipolytic effect of GH involves reduction of glycerol release from adipose tissue during a one hour incubation in the presence of a lipolytic agent (norepinephrine, or epinephrine in rats; glucagon in chickens). The sensitivity of the chicken system is slightly greater than that in the rat (Table 2). In both systems, GH appears to exert its inhibitory effect by acting between adenylate cyclase/cAMP and hormone-sensitive lipase (16, 32, 33).

**Structural requirements for the lipolytic and insulin-like effects of GH**

Growth hormone has many biological actions including growth promotion, lipolysis, insulin-like, and diabetogenic. It is possible that these activities are exerted by separate domains on the GH molecule and hence through different growth hormone receptors (GH-R). This seems unlikely since there appears to be only one GH-R (34), although there are observations which are not consistent with this view. A corollary to the hypothesis of a single GH-R mediating all the actions of GH is that identical structure-activity relationships should exist for all the biological actions of GH. There are several situations where GH preparations induce differential lipolytic and insulin-like responses from rat adipose tissue *in vitro*, suggesting that these different responses to GH have different structural requirements. A cleaved form of GH (Da 1 - a non-covalent complex of $hGH_{1-134}$, $hGH_{141-191}$) retains full lipolytic activity (potency x 1.0 compared to hGH) but has reduced insulin-like activity (26, 31). The insulin-like responses of rat adipose tissue to cleaved GH (Da1) are x 0.16 compared to hGH for $^{14}$C-glucose → $^{14}CO_2$; x 0.61 relative to human GH for $^{14}$C leucine → $^{14}CO_2$; and x 0.64 relative to human GH for anti-lipolytic activity (recalculated from 26, 31). Conversely, a putative GH analogue, from tapeworm larvae, retains insulin-like and growth-promoting activity in obese mice but lacks lipolytic or diabetogenic activities (35, 36). Studies using chicken adipose tissue *in vitro* provide strong evidence that the lipolytic and insulin-like effects of GH have different structural requirements (Table 3). A comparison of the lipolytic and insulin-like potencies of a series of GH preparations shows that the lipolytic response to GH has a much stricter structural requirement than the anti-lipolytic/insulin-like effect (Table 3). Preparations of GH from lower vertebrates (reptiles, amphibians and fish) have very low, if any, lipolytic activity. A comparison of the amino-acid residue sequences of GH from different vertebrates suggests that $Pro_{132}$, which is present in avian and mammalian GHs but not in lower vertebrate GHs, is required for the lipolytic effect of GH. All lower vertebrate GH preparations tested have high insulin-like (anti-lipolytic) activities in chicken adipose tissue *in vitro* (Table 3).

A mutated form of bovine GH (M8 - bGH E117L, G119R, A122D) has been genetically engineered to contain an idealized amphiphilic structure in the third α helix (38). This mutated form of bGH acts as a functional antagonist for the growth promoting effect of GH *in vivo* (38) and has a very low potency in stimulating lipolysis by chicken adipose tissue *in vitro* (Table 3). It is also a competitive antagonist for the lipolytic effects of GH on this system (37). However, unexpectedly, M8 retains the full insulin-like activity of GH, inhibiting lipolysis in the presence of glucagon (Table 3). The very high ratios of lipolytic to insulin-like potencies of lower vertebrate GH preparations and M8 (Table 3) provide strong evidence that there are different structural requirements for lipolytic and insulin-like effects of GH on chicken adipose tissue.

Table 3 Comparison of the lipolytic and insulin-like (anti-lipolytic) potencies of vertebrate growth hormones on chicken adipose tissue *in vitro*.

| Source of GH preparation | Potencies (μg equivalent bovine GH per μg) Lipolytic | Anti-lipolytic | Ratio of potencies |
|---|---|---|---|
| bovine[a] | 1.00 | 1.00 | 1.0 |
| human GH[b] | ~1.00 | ~1.00 | ~1.0 |
| RCMhGH | ~1.00 | ~1.00 | ~1.0 |
| 20KhGH[b] | 0.10[c] | ~1.00 | ~10 |
| human PL[b] | <0.01 | <0.01 | NA |
| chicken GH[d] | ~1.00 | ~1.00 | ~1.0 |
| turtle[e] | <0.01 | 0.67 | >67 |
| bullfrog[e] | <0.01 | 0.22 | >22 |
| sturgeon[e] | <0.01 | 1.03 | >103 |
| tuna[e] | <0.01 | 0.53 | >58 |
| bonita[e] | 0.037 | 6.1 | 165 |
| yellowtail[e] | <0.01 | 14.8 | >148 |
| salmon[e] | 0.030 | 6.1 | 203 |
| shark[e] | <0.01 | 0.78 | >78 |
| M8 - mutated bGH E117L, G119R, A122D[f] | 0.02 | ~0.81 | >40 |

[a]By definition. [b]Data from 30. [c]Potency relative to human GH 22K form. While the 20K form had a lower potency than 22K GH in the first hour of incubation this was not the case in the fourth hour when the potency was 0.61 equivalents of the 22K hGH. [d]Data from 15, 16. [e]Data from 29. [f]Data from 37. NA, not appropriate.

There is evidence that different signal transduction systems are involved in the lipolytic and insulin-like effects of GH on chicken adipose tissue. For instance, the lipolytic *but not* the insulin-like (anti-lipolytic) responses to GH are blocked by inhibitors of protein/RNA synthesis, by low calcium in the media or by the calcium channel blocker, verapamil (20). Conversely, the insulin-like effect but not the lipolytic effect of GH is suppressed in the presence of difluoromethyl ornithine, an inhibitor of polyamine synthesis (20).

There is additional evidence for distinct domains on the GH molecule being responsible for its different activities. Peptides have been synthesized which correspond to specific amino-acid sequences within GH in an attempt to examine whether specific domains on the GH molecule may be responsible for specific biological effects of GH. For instance, $hGH_{1-15}$ enhances the effect of insulin (39). Similarly, chronic or acute administration of $hGH_{1-43}$ reduces the increment in circulating concentrations of glucose following a glucose load in hypophysectomized genetically obese mice ($A^{vy}/A$) (40). Conversely, $hGH_{172-191}$ induces an acute hyperglycaemic effect in normal rats (41). While these reports provide evidence that specific domains on the GH molecule are responsible for some biological actions, they did not compare the biological effectiveness of the peptides with full length GH or, where they did, GH itself was ineffective.

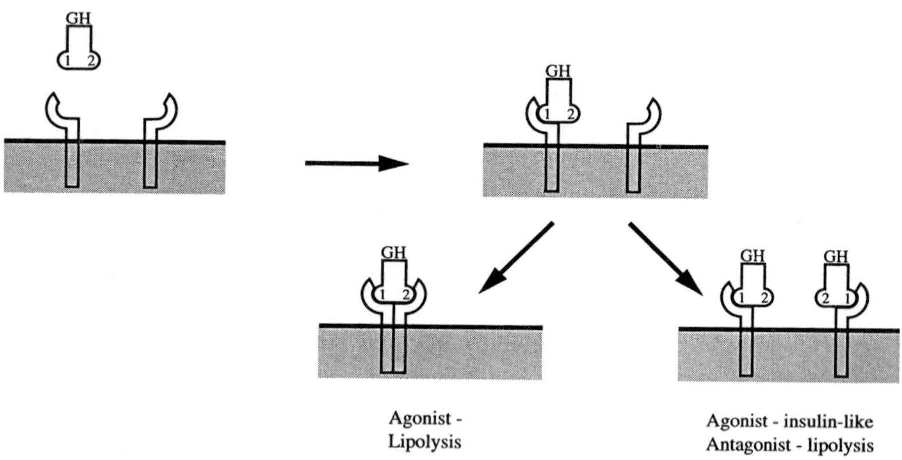


Fig. 2 Proposed model of GH interaction with the GH-R to evoke lipolytic and insulin-like effects on chicken adipose tissue and also to influence liver cells. [Figure based on 34 but modified in view of data summarized in the present communication]. Growth hormone (GH) molecule showing binding sites 1 and 2. Two univalent GH receptors (GH-R) bind sequentially to binding site 1 then 2 and ultimately dimerize such that GH evokes effects on lipolysis (chicken adipose tissue) and also on liver cells. Binding only to site 1 induces an insulin-like (anti-lipolytic) effect on chicken adipose tissue. Analogues of GH with only binding site 1 interact with the GHR but this will not result in dimerization and hence these act as antagonists for those effects of GH where dimerization of the GH-R is obligatory for mechanism of action of GH.

*Interactions between GH and its receptor*

A three dimensional structure of the growth hormone-growth hormone receptor complex has been proposed (42) based on X-ray crystallography of porcine GH (43) and the human GH-binding protein which consists of the extracellular domain of the GH-R (42). There are two sites on the GH molecule which bind to the GH-R. Binding site 1 consists of amino-acid residues located in α helix 4. Binding site 2 involves the N-terminal sequence, the α helix 1, the loop between α helix 1 and 2 and α helix 3. These sites on GH form hydrogen- and other close bonds with different amino-acid residues in the GH-R molecule. The GH-R can be viewed as univalent for GH because of the close proximity and overlap in loci in the GH-R binding to sites 1 and 2 on GH (34). It is probable that GH-R binds first to site 1 on GH due to its higher affinity and, subsequently, a second GH-R binds to the site 2. There is then dimerization of the GH-R which is thought to be obligatory for GH signal transduction (34). Consistent with this view is the finding that the GH analogue (M8-bGH E117L, G119R, A122D) with an altered α helix 3 to modify binding site 2 suppresses body growth in mice (38).

The potencies of a series of chemically derived GH fragments or GH analogues have been determined using a radioreceptor system which involves the inhibition of binding of $^{125}$I-bGH to liver membrane preparations from adult chickens in order to establish whether the chicken GH-R recognizes two binding sites on GH. Growth hormone analogues/fragments inhibit binding of $^{125}$I-bGH to GH-R on chicken liver membranes if they include either binding site 1 (e.g. $bGH_{1-60,153-191}$) or binding site 2 (e.g. $bGH_{1-132}$ or $bGH_{98-135}$) but not if they include neither binding site (e.g. $hGH_{1-43}$). Removal of most of binding site 1 (e.g. $bGH_{1-132}$) greatly reduces, while not totally eliminating, GH-R activity. This suggests that the GH-R binds preferentially or initially to binding site 1. These studies are consistent with the prevailing view that chicken GH-R binds to site 1 on GH followed by a second GH-R binding to site 2 (34).

The lack of lipolytic activity of M8 (bGH E117L, G119R, A122D) (Table 3), and its effectiveness as a competitive antagonist to GH, could be explained by the adipose GH-R binding to GH site 1 but the mutations in α helix 3 preventing binding to site 2 (Fig. 2). Similarly, the very low or non-existent lipolytic activities of lower vertebrate GH preparations is explicable by the absence of a proline residue in position 132 which is very close to α helix 3 and binding site 2. On the other hand, both M8 and lower vertebrate GH preparation retain full insulin-like activity, since they inhibit glucagon-induced lipolysis. This is consistent with the model (Fig. 2) in which binding of GH site 1 to a GH-R is sufficient to evoke at least one biological effect of GH - the insulin-like effect (anti-lipolytic in the presence of glucagon) without the requirement for GH-R dimerization. This model is consistent with the greater sensitivity of the tissue to the insulin-like rather than the lipolytic activity of GH (Table 2).

The lack of a marked reduction in biological (lipolytic and insulin-like) activity (Table 3) or radioreceptor activity (Table 4) of reduced and carboxymethylated hGH (RCM-hGH), suggests that disulphide bonds are not required for GH to maintain the structure essential for biological activity on chicken adipose tissue. While this may be the case for human GH where the tertiary structure may not be disrupted by reduction and carboxymethylation, it does not *per se* indicate that disulphide bonds are not required for the initial establishment of the crucial structure or its maintenance. Based

Table 4 Potency of GH preparations/fragments in displacing $^{125}$I-bGH from a (adult) chicken liver membrane preparation.

| **Hormone/analogue** | **Deficit in Structure of GH**[f] | **Potency**[a] **(µg bGH equivalents/µg)** | **Conclusions** |
|---|---|---|---|
| **cGH** | | 2.76 | |
| **bGH** | | 1.00 | |
| **GH fragments** | | | |
| des 1-4 bGH[d] | Amino-acid residues 1-4 close to binding site 2 | 0.68 | Amino-acid residues 1-4 are not important in the binding of GH to its receptor |
| ($bGH_{1\text{-}132}$) | Helix 4 and connection between helix 3 and 4 | 0.045 | Helix 4 (containing much of binding site 1 of GH[f]) is important for GH binding to the cGHR. Also GH binding site 2 has some role for GH binding |
| $hGH_{1\text{-}43}$ | Helix 2, 3, 4 | <0.002 | Helixes 1, 2, 3, 4 are essential for binding |
| $bGH_{98\text{-}135}$ | Helix 1, 2, 4 | 0.044 | Helix 3 (containing much of binding site 2[f]) has a role for GH in binding to the cGHR |
| $bGH_{1\text{-}166,\ 153\text{-}191}$ | Helix 2, 3 | 0.026 | Helix 1 and 4 containing respectively some of binding site 2[f] and 1 have a role in GH binding to the GHR |
| 22K hGH (native) | | 0.81 | Human GH binds to the cGHR |
| **Other GH analogues** | | | |
| 20K hGH (native) | Residues 32-46 | 0.38 | Residues 32-46 (between helix 1 and 2) have little importance in binding to the GHR |
| RCM hGH[e] (biosynthetic) | Disulphide bridges | 0.19 | There is little evidence for the importance of disulphide bridges for GH binding to the cGHR |
| C52S, C163S bGH[d] | One disulphide bridge | 0.10 | The maintenance of tertiary structure of GH by disulphide bonds is important to GH binding to the cGHR |
| C52S, C163S, C180SbGH[d] | Two disulphide bridges | 0.049 | |
| **Lactogens** | | | |
| oProlactin (native) | | 0.48[b] | Ovine prolactin binds to the cGHR |
| hPL | | 0.25[c] | |

Data from M Doscher, T A Peterla and C G Scanes, unpublished observations. [a]Potency based on ($ID_{50}$) 50% inhibition of binding of $^{125}$I-bGH. [b]Not due to GH contamination as oPRL preparation had a very low <0.020 potency in a sheep membrane radioreceptor assay. [c]Potency of 0.25 was based on 10% inhibition, higher doses of hPL did not inhibit $^{125}$I-bGH binding further. [d]Kindly provided by G Krivi (Monsanto, St. Louis, MO). [e]Reduced and carboxymethylated human GH, kindly provided by Dr J Kostyo (University of Michigan). [f]Based 42, 43.

on the invariant presence of the cysteines responsible for the disulphide bonds in GH throughout the vertebrates, a strong case can be made for their critical importance. Moreover, specifically mutated bovine GHs lacking either two or four cysteines showed much reduced radioreceptor activity with chicken liver membrane preparations (Table 4).

The use of chicken adipose and liver tissue *in vitro* is already providing intriguing data on the structural activity relationships of GH and the associated signal transduction mechanism(s). It is hoped that the potential of the systems will be recognized and utilized to gain further understanding of these mechanisms. The resulting research may be of benefit to basic and biomedical research and also provide insights ultimately for the application of knowledge to improve the efficient production of poultry and other agricultural animals.

**Acknowledgements**

Paper of the Journal Series, New Jersey Agricultural Experiment Station. Unpublished data are from studies supported by Hatch Act and State funds. We gratefully acknowledge the help of M Doescher and T Peterla Aust in some of the unpublished studies. Supported partially by a Rockefeller Foundation and NIH-Fogarty fellowship to CA.

**References**

1 Aramburo C, Campbell R M & Scanes C G (1989) Heterogeneity of chicken growth hormone (cGH). Identification of lipolytic and non-lipolytic variants. *Life Sciences* **45** 2201-2207

2 Aramburo C, Carranza M, Sanchez R & Perera G (1989) Partial biochemical and biological characterization of purified chicken growth hormone (cGH). Isolation of cGH charge variants and evidence that cGH is phosphorylated. *General and Comparative Endocrinology* **76** 330-339

3 Houston B & Goddard C (1988) Molecular forms of growth hormone in the chicken pituitary gland. *Journal of Endocrinology* **116** 35-41

4 Aramburo C, Montiel J L, Perera G, Navarrete S & Sanchez R (1990) Molecular isoforms of chicken growth hormone (cGH): different bioactivities of cGH charge variants. *General and Comparative Endocrinology* **80** 59-67

5 McCann-Levorse L M, Radecki S R, Donoghue D J, Malamed S, Foster D N & Scanes C G (1993) Ontogeny of pituitary growth hormone (GH) and GHmRNA in the chicken. *Proceedings of the Society for Experimental Biology and Medicine* **202** 109-130

6 Berghman L R, Lens P, Decuypere E, Kuhn E R & Vandesande F (1987) Glycosylated chicken growth hormone. *General and Comparative Endocrinology* **68** 408-414

7 Aramburo C, Navarrete S, Montiel J L, Sanchez R & Berghman L R (1991) Purification and electrophoretic analysis of glycosylated chicken growth hormone (G-cGH): evidence of G-cGH isoforms. *General and Comparative Endocrinology* **84** 135-146

8 Aramburo C & Scanes C G (1991) Endogenous cleavage of chicken growth hormone. Abstract 232 *73rd Annual Meeting of the Endocrine Society* Washington, DC, USA

9 Aramburo C, Montiel J L, Proudman J A, Berghman L R & Scanes C G (1992) Phosphorylation of prolactin and growth hormone. *Journal of Molecular Endocrinology* **8** 183-191
10 Aramburo C, Donoghue D, Montiel J L, Berghman L R & Scanes C G (1990) Phosphorylation of chicken growth hormone. *Life Sciences* **47** 945-952
11 Aramburo C, Montiel J L & Scanes C G (1992) Ontogenic changes in the size variants of growth hormone. Abstract 611 *9th International Congress of Endocrinology*. Nice, France
12 Houston B, O'Neill I E, Mitchell M A & Goddard C (1990) Purification and biological activity of a single charge isomer of pituitary-derived chicken growth hormone. *Journal of Endocrinology* **125** 207-215
13 Scanes C G, Aramburo C & Campbell R M (1990) Hormonal involvement in avian growth and development: growth hormone and insulin-like growth factor I. *Endocrinology of Birds: Molecular to Behavioral* pp 92-110. Eds M Wada, S Ishii & C G Scanes. Tokyo: Japan Scientific Societies Press. Berlin: Springer-Verlag
14 Montiel J L, Berghman L R & Aramburo C (1992) Identification of growth hormone molecular variants in chicken serum. *General and Comparative Endocrinology* **88** 298-306
15 Campbell R M & Scanes C G (1985) Lipolytic activity of purified pituitary and bacterially-derived growth hormone on chicken adipose tissue *in vitro*. *Proceedings of the Society for Experimental Biology and Medicine* **180** 513-517
16 Campbell R M & Scanes C G (1987) Growth hormone inhibition of glucagon and cAMP-induced lipolysis by chicken adipose tissue *in vitro*. *Proceedings of the Society for Experimental Biology and Medicine* **184** 456-460
17 Rudas P & Scanes C G (1983) Influences of growth hormone on glucose uptake by avian adipose tissue. *Poultry Science* **62** 1838-1845
18 Hall T R, Cheung A & Harvey S (1987) Some biological activities of recombinant DNA-derived growth hormone on plasma metabolite concentrations in the domestic fowl. *Comparative Biochemistry and Physiology* **86A** 29-34
19 Scanes C G (1992) Lipolytic and diabetogenic effects of native and biosynthetic growth hormone in the chicken: a re-evaluation. *Comparative Biochemistry and Physiology* **101A** 871-878
20 Campbell R M & Scanes C G (1988) Pharmacological investigations of the lipolytic and antilipolytic effects of growth hormone (GH) on chicken adipose tissue *in vitro*: evidence for the involvement of calcium and polyamines. *Proceedings of the Society for Experimental Biology and Medicine* **188** 177-184
21 Frigeri L G (1980) Absence of *in vitro* dexamethasone-dependent lipolytic activity from highly purified growth hormone. *Endocrinology* **107** 738-743
22 Frigeri L G, Robel G & Stebbins N (1982) Bacteria-derived human growth hormone lacks lipolytic activity in rat adipose tissue. *Biochemical and Biophysical Research Communications* **103** 1041-1046
23 Goodman H M & Grichting G (1983) Growth hormone and lipolysis: a re-evaluation. *Endocrinology* **113** 1697-1702
24 Goodman H M (1984) Biological activity of bacterial derived human growth hormone in adipose tissue of hypophysectomized rats. *Endocrinology* **114** 131-135
25 Goodman H M (1968) Multiple effects of growth hormone on lipolysis. *Endocrinology* **83** 300-308
26 Goodman H M & Kostyo J L (1981) Altered profiles of biological activity of growth hormone fragments on adipocyte metabolism. *Endocrinology* **108** 553-558

27 Fain J N, Kovacev V P & Scow R O (1965) Effect of growth hormone and dexamethasone on lipolysis and metabolism in isolated fat cells of the rat. *Journal of Biological Chemistry* **240** 3522-3529

28 Gorin E, Tai L-R, Honeyman T W & Goodman H M (1990) Evidence for a role of protein kinase C in the stimulation of lipolysis by growth hormone and isoproterenol. *Endocrinology* **126** 2973-2982

29 Campbell R M, Kawauchi H, Lewis U J, Papkoff H & Scanes C G (1991) Comparison of lipolytic and antilipolytic activities of lower vertebrate growth hormones on chicken adipose tissue *in vitro*. *Proceedings of the Society for Experimental Biology and Medicine* **197** 409-415

30 Campbell R M, Kostyo J L & Scanes C G (1990) Lipolytic and antilipolytic effects of human growth hormone, its 20-kilodalton variant, a reduced and carboxymethylated derivative and human placental lactogen on chicken adipose tissue *in vitro*. *Proceedings of the Society for Experimental Biology and Medicine* **193** 269-273

31 Chipkin S R, Szecowka J, Tai L-R, Kostyo J L & Goodman H M (1989) Different growth hormone receptor interactions mediate insulin-like and lipolytic responses of rat adipose tissue. *Endocrinology* **125** 450-458

32 Birnbaum R S & Goodman H M (1976) Studies on the mechanism of the antilipolytic effects of growth hormone. *Endocrinology* **99** 1336-1345

33 Bjorjell P, Rosberg S, Isaksson O & Belfrage P (1984) The antilipolytic insulin-like effect of growth hormone is caused by a net decrease of hormone-sensitive lipase phosphorylation. *Endocrinology* **115** 1151-1156

34 Fuh G, Cunningham B C, Fukunaga R, Nagata S, Goeddel D V & Wells J A (1992) Rational design of potent antagonists to the human growth hormone receptor. *Science* **256** 1679-1680

35 Salem M A M & Phares C K (1989) The growth factor from pleroceroid larvae of the tapeworm, *Spirometra mansonoides*, stimulates growth but is not diabetogenic. *Proceedings of the Society for Experimental Biology and Medicine* **191** 187-192

36 Salem M A M & Phares C K (1989) *In vitro* insulin-like actions of the growth factor from the tapeworm, *Spirometia mansonoides*. *Proceedings of the Society for Experimental Biology and Medicine* **190** 203-210

37 Campbell R M, Chen W Y, Wiehl P, Kelder B, Kopchick J J & Scanes C G (1993) A growth hormone (GH) analog which antagonizies the lipolytic effect but retains full insulin-like (anti-lipolytic) activity. *Proceedings of the Society for Experimental Biology and Medicine* **203** 311-316

38 Chen W Y, Wight D C, Wagner T E & Kopchick J J (1990) Expression of a mutated bovine growth hormone gene suppresses growth of transgenic mice. *Proceedings of the National Academy of Sciences of the USA* **87** 5061-5065

39 Ng F M & Bornstein J (1979) Insulin-potentiating action of a synthetic amino-terminal fragment of human growth hormone (hGH 1-15) in streptozotocin-diabetic rats. *Diabetes* **28** 1126-1130

40 Salem M A M, Lewis U J, Haro L S, Kishi K, McAllister D L, Seavey B K, Bee G & Wolff G L (1989) Effects of hypophysectomy and the insulin-like and anti-insulin pituitary peptides on carbohydrate metabolism in yellow $A^y/A$ (BALB/cxVY)$F_1$ hybrid mice. *Proceedings of the Society for Experimental Biology and Medicine* **191** 408-419

41 Wade J D, Pullin C D, Ng F M & Bornstein J (1977) The synthesis and hyperglycaemic activity of the amino acid sequence 171-191 of human growth hormone. *Biochemical and Biophysical Research Communications* **78** 827-832

42 deVos A M, Ultsch M & Kossiakoff A A (1992) Human growth hormone and its extracellular domain of its receptor: crystal structure of the complex. *Science* **225** 306-312

43 Abdel-Meguid S S, Shieh H-S, Smith W W, Dayringer H E, Violand B N & Bentle L A (1987) Three-dimensional structure of a genetically engineered variant of porcine growth hormone. *Proceedings of the National Academy of Sciences of the USA* **84** 6434-6437

Avian Endocrinology
Ed. P J Sharp, pp 275-284
Journal of Endocrinology Ltd, Bristol (1993)

# Insulin-like growth factors and IGF binding proteins

C Goddard[1], S C Butterwith[1], R D Roberts[1] and M J Duclos[2]
[1]Department of Cellular and Molecular Biology, AFRC Roslin Institute (Edinburgh), Roslin, Midlothian EH25 9PS, UK; [2]Station de Recherches Avicoles, INRA, Nouzilly, 37380 Monnaie, France

## Introduction

The literature on the insulin-like growth factors (IGF-I and IGF-II) in the chicken has expanded rapidly since the IVth International Symposium on Avian Endocrinology in Tokyo four years ago, and the aim of this article is to review the progress made. In contrast, much less is known about the IGF binding proteins in avian species, despite the almost exponential output of papers in the mammalian field during that period. We have some new data on IGF binding proteins in primary cell cultures prepared from different chicken tissues and this is also described in this review.

## Insulin-like growth factors

Both IGF-I and IGF-II have been identified in the chicken and the major advances have been: (i) the isolation of cDNAs, nucleotide sequencing, determination of the gene structure; (ii) peptide purification and amino acid sequencing; (iii) details of mRNA expression, tissue distribution, and regulation of plasma concentration; (iv) the production of recombinant derived chicken IGF-I in *E. coli*; and (v) administration of exogenous IGF-I.

### *Complementary DNA cloning and gene structure*

The cloning, isolation and characterisation of chicken IGF-I cDNAs has been described (1, 2) and although the nucleotide sequence encoding the mature peptide is firmly established, there still remains a discrepancy in exon number. A full length chicken IGF-I cDNA was isolated by Kajimoto and Rotwein (1). The 814 nucleotide cDNA contained 246 bp 5′-untranslated sequence, 459 nucleotides of coding information followed by a TAA stop codon and a 3′-untranslated region of 89 bp ending in a poly A tail of 17 nucleotides. The cDNA encoded a 153 amino acid peptide precursor using an ATG start site in exon 1, the only methionine common among primary translation products from different species. Northern blotting suggested transcription and processing into mRNAs of 1.9 and 2.6 kb.

Fawcett and Bulfield (2) isolated a number of chicken IGF-I cDNAs. One had sequence homologous to exons 1, 2, 3 and 5 of human IGF-I cDNA but had an additional exon (1A) spliced between exons 1 and 2. A second clone contained sequence from exons 1 and 2, without exon 1A, and was similar to Kajimoto and

Rotwein's clone (1). The sequence derived from the cDNA containing exon 1A predicted a 130 amino acid peptide comprising a 25 amino acid leader peptide, putative mature chicken IGF-I (70 amino acids) and an E peptide (35 amino acids). Multiple IGF-I mRNA species were reported in Northern blots probed with a full length chicken IGF-I cDNA probe.

IGF-I clones were isolated from a chicken genomic library and used to sequence the intron/exon boundaries. Southern blotting revealed that the chicken IGF-I gene is at least 23 kb in length (2). Recently, 82 kb of genomic DNA comprising the chicken IGF-I locus was isolated and mapped, and the data suggested that the chicken IGF-I gene is composed of 4 exons distributed over 50 kb of chromosomal DNA (3). Within the gene are 7 regions which contain repetitive elements. The promoter was sequenced, and transcription start sites were mapped and shown to be functional. Two poly(A) addition regions were identified within the 3′ untranslated sequence of exon 4 and the distal region contained 3 closely spaced polyadenylation sites. The apparent discrepancy between these studies concerning exon 1A remains to be resolved.

Mammalian IGF-II genes are complex transcription units containing multiple promoters and polyadenylation sites, and they exhibit alternate splicing of their primary transcripts. Recently, a cDNA clone isolated from a chick embryo cDNA library with a human IGF-II probe was sequenced and found to be homologous to the second coding exon of the human, mouse and rat IGF-II genes (4). Comparison of the nucleotide sequence with that of the corresponding genomic clone indicated that it was derived from a spliced antisense transcript of the chicken IGF-II gene which overlapped the second coding exon. Northern blotting revealed both sense and antisense transcripts in stage 22 and 36 chick embryos and suggest an even higher level of complexity in the IGF-II transcription unit.

*Purification and amino acid sequence*

The molecular biology of chicken IGFs has been complemented by the purification and amino acid sequencing of IGF-I and IGF-II. An initial report of partial sequences and properties (5) was confirmed (6) and then followed up by purification and sequencing of IGF-I (7) and IGF-II (8) from chicken serum and determination of their biological activity in a number of *in vitro* systems.

Both IGF peptides are highly conserved between species. Chicken IGF-I has 70 amino acids and shows 88.5% homology to human IGF-I. There are 8 amino acid substitutions between the chicken and human peptides but their physiological significance is unknown. The amino acid sequence of IGF-I purified from serum corresponds to that predicted from the nucleotide sequence of IGF-I cDNA (1, 2). Chicken IGF-I is biologically active in a number of cell culture systems and binds to IGF receptors from both avian and mammalian tissues (7). Similarly, chicken IGF-II, a 67 amino acid peptide, shares about 85% sequence with human IGF-II although there are marked differences at the N-terminus, and between residues 32 and 40. In particular, the N-terminal sequence of chicken IGF-II (Tyr-Gly-Thr-Ala-Glu) is different from the sequence (Ala-Tyr-Arg-Pro-Ser-Glu) present in most mammals but these changes in structure have a relatively small effect on receptor binding or activity in a number of bioassays.

*Expression of mRNA, tissue distribution, and regulation of plasma concentration*

The pattern of IGF-I mRNA expression and distribution of the peptide in different tissues throughout development and in the post-hatch rapid growth phase in the chicken is clear evidence for the importance of the IGFs in the regulation of growth and development. IGF-I has been partially purified from unfertilized chicken egg-yolk. The peptide is immunoreactive in an IGF-I radioimmunoassay and is biologically active in a lens cell bioassay (9) and this suggests that IGF-I of maternal origin is important at the earliest stages of embryogenesis. The use of reverse transcriptase-polymerase chain reaction (RT-PCR) demonstrates low levels of IGF-I mRNA expression in the blastoderm and gastrula with a tendency for an increase in the embryo throughout the first week of development. In mid and late embryogenesis, IGF-I mRNA is easily detectable in pancreas and brain but is low in liver. Liver IGF-I mRNA is increased markedly at 50 days post-hatch (10).

The distribution of IGF peptide immunoreactivity between stages 20 and 36 of chick embryo development has been described (11). Muscle, peripheral nerve and notochord are labelled whenever present. Insulin-like growth factor peptides are associated with myotubes in muscle and with neurones rather than Schwann cells in nerve tissue. Tissue distribution changes during differentiation, suggesting a correlation between IGFs and cell division in early mesenchyme, but in other tissues the association is with differentiating rather than dividing cells.

Solution hybridization/nuclease protection assays have shown IGF-I expression in whole embryos from days 6-15 of incubation (12), with IGF-I mRNA detectable in the eye, skeletal muscle and brain on days 17 and 20, but not in liver or heart until after hatching. Expression of IGF-I mRNA by RT-PCR has been found in liver, adipose tissue, ovary, heart, skeletal muscle, kidney and brain (in order of relative expression) in 6 week-old broiler chickens (13).

Circulating concentrations of IGF-I show an interesting parallel with tissue expression throughout growth and development. IGF-I is detectable in serum from embryos between day 6 and 9 of incubation, increases to a maximum between day 15 and 17 and then declines before hatching (12, 14-16). The increase is growth hormone (GH)-independent because serum GH concentrations do not increase in the embryo until after the peak of IGF-I (12, 16) and plasma IGF-I concentrations are unchanged in embryos which carry the sex-linked dwarf gene (14) a mutation of the GH receptor gene (17). The mid embryogenesis increase in IGF-I is absent in growth-retarded embryos in culture (15).

There is a post-hatch increase in plasma IGF-I concentrations which remain relatively constant from about 4-6 weeks of age. This could be GH-dependent regulation of IGF-I because the initial increase in IGF-I is accompanied by an increase in GH concentration, although it is well established that the IGF-I concentration is maintained or even increased despite a later decline in GH (7, 12, 16, 18, 19). All these data support the idea of an important role for IGF-I throughout growth and development in the chicken.

Targeted disruption of the IGF-II gene in transgenic mice to produce a dwarf phenotype has clearly shown the importance of IGF-II in embryo development (20), but the role of IGF-II in the chick has not been characterised. Plasma IGF-II concentrations are higher than IGF-I concentrations in the embryo between days 11 and 20 of incubation but they do not change until hatch when there is an increase in normal chicks and a decrease in those carrying the sex-linked dwarf gene (14). Plasma IGF-II concentrations have been determined in normal and dwarf chickens from lines selected for high or low body weight. Plasma IGF-II is marginally lower in adults from the low growth line and are also reduced in the dwarf (21) suggesting some GH regulation of IGF-II.

Attempts have recently been made to determine if the addition of exogenous IGF-I improves growth rate, body weight or body composition in broiler chickens. The results have not been encouraging. In one study, broiler cockerels were given daily intramuscular injections of human IGF-I (100 or 200 μg/kg body weight) from 11 to 24 days of age. There was no effect on the average daily gain or feed consumption or the gain to feed ratio and despite the highest dose of IGF-I causing a reduction in body ash content, there was no effect on fat or protein content. Plasma GH concentrations are depressed by IGF-I but there was no effect on $T_3$ or $T_4$ concentrations (22). In a similar experiment, human IGF-I was given to 4 week-old female broilers in a 14-day continuous infusion using osmotic minipumps. No differences in body weight were detected although abdominal fat was reduced in the highest dose group. The infusion did not alter plasma GH concentration but decreased $T_4$ and increased the $T_3$ concentrations (23). One reason for the lack of effect may be the use of human IGF-I. The production and characterization of recombinant chicken insulin-like growth factor-I from *E. coli* or variants which do not bind to IGF binding proteins will allow that to be tested (24).

## Insulin-like growth factor receptors

Distinct receptors which bind IGF peptides have been characterized in mammalian tissues. The type I receptor is structurally related to the insulin receptor, usually binds IGF-I, IGF-II and insulin in order of decreasing affinities and mediates the biological effects of IGF peptides. The type II receptor is a transmembrane molecule with a large extracellular domain, binds IGF-II with a high affinity and is the cation-independent mannose 6-phosphate receptor. It is not clear if it is a transmembrane signalling protein or has perhaps a structural or transport function. De Pablo and co-workers have clearly demonstrated the presence of the type I receptor in different tissues of the chick embryo using ligand binding and affinity labelling techniques. The receptor was identified in membranes prepared from 2 day-old embryos and there is a distinct tissue-specific developmental regulation of the receptor corresponding to well-defined patterns of organogenesis (25). There was no evidence for binding of IGF peptides to a type II receptor in these embryos. They also reported the presence of a cDNA sequence (CTK-2) that, by comparison with known tyrosine kinases, appeared to correspond to the tyrosine kinase domain of the type I receptor. Using RT-PCR they showed that CTK-2 was expressed in embryos as early as the blastoderm, during neurulation (day 1), and in early (days 2 and 3) and late (day 9) organogenesis (26). These data would suggest that the biological action of the IGF peptides, which are expressed throughout embryogenesis, is mediated by a type I receptor.

Using a similar experimental approach, we have shown the presence of a type I receptor in cell membranes prepared from tissues obtained from post-hatch chicks (27). Developmental regulation also occurs post-hatch. For example, in the liver, binding is low in newly hatched chicks, increases three-fold by day 3 and remains high until day 21 before declining to lower levels by 70 days of age. This is a result of a change in receptor number and not affinity. In common with De Pablo's studies on receptors in the embryo, no evidence was found for the presence of a receptor related to the mammalian type II receptor (27). A chicken IGF type I receptor cDNA has recently been cloned and shows 86% and 97% homology to human type I receptor α-subunit at the nucleotide and amino acid level, respectively (28). The clone was used to generate an antisense RNA probe for the detection of type I receptor mRNA transcripts in the liver of post-hatch chicks. The results showed a decrease in the amount of mRNA between 1 and 4 weeks of age and were consistent with the ligand binding studies (27). A major type I receptor mRNA transcript of 11 kb was detected by Northern blotting, with less abundant transcripts between 8 and 9 kb also present (28).

The major advance in the receptor field has been proof that, although the cation-independent mannose 6-phosphate receptor is present in the chick, it does not bind IGF-II. The chicken liver cation-independent mannose 6-phosphate receptor was purified to homogeneity and tested for its ability to bind iodinated human IGF-I, human IGF-II or chicken IGF-II. None of them bound even at a concentration of 1 μmol/l (29). Similarly, when the cation-independent mannose 6-phosphate receptor was purified from *Xenopus*, chicken and rat and then cross-linked to iodinated IGF-II, only the receptor from the rat was affinity labelled (30). These data suggest that the emergence of a high affinity IGF-II binding site on the cation-independent mannose 6-phosphate receptor occurred in evolution after the divergence of mammals from other vertebrates, and that the biological actions of IGF-II in non-mammalian vertebrates are initiated by the type I or related receptor. These results have recently been confirmed and extended to show that chick embryo fibroblasts possess a cation-independent mannose 6-phosphate receptor that is similar in size to the mammalian receptor which binds and internalizes lysosomal enzymes but does not bind IGF-II (31).

**Insulin-like growth factor binding proteins**

In mammals, there are at least six IGF binding proteins (IGFBPs) which have a major influence on the biological actions of IGF peptides (32). In contrast only three papers have been published in the last 4 years which deal directly with IGFBPs in chickens. The presence of IGFBPs has been demonstrated in chicken serum by chromatography under neutral and acidic conditions (33). In neutral conditions, about 70% of the immunoreactive IGF-I is associated with a large protein complex with a molecular weight of 150 Kd, 20-25% is associated with a smaller protein complex (45

Kd) with up to 6% free IGF-I. Chromatography under acid conditions dissociates the complexes and leaves an acid-stable binding protein of about 50-60 Kd. Ligand blotting with iodinated IGF-I detects three specific IGF binding proteins with molecular weights of about 40 Kd, 33 Kd and 28 Kd which we thought were IGFBP-3, 2 and 1 respectively. Further chromatographic studies confirmed the presence of IGF binding proteins in chicken serum with molecular weights similar to those found in human serum (34). The third paper also reported the presence of IGFBPs in chicken serum (35). However, the major binding protein found by neutral chromatography had a molecular weight of 55 Kd, much smaller than that reported earlier (33) or that found in human serum. The major IGFBP in ligand blots was about 32 Kd with only small amounts found corresponding to IGFBP-3, the major species, in human serum (35). The reason for the difference between the studies is not clear.

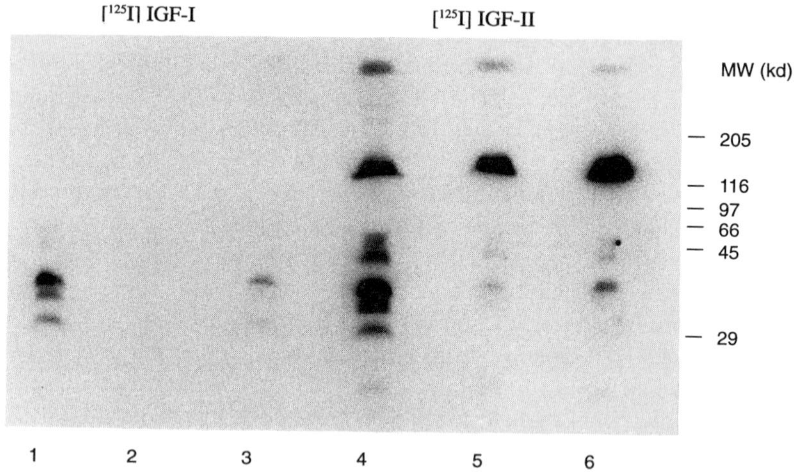


Fig. 1 Western Ligand blot of IGF binding proteins in serum from a six week-old broiler chicken. Serum (25 μl) was separated on 10% SDS-PAGE, transferred to nitrocellulose and incubated overnight with 400 000 c.p.m. $^{125}$I-labelled IGF-I (lanes 1-3) or $^{125}$I-labelled IGF-II (lanes 4-6) in the absence (lanes 1 and 4) or presence of unlabelled IGF-I (500 ng/ml; lanes 2 and 5) or unlabelled IGF-II (500 ng/ml; lanes 3 and 6). The blot was extensively washed, dried and developed by autoradiography. Molecular weights are indicated. (C Goddard, unpublished observation).

The importance of the IGFBPs in regulating the functions of the IGF peptides, particularly by autocrine or paracrine mechanisms, has led us to investigate their regulation at the tissue level. Ligand blotting was used to establish the pattern of IGFBPs in adipocyte precursor cells, skeletal muscle satellite cells, and ovarian follicular granulosa and thecal cells in primary culture. These have been compared with the serum IGFBP pattern. Up to six specific IGFBPs have been identified in ligand blots of serum from broiler and layer chickens (Fig. 1). These have molecular weights of about 44.5, 39.8, 37.2, 32.4, 29.5 and 24 Kd (Fig. 1). Four IGFBPs were found in ovarian follicular theca conditioned medium with molecular weights of 32.5, 31.5, 29.5 and 24.5 Kd. A similar pattern was found in ovarian follicular granulosa conditioned medium plus another IGFBP with a molecular weight of 30.5 Kd. Adipocyte precursor cells and skeletal muscle satellite cells produce similar IGFBPs with major components of 32.4, 29.5 and 24 Kd and with minor amounts of the 44.5 and 39.8 Kd proteins. Addition of IGF-I to almost all of these cell types in monolayer culture results in an increase in IGFBPs below 35 Kd, which in mammals have been shown to modulate the cellular response to IGF peptides. The ubiquitous presence of the IGF binding proteins and their regulation by IGF peptides in completely separate tissues indicates their importance in avian growth and development and highlights the need for more research.

**Acknowledgements**

We would like to thank Ross Breeders, the Ministry of Agriculture, Fisheries and Food and the Department of Education and Science for support.

**References**

1 Kajimoto Y & Rotwein P (1989) Structure and expression of a chicken insulin-like growth factor I precursor. *Molecular Endocrinology* **3** 1907-1913

2 Fawcett D H & Bulfield G (1990) Molecular cloning, sequence analysis and expression of putative chicken insulin-like growth factor I cDNAs. *Journal of Molecular Endocrinology* **4** 201-211

3 Kajimoto Y & Rotwein P (1991) Structure of the chicken insulin-like growth factor I gene reveals conserved promotor elements. *Journal of Biological Chemistry* **266** 9724-9731

4 Taylor E R, Selerio E A P & Brickell P M (1991) Identification of antisense transcripts of the chicken insulin-like growth factor-II gene. *Journal of Molecular Endocrinology* **7** 145-154

5 Dawe S R, Francis G L, McNamara P J, Wallace J C & Ballard F J (1988) Purification partial sequences and properties of chicken insulin-like growth factors. *Journal of Endocrinology* **117** 173-181

6 Armstrong D G, Duclos M J & Goddard C (1990) Biological activity of insulin-like growth factor I purified from chicken serum. *Domestic Animal Endocrinology* **7** 383-393

7 Ballard F J, Johnson R J, Owens P C, Francis G L, Upton F M, McMurtry J P & Wallace J C (1990) Chicken insulin-like growth factor-I: amino-acid sequence, radioimmunoassay and plasma levels between strains and during growth. *General and Comparative Endocrinology* **79** 459-468

8 Kallincos N C, Wallace J C, Francis G L & Ballard F J (1990) Chemical and biological characterisation of chicken insulin-like growth factor-II. *Journal of Endocrinology* **124** 89-97

9 Scavo L, Alemany J, Roth J & De Pablo F (1989) Insulin-like growth factor-I activity is stored in the yolk of the avian egg. *Biochemical and Biophysical Research Communications* **162** 1167-1173

10 Serrano J, Shuldiner A R, Roberts Jr C T, LeRoith D & De Pablo F (1990) The insulin-like growth factor-I gene is expressed in chick embryos during early organogenesis. *Endocrinology* **127** 1547-1549
11 Ralphs J R, Wylie L & Hill D J (1990) Distribution of insulin-like growth factor peptides in the developing chick embryo. *Development* **109** 51-58
12 Kikuchi K, Buonomo F C, Kajimoto Y & Rotwein P (1991) Expression of insulin-like growth factor-I during chicken development. *Endocrinology* **128** 1323-1328
13 Goddard C & Boswell J M (1991) Molecular biology and the growth of poultry. *Critical Reviews in Poultry Biology* **3** 325-340
14 Huybrechts L M, Michielsen R, Darras V M, Buonomo F C, Kühn E R & Decuypere E (1989) Effect of the sex-linked dwarf gene on thyrotrophic and somatotrophic axes in the chick embryo. *Reproduction Nutrition and Development* **29** 219-226
15 Robcis H L, Caldes T & De Pablo F (1991) Insulin-like growth factor-I serum levels show a mid-embryogenesis peak in the chicken that is absent in growth-retarded embryos cultured *ex ovo*. *Endocrinology* **128** 1895-1901
16 McGuinness M C & Cogburn L A (1990) Measurement of developmental changes in plasma insulin-like growth factor-I levels of broiler chickens by radioreceptor assay and radioimmunoassay. *General and Comparative Endocrinology* **79** 446-458
17 Burnside J, Liou S S & Cogburn L A (1991) Molecular cloning of the chicken growth hormone receptor complementary deoxyribonucleic acid: mutation of the gene in sex-linked dwarf chickens. *Endocrinology* **128** 3183-3192
18 Goddard C, Wilkie R S & Dunn I C (1988) The relationship between insulin-like growth factor, growth hormone, thyroid hormone and insulin in chickens selected for growth. *Domestic Animal Endocrinology* **5** 165-176
19 Johnson R J, McMurtry J P & Ballard F J (1989) Ontogeny and secretory patterns of plasma insulin-like growth factor-I concentrations in meat-type chickens. *Journal of Endocrinology* **124** 81-87
20 DeChiara T M, Robertson E J & Efstratiadis A (1991) Parental imprinting of the mouse insulin-like growth factor II gene. *Cell* **64** 849-859
21 Scanes C G, Dunnington E A, Buonomo F C Donoghue D J & Siegel P B (1989) Plasma concentrations of insulin-like growth factors (IGF-) I and IGF-II in dwarf and normal chickens of high and low weight selected lines. *Growth, Development & Aging* **53** 151-157
22 McGuinness M C & Cogburn L A (1991) Response of young broiler chickens to chronic injection of recombinant-derived human insulin-like growth factor-I. *Domestic Animal Endocrinology* **8** 611-620
23 Huybrechts L M, Decuypere E, Buyse J, Kühn E R & Tixier-Boichard M (1992) Effect of recombinant human insulin-like growth factor-I on weight gain, fat content, and hormonal parameters in broiler chickens. *Poultry Science* **71** 181-187
24 Upton F Z, Francis G L, Ross M, Wallace J C & Ballard F J (1992) Production and characterisation of recombinant insulin-like growth factor-I from *Escherichia coli*. *Journal of Molecular Endocrinology* **9** 83-92
25 Bassas L, Lesniak M A, Serrano J, Roth J & De Pablo F (1988) Developmental regulation of insulin and type I insulin-like growth factor receptors and absence of type II receptors in chicken embryo tissues. *Diabetes* **37** 637-644
26 Scavo L, Serrano J, Roth J & De Pablo F (1991) Genes for the insulin receptor and the insulin-like growth factor I receptor are expressed in the chicken embryo

blastoderm and throughout organogenesis. *Biochemical and Biophysical Research Communications* **176** 1393-1401

27 Duclos M J & Goddard C (1990) Insulin-like growth factor receptors in chicken liver membranes: binding properties, specificity, developmental pattern and evidence for a single receptor type. *Journal of Endocrinology* **125** 199-206

28 Armstrong D G & Hogg C O (1992) The expression of a putative insulin-like growth factor-I receptor gene in the liver of the developing chick. *Journal of Molecular Endocrinology* **8** 193-201

29 Canfield W M & Kornfeld S (1989) The chicken liver cation-independent mannose 6-phosphate receptor lacks the high affinity binding site for IGF-II. *Journal of Biological Chemistry* **264** 7100-7103

30 Clairmont K B & Czech M P (1989) Chicken and *Xenopus* mannose 6-phosphate receptors fail to bind insulin-like growth factor II. *Journal of Biological Chemistry* **264** 16390-16392

31 Yang Y W-H, Robbins A R, Nissley S P & Rechler M M (1991) The chick embryo fibroblast cation-independent mannose 6-phosphate receptor is functional and immunologically related to the mammalian insulin-like growth factor-II (IGF-II)/man 6-P receptor but does not bind IGF-II. *Endocrinology* **128** 1177-1189

32 Shimasaki S & Ling N (1991) Identification and molecular characterization of insulin-like growth factor binding proteins (IGFBP-1, -2, -3, -4, -5 and 6). *Progress in Growth Factor Research* **3** 243-266

33 Armstrong D G, McKay C O, Morrell D J & Goddard C (1989) Insulin-like growth factor-I binding proteins in serum from the domestic fowl. *Journal of Endocrinology* **120** 373-378

34 Lee P D K, Peacock A, Roessler M K, Hester J & Reeves J T (1989) Insulin-like growth factor (IGF)-I and IGF-binding activity in normal and fast growing chickens. *Life Sciences* **45** 2465-2470

35 Francis G L, McMurtry J P, Johnson R J & Ballard F J (1990) Plasma clearance of chicken and human insulin-like growth factor-I and their association with circulating binding proteins in chickens. *Journal of Endocrinology* **124** 361-370

Avian Endocrinology
Ed. P J Sharp, pp 285-296
Journal of Endocrinology Ltd, Bristol (1993)

# Metabolic and endocrine control of genetic variation in fat deposition in growing chickens

H D Griffin
Department of Cellular and Molecular Biology, AFRC Roslin Institute (Edinburgh), Roslin, Midlothian EH25 9PS, UK

## Introduction

Commercial strains of broiler chicken in the UK currently contain 13-18% of their body weight as ether-extractable fat and higher levels are often reported from the US. Ducks contain 25-30% of their body weight as fat and turkeys 7-16%, depending on age and sex. These high levels of fat are cause for concern because they decrease lean meat yield and value and could compromise the image of poultry as an important contributor to 'lean cuisine'.

Fatness is a moderately heritable trait in poultry and experimental lines of leaner chicken have been produced by selection against abdominal fat, and for feed efficiency or low plasma very low density lipoprotein (VLDL) concentration (see reference 1 for details). Several of the major poultry breeding companies have already incorporated one or more of these criteria into their selection programmes and in the longer term substantially leaner strains of broilers should become available to commercial growers.

The availability of divergent lines has encouraged comparative studies to determine the physiological basis for genetic differences in fattening and feed efficiency in broiler chickens and this paper reviews how far these have contributed to the understanding of the metabolic and endocrine mechanisms involved. The control of lean and fat deposition in birds has been the subject of a number of recent reviews (2-5): this review summarises the more recent findings and/or those that appear to be particularly relevant in the present context.

## Mechanisms involved in fat deposition in birds

Avian adipose tissue has a very limited capacity for *de novo* fatty acid synthesis and most of the fatty acids that accumulate in avian adipose tissue are derived from the diet or synthesised in the liver. Transport of fat between tissues takes place in the plasma of birds in the form of triglyceride-rich lipoproteins and as non-esterified fatty acids in much the same way as in mammals, although dietary fat is secreted directly into the portal vein as portomicrons rather than *via* the lymphatic system as chylomicrons (6). The avian liver exports surplus fatty acids as very low density lipoproteins (VLDL), which are a little smaller than portomicrons but have very similar lipid and apoprotein composition, at least in immature birds.

## Hepatic lipogenesis

Early studies on lipid metabolism in birds identified the liver as the major site of lipogenesis. In 1969, for example, O'Hea and Leveille (7) estimated that the liver accounted for about 95% of *de novo* fatty acid synthesis in young chicks. More recent investigations on broiler chickens close to commercial killing ages indicate that the liver accounts for only about 65% of whole body lipogenesis, with the remainder distributed throughout a variety of tissues, including skin, bone and adipose tissue (8, 9).

Hepatic fatty acid metabolism is influenced by a large number of hormones, including insulin, glucagon, corticosteroids, thyroid hormones, prolactin and oestrogen, and their possible role in regulating hepatic lipogenesis in birds has been considered in detail in a recent series of reviews (1). It is important to appreciate that some of these are more involved in particular physiological states or stages of development than in others. Triiodothyronine ($T_3$) and corticosterone, for example, appear to play the major roles in the induction of lipogenic enzyme activity in the liver of the neonatal chick as it makes the transition to a carbohydrate-rich diet, with insulin and IGF-I having minor roles at this stage of development. In contrast, the induction of lipogenic enzymes during the fasted-fed transition appears to be largely controlled by insulin, with $T_3$ simply accelerating the induction of activity (10). However, these responses are not obviously related to fattening in young growing broilers reared under typical commercial conditions in which the availability of feed and extended photoperiod ensures birds are fully fed throughout the 24 h of each day.

Most of the studies on hepatic lipid metabolism in birds have focused on regulation of lipogenesis and other aspects of hepatic fatty acid metabolism have tended to be neglected. However, plasma concentrations of β-hydroxybutyrate are about 5-fold greater in *ad libitum*-fed broilers than in fed rats, and this suggests that oxidation is a quantitatively important fate of fatty acids in avian liver, even in the fully fed state. This may be a reflection of the relatively low levels of malonyl coA in avian liver (11) and/or the apparently high glucagon concentrations in avian plasma (12). A combination of high rates of β-oxidation and lipogenesis would provide a particularly sensitive mechanism for the control of the supply of fatty acids for VLDL synthesis and secretion and explain how relatively small changes in hepatic lipogenesis can have a large effect on rate of VLDL secretion (13).

Insulin has a key role in regulating hepatic lipogenesis in mammals, but one of the problems in assigning it a similar role in birds has been the difficulty in demonstrating appropriate responsiveness to insulin *in vitro*. Lipogenesis in liver slices and freshly isolated hepatocytes, for example, is only slightly elevated by short term incubation with insulin (14). Recently we have demonstrated that the rate of lipogenesis in hepatocytes maintained in primary culture for 1-2 days is very sensitive to porcine insulin, with an $ED_{50}$ of between 1-2 ng/ml. Chicken insulin has approximately 2-fold higher affinity for the chicken insulin receptor than porcine insulin and insulin concentrations in the plasma of growing broilers range from 0.5-1.0 ng/ml and our evidence appears to demonstrate for the first time that chicken hepatocytes are responsive to insulin over normal physiological concentrations. Human recombinant IGF-I also stimulates lipogenesis in chick hepatocytes in primary culture, but over concentrations which suggest that it is acting through the insulin receptor rather than the IGF-receptor. This is distinct from the situation in hepatocytes from chick embryos, where IGF-I is more effective than insulin in potentiating the induction of lipogenic enzyme activity by $T_3$ and where both insulin and IGF-I are thought to act through the IGF-receptor (10). Locally high concentrations of IGF-I *in vivo* may therefore stimulate hepatic lipogenesis, although in doing so they may decrease its responsiveness to changes in plasma insulin concentration.

Glucagon is a very potent inhibitor of hepatic lipogenesis *in vitro* and has been proposed to be a much more significant partner in the insulin/glucagon ratio in birds than in mammals (15). Although reported plasma concentrations of glucagon in birds are relatively high, there is considerable doubt about the specificity of the heterologous radioimmunoassays that have been used and such data must be viewed with caution. Triiodothyronine does not have any effect on lipogenesis in hepatocytes from fully-fed birds in which lipogenic enzyme activity appears to be already fully induced (16). In contrast, the feeding of $T_3$ in the diet reduces fatness in broilers and inhibits hepatic lipogenesis, but this effect is most probably mediated through changes in plasma insulin/glucagon ratio (12).

Chickens are not as responsive to exogenous growth hormone as, for example, the pig, and its effects appear to be very dependent on age and sex of the birds, dosage and method of administration (3, 4). Pulsatile, but not continuous, administration of GH has been shown to decrease fat content and inhibit hepatic lipogenesis by over 70% in 8-11 week old female broilers (17, 18). The difference in response to different modes of administration was attributed to differential effects on plasma concentrations of insulin and corticosterone rather than to any direct effect of GH on the liver. High concentrations of GH (1 μg/ml) have been reported to inhibit the small increase in lipogenesis caused by insulin in freshly isolated hepatocytes (19), but we have been unable to repeat these observations in either freshly isolated hepatocytes or in hepatocytes in primary culture. Many of the fatty acids released from adipose tissue as a consequence of any lipolytic effect of GH will be taken up by the liver and this could provide an additional mechanism for inhibition of hepatic lipogenesis.

**Plasma lipoprotein metabolism**

Very low density lipoproteins secreted from the liver and portomicrons secreted from the intestine are hydrolysed in the peripheral circulation by lipoprotein lipase (LPL) and the fatty acids released are taken up into surrounding tissues to be re-esterified (in adipose tissue) or oxidised (in heart or skeletal muscle). In rats, up to 40% of plasma lipoprotein triglyceride is taken up and oxidised by muscle. Only 20-25% of [$^{14}C$]-labelled VLDL triglyceride was found to be oxidised to [$^{14}CO_2$] in young broiler chickens (20) and this suggests that efficiency of transfer of VLDL triglyceride from the liver to adipose tissue is relatively high in birds. The distribution of circulating triglyceride between different depots has been linked to differences in tissue lipoprotein lipase activity (21), but the major problem in determining the specific contribution of LPL to differences in fattening is the present inability to measure functional activity in tissues (i.e. the proportion of LPL at the capillary bed). We have found that only about 2% of the total tissue LPL activity in chicken heart is released by perfusing with heparin and a similarly low functional LPL activity in skeletal muscle could account for the relatively high proportion of lipoprotein triglyceride taken up into adipose tissue in birds.

**Adipose tissue metabolism**

Adipose tissue grows in young animals by a combination of an increase in cell size and number. Hyperplasia appears to make the major contribution to the growth of the abdominal fat pad in younger broilers, but fat cell numbers do not increase much after about 14 weeks of age and further growth of the fat pad occurs mainly by hypertrophy (22). Adipose tissue accounts for an estimated 95% of whole body lipogenesis in the pig and in this species an increase in fat cell number would be expected to have a direct effect on capacity for *de novo* fatty acid synthesis. In contrast, it is not at all clear how an increase in fat cell number *per se* would effect fat deposition in birds, since adipose tissue lipogenesis accounts for perhaps only 10-15% of the total fatty acids accumulating in avian adipose tissue, even in leaner birds (13). Adipocyte hyperplasia might influence functional LPL activity or blood flow, but there is no evidence of this.

Glucagon and GH are the major lipolytic hormones in birds. *In vitro*, the initial exposure of adipocytes to GH reduces their sensitivity to glucagon, but in longer term incubations the effects of GH and glucagon are additive (3). The sensitivity of chicken adipocytes to glucagon declines with age, although this may be simply a reflection of the accompanying increase in adipocyte size.

**Why are broiler chickens fat?**

Young chicks of strains selected for egg-laying performance contain less than half the proportion of carcass fat of rapidly growing broiler strains and differences in the abdominal fat pad of broiler and layers are even greater (23). The relative fatness of broiler strains is generally believed to be a direct consequence of their intense selection for rapid growth for 40 or more generations, although across livestock species the evidence that an increase in fatness is an inevitable consequence of selection for growth is not conclusive.

Recent comparisons of lipid metabolism in broiler and layer strains have demonstrated that differing rates of lipolysis play the major part in determining their relative fatness (24). The rate of lipoprotein synthesis and secretion from the liver and the proportion of circulating plasma triglyceride taken up into the abdominal fat pad are both about 50% higher in broilers than in layers. Nevertheless, because of the lower rate of growth and relative leanness of layers, uptake of lipoprotein fatty acids in to the abdominal fat pad of layers occurs at about 4-fold the net rate of accumulation of fat by the depot. In layers, adipose tissue triglyceride shows a very rapid turnover, with a half-life of 2-3 days but there is no turnover of adipose tissue triglyceride in broilers. There are no significant differences in plasma immunoreactive glucagon concentrations between broilers and layers (24) and the differences between strains in turnover of adipose tissue triglyceride are almost certainly a consequence of the effects of selection for growth on plasma GH levels. Plasma GH concentrations rise rapidly after hatch in both broilers and layers, reaching a peak after about 2 weeks of age and declining rapidly thereafter. However, plasma GH concentrations are 5-10 fold higher in layer strain chicks than in broilers (25). Plasma insulin concentrations tend to be higher in broilers than layers and although these differences are not very large or consistent, they may be sufficient to account for the difference in hepatic lipogenesis between strains implied by their different rates of VLDL synthesis and secretion.

Apparent confirmation of the relationship between growth, GH and body composition is also provided by comparisons of low- and high-body weight lines (LW and HW) produced by more than 25 generations of divergent selection for body weight by Seigel and colleagues (26, 27). These lines differ by more than 6-fold in body weight at 56 days of age (the age of selection) and the heavier line has 2-fold greater percentage of abdominal fat. As in broilers and layers, differences in body composition between LW and HW lines are greater at an earlier age. Endocrinological comparisons have shown that the leaner LW line has a 4-fold greater concentration of plasma GH, whereas plasma IGF-I concentrations are almost 2-fold greater in the HW line (26, 27). In contrast, there are no significant differences between LW and HW lines in plasma thyroid hormone concentrations. The leaner LW line is more able to clear glucose from the plasma after intubation of glucose after an overnight fast, and plasma concentrations of insulin were elevated in this test in the HW relative to the LW line (38). Hepatic lipogenesis, as assessed by measurement of lipogenic enzyme activity and the incorporation of [$^{14}$C]-acetate into liver slices *in vitro*, was reported to be significantly higher in the leaner, LW line (28) but plasma non-esterified fatty acid concentrations were greater and fatty acid release from adipose tissue pieces slightly (though not statistically) higher in the LW line. It was concluded (28) from this limited evidence that the relative leanness of the LW line chicks was a consequence of rapid turnover of adipose tissue triglyceride, most likely mediated by their higher plasma GH concentration.

**Fat and lean lines of broiler**

Lines of broiler chicken differing substantially in fatness have been produced by selecting against fat using sib analysis or plasma very low density lipoprotein concentration or indirectly using feed conversion efficiency (see 1). The endocrine and metabolic mechanisms responsible have been studied in greatest detail in the lean and fat lines produced in France by Leclercq by divergent selection for abdominal fat weight (20) and by Whitehead and Griffin (1, 30) by divergent selection for plasma VLDL concentration. In both cases, differences in body composition between lines persist when they are pair fed and this indicates that they are primarily of metabolic origin rather than a consequence of differences in appetite.

*French lean and fat lines*

The French lean (LL) and fat (FL) lines were produced from a synthetic broiler population by divergent selection for abdominal fat pad weight in males. The response to selection was relatively rapid and after 7 generations of selection the lines differed by over 3-fold in abdominal fat pad weight in both males and females (20). The rate of hepatic lipogenesis in the FL line is only about 50% greater than in the LL line (8, 9) and although the activity (31) and mRNA levels (32) of lipogenic enzymes tend to be higher in the fatter line, differences between lines are small. A consistent difference in the activity and mRNA levels of Δ-9 desaturase between lines has led to the suggestion (31) that the major effect of selection on hepatic lipid metabolism was to alter the partition of fatty acids between oxidation and VLDL secretion. However there is no direct evidence of this and indeed the rate of VLDL secretion in the fat line is only about 1.5 fold greater than that in the lean line (33). It should be recognised that there is not a simple relationship between rate of hepatic lipogenesis and body fat content in lean and fat lines. Differences in body composition persist when birds are fed diets containing sufficiently high level of fat to almost totally suppress hepatic lipogenesis (9) and any explanation of the underlying genetic differences between lines must account for this.

Comparison of plasma lipoprotein metabolism in the LL and FL lines has shown that there is a 2-fold greater uptake of circulating VLDL triglyceride into the fat pad of the fat line. This was accompanied by a 2-fold greater activity of LPL in the abdominal fat pad and this was suggested (34) to be important in determining the overall body compositions of the two lines. Because only about 20% of circulating VLDL triglyceride is oxidised, changes in LPL activity in response to selection are likely to be more important in determining the distribution of circulating triglyceride between the abdominal fat pad and other fat depots than its partition between fat deposition and oxidation.

The possible importance of adipose tissue triglyceride turnover has been examined in LL and FL birds by feeding a diet containing isomerised rapeseed oil between 4 and 10 weeks of age and following the decrease in the elaidic acid content of their abdominal fat pads over the following 10 weeks (35). The results indicated very similar half lives of 29 and 32 days in the two lines and this suggests that the difference in fatness between the two lines was not related to any difference in lipolysis. No difference has been found in the lipolytic response of adipose tissue from FL and LL birds to glucagon *in vitro* (36).

Many of the endocrinological studies on FL and LL lines have been directed at identifying a possible role for insulin (37). Plasma concentrations of glucose tend to be slightly lower in the fat line and this is associated with similar or slightly higher plasma insulin concentrations in both the fed and fasted states. Plasma insulin concentrations in the fat line following refeeding after a fast or a glucose tolerance test were found to be significantly elevated after 4 generations of selection. However, in subsequent generations refeeding after a fast was associated with higher plasma glucose and insulin levels in the lean line. The variation in results from these and similar studies on the HW and LW lines (38) appear to substantiate an earlier conclusion that there is no relationship between glucose tolerance and fattening in broilers (39).

Decreased hepatic responsiveness to insulin in obese rats have been linked to a decreased number and activity of insulin receptors, but no difference in these parameters was found in the LL and FL lines (40). Between line differences in insulin/glucagon ratio have been reported (41) and although these are small, they may be sufficient to contribute to the relative small differences in their rates of hepatic lipogenesis and VLDL secretion. No significant differences in plasma levels of corticosterone, growth hormone or oestrogen have been found between the LL and FL lines and although the plasma concentrations of luteinizing hormone were significantly higher in the fatter line (42), the relevance of this to differences in body composition is unclear.

*High- and low-VLDL lines*

The possibility of using plasma very low density lipoprotein concentrations as a selection parameter arose from observations that plasma VLDL concentrations and body fat content were moderately highly correlated in fully-fed broilers. Subsequent divergent selection created lean and fat lines with substantial differences in body composition and feed efficiency (1, 30). The greater than 10-fold difference in plasma VLDL concentration between high- and low-VLDL lines after 10 generations of selection is largely a consequence of a 3-4-fold increase in the rate of VLDL synthesis and secretion by the liver (20, 43). This is accompanied by an approximately 2-fold difference in *de novo* fatty acid synthesis (11) and smaller, but significant, differences in hepatic lipogenic enzyme activity (44). These differences are greater than found in the FL and LL lines and these lines may be a better model for studying the effects of selection on hepatic lipid metabolism.

Divergent selection for plasma VLDL has produced a 2.5-fold greater uptake of circulating VLDL triglyceride into the abdominal fat pad of the high-VLDL line and this occurred despite a somewhat lower total LPL activity in the fat pads of the fatter high-VLDL line (44). This observation emphasises the uncertain relationship between total tissue LPL activity and lipoprotein metabolism, an issue that will only be resolved if methods can be developed to measure functional LPL activity. There are no differences in proportion of [$^{14}$C]-labelled VLDL triglyceride oxidised to $CO_2$ in the high- and low-VLDL lines (20).

Endocrine comparisons of the high-and low-VLDL lines have so far been limited to measurement of plasma hormone concentrations. No significant differences between lines have been found for plasma corticosterone, insulin, glucagon or oestrogen. Triiodothyronine tends to be negatively associated with genetic differences in fatness, but although a significantly high level of $T_3$ was found in the low-VLDL line in an initial study, this has not been confirmed in later generations. We have found consistently higher levels of GH in young males of the low VLDL line and this appears to be due to differences in pulse amplitude rather than frequency. At the metabolic level, divergent selection for plasma VLDL concentration appears to have had greater effect upon hepatic lipid metabolism than on adipose tissue, and the significance of higher GH levels on VLDL secretion rate is unclear since there is as yet little evidence that GH has a direct effect on lipid metabolism in avian liver.

*Other selected lines*

There have been relatively few studies on the mechanisms underlying differences in body composition in other selected lines, although when they have been made they tend to reinforce some of the observations already described. For example, Hood and Pym (45) examined differences in hepatic lipogenesis and lipogenic enzyme activity in lines selected for improved feed efficiency, food consumption or leanness and found significant but small differences between lines. Comparison of lines selected for growth or feed efficiency differ significantly in body composition and the leaner, more efficient line had greater circulating GH concentrations (46), although in this case it is not clear to what extent this is due to selection for growth or feed efficiency. Differences in the *in vitro* sensitivity of adipose tissue to GH and glucagon have been reported in these lines (47), but there is as yet no corroborating *in vivo* evidence that this is an important mechanism in determining their relative fatness.

## Conclusions

The evidence for the involvement of GH in the relative fatness of faster growing birds is convincing, but the involvement of any single hormone in the response of broilers to selection for leanness has yet to be demonstrated. This is perhaps not too surprising considering the number of potential candidates and the lack of any evidence of the involvement of a major gene from the analysis of crosses between both the FL and LL (29) and high and low-VLDL lines. In contrast, significant progress has been made in developing approaches to identifying the major metabolic responses to selection for leanness and these might fruitfully be used to target areas for more focused studies of the hormonal mechanisms involved. What is clear is that the mechanisms controlling genetic variation in body composition will vary according to the selection criteria used to create it.

## Acknowledgements

Work in the author's laboratory was supported by a commission from the Ministry of Agriculture, Fisheries and Food. The contributions of Dawn Windsor, Chris Goddard and Ian George to the unpublished studies referred to in the text are gratefully acknowledged.

## References

1 Leclercq B & Whitehead C C (1988) *Leanness in Domestic Birds: genetic, metabolic and hormonal aspects.* London: Butterworths

2 Leclercq B (1984) Adipose tissue metabolism and its control in birds. *Poultry Science* **63** 2044-2054

3 Scanes C G (1987) The physiology of growth, growth hormone, and other growth factors in poultry. *CRC Critical Reviews in Poultry Biology* **1** 51-105

4 Vasilatos-Younken R & Scanes C G (1991) Growth hormone and insulin-like growth factors in poultry growth: required, optimal, or ineffective? *Poultry Science* **70(8)** 1764-1780

5 Griffin H D, Goddard C & Butterwith S C (1992) Control of fat deposition in birds. *Proceedings of the 51st University of Nottingham Easter School in Agricultural Science 1991* 231-248

6 Griffin H D & Hermier D (1988) Plasma lipoprotein metabolism in poultry. *Leanness in Domestic Birds: genetic, metabolic and hormonal aspects* pp 175-201. Eds B Leclercq & C C Whitehead. London: Butterworths

7 O'Hea E K & Leveille G A (1969) Lipid biosynthesis and transport in the domestic chick (*Gallus domesticus*). *Comparative Biochemistry and Physiology* **30** 149-159

8 Saadoun A & Leclercq B (1986) *In vivo* lipogenesis in genetically fat and lean chickens of various ages. *Comparative Biochemistry and Physiology B* **83** 607-611

9 Saadoun A & Leclercq B (1987) *In vivo* lipogenesis of genetically lean and fat chickens: effects of nutritional state and dietary fat. *Journal of Nutrition* **117** 428-435

10 Goodridge A G, Fantozzi D A, Klautky S A, Ma X J, Roncero C & Salati L M (1991) Nutritional and hormonal regulation of genes for lipogenic enzymes. *Proceedings of the Nutrition Society* **50(2)** 115-122

11 Griffin H D, Windsor D & Zammit V (1990) Regulation of palmitoylacyltransferase I in chick liver. *Biochemical Society Transactions* **18** 981-982

12 Cogburn L A, Liou S S, Alfonso C P, McGuinness M C & McMurtry J P (1989) Dietary thyrotropin-releasing hormone stimulates growth rate and increases insulin:glucagon molar ratio of broiler chickens. *Proceedings of the Society for Experimental Biology and Medicine* **192** 127-134

13 Griffin H D, Guo K, Windsor D & Butterwith S C (1992) Adipose tissue lipogenesis and fat deposition in leaner broiler chickens. *Journal of Nutrition* **122** 363-368

14 Vasilatos-Younken R (1986) Age-related changes in tissue metabolic rates and sensitivity to insulin in the chicken. *Poultry Science* **65** 1391-1399

15 Hazelwood R L (1984) Pancreatic hormones, insulin/glucagon molar ratios, and somatostatin as determinants of avian carbohydrate metabolism. *Journal of Experimental Zoology* **232** 647-652

16 Tarlow D M, Watkins P A, Reed R E, Miller R S, Zwergel E E & Lane D (1977) Lipogenesis and the synthesis and secretion of very low density lipoproteins by avian liver cells in non-proliferating monolayer culture. Hormonal effects. *Journal of Cell Biology* **73** 332-353

17 Vasilator-Younken R, Cravener T L, Cogburn L A, Mast M G & Wellenreiter R (1988) Effect of pattern of administration on the response to exogenous, pituitary-derived chicken growth hormone by broiler-strain pullets. *General and Comparative Endocrinology* **71** 268-283

18 Rosebrough R W, McMurtry J P & Vasilatos-Younken R (1991) Effect of pulsatile or continuous administration of pituitary-derived chicken growth hormone (p-cGH) on lipid metabolism in broiler pullets. *Comparative Biochemistry and Physiology* **99A** 207-214

19 Harvey S, Scanes C G & Howe T (1977) Growth hormone effect on *in vitro* metabolism of avian adipose and liver tissue. *General and Comparative Endocrinology* **33** 322-328

20 Griffin H D, Windsor D & Whitehead C C (1991) Changes in lipoprotein metabolism and body composition in chickens in response to divergent selection for plasma lipoprotein metabolism. *British Poultry Science* **32** 195-201

21 Butterwith S C (1989) Contribution of lipoprotein lipase activity to the differential growth of three adipose tissue depots in young broiler chickens. *British Poultry Science* **30** 927-933

22 Hood R L (1984) Cellular and biochemical aspects of fat deposition in the broiler chicken. *World's Poultry Science Journal* **40** 160-169

23 Griffin H D, Butterwith S C & Goddard C (1987) Contribution of lipoprotein lipase to differences in fatness between broiler and layer-strain chickens. *British Poultry Science* **28** 197-206

24 Griffin H D, Windsor D & Goddard C (1991) Why are young broiler chickens fatter than layer-strain chicks? *Comparative Biochemistry and Physiology* **100A** 205-210

25 Goddard C, Wilkie R & Dunn I C (1988) The relationship between insulin-like growth factor, growth hormone, thyroid hormone and insulin in chickens selected for growth. *Domestic Animal Endocrinology* **5** 165-176

26 Nir I, Harvey S, Cherry J A, Dunnington E A, Klandorf H and Siegel P B (1987) Growth associated traits in parental and F1 populations of chickens under different feeding programmes. 4. Growth and thyroid hormone. *Poultry Science* **66** 32-37

27 Scanes C G, Dunnington E A, Buonomo F C, Donoghue D J & Siegel P B (1989) Plasma concentrations of insulin-like growth factors (IGF)-I and IGF-II in dwarf and normal chickens of high and low weight selected lines. *Growth, Development and Aging* **53** 151 157

28 Calabotta D F, Cherry J A, Siegel P B & Jones D E (1985) Lipogenesis and lipolysis in fed and fasted chicks from high and low body weight lines. *Poultry Science* **64** 700-704

29 Leclercq B (1988) Genetic selection of meat-type chickens for high or low abdominal fat. *Leanness in Domestic Birds: genetic, metabolic and hormonal aspects* pp 25-40. Eds B Leclercq & C C Whitehead. London: Butterworths

30 Whitehead C C & Griffin H D (1984) Development of divergent lines of lean and fat broilers using plasma very low density lipoprotein concentration as selection criterion: the first three generations. *British Poultry Science* **25** 573-582

31 Legrand P, Mallard J, Bernard-Griffiths M A, Douaire M & Lemarchal P (1987) Hepatic lipogenesis in genetically lean and fat chickens. *In vitro* studies. *Comparative Biochemistry and Physiology B* **4** 789-792

32 Douaire M, Le Fur N, El Khadir-Mounier C, Langlois P, Flamart F & Mallard J (1992) Identifying genes involved in the variability of genetic fatness in the growing chickens. *Poultry Science* **71** 1911-1920

33 Leclercq B, Hermier D & Guy G (1990) Metabolism of very low density lipoprotein in genetically lean or fat lines of chicken. *Reproduction Nutrition and Development* **30** 701-715

34 Hermier D, Quignard-Boulange A, Dugail I, Guy G, Salichon M R, Brigant L, Ardouin B & Leclercq B (1989) Evidence of enhanced storage capacity in adipose tissue of genetically fat chickens. *Journal of Nutrition* **119** 1369-1375

35 Lemarchal P, Beaunez P, Roger C & Legrand P (1988) *In vivo* turnover of triglyceride fatty acids in adipose tissue of chickens (*Gallus domesticus*) selected for low or high adiposity. *Comparative Biochemistry and Physiology B* **89** 227-231

36 Leclercq B, Chevalier B, Derouet M & Simon J (1988) *In vitro* sensitivity of adipocytes from lean or fat chickens to glucagon and to an analogue of adenosine. *Leanness in Domestic Birds: genetic, metabolic and hormonal aspects* pp 239-242. Eds B Leclercq & C C Whitehead. London: Butterworths

37 Simon J (1988) Insulin in birds: metabolic effects and possible implications in genetically fat and lean chickens. *Leanness in Domestic Birds: genetic, metabolic and hormonal aspects* pp 258-268. Eds B Leclercq & C C Whitehead. London: Butterworths

38 Sinsigalli N A, McMurtry J P, Cherry J A & Siegel P B (1987) Glucose tolerance, plasma insulin and immunoreactive glucagon in chickens selected for high and low body weight. *Journal of Nutrition* **117** 941-947

39 Simon J (1980) Apparent independence of glucose tolerance and body fat content in 8-week-old broilers. *British Poultry Science* **21** 309-314

40 Simon J, Chevalier B, Derouet M & Leclercq B (1991) Normal number and kinase activity of insulin receptors in liver of genetically fat chickens. *Journal of Nutrition* **121** 379-385

41 Dimock H D, Marliss E B & Touchburn S P (1988) Plasma pancreatic hormones in genetically fat and lean chickens. *Leanness in Domestic Birds: genetic, metabolic and hormonal aspects* pp 323-325. Eds B Leclercq & C C Whitehead. London: Butterworth

42 Williams J, Harvey S & Leclercq B (1986) Plasma levels of luteinizing hormone, growth hormone and estradiol from six weeks of age to sexual maturity in two lines of chickens selected for low and high abdominal fat content. *British Poultry Science* **65** 1782-1786

43 Griffin H, Acamovic F, Guo K & Peddie J (1989) Plasma lipoprotein metabolism in lean and in fat chickens produced by divergent selection for plasma very low density lipoprotein concentration. *Journal of Lipid Research* **30** 1243-1250

44 Bannister D W, Lee A, Whitehead C C & Griffin H D (1984) Lipogenic enzyme activity and fructose: 2-6 bisphosphate concentration in livers of two lines of domestic fowl (*Gallus domesticus*) selected for different body fat content. *International Journal of Biochemistry* **16** 1301-1305

45 Hood R L & Pym R A E (1982) Correlated responses for lipogenesis and adipose tissue cellularity in chickens selected for body weight gain, food consumption, and food conversion efficiency. *Poultry Science* **61** 122-127

46 Decuypere E, Leenstra F, Buyse J, Beuving G & Berghman L (1991) Temporal secretory patterns of growth hormone in male meat-type chickens of lines selected for body weight gain or food conversion. *British Poultry Science* **32** 1121-1128

47 Buyse J, Decuypere E, Leenstra F & Scanes C G (1992) Abdominal adipose tissue from broiler chickens selected for body weight or for food conversion efficiency differ in *in vitro* responsiveness to glucagon and to chicken growth hormone, but not to dibutyryl cAMP. *British Poultry Science* **33** 1069-1076

# Part Six
# Endocrine Control of Ovarian and Oviduct Function

Avian Endocrinology
Ed. P J Sharp, pp 297-308
Journal of Endocrinology Ltd, Bristol (1993)

# The molecular biology and endocrinology of inhibin in the domestic hen

P A Johnson and S-Y Wang
Department of Animal Science, 202 Morrison Hall, Cornell University, Ithaca, NY 14853, USA

## Introduction

Inhibin is a gonadal protein hormone which has been shown to be involved in the negative feedback regulation of pituitary follicle stimulating hormone (FSH) in mammals (1, 2). More recently, the local or paracrine role of inhibin in the ovary has been defined (3). Mammalian inhibin is a dimeric protein hormone produced by the ovary and testis and composed of an α- (approximate mol. wt 18,000) and a ß-subunit (approximate mol. wt. 14,000). The ß-subunit exists in two forms ($ß_A$ and $ß_B$) which are distinct but quite similar. Therefore, depending upon the ß-subunit involved, inhibin exists as inhibin-A or inhibin-B. Both subunits, when joined by disulphide bonds, are required for biological activity in suppressing FSH secretion from the pituitary (4). Interestingly, combination of two ß-subunits gives rise to the hormone activin, a substance found to release FSH from the pituitary (4).

In mammals, the mRNAs for the inhibin α- and ß-subunits have been examined in follicular tissue by *in situ* hybridization, solution hybridization, as well as by Northern analysis (5, 6). Both of these studies show an increase of expression of the two subunits (α- and $ß_A$-subunit) associated with follicular development, with a 10-25 fold greater abundance of the α-subunit compared to the $ß_A$-subunit. This suggests that the inhibin $ß_A$-subunit may be limiting in the formation of biologically active inhibin (5, 6). It also suggests that the α-subunit may have an independent physiological role (7). Support for differential regulation and action of the subunits is given by differences observed in the regulatory sequences at the 5′ flanking sequences of the subunits (8).

There are few reports characterizing inhibin in birds. Initial studies indicated that chicken testes preparations were effective in selectively reducing the postcastration rise in FSH in adult male rats (9, 10). Lewis *et al.* (11) reported that inhibin activity (as measured by heterologous RIA) decreased in roosters after castration and that the messenger RNA for the α-subunit of inhibin cross-hybridized with a human α-inhibin complementary DNA. More recently, immunoreactive inhibin concentration has been estimated in the plasma, amniotic fluid and gonads of chick embryos (12) using a heterologous assay system. An immunohistochemical study (using antibodies directed against human peptides) examined the presence and distribution of inhibin α- and $ß_A$-

subunits in chicken tissues during the first week of embryonic development (13). A cDNA fragment (270 bp) coding for the inhibin $\beta_B$-subunit has been cloned from chick embryo hypoblast and a fragment of the $\beta_A$-subunit cloned from chicken genomic DNA (14).

Two studies have investigated the bioactivity of chicken ovarian inhibin. Both of these studies demonstrated a selective suppression of FSH when granulosa cell conditioned medium (charcoal-stripped) was incubated with rat (15) or sheep (16) anterior pituitary cells. Akashiba *et al.* (15) demonstrated a greater amount of biologically active inhibin in the granulosa of the largest ($F_1$) follicle as compared to that of the third largest ($F_3$) follicle. In the study by Tsonis *et al.* (16), both granulosa and theca were tested and although theca cell conditioned medium suppressed FSH, it was less potent than granulosa cell-conditioned medium. Using a heterologous RIA, Vanmontfort *et al.* (17) demonstrated the presence of immunoreactive inhibin in the granulosa cells of the hen as well as *in vitro* production of inhibin from granulosa cells. Additionally, they showed that immunoreactive inhibin production was inversely correlated with follicle size (among the hierarchical follicles).

In order to understand further the function of inhibin in the hen, we have primarily taken a molecular approach. Results of the cloning and sequence analysis of the chicken inhibin $\alpha$-subunit are summarized in this manuscript. In addition, results from Northern and slot blot analysis of the RNA for the $\alpha$-subunit from the largest preovulatory follicles are presented. Finally, we discuss plasma immunoreactive inhibin data, obtained using a heterologous RIA system, validated for use in the chicken.

## Cloning and nucleotide sequence of the complementary DNA (cDNA) for the inhibin $\alpha$-subunit in the hen

Prior to initiating this aspect of the work, it was necessary to determine that a mammalian probe for the inhibin $\alpha$-subunit could be used to detect the chicken messenger RNA (mRNA) for the $\alpha$-subunit of inhibin. We began our studies with the $\alpha$-subunit because this subunit is unique to inhibin as opposed to the $\beta$-subunit which is also present in the molecule activin. We believed that a mammalian probe would hybridize with the chicken mRNA because of the great similarity in sequence of the $\alpha$-subunit of inhibin among various mammalian species. The amino acid sequence of the $\alpha$-subunit of inhibin exhibits approximately 80% similarity among the porcine, bovine, human and murine species (18). A porcine cDNA probe for the inhibin $\alpha$-subunit (19) was obtained from Dr Kelly Mayo (Northwestern University). This probe was used in Northern hybridization with chicken RNA extracted from the granulosa layers of the large preovulatory follicles. A single, main band of hybridization was detected at approximately 1.7 kb. This indicated that the mammalian inhibin $\alpha$-subunit was sufficiently similar in sequence to chicken inhibin $\alpha$-subunit to permit the use of the porcine cDNA to screen a chicken cDNA library.

Granulosa layers of the largest preovulatory follicles ($F_1$ through $F_4$) from laying hens were used as a source of mRNA for the commercial preparation of a cDNA library in lambda gt10. We used the porcine cDNA probe (19) labelled by the random primer method to screen the cDNA library. Out of many positive clones (approximately 40), 6 were plaque-purified after tertiary screening and four of the longest (1.7, 1.3, 1.0 and 0.8kb) were selected for subcloning and further analysis. On the basis of restriction enzyme analysis of the clones, a subcloning and sequencing strategy was devised. All four clones contained overlapping inserts and shared a common 3′ end. The positive clones were subcloned into the Bluescript plasmid vector. Some were digested with PstI and portions of the inserts were subcloned into Bluescript plasmid vector at the PstI site. These subclones were used to determine the sequence of the putative full-length cDNA. Dideoxy-mediated chain termination sequencing (20) was performed on denatured double-stranded plasmid DNA by extending primers hybridized to the T3 or T7 promoters of the vector.

Sequence analysis showed that the longest clone contained a complete open reading frame encoding the inhibin α-subunit (21). Beginning with a start (ATG) and ending with a stop (TAG) codon, the open reading frame consisted of 987 nucleotides. This predicted a precursor protein of 328 amino acids from methionine (-60) to valine (+268). Similarly to the mammalian inhibin α-subunits, the chicken proinhibin α-subunit is predicted to be derived from its precursor by cleavage at Arg-Arg (-2,-1). Cleavage at Arg-Arg (+154, +155) gives rise to an N-terminal and a C-terminal portion. The C-terminal portion presumably functions as the mature α-subunit. A stop codon (TAG) follows the code for the amino acid valine (position 269) and suggests that a long 3′ untranslated region of approximately 500 bases follows the open reading frame. Although a classical poly(A) tail was not identified, a polyadenylation signal (AATAAA) was found.

Figure 1 schematically depicts the full-length amino acid sequence for the chicken inhibin α-subunit derived from the cDNA sequence. It is shown with reference to bovine pre-proinhibin as published by Forage *et al.* (22). Substantial similarities between the chicken and bovine inhibin α-subunit are evident. First, a large precursor protein is synthesized, followed by proteolytic processing forming polypeptides of different lengths, of which the C-terminal fragment is the mature α-subunit. In the chicken, the mature α-subunit is 113 amino acids compared to 134 amino acids in the bovine species. Secondly, both the chicken and the bovine inhibin α-subunit possess a potential glycosylation site at a similar position in the molecule (amino acid 32 in the chicken; amino acid 36 in the bovine α-subunit). Finally, the number of cysteine residues is the same and their corresponding positions in the amino acid sequences are almost identical. There are four cysteine residues in the N-terminal fragment of the amino acid sequence suggesting possible internal bonding sites which permit the separation of the N-terminal fragment from the C-terminal fragment. The remaining odd number of seven cysteine residues in the C-terminal fragment suggests that at least one cysteine could be linked to the ß-subunit by forming a disulphide bond.

The full length cDNA sequence for the inhibin α-subunit was scanned for homology to the bovine cDNA sequence in GenBank (National Institute of Health genetic sequence databank: GenBank tape release 66.0 [12/90]) and EMBL (European Molecular Biology Laboratory EMBL data library, tape release 25.0 [11/90]) data bases. The percent identity of the chicken inhibin α-subunit cDNA to the bovine sequence is 89%. The PIR-Protein data base (National Biomedical Research Foundation protein sequence data base of the Protein Identification Resource, release 26.0 [9/90]) indicated that the percent identity for the predicted amino acid sequence of the chicken inhibin α-subunit to the bovine sequence is 60%.

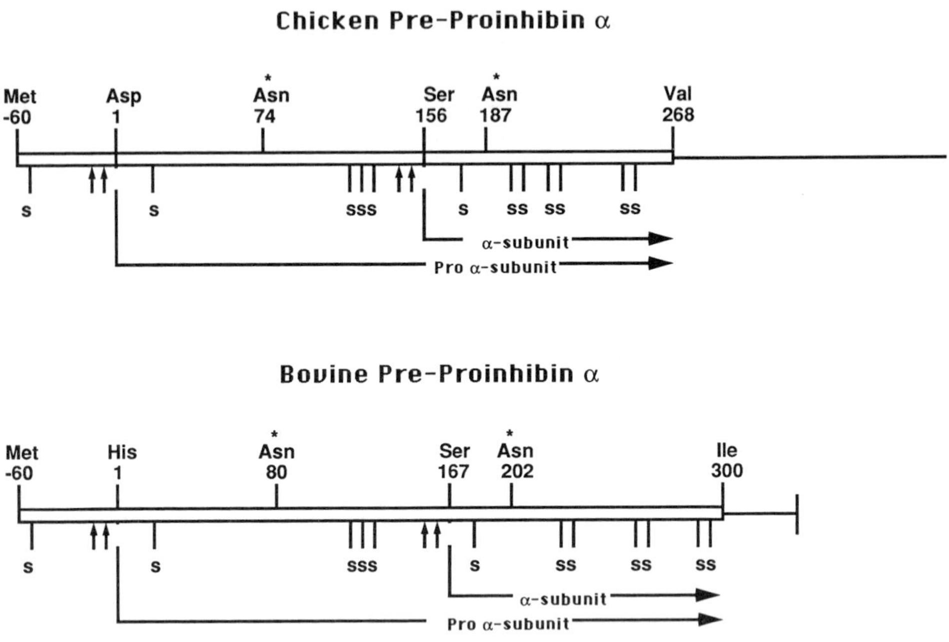


Fig. 1 Schematic version of the chicken cDNA for the inhibin α-subunit as compared to the bovine inhibin α-subunit reported by Forage *et al.* (22). The numbers refer to the amino acid residues and the positions of cysteine residues are indicated by (S). Asterisks (*) represent potential glycosylation sites.

The specificity of the clone was tested by Northern analysis with RNA prepared from several different tissues of the hen (21). RNA was prepared from hen brain, spleen, liver, kidney and the ovarian granulosa layer. The RNA was separated on a denaturing formaldehyde gel, transferred to a nylon membrane and hybridized with the two longest clones (1585 and 942 bp) labelled by the random primer method. With both probes, a single band of hybridization (at approximately 1.7 kb) was only observed in the lane containing RNA extracted from the ovarian granulosa layer. These results suggested that, of the tissues tested, inhibin α-subunit mRNA was unique to the granulosa layer. In addition, we have performed subsequent Northern hybridizations with the full length inhibin α-subunit clone using RNA extracted from the theca layer and have found no evidence for inhibin α-subunit mRNA expression in this layer.

**Regulation of the messenger RNA for the inhibin α- and ß-subunits**

In order to begin to understand the regulation of inhibin production by the hen's ovary, expression of the mRNA for the inhibin subunits by the preovulatory follicles has been estimated. The hen has no follicular fluid from which to isolate inhibin and therefore, the location and quantification of inhibin by mRNA measurements seems an appropriate method to follow. In these studies of mRNA quantification, the concentration of inhibin mRNA was determined; that is, the amount of inhibin mRNA per microgram of RNA. The preovulatory follicles of the hen differ markedly in size and therefore, the contribution that an individual follicle may make to plasma inhibin is a reflection of concentration as well as follicle size.

As a first approach to the quantitative analysis of expression of the inhibin α-subunit among follicles, the porcine inhibin α-subunit cDNA was used as the probe in Northern and slot blot analysis (23). RNA was isolated from the granulosa layer of the largest ($F_1$-$F_4$) preovulatory follicles, and the size of the mRNA was estimated by Northern hybridization at approximately 1.7 kb for all follicle sizes. Slot blot analysis was used for quantitative estimation. No difference in mRNA concentration for the inhibin α-subunit was detected in the granulosa layer among the four largest follicles. These studies are currently being repeated using our chicken cDNA probe for the inhibin α-subunit. As mentioned earlier, it appears that the inhibin ß-subunit is limiting in mammals with respect to formation of biologically active inhibin (5, 6). We are also examining the expression of the inhibin $ß_A$-subunit mRNA in hen granulosa tissue using a rat cDNA probe for the inhibin $ß_A$-subunit (24).

**Plasma concentrations of immunoreactive inhibin in the hen**

Using a bovine inhibin radioimmunoassay kit (25; obtained from the NICHHD of NIH) validated for chicken plasma, we have measured circulating concentrations of immunoreactive inhibin in the hen (26). No significant peak in plasma immunoreactive inhibin was detected throughout the cycle, in spite of a significant preovulatory rise in plasma LH. The highest level of plasma immunoreactive inhibin was attained, however, at 2200 h. This is the time when the largest ($F_1$) follicle is secreting maximal progesterone and during which the preovulatory surge of LH is induced. Figure 2 depicts the plasma inhibin and FSH values obtained from this experiment using the homologous chicken FSH RIA developed by Krishnan *et al.* (27). Similar to the inhibin data, no marked fluctuations occurred in plasma FSH throughout the cycle. In addition, no reciprocal relationship between the two hormones is observed. In sheep, the association between increased inhibin level and preovulatory follicular development has been somewhat controversial. Some workers have shown increased inhibin secretion during the follicular phase (28) and others have demonstrated little change (29, 30). This observation of minimal change suggests that inhibin may function as a longer-term regulator of FSH levels when compared to oestradiol (29). This may be the case in the hen.

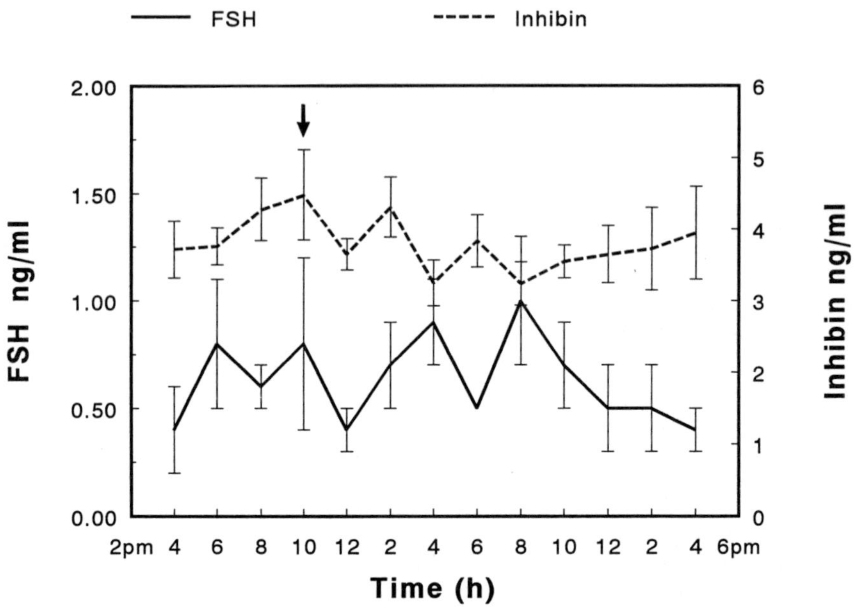


Fig. 2 Concentrations of plasma FSH and inhibin (ng/ml) throughout the ovulatory cycle of the domestic hen. Blood samples were collected at 2 h intervals throughout a complete ovulatory cycle (n=5). Blood sampling was begun at 4 pm on the day of terminal oviposition of the sequence and continued through 4 pm on the pause day (day of first ovulation of the sequence). The arrow shows the time of the predicted peak of the preovulatory surge of LH.

In a preliminary effort to elucidate the source of plasma immunoreactive inhibin, we removed 3-4 of the largest preovulatory follicles and assayed plasma immunoreactive inhibin concentrations at 0, +6 and +24 h. Removal of these largest follicles led to a significant decline in plasma immunoreactive inhibin concentrations (26) which were significantly depressed at both +6 and +24 h. Although plasma immunoreactive inhibin was not totally eliminated, this study demonstrated that more than 50% of the circulating immunoreactive inhibin was derived from the 3-4 largest preovulatory follicles. This identified the ovary, and more specifically, the largest follicles as a primary source of secreted immunoreactive inhibin in the hen.

An additional experiment was done in which we removed all the hierarchical follicles as well as the small yellow follicles (SYF) down to approximately 5 mm. As seen in Figure 3, when the hierarchical follicles as well as the SYF were removed, plasma immunoreactive inhibin was reduced by 86% within 3 h. Associated with this was a marked rise in plasma FSH with minimal change in LH. Although steroids are also removed in this type of experiment, most oestradiol is produced by the small (<1 mm) white follicles (31) and LH values were minimally changed in this experiment. Hormone values were relatively unchanged in the sham-operated animals.

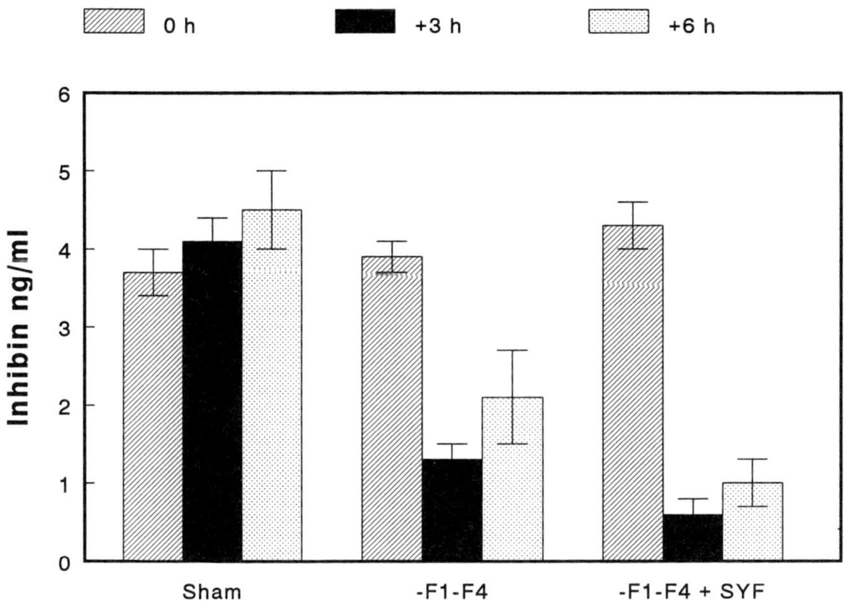


Fig. 3 Concentrations of plasma immunoreactive inhibin in a laying hen after removal of the largest yellow yolky follicles ($F_1$-$F_4$) and the small yellow follicles (SYF). Follicles were surgically removed from hens (n=5 per group) and blood samples taken at time 0, +3 h and +6 h. Removal of the largest yellow yolky follicles as well as the SYF down to approximately 5 mm caused an 86% reduction in plasma inhibin concentration within 3 h. Sham operated hens showed no change in concentrations of plasma inhibin throughout the sampling period.

We performed a subsequent experiment to determine the contribution of the individual hierarchical follicles to plasma immunoreactive inhibin (32, 33). In this experiment, we selectively removed increasing numbers of preovulatory follicles: $F_1$, $F_1$+$F_2$, $F_1$+$F_2$+$F_3$ or $F_1$+$F_2$+$F_3$+$F_4$. Removal of an increasing number of follicles caused a progressive decline in immunoreactive inhibin concentrations as compared to sham-operated animals. In addition, an increased length of time was required for inhibin level to return to baseline with increasing number of follicles removed. These data suggest that all the preovulatory follicles contribute to plasma immunoreactive inhibin. In this experiment we also examined plasma concentrations of FSH and LH. Plasma FSH was significantly elevated at +6 h relative to follicle removal in all groups except for the sham-operated group, while plasma LH was unchanged. We observed that plasma FSH was not sequentially elevated by removal of increasing numbers of follicles (after the $F_1$), in spite of a progressive fall in inhibin level. This suggested the possibility that the $F_1$ follicle may be the only follicle producing biologically active inhibin capable of causing an acute change in FSH concentration.

We have attempted to assess the relationship between plasma immunoreactive inhibin and laying rate as an indication of the biological function of inhibin in the hen (34, 35). Repeated blood samples (at 2 h intervals starting at a time 6 h after an LH surge) were taken at the beginning of the sequence from hens laying short sequences as well as hens laying long sequences. Plasma immunoreactive inhibin levels were significantly higher in hens laying short sequences (3-7 eggs) compared to that in hens laying long sequences (>20 eggs). Likewise, expression of the mRNA for the inhibin α-subunit was greater in the granulosa layer of hens laying short sequences compared to those laying long sequences. Plasma FSH was not different between the two groups. It is possible that the time of blood sampling did not coincide with a possible alteration in plasma FSH concentration or that inhibin may be primarily exerting a paracrine effect within the ovary that has importance with respect to follicular development (3).

A related study recently reported on the concentration of immunoreactive inhibin in the follicular tissues of the hierarchical follicles (17). The granulosa layer was found to be the major source and, in addition, the concentration of inhibin decreased with increasing follicle size; that is, the concentration varied ($F_4 > F_3 > F_2 = F_1$) among these follicles although the total inhibin content per granulosa layer did not differ. These authors also showed that granulosa cell cultures from follicles of different sizes had decreased production of immunoreactive inhibin with increased follicle maturity. These results conflict with our demonstration of no marked difference in mRNA expression of the inhibin α-subunit among the granulosa layers of the four largest follicles (23). In addition, it is difficult to reconcile these data with our observation of an increase in FSH after removal of the $F_1$ follicle but no further increase in FSH after removal of increasing numbers of follicles up through the $F_4$ follicle (33). Since it has been reported that the ß-subunit is limiting in the formation of biologically active inhibin in mammals (5, 6), we are currently examining the expression of the mRNA for the $ß_A$-subunit in the granulosa layer of the preovulatory follicles of the hen. These studies may enhance our understanding of the potential for formation of biologically active, intact inhibin.

In assessing plasma immunoreactive inhibin in the hen, we have used the conventional endocrine techniques of serial blood sampling, surgical follicular removal and hormone assay to begin to understand the feedback relationship between the immunoreactive inhibin produced by individual follicles and the gonadotrophins. It is important to point out that the inhibin RIA (25) which our group and others (12, 17) commonly use for immunoreactive inhibin measurement in the hen has the problem of cross-reaction of the antibody with free inhibin α-subunit (36, 37). The contribution of free secreted inhibin α-subunit to total plasma immunoreactive inhibin in the hen is not known at this time. Therefore, studies assessing immunoreactive inhibin in the hen should be interpreted with caution. At this time, however, this assay system permits inroads into our knowledge of inhibin in the hen since two-site immunoassays, which measure intact inhibin, are not sensitive enough to determine plasma values (38, 39).

## Conclusion

Results from the cloning and sequencing of the inhibin α-subunit of the hen indicate that there is substantial similarity of the molecule to the mammalian (bovine) α-subunits. Expression of the mRNA for the inhibin α-subunit demonstrates tissue specificity localized to the granulosa layer in the hen. This clone will provide a sensitive probe for the study of inhibin with respect to its role in the regulation of follicular development. Studies assessing the plasma concentrations of inhibin in the hen have demonstrated that the ovary is the major source of immunoreactive inhibin. In addition, associated with follicle removal and a fall in inhibin levels, there is an acute rise in FSH with no change in LH.

## Acknowledgments

We thank Dr Kelly Mayo for the porcine cDNA clone for the inhibin α-subunit. We are grateful to Dr Gabriel Bialy of NICHHD at the NIH (Bethesda, MD) for supplying us with the materials for the inhibin RIA. We also thank Dr Peter Sharp (AFRC Roslin Institute (Edinburgh), Roslin, Midlothian, UK) and Dr John Proudman (USDA, Beltsville, MD) for the reagents to perform the chicken LH and FSH RIAs, respectively. We are grateful to Mr C-C Chen for comments on the manuscript and we gratefully acknowledge the skilled assistance of Ms Carrie Brooks and Ms Sophia Ramlal.

## References

1 Hasegawa Y, Miyamoto K, Igarashi M, Yanaka T, Sasaki K & Iwamura S (1987) Changes in serum concentrations of inhibin during the estrous cycle of the rat, pig and cow. *Inhibin-Non-Steroidal Regulation of Follicle Stimulating Hormone Secretion* pp 119-133. Eds H G Burger, D M deKretser, J K Findlay & M Igarashi. New York: Raven Press

2 Rivier C, Meunier H, Roberts V & Vale W (1990) Inhibin: role and secretion in the rat. *Recent Progress in Hormone Research* **46** 231-259

3 Findlay J K (1993) An update on the roles of inhibin, activin, and follistatin as local regulators of folliculogenesis. *Biology of Reproduction* **48** 15-23

4 Ling N, Ueno N, Ying S-Y, Esch F, Shimasaki S, Hotta M, Cuevas P & Guillemin R (1988) Inhibins and activins. *Vitamins and Hormones* **44** 1-46

5 Woodruff T, D'Agostino J, Schwartz N & Mayo K (1988) Dynamic changes in inhibin messenger RNAs in rat ovarian follicles during the reproductive cycle. *Science* **239** 1296-1299

6 Rodgers R J, Stuchbery S J & Findlay J K (1989) Inhibin mRNAs in ovine and bovine ovarian follicles and corpora lutea throughout the estrous cycle and gestation. *Molecular and Cellular Endocrinology* **62** 95-101

7 Schneyer A L, Sluss P M, Whitcomb R W, Martin K A, Sprengel R & Crowley Jr W F (1991) Precursors of α-inhibin modulate follicle stimulating hormone receptor binding and biological activity. *Endocrinology* **129** 1987-1999

8 Feng Z-M, Li Y-P & Chen C-L C (1989) Analysis of the 5′-flanking regions of rat inhibin α- and $ß_B$-subunit genes suggests two different regulatory mechanisms. *Molecular Endocrinology* **3** 1914-1925

9 Bandivdekar A H, Vijayalakshmi S, Moodbidri S B & Sheth A R (1982) Low-molecular-weight inhibin from sheep, human, rat and chicken testes. *Journal of Andrology* **3** 140-143

10 Bandivdekar A H, Moodbidri S B & Sheth A R (1984) FSH receptor binding inhibiting activity associated with low molecular weight inhibin from sheep, human, rat and chicken. *Experientia* **40** 994-995

11 Lewis W M, Muster K M, Martin T L, Ireland J L H, Bursian S, Balander R, Mason A J & Ireland J J (1989) Evidence for inhibin in roosters. *Biology of Reproduction* **40** (Supplement 1) 109

12 Rombauts L, Vanmontfort D, Verhoven G & Decuypere E (1992) Immunoreactive inhibin in plasma, amniotic fluid, and gonadal tissue of male and female chick embryos. *Biology of Reproduction* **46** 1211-1216

13 Blauer M, Kohonen J, Leivonen I, Vilja P & Touhimaa P (1992) Inhibin α- and $ß_A$-subunit immunoreactivity in the chicken embryo during morphogenesis. *Development* **115** 1079-1086

14 Mitrani E, Ziv T, Thomsen G, Shimoni Y, Melton D A & Bril A (1990) Activin can induce the formation of axial structures and is expressed in the hypoblast of the chick. *Cell* **63** 495-501

15 Akashiba H, Taya K, Sasamoto S, Goto H, Kamiyoshi M & Tanaka K (1988) Secretion of inhibin by chicken granulosa cells *in vitro*. *Poultry Science* **67** 1625-1631

16 Tsonis C G, Sharp F J & McNeilly A S (1988) Inhibin bioactivity and pituitary cell mitogenic activity from cultured chicken ovarian granulosa and thecal/stromal cells. *Journal of Endocrinology* **116** 293-299

17 Vanmontfort D, Rombauts L, Decuypere E & Verhoeven G (1992) Source of immunoreactive inhibin in the chicken ovary. *Biology of Reproduction* **47** 977-983

18 Vale W, Rivier C, Hsueh A, Campen C, Meunier H, Bicsak T, Vaughan J, Corrigan A, Bardin W, Sawchenko P, Petraglia F, Yu J, Plotsky P, Speiss J & Rivier J (1988) Chemical and biological characterization of the inhibin family of protein hormones. *Recent Progress in Hormone Research* **44** 1-34

19 Mayo K E, Cerelli G M, Speiss J, Rivier J, Rosenfeld M G, Evans R M & Vale W (1986) Inhibin A-subunit cDNAs from porcine ovary and human placenta. *Proceedings of the National Academy of Sciences of the USA* **83** 5849-5853

20 Sanger F, Nicklen S & Coulson A R (1977) DNA sequencing with chain-terminating inhibitors. *Proceedings of the National Academy of Sciences of the USA* **74** 5463-5467

21 Wang S-Y & Johnson P A (1993) Complementary deoxyribonucleic acid (cDNA) cloning and sequence analysis of the α-subunit of inhibin from chicken ovarian granulosa cells. *Biology of Reproduction* (in press)

22 Forage G R, Ring J M, Brown R W, McInerney B V, Cobob G S, Gregson R P, Robertson D M, Morgan F J, Hearn M T W, Findlay J K, Wettenhall R E H, Burger H G & deKretser D M (1986) Cloning and sequence analysis of cDNA species coding for the two subunits of inhibin from bovine follicular fluid. *Proceedings of the National Academy of Sciences of the USA* **83** 3091-3095

23 Johnson P A & Wang S-Y (1993) Characterization and quantitation of mRNA for the inhibin α-subunit in the granulosa layer of the domestic hen. *General and Comparative Endocrinology* **90** 43-50

24 Woodruff T K, Meunier H, Jones P B C, Hsueh A J W & Mayo K E (1987) Rat inhibin: molecular cloning of α- and ß-subunit complementary deoxyribonucleic acids and expression in the ovary. *Molecular Endocrinology* **1** 561-568

25 Robertson D M, Tsonis C G, McLachlan R I, Handelsman D J, Leask R, Baird D T, McNeilly A S, Hayward S, Healy D L, Findlay J K, Burger H G & de Kretser D M (1988) Comparison of inhibin immunological and *in vitro* biological activities in human serum. *Journal of Clinical Endocrinology and Metabolism* **67** 438-443

26 Johnson P A, Wang S-Y & Brooks C (1993) Characterization of a source and levels of plasma immunoreactive inhibin during the ovulatory cycle of the domestic hen. *Biology of Reproduction* **48** 262-267

27 Krishnan K A, Proudman J A, Bolt D J & Bahr J M (1993) Development of an homologous radioimmunoassay for chicken follicle-stimulating hormone and measurement of plasma FSH during the ovulatory cycle. *Comparative Biochemistry and Physiology* (in press)

28 Findlay J K, Clarke I J & Robertson D M (1990) Inhibin concentrations in ovarian and jugular venous plasma and the relationship of inhibin with follicle-stimulating hormone and luteinizing hormone during the ovine estrous cycle. *Endocrinology* **126** 528-535

29 Mann G E, McNeilly A S & Baird D T (1989) Source of ovarian inhibin secretion during the estrous cycle of the sheep. *Journal of Endocrinology* **123** 181-188

30 Campbell B K, Mann G E, McNeilly A S & Baird D T (1990) The pattern of ovarian inhibin, estradiol and androstenedione secretion during the estrous cycle of the ewe. *Endocrinology* **127** 227-235

31 Robinson F E & Etches R J (1986) Ovarian steroidogenesis during follicular maturation in the domestic fowl (*Gallus domesticus*). *Biology of Reproduction* **35** 1096-1105

32 Johnson P A, Brooks C & Wang S-Y (1992) Contribution of preovulatory follicles to plasma immunoreactive inhibin level in the domestic hen. Abstract 155. *Biology of Reproduction* **46** (Supplement 1)

33 Johnson P A (1993) Inhibin in the hen. *Poultry Science* **72** 955-958

34 Wang S-Y & Johnson P A (1991) Decreased ovulation rate is associated with an increased amount of messenger RNA for inhibin α-subunit in the domestic hen. *Signaling Mechanisms and Gene Expression in the Ovary* pp 317-322. Ed. G Gibori. New York: Springer-Verlag

35 Wang S-Y & Johnson P A (1993) Increase in ovarian α-inhibin gene expression and plasma immunoreactive inhibin level is correlated with a decreased rate in ovulation in the domestic hen. *General & Comparative Endocrinology* (in press)

36 Robertson D M, Giacometti M, Foulds L M, Lahnstein J, Goss N K, Hearn M T W, de Kretser D M (1989) Isolation of inhibin α-subunit precursor proteins from bovine follicular fluid. *Endocrinology* **125** 2141-2149

37 Schneyer A L, Mason A J, Burton L E, Zeigner J R, Crowley Jr W F (1990) Immunoreactive inhibin α-subunit in human serum: implications for radioimmunoassay. *Journal of Clinical Endocrinology and Metabolism* **70** 1200-1212

38 Betteridge A & Craven R P (1991) A two-site enzyme-linked immunosorbent assay for inhibin. *Biology of Reproduction* **45** 748-754

39 Knight P G, Groome N & Beard A J (1991) Development of a two-site immunoradiometric assay for dimeric inhibin using antibodies against chemically synthesized fragments of the α and ß subunit. *Journal of Endocrinology* **129** R9-R12

Avian Endocrinology
Ed. P J Sharp, pp 309-320
Journal of Endocrinology Ltd, Bristol (1993)

# Involvement of plasminogen activator and collagenase and changes in extracellular matrix during the proliferative and ovulatory phases of the chicken ovary

J A Jackson[1], S A Tischkau[1], P Zhang[2], Y Yoshimura[3] and J M Bahr[1]
[1]Departments of Animal Sciences and Physiology, University of Illinois at Urbana-Champaign, Animal Genetics Laboratory, Urbana, IL 61801, USA; [2]Departments of Microbiology and Immunology, Albert Einstein College of Medicine, Bronx, NY 10461, USA; [3]Faculty of Applied Biological Science, Hiroshima University, Japan

## Introduction

The chicken ovulates and oviposits an egg every 25-26 h with the exception of a pause day on which no egg is laid. This pattern of frequent ovulations requires selection and rapid growth of follicles. During the last 5-7 days preceding ovulation, the follicle is in a proliferative phase, increasing in diameter from 10-12 mm to 30 mm or more. Once the follicle has reached sufficient maturity and is competent to ovulate, it enters the ovulatory phase. This brief phase initiated by the luteinizing hormone (LH) surge is of 4-6 h duration and is terminated by ovulation.

During the proliferative phase the follicle undergoes extensive growth and tissue remodelling. Ovulation, on the other hand, is dependent upon degradation and/or dissociation of tissue in the stigma region of the follicle. Both of these tissue responses require the controlled action of proteases, specifically plasminogen activator and collagenase. Plasminogen activator, a serine protease, converts plasminogen to plasmin, which then acts to cause tissue remodelling or to convert latent collagenase to active collagenase. Plasmin and collagenase may cause degradation and/or dissociation of theca connective tissue specifically in the stigma region resulting in ovulation.

This review describes recent studies on the chicken ovary of the role and regulation of plasminogen activator and collagenase, and measurements of the components of the extracellular matrix, collagen and proteoglycans/glycosaminoglycans. The role of the germinal disc region in regulating the proliferative phase of follicle growth and the interactions between the granulosa and theca layers in the production of plasminogen activator are discussed. Finally, the involvement of plasminogen activator and collagenase is considered in relation to changes in the extracellular matrix of the stigma region prior to ovulation.

## The germinal disc region of the ovarian follicle - the primary proliferative centre

Evidence is accumulating that suggests the germinal disc region, comprising the germinal disc plus granulosa cells overlying the germinal disc, may be the primary proliferative centre for the growing follicle. Early histological studies described regional differences in morphology and growth within the granulosa layer of pre-ovulatory follicles of the hen. The columnar cells in the granulosa layer overlying the germinal disc are thought to be less mature than the cuboidal cells observed in peripheral regions of the granulosa layer (1, 2). In flow cytometry studies, a higher percentage of granulosa cells in the germinal disc region are found in proliferative stages of the cell cycle (3). In the mouse, the oocyte secretes a protein, termed enabling factor, which stimulates proliferation of the surrounding granulosa cells (4). Finally, $^{3}$H-thymidine incorporation is greatest in granulosa cells nearest the germinal disc (5).

While the preceding evidence suggests that the germinal disc may be important in directing pre-ovulatory growth of the follicle in the hen, the mechanism by which this occurs has not been defined. A series of simple experiments has been carried out to investigate the role of the germinal disc in follicular growth and ovulation (Table 1).

Table 1 The effect of local freezing of the germinal disc region of the largest pre-ovulatory ovarian follicle ($F_1$), on ovulation in the hen (n=4). Unpublished data.

| **Region of the $F_1$ follicle destroyed by local freezing** | **Hours before predicted ovulation** | **Response** | **Pre-ovulatory increases in plasma progesterone and LH** |
|---|---|---|---|
| Germinal disc region | 12 | Ovulation | YES |
| Germinal disc region | 24 | Atresia | NO |
| Non-germinal disc region | 12 | Ovulation | YES |
| Non-germinal disc region | 24 | Ovulation | YES |

At 24 h and 12 h before a predicted ovulation the germinal disc region of the largest pre-ovulatory follicle ($F_1$) was destroyed by localized freezing and, in controls, an area of similar size was frozen on the side of the $F_1$ follicle opposite to the germinal disc region. Observations were made on the occurrence of ovulation, pre-ovulatory increases in concentrations of plasma LH and progesterone ($P_4$) and the ability of cultured granulosa cells to secrete $P_4$ in response to LH. Destruction of the germinal disc region at 24 h, but not at 12 h before ovulation, resulted in atresia and an absence of pre-ovulatory increases in plasma $P_4$ and LH (Table 1). Short term incubation of granulosa cells isolated from follicles in which the germinal disc regions had been destroyed at 24 h and 12 h before a predicted ovulation indicated that the granulosa cells were able to respond to LH stimulation and secrete $P_4$ despite the fact that the follicles in which the germinal disc regions were destroyed at 24 h were atretic. This experiment suggests that the germinal disc region is essential for follicular growth during the proliferative phase, but once the follicle is sufficiently mature to ovulate, the germinal disc region is no longer required.

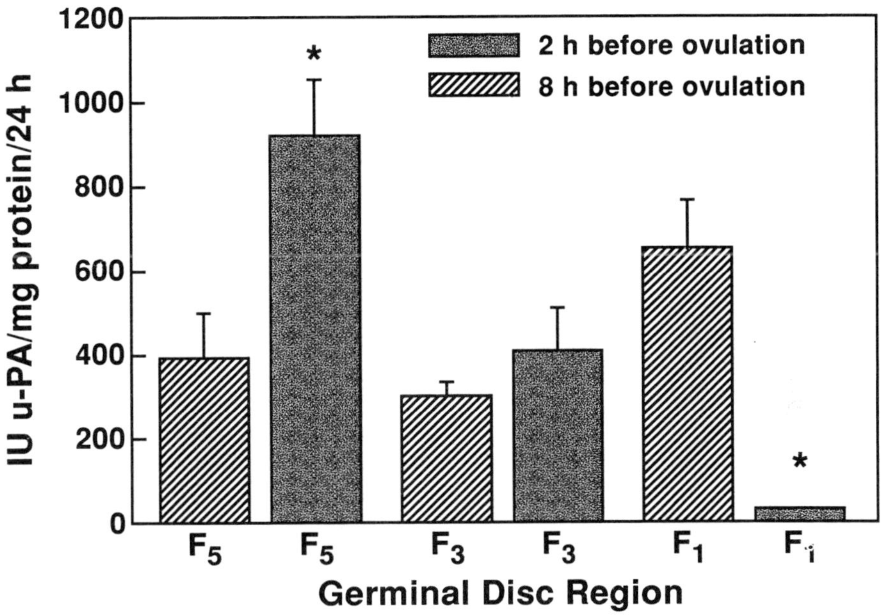


Fig. 1 A comparison of the production of plasminogen activator by germinal disc regions from hen pre-ovulatory follicles ($F_1$, $F_3$, $F_5$) collected 8 h (before LH surge) and 2 h (after LH surge), prior to ovulation. The $F_1$ follicle is the most mature, and ready for ovulation while the $F_3$ and $F_5$ follicles are approximately 60-50 h and 110-125 h before ovulation. The germinal disc regions from the $F_5$, $F_3$, and $F_1$ follicles were incubated in 1 ml medium for 24 h at 37 °C. Medium was assayed for plasminogen activator activity using the chromogenic substrate, S-2251. Values are means ± s.e. (n=3-5). IU u-PA, international units of urokinase-type plasminogen activator. *$P$<0.001 is significantly different compared with the same follicle taken 8 h before ovulation. Unpublished data.

Experiments have been carried out to establish whether plasminogen activator plays a role in proliferative processes regulated by the germinal disc region (6). Germinal disc regions were removed 8 h and 2 h before ovulation to measure the content and production of plasminogen activator. Extremely high concentrations of plasminogen activator activity were present in the germinal disc region of the largest pre-ovulatory follicle ($F_1$) when the follicle was obtained 8 h before ovulation and before the initiation of the pre-ovulatory LH surge. However, the germinal disc region from the $F_1$ follicle obtained 2 h before ovulation and after the LH surge, contained low amounts of plasminogen activator activity. This activity remained low immediately prior to ovulation (6). *In vitro* measurements of plasminogen activator activity produced by germinal disc regions from pre-ovulatory follicles ($F_5$, $F_3$, and $F_1$), obtained at 8 h (before LH surge) and 2 h (after LH surge) before ovulation, indicated that production increases with follicular maturation, and that plasminogen activator production by the $F_1$ is suppressed by LH whereas that of the $F_5$ is increased (Fig. 1).

These observations suggest that plasminogen activator activity in the germinal disc region is differentially regulated by LH as follicular maturation progresses and is an important component of the mechanisms controlling the proliferative phase of follicular growth. In contrast, plasminogen activator production by the cumulus-oocyte complex of the rat pre-ovulatory follicle is always stimulated by LH with the highest production occurring immediately before ovulation (7).

**Interaction of the granulosa and theca layers in plasminogen activator production**

The role of the granulosa and theca layers in the production of plasminogen activator has been investigated during the proliferative phase of the ovulatory cycle (6, 8). Although the germinal disc may be the proliferative centre of the follicle, tissues which comprise the follicular wall must undergo significant remodelling to participate in the rapid growth of the follicle. Plasminogen activator activity is present in granulosa and theca layers obtained during the proliferative phase before the pre-ovulatory surge of LH, and the theca layer contains higher amounts of plasminogen activator activity than the granulosa layer (6, 8). Plasminogen activator activity in granulosa and theca layers obtained during the proliferative phase increase with maturation of the $F_6$-$F_1$ follicles (8).

The presence of plasminogen activator activity in granulosa and theca layers does not necessarily indicate that both cell layers produce it. An explant system in which equal proportions of granulosa and theca layers are cultured alone or in combination has been developed to establish the mechanisms controlling the production of ovarian plasminogen activator (9). Plasminogen activator activity is nearly non-detectable in the medium when granulosa and theca layers were cultured alone, whereas co-incubation of the 2 cell layers results in significant production of plasminogen activator activity (Fig. 2A). Tilly and Johnson have previously reported that granulosa (10) and theca (11) cells cultured separately produce very low amounts of plasminogen activator activity. In subsequent studies, we found that the theca layer produced a factor(s) that stimulates the granulosa layer to produce plasminogen activator (Fig. 2B). The theca factor does not appear to be a prostaglandin because addition of indomethacin to co-cultures of granulosa and theca layers does not suppress the stimulatory effect of the theca layer on production of plasminogen activator activity by granulosa cells. Moreover, addition of prostaglandins $E_2$ or $F_{2\alpha}$, which are present in chicken follicles during the proliferative and ovulatory phases, does not stimulate granulosa cells to produce plasminogen activator activity.

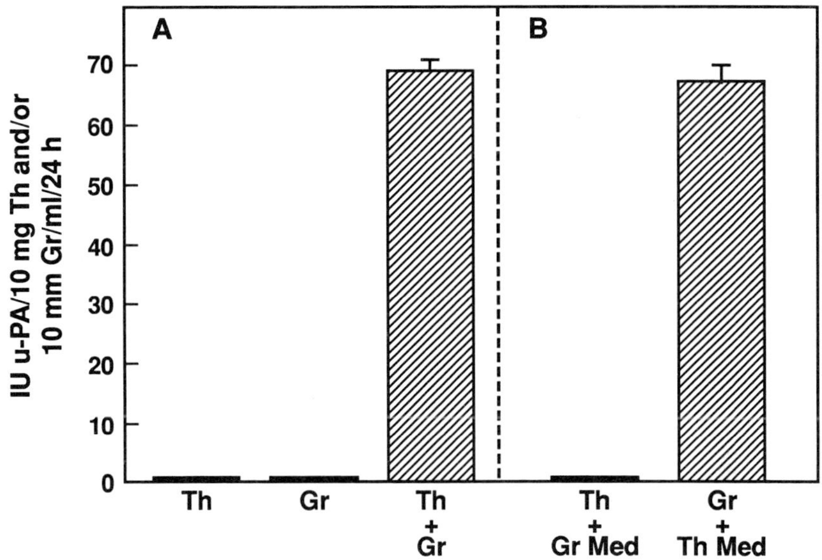


Fig. 2 The production of plasminogen activator activity by cultured theca (Th) or granulosa (Gr) layers from the $F_1$ follicle collected 8 h before ovulation. Equal proportions of Th (10 mg wet weight) or the Gr (10 mm in diameter) were (A) incubated alone or in combination for 24 h or (B) incubated (18 h) in medium (med) in which Gr or Th layers had previously been incubated for 12 h, respectively. Values are means ± s.e. (n=3-5). IU u-PA, international units of urokinase-type plasminogen activator. Data from (9).

Other experiments have investigated the role of LH in regulating plasminogen activator production by the granulosa layer (9). Co-incubation of theca and granulosa layers removed during the proliferative phase and before the LH surge (8 h before ovulation) results in the production of large amounts of plasminogen activator activity (Fig. 3). In contrast, co-incubation of theca and granulosa layers removed after the LH surge (2 h before ovulation) or addition of LH to co-cultures of granulosa and theca layers obtained 8 h before ovulation results in a suppression of plasminogen activator activity (Fig. 3). Therefore the LH surge required for ovulation may inhibit plasminogen activator production by the granulosa layer.

Finally, since two types of plasminogen activator exist, namely tissue type plasminogen activator (tPA) and urokinase (uPA), the type of plasminogen activator produced by the pre-ovulatory follicle has been investigated (9). The molecular size of tPA is approximately 70,000 kDa while uPA is approximately 50,000 kDa. Substrate gel electrophoresis indicates that uPA is the predominant plasminogen activator produced by the germinal disc region and the granulosa layer before exposure to LH (9).

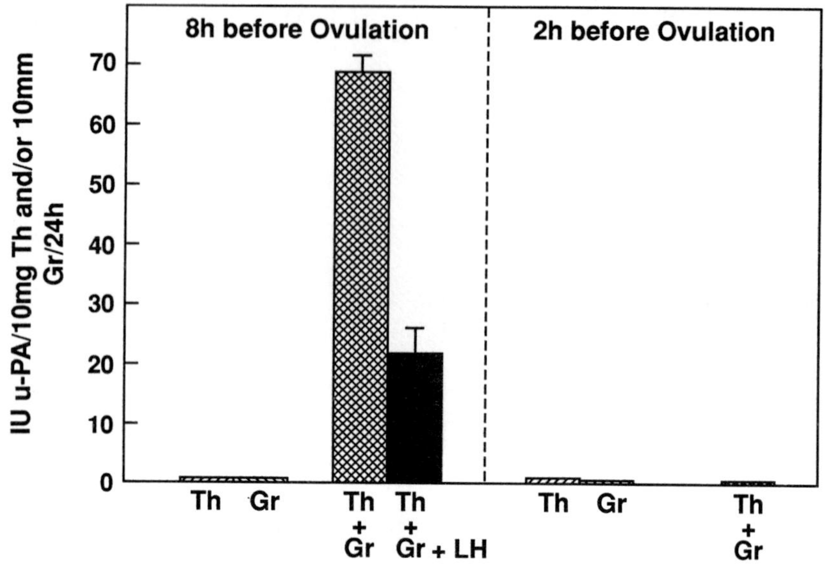


Fig. 3 The effect of LH on the production of plasminogen activator activity by cultured theca (Th) and granulosa (Gr) layers from the $F_1$ follicles collected 8 h and 2 h before ovulation. Equal proportions of Th (10 mg wet weight) or Gr (10 mm in diameter) were cultured alone, or together with or without LH (25 ng/ml) for 24 h. IU u-PA, international units of urokinase-type plasminogen activator. Values are means ± s.e. (n=3-5). Data from (9).

## The involvement of plasminogen activator and collagenase in ovulation

The final stage in the maturation of the pre-ovulatory follicle is the ovulatory phase. This phase is initiated by the LH surge and is terminated by the rupture of the stigma releasing the egg into the infundibulum. The theca layer is composed of connective tissue with an extracellular matrix comprised of collagen and interfibrillar matrix. Connective tissue interfibrillar matrix is usually composed of proteoglycans, which consist of long carbohydrate chains called glycosaminoglycans covalently attached to a protein backbone. Prior to ovulation there are pronounced changes in the extracellular matrix of the theca layer, specifically in the stigma region, the site of follicular rupture. Morphological studies of the follicular wall indicated there was a decrease in the interfibrillar matrix and degradation and/or dissociation of collagen fibres in the stigma region prior to ovulation in both mammals (12, 13) and hens (14, 15). Collagen fibres of the theca externa are extensively intertwined and form a complicated network in the nonstigma region, whereas the fibres are arranged parallel to the long axis in the stigma region of the hen (16). The collagen concentrations of the stigma and nonstigma regions of the $F_1$ follicle have been determined by measuring hydroxyproline (Fig. 4A). Immediately before ovulation the theca layer of the stigma region has a lower collagen concentration than the nonstigma region (Fig. 4A). The lower collagen concentration and longitudinal arrangement of the collagen fibres suggests that the stigma area has less intrinsic tensile strength than the nonstigma region (16, 17).

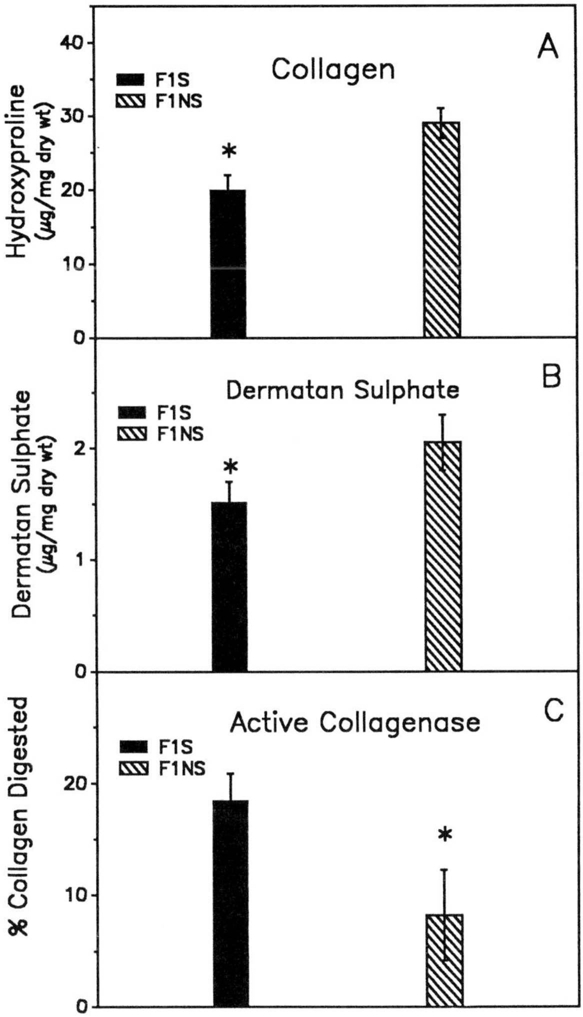


Fig. 4 A comparison of the concentration of (A) collagen and (B) the glycosaminoglycan, dermatan sulphate and (C) collagenase activity in the stigma ($F_1S$) and non-stigma ($F_1NS$) areas of the $F_1$ follicle immediately before ovulation. Collagen was measured as hydroxyproline. Values are means ± s.e. (n=3-5) **P*<0.05, *P*<0.05, significantly different compared with the non-stigmal area. Data from (17, 18).

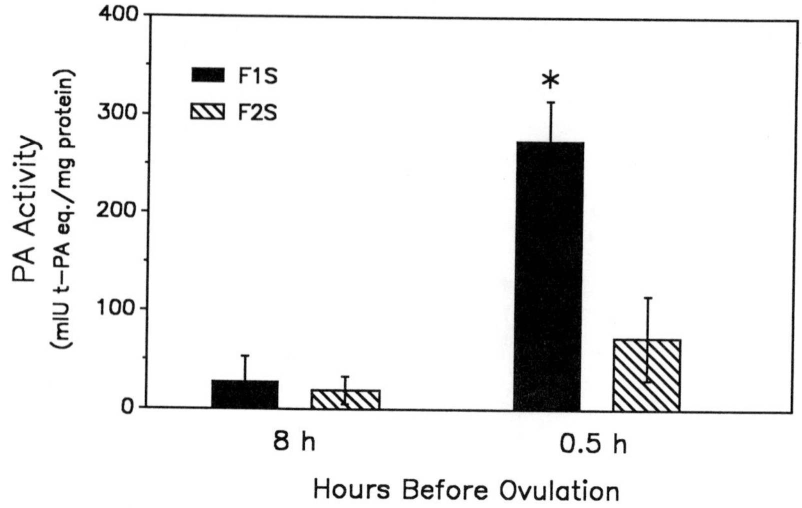


Fig. 5 A comparison of plasminogen activator (PA) activity in the stigma region of a $F_1$ ($F_1S$) or $F_2$ ($F_2S$) pre-ovulatory follicle removed 0.5 h or 8 h before a predicted ovulation: $^*P<0.001$, significantly different compared with the stigmal region of the $F_2$ follicle and the $F_1S$ at 8 h before ovulation. Values are means ± s.e. (n=3-5). Data from (6). mIU-t-PA, milli-international units of tissue type plasminogen activator.

The glycosaminoglycans, dermatan sulphate, heparin sulphate and hyaluronic acid, are components of the interfibrillar matrix of the theca layer and the predominant glycosaminoglycan is dermatan sulphate (17). There is a concomitant decrease in the concentration of dermatan sulphate and hyaluronic acid specifically in the stigma region of the theca layer compared to the nonstigma region prior to ovulation (Fig. 4B). Dermatan sulphate proteoglycans may need to be degraded to allow better access of proteases for degradation of collagen fibres. Alternatively, interfibrillar matrix components may be decreased to dissociate collagen fibres into individual fibrils. The decrease in dermatan sulphate proteoglycans and hyaluronic acid may further diminish the tensile strength of the stigma region, thereby facilitating rupture of the follicular wall for ovulation.

The decrease in collagen and dermatan sulphate concentrations in the stigma region immediately before ovulation is correlated with the amount of active collagenase (Fig. 4C) and plasminogen activator (Fig. 5) activity. Active collagenase is a measurement of collagenase activity after destroying the putative collagenase inhibitors with dithiothreitol and iodoacetamide, but without activation of latent (inactive) collagenase. Active collagenase activity is greater in the stigma region compared to nonstigma region of the theca layer (Fig. 4C). Moreover, there is a nine-fold increase in plasminogen activator activity specifically in the stigma region of the $F_1$ follicle immediately before ovulation (0.5 h) compared to 8 h (before LH surge) prior to ovulation (Fig. 5). There was no increase in plasminogen activator activity in the stigma region of the less mature $F_2$ follicle which will not ovulate until 26 h later (Fig. 5).

Studies in the rat have demonstrated a decrease in ovarian collagen concentrations (19, 20) and an increase in ovarian collagenase (20, 21) and plasminogen activator (22, 23) activities at the time of follicular rupture. Plasmin produced by the activation of plasminogen by plasminogen activator, is capable of activating collagenase in the rat ovary (20). Data from our studies suggest that a specific increase in plasminogen activator activity in the stigma region may result in an increase in active collagenase and plasmin which, in turn, may cause degradation and/or dissociation of collagen fibres and degradation of interfibrillar dermatan sulphate proteoglycans (24, 25), respectively, which are required for ovulation. The structural difference of the stigma region in the arrangement of collagen fibres may facilitate the degradation and/or dissociation of collagen and the proteoglycan interfibrillar matrix by collagenase and plasmin.

## Conclusion

The dynamic growth of the pre-ovulatory follicle during the proliferative phase and tissue and site specific changes in the extracellular matrix during the ovulatory phase are dependent upon local and regulated production of plasminogen activator and collagenase. During the proliferative stage of follicular growth, the germinal disc is required for continued growth of the follicle and the germinal disc region is a rich source of plasminogen activator. The differential response of the germinal disc region

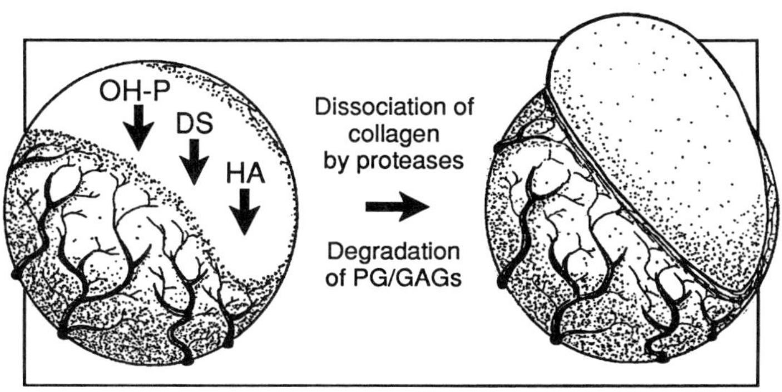


Fig. 6 Diagrammatic representation describing changes in the ovarian follicular wall in the domestic hen which result in ovulation. Collagen fibres are degraded and/or dissociated by proteases. There is a degradation of proteoglycans (PG) and glycosaminoglycans (GAGs). Dermatan sulphate (DS) and hyaluronic acid (HA) decrease specifically in the stigma region (clear area) of the follicle. Concentrations of collagen, as determined by hydroxyproline (OH-P) are inherently lower in the stigma region of the follicle. Reproduced with permission from Bahr (26).

to LH, e.g. stimulation of plasminogen activator activity in the germinal disc region of the $F_5$ follicle and suppression of plasminogen activator activity in the germinal disc region of the $F_1$ follicle is an intriguing observation. Interaction of granulosa and theca layers results in the production of plasminogen activator required for growth and tissue remodelling of the pre-ovulatory follicle. Finally, the impingement of the LH surge on the $F_1$ follicle causes the follicle to make the transition from the proliferative phase to the ovulatory phase. This latter stage is characterized by localized increases in proteases and highly specific structural changes in the stigma region of the pre-ovulatory follicle terminating in ovulation (Fig. 6).

## Acknowledgements

Supported in part by USDA-Ag89-37240-4769, NSF-DCB 91-08106, NSF DCB 92-07535, and NIH PHS-5T32-HDO7028.

## References

1 Perry M M, Gilbert A B & Evans A J (1978) The structure of the germinal disc region of the hen's ovarian follicle during the rapid growth phase. *Journal of Anatomy* **127** 379-392

2 Bakst M R (1979) Scanning electron microscopy of hen granulosa cells before and after ovulation. *Scanning Electron Microscopy* **3** 307-312

3 Marrone B L, Jamaluddin M & Hertelendy F (1990) Regional patterns of cell maturation and progesterone biosynthesis in the avian granulosa layer. *Biology of Reproduction* **42** 405-412

4 Vanderhyden B C, Telfer E E & Eppig J J (1992) Mouse oocytes promote proliferation of granulosa cells from preantral and antral follicles *in vitro*. *Biology of Reproduction* **46** 1196-1204

5 Tilly J L, Kowalski K I, Li Z, Levorse J M & Johnson A L (1992) Plasminogen activator activity and thymidine incorporation in avian granulosa cells during follicular development and the periovulatory period. *Biology of Reproduction* **46** 195-200

6 Jackson J A, Zhang P & Bahr J M (1993) Plasminogen activator activity in preovulatory follicles during the ovulatory cycle of the chicken. *Biology of Reproduction* (in press)

7 Liu Y X & Hsueh A J W (1987) Plasminogen activator activity in cumulus-oocyte complexes of gonadotropin-treated rats during the periovulatory period. *Biology of Reproduction* **36** 1055-1062

8 Politis I, Wang L, Turner J D & Tsang B K (1990) Changes in tissue-type plasminogen activator-like and plasminogen activator inhibitor activities in granulosa and theca layers during ovarian follicle development in the domestic hen. *Biology of Reproduction* **42** 747-754

9 Jackson J A, Tischkau S A, Zhang P & Bahr J M (1993) Plasminogen activator production by granulosa layer is stimulated by factor(s) produced by the theca layer and inhibited by the LH surge in the chicken. *Biology of Reproduction* (in press)

10 Tilly J L & Johnson A L (1987) Presence and hormonal control of plasminogen activator in granulosa cells of the domestic hen. *Biology of Reproduction* **37** 1156-1164

11 Tilly J L & Johnson A L (1990) Control of plasminogen activator activity in the thecal layer of the largest preovulatory follicle in the hen ovary. *Endocrinology* **126** 2079-2087
12 Espey L L (1967) Ultrastructure of the apex of the rabbit Graafian follicle during the ovulatory process. *Endocrinology* **81** 267-276
13 Bjersing L & Cajander S (1974) Ovulation and the mechanism of follicle rupture. V. Ultrastructure of tunica albuginea and theca externa of rabbit Graafian follicles prior to induced ovulation. *Cell and Tissue Research* **153** 15-30
14 Yoshimura Y & Koga O (1982) Ultrastructural changes of the stigma of the follicle during the process of ovulation in the hen. *Cell and Tissue Research* **224** 349-359
15 Yoshimura Y & Koga O (1983) Ultrastructural changes in the follicular stigma during the ovulation process in the hen. *Avian Endocrinology: environmental and ecological perspectives* pp 107-115. Eds S Mikami, K Homma & M Wada. Tokyo: Japanese Scientific Societies Press
16 Fujii S & Yoshimura Y (1979) Morphological observations on the stigma of the follicular wall concerning the mechanism of ovulation in hens. *The Faculty of Applied Biological Science, Hiroshima University, Japan* **18** 185-196
17 Jackson J A, Friberg A C & Bahr J M (1991) Preovulatory changes in glycosaminoglycans and collagen content in the stigma region of the follicle of the domestic hen. *Biology of Reproduction* **45** 301-307
18 Jackson J A & Bahr J M (1993) Preovulatory collagenase activity in the follicle of the chicken. *Molecular Basis of Reproductive Endocrinology* pp 210-214. Eds P E K Leung, A J W Hsueh & H G Frieson. New York: Springer-Verlag
19 Morales T I, Woessner J F, Marsh J M & LeMaire W J (1983) Collagen, collagenase and collagenolytic activity in rat Graafian follicles during follicular growth and ovulation. *Biochimica et Biophysica Acta* **756** 119-122
20 Reich R, Tsafriri A & Mechanic G L (1985) The involvement of collagenolysis in ovulation in the rat. *Endocrinology* **116** 522-527
21 Curry T E, Dean D D, Woessner J F & LeMaire W J (1985) The extraction of a tissue collagenase associated with ovulation in the rat. *Biology of Reproduction* **33** 981-991
22 Beers W H, Strickland S & Reich E (1975) Ovarian plasminogen activator: relationship to ovulation and hormonal regulation. *Cell* **6** 387-394
23 Liu Y X, Cajander S B, Ny T, Kristensen P & Hsueh A J W (1987) Gonadotropin regulation of tissue-type and urokinase-type plasminogen activators in rat granulosa and theca-interstitial cells during the periovulatory period. *Molecular and Cellular Endocrinology* **54** 221-229
24 Yanagishita M & Hascall V C (1979) Biosynthesis of proteoglycans by rat granulosa cells cultured *in vitro*. *Journal of Biological Chemistry* **254** 12355-12364
25 Yanagishita M, Rodbard D & Hascall V C (1979) Isolation and characterization of proteoglycans from porcine ovarian follicular fluid. *Journal of Biological Chemistry* **254** 911-920
26 Bahr J M (1991) The chicken ovary as a model of follicular development. *Seminars in Reproductive Endocrinology* **9** 352-359

Avian Endocrinology
Ed. P J Sharp, pp 321-330
Journal of Endocrinology Ltd, Bristol (1993)

# The role of epidermal growth factor and other factors in the paracrine and autocrine control of ovarian follicular development in the domestic hen

M Peddie, M Onagbesan and I Woolveridge
Department of Physiology, School of Biomedical Sciences, University of Southampton, Southampton SO9 3TU, UK

## Introduction

The ovulatory cycle of the hen is based on a 24-27 h rhythm with successive ovulations occurring, in older birds, at progressively later times on each successive day until, finally, there is a pause in the sequence and the 'clutch' is completed (1). In younger birds the clutch sequence is not obvious, with individuals attaining laying rates of 95% at peak lay. In order to sustain these exceptionally high laying rates, the ovary of the domestic hen contains a series of yolk-filled follicles, arranged in a precise hierarchy so that, at any time, the largest follicle is the next destined to ovulate and the sequentially smaller ones are destined to ovulate on the subsequent successive days (2). Atresia of follicles during this period is very rare. The yolk-filled follicles are recruited, by processes which are not known, from the pool of white follicles in the ovary, so that a hierarchy of maturing follicles is sustained. At any time, there are approximately 5 or 6 yolk-filled follicles in the ovary with, conventionally, F1 being the largest and F6 the smallest.

During follicular growth, there is not only a substantial accumulation of yolk, but the number of granulosa and theca cells also increases rapidly (3). Furthermore, during this period there are marked changes in the profile of steroid production by the follicles. Thus there is:

a) a marked increase in basal and LH stimulated progesterone production by the granulosa cells from F5 through to F1 (4);
b) a dramatic reduction in progesterone metabolism to androstenedione by the theca of F2/F1 in the few hours preceding ovulation (5);
c) a marked decline in aromatase activity in the theca from F4 to F3, so that no aromatase activity persists in F2, 48h before ovulation (6, 7).

Many aspects of these cardinal features of follicular development in birds have been reviewed recently, (1, 8-10). Therefore, only brief mention of the most salient points in the basic control of steroid production by the follicles will be made, while this review will focus on recent work, which demonstrates that peptide growth factors and steroids participate in the local regulation of steroidogenesis and LH-sensitivity.

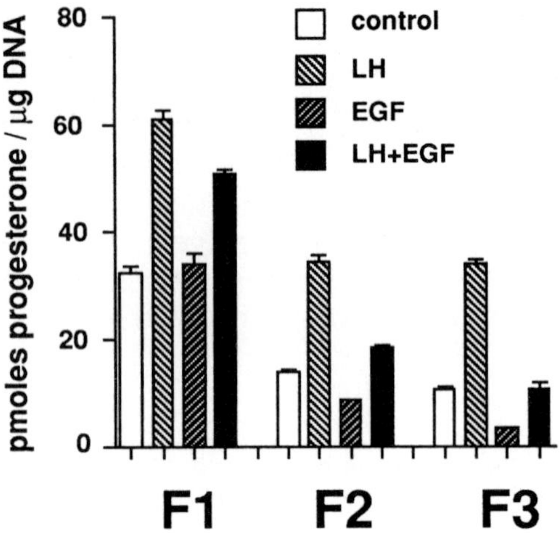


Fig. 1 The effect of EGF on progesterone production by granulosa cells. The cells were dispersed and plated in 2.5% charcoal stripped fetal calf serum in M199 with the addition of LH (100 ng/ml) EGF (20 ng/ml) or a combination of the two.

## Epidermal growth factor

Epidermal growth factor, in mammals, is a 53 amino acid peptide with close structural homology to another growth factor, transforming growth factor α (TGFα), especially in the region of the receptor-binding domain (11). Generally, they bind to a common receptor, epidermal growth factor-receptor (EGF-R), with the same affinity and broadly similar biological effects (12-14). However, in birds, the homologous peptides have not been characterized. Nevertheless, the EGF-R from avian tissues has been partially sequenced and shown to contain an extracellular domain which is dissimilar to the mammalian sequence (15). Since there is considerable homology between cErb-B and the human EGF-receptors (16-18), and there are common epitopes shared by the avian and human EGF-receptors (19), it has been possible to predict the structure of the avian EGF-R (15). The predicted structure of avian EGF-R is broadly similar to that in other species, with a glycosylated extracellular domain and a cytoplasmic domain which possesses both tyrosine kinase and a threonine kinase activity, and other potential phosphorylation sites. The affinity of the avian EGF-R and its ability to discriminate between EGF and TGFα as ligands have also been investigated. Thus the isolated avian EGF-R from fibroblasts has an affinity of $0.7 \times 10^{-9}$ M for human TGFα but a much lower affinity ($1.4 \times 10^{-7}$ M) for murine EGF (15, 20).

Epidermal growth factor receptors which have been localised immunocytochemically in both the theca and granulosa cells of the ovary of the domestic hen (21) probably mediate the inhibitory effect of EGF on progesterone production by granulosa cells (22, 23). Inhibition of LH-stimulated progesterone production by EGF may be transduced by a protein kinase C-dependent mechanism (24). Epidermal growth factor also causes an increase in intracellular pH in granulosa cells by a mechanism mediated by protein kinase C (25, 26). These observations are consistent with the effects of EGF on mammalian cells (27), and might also be associated with alterations in other components of second messenger systems known to be associated with EGF receptor activation in mammals, including increased phosphoinositide hydrolysis and intracellular calcium (28). The involvement of $Ca^{2+}$ as a component of the LH-induced steroidogenic response in granulosa cells in birds is well established (29), and calcium ionophores also elevate intracellular pH in granulosa cells (30). Ultimately, EGF and TGFα appear to inhibit progesterone production by reducing LH activated adenylate cyclase (9).

In addition to its inhibitory effect on progesterone production, EGF also stimulates the proliferation of avian granulosa cells in media containing serum in a dose dependent manner (31). In serum-free culture systems too, EGF and TGFα exert mitogenic effects on both avian granulosa and thecal cells (Fig. 2; 33, 35, 36). The ability of EGF to inhibit basal and LH-stimulated progesterone production by the granulosa cells and to stimulate proliferation of granulosa cells decreases with follicular maturity (Figs. 1 and 2). Similarly, EGF and TGFα inhibit LH-induced aromatase activity and androstenedione production by theca cells (34, 36, 37). These observations are consistent with the effects of EGF or TGFα on the mammalian ovary (37-39).

The physiological relevance of these observations cannot be fully evaluated until it is established whether EGF or TGFα are produced by the ovarian cells. However, chicken ovarian thecal cell-conditioned medium, but not granulosa-conditioned medium, contains biological activity similar to that of EGF or TGFα since it suppresses both LH-induced progesterone production and LH-induced aromatase activity in the theca cells, in a dose-dependent manner, and also induces proliferation of thecal cells (21, 34). Likewise, theca-conditioned medium stimulates pituitary cell proliferation in an EGF-like fashion (40). Similar observations were made initially for pig (41) and rat (42) thecal tissues which were subsequently confirmed when the expression of the TGFα gene was localized exclusively to the theca (43). However, more substantive data are required to confirm a similar localization of production in avian tissues.

**Insulin-like growth factor-I**

In some respects, more is known about IGF-I in the avian ovary than about EGF. The avian IGF-I gene is expressed in the avian ovary (44), while IGF-I ligand binding to granulosa cells has also been reported (45). The avian IGF-I receptor is dissimilar to the mammalian, in that there is only one subtype which binds IGF-II and insulin as well (46). Some of the reported discrepancies in the literature on the effects of IGF-I on avian cells may relate to this.

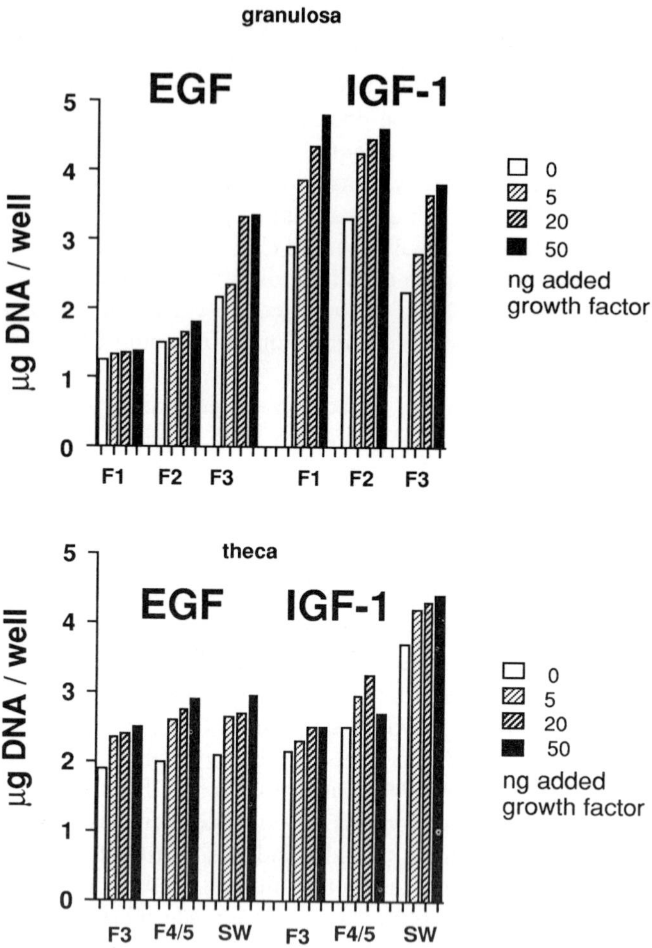


Fig. 2 The effects of EGF or IGF-I on the DNA content of cultured granulosa or theca cells. The cells were dispersed and plated in M199 with or without the addition of EGF (5-50 ng/ml) or IGF-I (5-50 ng/ml). DNA was determined in the culture cells after 72 h.

Insulin-like growth factor-I has a mitogenic action *in vitro* on both granulosa and theca cells of hens (47), but information on the effects of IGF-I on steroidogenesis is more elusive. In mammals IGF-I potentiates progesterone secretion from granulosa cells (48). The effects of IGF-I are modulated by IGF-binding proteins (BP), but little is known about IGF-BPs in birds (43, 49), especially about those which might be produced in the ovary, and perhaps this component is crucial. Nevertheless, IGF-I stimulates proliferation of both theca and granulosa cells, in the absence of serum, in a dose-dependent manner (Fig. 2). Insulin-like growth factor-I also potentiates basal progesterone production and enhances LH-stimulated progesterone production (Fig. 3). Insulin-like growth factor-I also inhibits LH-induced aromatase activity in cultured theca cells (33, 35), an effect which is opposite to that reported in mammals (38).

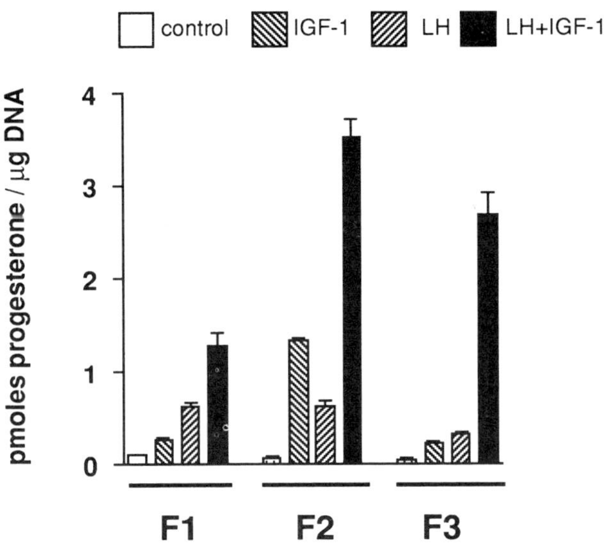


Fig. 3 The effects of IGF-I on progesterone secretion by cultured granulosa cells. The granulosa cells were prepared and plated into collagen-coated tissue culture dishes, in serum-free medium M199 with 50ng/ml insulin, transferrin, and selenium. The IGF-I (20ng/ml) was added at the beginning of the experiment; after 72 h, the cells were rinsed, the medium replaced with fresh medium but with no further IGF, while LH (100 ng/ml) was added for the subsequent 3 h. A control group (control) was not treated with IGF-I or LH, while a second control group (LH) was treated with LH only during the last 3 h of the experiment. The medium was collected at the end of this period and subsequently assayed for progesterone by radioimmunoassay.

**Role of steroids as paracrine modulators of ovarian function**

Steroids may have an autocrine or paracrine role in the ovary (10, 51). Both theca and granulosa cells possess progesterone receptors (52), but a role for these has not emerged until recently. Progesterone inhibits androstenedione production by theca from the F2 ovarian follicles with a similar but less marked effect on theca from F3 follicles (53). This effect seems to be predominantly by an inhibition of C17-20 lyase, and involves the synthesis of new proteins, since it is ameliorated by the addition of cycloheximide (Fig. 4).

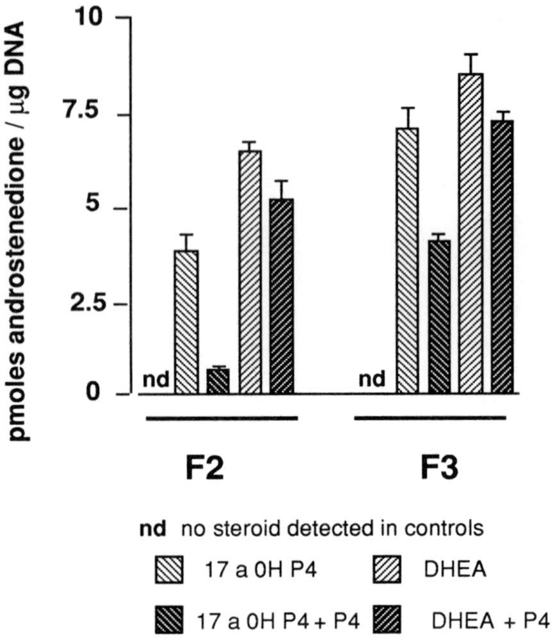


Fig. 4 Progesterone (P4) inhibition of androstenedione production by avian theca cells in culture. F2 and F3 theca cells were dispersed and cultured in 48 well plates in the presence of 5% fetal calf serum in M199. Progesterone was added ($10^{-6}$ M) for an 8-hour period, the cells washed, fresh medium containing 17α hydroxyprogesterone, (17αOH-P4), or dehydroepiandrosterone (DHEA) added, for a further 2 h. Subsequently, the medium was assayed for androstenedione by radioimmunoassay.

Androgens and oestrogens may play a physiological role in regulating the granulosa cell progesterone production (51), particularly in the less mature follicles. This local influence is likely to disappear following the dramatic reduction in C17-20 lyase activity in the F2/F1 follicles. Since oestrogens interact with EGF in many systems, it is possible EGF and oestradiol might potentiate or inhibit each other's actions on the granulosa cells. However, an attempt to demonstrate that oestradiol ($10^{-7}$ M or $10^{-8}$ M) inhibits granulosa cell progesterone production was unsuccessful (M Peddie, unpublished observation), rather it tended to increase LH responsiveness of the cultured cells. Also, the addition of oestradiol to cultured granulosa cells potentiates the ability of EGF to cause proliferation and inhibit progesterone production (50). This observation requires confirmation since other reports show that $10^{-6}$ M oestradiol inhibits granulosa cell progesterone production (51).

## Conclusions

New information is emerging about the actions of growth factors as modulators of ovarian function in birds, but substantially more is required before a complete understanding of the integrated control of follicular development is achieved. However, the following points have been established:

1) EGF (and TGFα) have inhibitory effects on progesterone production by the granulosa cells;
2) EGF and IGF-I cause proliferation of granulosa and theca cells;
3) IGF-I and an EGF-like factor are synthesized in the ovary;
4) EGF and IGF-I inhibit LH-induced aromatase activity in the theca;
5) EGF and TGFα inhibit C17-20 lyase activity in the theca;
6) Progesterone inhibits C17-20 lyase activity;
7) Oestrogens potentiate EGF-suppression of progesterone production from granulosa cells;
8) Other factors which might act as local modulators of ovarian function, such as neuropeptides, cytokines, prostanoids or neurotransmitters remain to be investigated.

## References


1 Cunningham F J (1987) The ovulatory cycle of the hen. *Oxford Reviews of Reproductive Biology* **9** 96-136

2 Warren D C & Scott H M (1935) The time factor in egg formation. *Poultry Science* **14** 195-207

3 Etches R J, MacGregor H E, Morris T F & Williams J B (1983) Follicular growth and maturation in the domestic hen. *Journal of Reproduction and Fertility* **67** 351-358

4 Hammond R W, Burke W H & Hertelendy F (1981) Influence of follicular maturation on progesterone release in chicken granulosa cells in response to turkey and ovine gonadotrophins. *Biology of Reproduction* **24** 1048-1055

5 Marrone B L & Hertelendy F (1985) Decreased androstenedione production with increased follicular maturation in theca cells from the domestic hen. *Journal of Reproduction and Fertility* **74** 543-550

6 Armstrong D G (1984) Ovarian aromatase activity in the domestic fowl (*Gallus domesticus*). *Journal of Endocrinology* **100** 81-84

7 Wang S C & Bahr J M (1983) Estradiol secretion by theca cells of the domestic hen during the ovulatory cycle. *Biology of Reproduction* **28** 618

8 Bahr J M & Palmer S S (1989) The influence of ageing on ovarian function. *Critical Reviews in Poultry Biology* **2** 103-110

9 Etches R J (1990) The ovulatory cycle of the hen. *Critical Reviews in Poultry Biology* **2** 293-318

10 Johnson A L (1990) Steroidogenesis and actions of steroids in the hen ovary. *Critical Reviews in Poultry Biology* **2** 319-346

11 Marquardt H, Hunkapiller M W, Hood L E & Todaro G J (1984) Rat TGFα type 1: structure and relation to EGF. *Science* **223** 1079

12 Dernyck R (1986) TGFα: structure and biological activities. *Journal of Cellular Biochemistry* **32** 293

13 Massague J (1983) Epidermal growth factor-like transforming factor. *Journal of Biological Chemistry* **258** 13614-13620

14 Todaro G J, Fryling C & Delarco J E (1980) Transforming growth factors produced by certain human tumor cells: polypeptides that interact with epidermal growth factor receptors. *Proceedings of the National Academy of Sciences of the USA* **77** 5258-5262

15 Lax I, Johnson A, Howk R, Sap J, Bellot F, Winkler M, Ullrich A, Vennstrom B, Schlessinger J & Givol D (1988) Chicken epidermal growth factor (EGF) receptor: cDNA cloning, expression in mouse cells, and differential binding of EGF and transforming growth factor alpha. *Molecular and Cellular Biology* **8** 1970-1978

16 Downward J, Yarden Y, Mayes E, Scrace G, Totty N, Stockwell P, Ullrich A, Schlessinger J & Waterfield M D (1984) Close similarity of epidermal growth factor receptor and vErb B oncogene protein sequences. *Nature* **367** 521-527

17 Nilsen T W, Maroney P A, Goodwin R G, Rottman F M, Crittenden L B, Raines M A & Kung H J (1985) CerbB activation in ALV-induced erythroblastosis: novel RNA processing and promoter insertion result in expression of an amino truncated EGF receptor. *Cell* **41** 719-726

18 Ullrich A, Coussens L, Hayflick J S, Dull T J & Gray A (1984) Human epidermal growth factor receptor, cDNA sequence and aberrant expression of the amplified gene in A431 epidermoid carcinoma cells. *Nature* **309** 418-425

19 Kris R M, Lax I, Gullick W, Waterfield M D, Ullrich A, Fridkin M & Schlessinger J (1985) Antibodies against a synthetic peptide as a probe for the kinase activity of the avian EGF-receptor and V-erbB protein. *Cell* **40** 619-625

20 Lax I, Bellot F, Howk R, Ullrich A, Givol D & Schlessinger J (1989) Functional analysis of the ligand binding site of EGF-receptor luteinizing chimeric chicken/human receptor molecules. *The EMBO Journal* **8** 421-427

21 Peddie M J (1992) The role of EGF and other growth factors in the local regulation of steroid production in ovarian follicles. *Proceedings of the Fifth International Symposium on Avian Endocrinology*. Abstract S10/3. September 1992. Edinburgh: AFRC Roslin Institute

22 Pulley D D & Marrone B L (1986) Inhibitory action of epidermal growth factor on progesterone biosynthesis in hen granulosa cells during short-term culture: two sites of action. *Endocrinology* **118** 2284-2291

23 Tilly J L & Johnson A L (1990) Modulation of hen granulosa cell steroidogenesis and plasminogen activator-activity by transforming growth factor alpha. *Growth Factors* **3** 246-255

24 Tilly J L & Johnson A L (1991) Protein kinase C in preovulatory follicles from the hen ovary. *Domestic Animal Endocrinology* **8** 1-13

25 Li M, Morley P, Asem E K & Tsang B K (1991) Epidermal growth factor elevates intracellular pH in chicken granulosa cells. *Endocrinology* **129** 656-662

26 Li M, Morley P & Tsang B K (1991) Epidermal growth factor elevates intracellular pH in chicken granulosa cells by activating protein kinase C. *Endocrinology* **129** 2957-2964

27 Molenaar W H, Tsien R Y, Van Der Saag O T & de Laat S W (1983) $Na^+/H^+$ exchange and cytoplasmic pH in the action of epidermal growth factor in human fibroblasts. *Nature* **304** 645-648
28 Cheyette T E & Cross D J (1991) Epidermal growth factor-stimulated calcium ion transients in individual A431 cells: initiation kinetics and ligand concentration dependence. *Cell Regulation* **2** 827-840
29 Asem E K & Hertelendy F (1986) Role of calcium in luteinizing hormone-induced progesterone and cyclic AMP production in granulosa cells of the hen (*Gallus domesticus*). *General and Comparative Endocrinology* **62** 120-128
30 Asem E K, Li M & Tsang B K (1992) Calcium ionophores increase intracellular pH in chicken granulosa cells. *Journal of Molecular Endocrinology* **9** 1-6
31 Yoshimura Y & Tamura T (1988) Effects of gonadotrophins, steroid hormones and epidermal growth factor on the *in vitro* proliferation of chicken granulosa cells. *Poultry Science* **67** 814-818
32 Onagbesan O M, Peddie M & Williams J (1993) Unpublished observations
33 Onagbesan O M, Peddie M & Williams J (1993) Unpublished observations
34 Onagbesan O M, Peddie M & Williams J (1993) Unpublished observations
35 Onagbesan O M, Peddie M & Williams J (1991) Effects of growth factors, serum and gonadotrophins on cell morphology and estrogen production by chicken theca cells cultured in defined medium. *Journal of Endocrinology* **129** Supplement Abstract 190
36 Onagbesan O M, Peddie M & Williams J (1992) Effects of EGF or TGFα on androstenedione production by the theca cells of the chickens *in vitro*. *Journal of Endocrinology* **132** Supplement Abstract 283
37 Franchimont P, Hazec-Hagelstein M T, Chalet-Renard Ch & Jaspar J M (1986) Effect of epidermal growth factor on DNA and protein synthesis, progesterone and inhibin production by bovine granulosa cells in culture. *Acta Endocrinologica* **111** 122-127
38 Steinkampf M P, Mendelson C R & Simpson E R (1988) Effects of epidermal growth factor and insulin-like growth factor 1 on the levels of mRNA encoding aromatase cytochrome P-450 of human ovarian granulosa cells. *Molecular and Cellular Endocrinology* **59** 93-99
39 Trzaciak W H, Duda T, Waterman M R & Simpson E R (1987) Effects of epidermal growth factor on the synthesis of the cholesterol side-chain cleavage enzyme complex in rat ovarian granulosa cells in primary culture. *Molecular and Cellular Endocrinology* **52** 42-50
40 Tsonis C G, Sharp P J & McNeilly A S (1988) Inhibin bioactivity and pituitary cell mitogenic activity from cultured chicken ovarian granulosa and thecal/stromal cells. *Journal of Endocrinology* **116** 293-299
41 Hsu C-J, Holmes S D & Hammond J M (1987) Ovarian epidermal growth factor-like activity. Concentrations in porcine follicular fluid during follicular enlargement. *Biochemical and Biophysical Research Communications* **147** 242-247
42 Skinner M K, Lobb D & Dorrington J H (1987) Ovarian thecal/interstitial cells produce an epidermal growth factor-like substance. *Endocrinology* **121** 1892-1899
43 Kudlow J E, Kobrin M S, Purchio A F, Twardzik D R, Hernandez E R, Asa S L & Adashi E Y (1987) Ovarian transforming growth factor-α gene expression: immunohistochemical localization to the theca-interstitial cells. *Endocrinology* **121** 1577-1579

44 Roberts R D, Boswell J, Burt D, Sharp P J & Goddard C (1990) IGF-I in the avian ovary. *Journal of Reproduction and Fertility* Abstract Series 6, Abstract 18

45 Huybrechts L M, Decuypere E & Kühn E R (1991) Granulosa cells of the preovulatory follicles of the domestic fowl contain an insulin-like growth factor (IGF) receptor. *Medical Science Research* **19** 699-700

46 Duclos M J, Wilkie R S & Goddard C (1991) Stimulation of DNA synthesis in chicken muscle satellite cells by insulin and insulin-like growth factors: evidence for exclusive mediation by a type-1 insulin-like growth factor receptor. *Journal of Endocrinology* **128** 35-42

47 Roberts R D (1991) The mitogenic effects of insulin-like growth factor 1 (IGF-I) on chicken preovulatory granulosa and theca. *Journal of Reproduction and Fertility* Abstract Series 8, Abstract 76

48 Schams D, Koll R & Li Ch (1988) Insulin-like growth factor-1 stimulates oxytocin and progesterone production by bovine granulosa cells in culture. *Journal of Endocrinology* **116** 97-100

49 Francis G L, McMurtry J P, Johnson R J & Ballard F J (1990) Plasma clearance of chicken and human insulin-like growth factor-1 and their association with circulatory binding proteins in the chicken. *Journal of Endocrinology* **124** 361-370

50 Peddie M J (1992) Interactions of EGF and oestradiol on cell replication and progesterone production by granulosa cells of the chicken. *Journal of Endocrinology* **132** Supplement Abstract 284

51 Lee H T & Bahr J M (1989) Inhibitory sites of androgens and estradiol in progesterone biosynthesis in granulosa cells of the domestic hen. *Endocrinology* **125** 760-765

52 Yoshimura Y & Bahr J M (1991) Localisation of progesterone receptors in pre- and post-ovulatory follicles of the domestic hen. *Endocrinology* **128** 323-330

53 Woolveridge I, Onagbesan O M & Peddie M (1992) Effects of cycloheximide on androstenedione secretion by the theca cells of chickens, *in vitro*. *Journal of Endocrinology* **132** Supplement Abstract 282

Avian Endocrinology
Ed. P J Sharp, pp 331-344
Journal of Endocrinology Ltd, Bristol (1993)

# Signal transduction in avian granulosa cells

F Hertelendy, M Jamaluddin and M Molnár
Department of Obstetrics and Gynecology, St Louis University Medical Center,
3635 Vista at Grand, PO Box 15250, St Louis, MO 63110-0250, USA

## Introduction

The concept of signal transduction by second messengers, originally introduced by Sutherland *et al.* (1) more than a quarter century ago, has been, over the intervening years, one of the most actively pursued areas in cellular and molecular endocrinology. Signals, generated by many hormones, upon association with specific receptors located on the cell surface, are transmitted across the plasma membrane by a variety of mechanisms, producing messenger molecules which in turn may activate key enzymes involved in evoking hormone-specific physiological responses. Such transmembrane signalling, generally referred to as signal transduction, has been documented in a wide variety of endocrine cells. In some cells, a single mechanism (e.g. the activation of the adenylate cyclase/cyclic AMP system) is responsible for transmitting the hormone-generated signal, whereas other cells are regulated by multiple signal transduction mechanisms operating in parallel or, more often, via 'crosstalk' (2). There is abundant evidence that a superfamily of guanine nucleotide binding proteins, (G-proteins) are intimately involved in signal transduction by coupling receptors to key enzymes, thereby stimulating or inhibiting enzyme activity, as well as regulating the operation of ion channels (3). Yet another group of hormones (insulin and various growth factors) elicit their action by autophosphorylation of receptor subunits to transmit the signal (4). The purpose of this brief review is to summarize the available evidence for the operation of signal-transducing pathways in avian granulosa cells and to propose a unified working model for the interaction of these pathways.

## The adenylate cyclase-cyclic AMP signalling pathway

*General considerations*

It is now well established that the adenylate cyclase enzyme is coupled to receptors on the surface of the plasma membrane via G-proteins. In some cell types, the coupling is by both stimulatory (Gs) and inhibitory (Gi) proteins. These, and all other known members of the superfamily of GTP-binding proteins (5) consist of $\alpha$-, ß- and $\tau$-subunits, of which the latter two are apparently identical, whereas the $\alpha$-subunit is the product of different genes (6). Under unstimulated conditions, the nucleotide binding site is occupied by GDP. Upon stimulation by the hormone-receptor complex, GDP is exchanged for GTP, the ß-, $\tau$-subunits dissociate from the heterotrimer G-protein and this allows the $\alpha$-GTP unit to activate the enzyme, generating from ATP, in the presence of $Mg^{2+}$, the intracellular signalling molecule, cyclic AMP. Conversely, the activation of Gi (Giα-GTP) inhibits the enzyme. The intracellular concentration of cyclic AMP is negatively regulated by phosphodiesterase enzyme, which breaks down cyclic AMP to the biologically inactive 5′-AMP. Cyclic AMP activates a group of protein kinases (protein kinase A), which phosphorylate a wide variety of proteins, thereby changing their molecular confirmation, rendering them active or inactive. In elucidating the actions of G-proteins, much use has been made of two bacterial toxins. Cholera toxin permanently activates the transducing system by transferring ADP-ribose to Gα, whereas ADP-rybosylation of Giα uncouples the system from the inhibitory hormone-receptor complex (7).

*Role of cyclic AMP in avian granulosa cells*

Although the involvement of the adenylate cyclase-cyclic AMP signalling pathway in ovarian cells of the rat and some other mammals had been unequivocally established by the mid-1970s (8), it was not until a few years later that experimental evidence was obtained implicating cyclic AMP in the steroidogenic action of LH in avian granulosa cells, by demonstrating that exogenous cyclic AMP analogues mimicked the effects of LH (9), and subsequently by direct measurements of LH-stimulated adenylate cyclase activity in membrane preparations of chicken granulosa cells (10-12). These early experiments using broken cell preparation, while achieving only modest, 2-3-fold increments in cyclic AMP production in $F_1$ follicles even in the presence of pharmacological doses (up to 100μg/ml) of LH (11) (particularly when viewed in the light of subsequent studies using intact cells (13, 14) which respond to near physiological concentrations of LH), indicated that the LH-sensitivity of the receptor-coupled adenylate cyclase increases during the last few days of follicular maturation, thus paralleling the progressive increase in the steroidogenic capacity of maturing granulosa cells (15-17). These studies implied that the LH-activatable adenylate cyclase (reflecting, perhaps, the availability of LH-receptors) represents the rate-limiting step in the biochemical events leading to increased progesterone synthesis (18). Several lines of evidence, however, suggest that this may not be the case. When the production of cyclic AMP and progesterone was measured in the same population of intact granulosa cells stimulated with LH or forskolin (a non-receptor-coupled activator of adenylate cyclase), there was no positive correlation between the well-documented progressive increase in progesterone output ($F_6 \rightarrow F_1$) and cyclic AMP production (14). On the contrary, granulosa cells from younger follicles ($F_3$) were more responsive in terms of LH-stimulated cyclic AMP generation than those isolated from $F_1$ follicles, indicating that the LH receptor-coupled adenylate cyclase signalling pathway is fully competent in maturing granulosa cells well before the steroid synthesizing biochemical machinery is equally capable of responding to the incoming signals. This was further corroborated by the demonstration that dibutyryl cyclic AMP was also less effective steroidogenically in developing rather than mature granulosa cells (19). Thus, the rate-limiting step must reside at a post-cyclic AMP site. Indeed, when the activity of cholesterol side chain cleavage enzyme (P450scc) was measured in a mitochondrial preparation from granulosa cells of mature and developing follicles, a significant positive correlation between enzyme activity and follicular development was obtained (20). In elegant experiments, Tilly *et al.* (21) confirmed these results recently, by demonstrating significantly more P450scc mMRA transcripts in granulosa tissue of $F_1$ than in $F_3$ and smaller follicles.

Luteinizing hormone-induced signal transduction mediated by the intracellular messenger cyclic AMP affects a broad spectrum of cellular functions such as glucose uptake and metabolism (22), as well as RNA and protein synthesis (23), all of which are interrelated with progesterone synthesis, representing the principal LH-regulated physiological function of mature granulosa cells. Inhibition of these processes also impairs steroidogenesis. How is the signal carried by cyclic AMP translated into RNA and protein synthesis? Although the activation of protein kinase A by cyclic AMP has been known for some time, the connection between protein kinase A-catalyzed phosphorylation and gene transcription has been elucidated only recently. Using recombinant DNA techniques, cyclic AMP response elements (CRE) have been found in DNA sequences of genes encoding peptide hormones (24). Further, a nuclear protein has been identified that binds CRE (CREB), which can be phosphorylated by cyclic AMP-dependent protein kinase catalytic subunit (25). The phosphorylated, dimerized CREB in turn enhances the transcription of genetic message (26). Is such a mechanism operative in chicken granulosa cells? Recent studies by Andrisani and Asem (27) using gel retardation (*in vitro* DNA binding) assay detected the presence of CREB in extracts of granulosa cells from mature, as well as developing, follicles, preincubated with LH or forskolin. Using a specific enzyme-linked immunoassay to quantitate CREB, these investigators found a modest increase with follicular maturation.

Is the adenylate cyclase-cyclic AMP signalling pathway the only transducing mechanism in avian granulosa cells? Several observations, to be discussed below, lead us to believe that in hen granulosa cells there exist multiple LH-triggered signalling mechanisms, operating in concert to maintain optimal intracellular conditions, and thereby allowing an appropriate physiological response to be elicited.

**The Ca-calmodulin system**

The widespread role of $Ca^{2+}$ in cellular regulation is well-recognized. The multiple actions of $Ca^{2+}$ are believed to be mediated in conjunction with various $Ca^{2+}$ binding proteins, of which calmodulin has been investigated most extensively (28). Several lines of evidence have suggested that the LH-induced rise in cyclic AMP is insufficient to explain fully the subsequent steroidogenic response in hen granulosa cells (13, 23, 29). For example, it was observed that forskolin-provoked cyclic AMP responses exceeded those obtained with LH stimulation, yet progesterone production was markedly lower under the influence of forskolin (13). In subsequent experiments, it was found that omission of $Ca^{2+}$ from the incubation medium reduced LH-inducible progesterone synthesis by about 50% (30). A more detailed analysis revealed that in addition to the significance of extracellular calcium, $Ca^{2+}$ derived from intracellular sources may also play an important functional role in LH-initiated signal transduction. It was observed that TMB-8, a putative inhibitor of intracellular calcium mobilization, suppressed LH-promoted progesterone synthesis in a dose-dependent manner, even in the presence of extracellular calcium. This suggests that calcium mobilization from intracellular compartments during LH stimulation is an integral part of LH action (31). Moreover, this study also suggested that extracellular calcium must first enter an intracellular compartment, probably in close association with the cytoplasmic side of the plasma membrane, from which it is then released in response to LH stimulation. Such an arrangement is concordant with the 'capacitative' model of $Ca^{2+}$ flux proposed by Putney (32). Direct evidence for LH-promoted $Ca^{2+}$ mobilization was provided by the use of digitonin-permeabilized granulosa cells (33). When these cells were stimulated with LH, after preloading them with $^{45}Ca$ in the presence of ATP, and ruthenium red to prevent mitochondrial uptake, there was a rapid, dose-related and biphasic release of $^{45}Ca$. The source of calcium was localized to the endoplasmic reticulum, since in the presence of vanadate, which blocks the uptake of $^{45}Ca$ into this compartment, LH was ineffective. Interestingly, the kinetics of $^{45}Ca$ efflux in response to forskolin and 8-bromo cyclic AMP was markedly different from that observed with LH stimulation, providing additional evidence that the signalling pathway initiated by LH is a complex, multifaceted mechanism. As in many other cell types, calcium probably acts in association with calmodulin, present in the cytoplasm of all eukaryotes (28). The Ca-calmodulin complex then interacts with target enzymes, enhancing their activity, or calmodulin may serve as a subunit of various enzyme proteins. Enzymes such as cyclic AMP phosphodiesterase and adenylate cyclase are Ca-calmodulin-dependent, and may thus represent one way by which the cyclic AMP and calcium signalling pathways interact. Indeed, phenothiazine antagonists of calmodulin inhibit not only LH-promoted progesterone production, but also cyclic AMP generation in hen granulosa cells (34). It should be pointed out, however, that the influence of calcium on adenylate cyclase is concentration-dependent. At submicromolar $[Ca^{2+}]_i$, calcium is stimulatory, whereas at higher concentrations it inhibits LH-supported cyclic AMP generation (35).

Ionophores, capable of transporting calcium across biological membranes, have been widely used to study the role and function of $[Ca^{2+}]_i$. Of these, A23187, ionomycin and lasalocid have been shown to stimulate basal production of progesterone in granulosa cells (36, 37), having a biphasic effect on cyclic AMP levels (35). However, in combination with LH, these ionophores were found to inhibit progesterone production in response to LH, forskolin, or cyclic AMP stimulation. This was attributed to morphological changes brought about by ionophores, as well as to the suppression of cellular ATP levels (36).

## The phosphoinositide cycle

The discovery of a second major transmembrane signalling pathway has led to the recognition that in a great many cells, including avian granulosa cells, $[Ca^{2+}]_i$ is modulated by receptor-coupled mechanisms. In this system the ligand-receptor complex is coupled by a G-protein (designated Go or Gq) to phospholipase C, which hydrolyses phosphatidylinositol-4,5-bisphosphate ($PIP_2$) to inositol-1,4,5-trisphosphate ($IP_3$) and diacylglycerol (DAG) (38). Inositol-1,4,5 trisphosphate binds to specific receptors on endoplasmic reticulum membranes, causing rapid release of $Ca^{2+}$, which serves, among its numerous other functions, together with DAG, to activate the ubiquitous enzyme protein kinase C (39). Inositol-1,4,5 trisphosphate can be further metabolized by phosphorylation to various inositol polyphosphates or sequentially dephosphorylated to inositol phosphate, which can reenter the cycle. Diacylglycerol localized both in the membrane and the cytoplasm, can be phosphorylated by DAG kinase to phosphatidic acid and then resynthesized to phosphatidylinositol, or further hydrolyzed by DAG lipase to release arachidonic acid, the precursor of prostanoids. The protein kinase C family of enzymes number at least eight members derived from several different genes (40), phosphorylating a wide array of protein substrates. Using tumour-promoting phorbol esters as activators and various enzyme inhibitors, a great deal of information has accumulated in recent years with respect to the functional role of protein kinases in cell regulation.

The first direct evidence that the phosphoinositide signalling pathway is operative in avian granulosa cells was provided by Hertelendy *et al.* (41), who observed a rapid, and dose-related, increase in $[Ca^{2+}]_i$ in LH-stimulated cells, as measured by the fluorescent dye, fura-2. Associated with this increase in $Ca^{2+}$-transient was an equally swift and significant loss of $[^3H]$ from $PIP_2$ in intact granulosa cells prelabelled with $[^3H]$myoinositol. Moreover, exposure of permeabilized granulosa cells to GTPτS, a stable analogue of GTP, mimicked these effects of LH by causing a rapid increase in $IP_3$, with a simultaneous drop in $PIP_2$ and a rise in $^{45}Ca$ efflux. These results can be interpreted to show an LH-triggered G-protein-coupled activation of phospholipase C, releasing the second messengers $IP_3$ and DAG in a fashion that is analogous to the cyclic AMP-generating system. Additionally, it was also found in this study that the stage of follicular maturity had a profound effect on the LH-activation of this signalling pathway with granulosa cells from the $F_1$ follicle exhibiting the highest phospholipase C activity and calcium mobilization. This novel observation supports the notion that during the last 2-3 days of follicular development not only does the activity of key steroidogenic enzymes increase (20, 42), but the LH-sensitive phospholipase C-$IP_3$-$Ca^{2+}$ signal transduction pathway also undergoes 'maturation'. Not surprisingly, the activity of protein kinase C, representing the second limb of this bifurcating signal transduction, also increases in a parallel fashion with follicular growth (43). A recent study by Canadian investigators (44) has demonstrated a rapid (within seconds) and transient oestrogen-induced increase in $[Ca^{2+}]_i$ in chicken granulosa cells from $F_1$ and $F_2$ follicles. This increase was independent of extracellular calcium, but was blocked by phospholipase C inhibitors, suggesting the involvement of $IP_3$ as an intracellular messenger, and a nongenomic action of oestrogens.

An additional and significant feature of the inositol lipid signalling pathway is the inhibitory effect of protein kinase C activation on LH-promoted cyclic AMP and progesterone production, representing perhaps the turn-off or negative feedback signal. Several laboratories have now shown that the bioactive ß-phorbol esters, and DAG analogs known to activate protein kinase C, inhibit LH actions in hen granulosa cells (37, 45, 46). Conversely, inhibitors of protein kinase C have been found to potentiate both the production of inositol phosphates and cyclic AMP in LH-stimulated chicken granulosa cells (47). Such 'crosstalk' between these two signalling systems may represent a fundamental aspect of granulosa cell regulation.

**Desensitization and signal coupling**

A sequential desensitization-resensitization phenomenon during the ovulation cycle of the domestic chicken has been suggested (19). Such desensitization can result from receptor down-regulation, uncoupling of the hormone-receptor complex from its effector system, or an interaction between dual signalling systems, as described above. Studies in Ishii's laboratory failed to detect significant differences in LH-receptor population in granulosa cells of the 3 largest follicles of Japanese quail (48). One may argue, therefore, that the acquisition of steroidogenic competence and, by implication, the loss of such, as in the case of desensitized granulosa cells, is not so much a function of receptor concentration as of the ability of the cell to transmit and translate the ligand-generated signal. Exposure of granulosa cells to LH for 1-4 h leads to significant attenuation of both cyclic AMP and inositol phosphate generation, as well as progesterone production in response to a second LH stimulation (49). This is not due to depletion of precursor stores, since it cannot be prevented by incorporation of lipoproteins in the incubation media during the 'desensitization' period. However, when staurosporine or H-7, two inhibitors of protein kinase C, were present during the first and second incubation periods, desensitization was completely prevented. These observations suggest that protein kinase=catalyzed phosphorylation of the coupling mechanism and/or the receptor itself, as has been shown for the avian erythrocyte ß-adrenergic receptor (50), is responsible for the desensitization. Moreover, the desensitization appears to be of the homologous type, since forskolin was fully effective in stimulating both cyclic AMP and progesterone production in LH-desensitized cells, suggesting, furthermore, that the locus of desensitization is upstream from the catalytic domain of adenylate cyclase. It would be of interest to see if a covalent modification of Gsα by cholera toxin could render the signalling mechanism immune to desensitization. This proposed mechanism does not rule out, of course, the 'classical' concept of desensitization by receptor internalization, followed by lysosomal degradation of receptor protein.

**Arachidonic acid: a second messenger in avian granulosa cells?**

Arachidonic acid has been implicated in signal transduction in a wide variety of cells, mainly because it serves as the principal precursor of many bioactive metabolites, the eicosanoids, which include the classic prostaglandins and leukotrienes. However, more recently, evidence based on observations that inhibitors of arachidonic acid metabolism do not influence the actions of exogenous arachidonic acid suggests that this fatty acid itself may act as an important intracellular messenger (51). For these reasons, the regulation of arachidonic acid release and its biomolecular actions have assumed importance. Arachidonic acid is a 20 carbon polyunsaturated fatty acid that is provided in the diet. It occurs chiefly as a component of glycerophospholipids, being attached at sn-2 position. Since free arachidonic acid levels of tissues and cells are believed to be very low (although direct quantitative evidence to support this notion is lacking; the assumption being based on the high affinity of acetyl transferases that catalyze the incorporation of arachidonic acid into glycerophospholipids), mobilization of arachidonic acid from endogenous stores is regarded as the rate-limiting step in the arachidonic acid cascade (52).

*Regulation of arachidonic acid release*

Although the mechanism usually considered for the control of free arachidonic acid levels involves the activation of phospholipase $A_2$, there is strong evidence that there exist in many types of cells multiple interconnected pathways of arachidonic acid release (53). These will now be briefly considered. Convincing evidence has been marshalled in support of the regulation of phospholipase $A_2$ by stimulatory and inhibitory G-proteins (Gs, Gi) coupled to receptors (53). Because phospholipase $A_2$ activation requires $Ca^{2+}$ (indeed, phospholipase $A_2$ can be activated by calcium ionophores, such as A23187 or ionomycin, in a receptor-G-protein-independent fashion), activation of phosphoinositol-specific phospholipase C generating $IP_3$ which raises $[Ca^{2+}]_i$ may also lead to, or contribute to, phospholipase $A_2$ activation and arachidonic acid release. It has also been proposed that activation of G-proteins may regulate ion channels (55), allowing the influx of $Ca^{2+}$ and leading to phospholipase $A_2$ activation. Thus the activity of phospholipase $A_2$ is intimately related to $[Ca^{2+}]_i$. Moreover, DAG - the second product of phospholipase C-catalyzed hydrolysis of inositol phospholipids - serves as substrate to DAG lipase and monoacylglycerol lipase, releasing free arachidonic acid. The crosstalk between these two pathways is further underscored by the demonstration that protein kinase C - activated by DAG in the presence of $Ca^{2+}$ - can promote arachidonic acid release by phosphorylating and thereby inactivating lipocortins, the endogenous inhibitors of phospholipase $A_2$ (56).

In addition to serving as substrate for phospholipase $A_2$, phosphatidylcholine may also provide arachidonic acid by the action of a phosphatidylcholine-specific phospholipase C and phospholipase D, releasing DAG and phosphatidic acid, respectively (57). G-protein-coupled activation of these enzymes has been demonstrated in a variety of cells. Once again, protein kinase C and $Ca^{2+}$ have been shown to activate both phosphatidylcholine specific phospholipase C and phospholipase D, representing key intracellular signal processors linking different pathways that regulate free arachidonic acid levels. Despite the convincing evidence obtained using various cell types, there is very little corresponding experimental data with respect to gonadal cells. In rat granulosa cells, a GnRH analogue activates phospholipase D (58). Moreover, preliminary evidence indicates that chicken granulosa cells, prelabelled with $[^3H]$arachidonic acid to apparent isotopic equilibrium and stimulated with arginine vasotocin, release not only free $[^3H]$arachidonic acid, but also $[^3H]$DAG, even when phosphoinositide-phospholipase C is inhibited and no increase in inositol phosphates is detected (59). The concomitant decrease in the labelled $[^3H]$phosphoinositide, $[^3H]$phosphatidylethanolamine, and $[^3H]$phosphatidylcholine provides further evidence that in addition to phospholipase $A_2$, phospholipase D and possibly phosphatidylcholine-phospholipase C are also activated.

Finally, arachidonic acid may also enter the intracellular pool bound to lipoproteins, using the LDL receptor pathway (60). Addition of both chicken and human lipoproteins, particularly LDL and VLDL, significantly stimulates progesterone synthesis in chicken granulosa cells, as well as potentiating LH- and $BU_2cAMP$-evoked responses (61). This study, and the earlier characterization of lipoprotein receptors in chicken granulosa membranes (62), support the suggestion that arachidonic acid esterified to cholesterol enters the cells by the LDL pathway.

*Mode of action of arachidonic acid*

As pointed out earlier, most of the studies involving the role of arachidonic acid in cellular functions have been linked to its conversion to one or more of the potent prostanoids or leukotrienes. There is growing evidence, however, that arachidonic acid itself acts as a second messenger (51). For example, arachidonic acid has been shown to mobilize intracellular $Ca^{2+}$ in a variety of cells (63), including bovine luteal cells (64). Recent studies have confirmed this effect of arachidonic acid in chicken granulosa cells and have uncovered a dual mode of action (65). At submicromolar concentrations, arachidonic acid releases $Ca^{2+}$ only from the endoplasmic reticulum, whereas at higher concentrations it also mobilizes $Ca^{2+}$ from the mitochondrial compartment. Arachidonic acid has also been implicated as an intracellular modulator of steroidogenesis in ovarian cells of the rat (66), chicken (67-69), and goldfish (70), as well as in rat Leydig tumour cells (71). However, whereas in rat granulosa cells the effect of arachidonic acid on progesterone production is completely blocked by an inhibitor of the lipoxyenase pathway, but not by indomethacin, arachidonic acid-stimulated testosterone production in goldfish preovulatory follicles is abolished by cyclooxygenase inhibitors. Both of these studies indicate that the bioactive substances are metabolites of arachidonic acid. On the other hand, in chicken granulosa and theca cells, exogenous arachidonic acid stimulates basal progesterone and androstenedione synthesis, while inhibiting LH-promoted steroidogenesis (67, 69). However, these effects of arachidonic acid are not inhibited by either cyclooxygenase or lipoxygenase inhibitors, supporting the idea that arachidonic acid itself has a modulatory role. Clearly, further studies will be necessary to establish or rule out the role of arachidonic acid as a second messenger in signal transduction.

## Conclusion

A hypothetical scheme for the interaction of signalling pathways in avian granulosa cells is presented in Figure 1.

Binding of LH to its receptor sets in motion a dual signal transducing mechanism: one generating cyclic AMP, and the other $IP_3$ and DAG, as second messengers. Cyclic AMP via protein kinase A is probably responsible for most of the steroidogenic response to LH stimulation. Phospholipase C activation is important in raising $[Ca^{2+}]_i$ above $10^{-7}$M via the agency of $IP_3$, as well as providing negative feedback signal by means of the DAG-protein kinase C mechanism. DAG derived from phosphoinositide breakdown may also furnish arachidonic acid after hydrolysis by neutral lipases. AVT may activate phospholipase $A_2$ via a G-protein-coupled mechanism and/or by opening

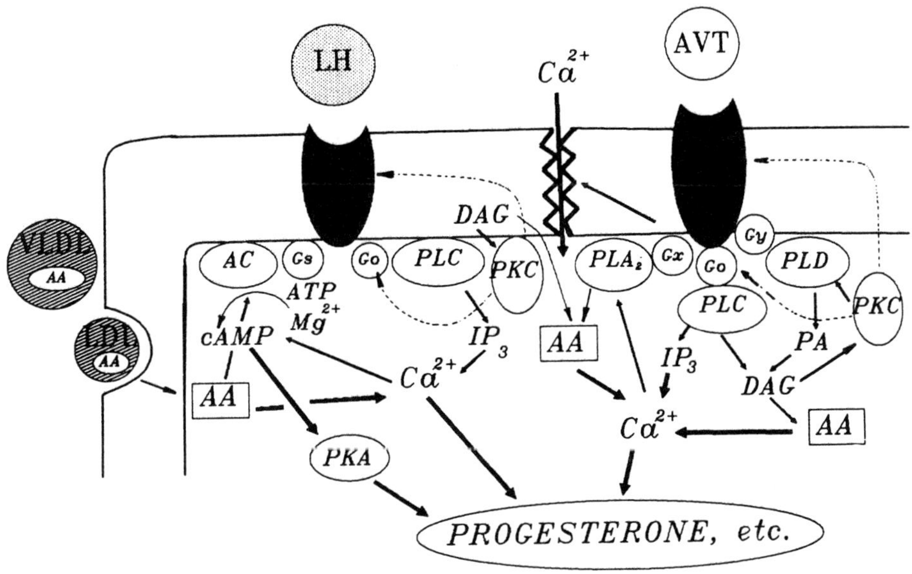


Fig. 1 A summary of LH and AVT signal transducing pathways and their interactions in the granulosa cell of the domestic chicken. LH: luteinizing hormone; AVT: arginine vasotocin; AC: adenylate cyclase; Gs: G-protein that couples LH receptor to AC; Go: G-protein that couples to PLC; $IP_3$: inositol-1,4,5-trisphosphate; PKC: protein kinase C; PKA: cyclic AMP-dependent protein kinase A; VLDL: very low density lipoprotein; LDL: low density lipoprotein; $PLA_2$: phospholipase $A_2$; PLD: phospholipase D; DAG: diacylglycerol; Gx, Gy: putative G-proteins; PA: phosphatidic acid. Solid lines indicate stimulation, broken ones inhibition.

Ca channels, thereby activating the Ca-dependent phospholipase $A_2$. AVT may also activate phosphatidylcholine-specific phospholipases C and D, also via G-proteins, or as a result of activation of protein kinase C by DAG generated in the phosphoinositide cycle. The sustained rise resulting from enhanced release of arachidonic acid, $IP_3$ and calcium influx, modulates, in association with calmodulin, key enzymes involved in a variety of cellular functions essential to support the steroidogenic mechanism. Finally, arachidonic acid attached to circulating lipoproteins may be delivered directly by receptor-mediated pinocytosis and released in the interior of the cell. Such an interactive signal transduction system would carry both positive and negative feedback signals, allowing for optimal execution of hormone-regulated functions.

## References

1 Sutherland E W, Oye I & Butcher R W (1965) The action of epinephrine and the role of the adenylyl cyclase system in hormone action. *Recent Progress in Hormone Research* **21** 623-646

2 Houslay MD (1991) 'Cross-talk': a pivotal role for protein kinase C in modulating relationships between signal transduction pathways. *European Journal of Biochemistry* **195** 9-27

3 Gilman A G (1987) G-proteins: transducers of receptor-generated signals. *Annual Review of Biochemistry* **56** 615-649

4 Taylor R (1991) Insulin action 1991. *Clinical Endocrinology* **34** 159-171

5 Bourne H R (1990) The GTPase superfamily: a conserved switch for diverse cellular functions. *Nature* **348** 125-132

6 Spiegel A M (1989) Receptor-effector coupling by G-proteins: implications for endocrinology. *Trends in Endocrinology and Metabolism* **1** 72-76

7 Davis J R E, Bidey S P & Tomlinson S (1992) Signal transduction in endocrine tissues. *Clinical Endocrinology* **36** 437-449

8 Marsh J M (1975) The role of cyclic AMP in gonadal function. *Advances in Cyclic Nucleotide Research* pp 137-199. Eds P Greengard & G A Robison. New York: Raven

9 Hammond R W, Todd H & Hertelendy F (1980) Effects of mammalian gonadotropins on progesterone release and cyclic nucleotide production by isolated avian granulosa cells. *General and Comparative Endocrinology* **41** 467-476

10 Hertelendy F, Takáts A & Lintner F (1981) Comparison between LH-induced adenyl cyclase activation and progesterone production in avian granulosa cells. *IRCS Medical Science* **9** 382

11 Calvo F O, Wang S-C & Bahr J M (1981) LH-stimulable adenylyl cyclase activity during the ovulatory cycle in granulosa cells of the three largest follicles and the postovulatory follicle of the domestic hen (*Gallus domesticus*). *Biology of Reproduction* **25** 805-812

12 Takáts A & Hertelendy F (1982) Adenylate cyclase activity of avian granulosa: effect of gonadotropin releasing hormone. *General and Comparative Endocrinology* **48** 515-524

13 Asem E K & Hertelendy F (1983) Effects of forskolin on progesterone and cyclic AMP production in avian granulosa cells. *Biology of Reproduction* **29** 1098-1104

14 Asem E K & Hertelendy F (1986) Steroidogenesis and cyclic AMP production in isolated avian granulosa cells during follicular maturation: lack of positive correlation. *Acta Endocrinologica* **113** 289-297

15 Huang E S-R, Kao K J & Nalbandov A V (1979) Synthesis of sex steroids by cellular components of chicken follicles. *Biology of Reproduction* **20** 454-461

16 Hammond R W, Burke W H & Hertelendy F (1981) Influence of follicular maturation on progesterone release in chicken granulosa cells in response to turkey and ovine gonadotropins. *Biology of Reproduction* **24** 1048-1055

17 Marrone B L & Hertelendy F (1983) Steroid metabolism by avian ovarian cells during follicular maturation. *Biology of Reproduction* **29** 953-962

18 Bahr J M & Calvo F O (1984) A correlation between adenylyl cyclase activity and responsiveness to gonadotrophins during follicular maturation in the domestic hen. *Reproductive Biology of Poultry* pp 75-87. Eds F J Cunningham, P E Lake & D Hewitt. Harlow: Longman

19 Hertelendy F & Asem E K (1984) Steroidogenesis in granulosa cells during follicular maturation: evidence for desensitization-resensitization during the ovulation cycle. *Journal of Experimental Zoology* **232** 513-520

20 Asem E K & Hertelendy F (1985) Influence of follicular maturation on luteinizing hormone-, cyclic 3′,5′-adenosine monophosphate-, forskolin- and cholesterol-stimulated progesterone production in hen granulosa cells. *Biology of Reproduction* **32** 257-268

21 Tilly J L, Kowalski K I & Johnson A L (1991) Cytochrome $P_{450}$ side-chain cleavage ($P_{450scc}$) in the hen ovary. II. $P_{450scc}$ messenger RNA, immunoreactive protein, and enzyme activity in developing granulosa cells. *Biology of Reproduction* **45** 967-974

22 Zakár T & Hertelendy F (1980) Effects of mammalian LH, cyclic AMP and phosphodiesterase inhibitors on steroidogenesis, lactate production, glucose uptake and utilization by avian granulosa cells. *Biology of Reproduction* **22** 810-816

23 Zakár T & Hertelendy F (1980) Steroidogenesis in avian granulosa cells: early and late kinetics of oLH and dibutyryl cyclic AMP-promoted progesterone production. *Biology of Reproduction* **23** 974-980

24 Montminy M R, Sevarino K A, Wagner J A, Mandel G & Goodman R H (1986) Identification of a cyclic AMP-responsive element within the rat somatostatin gene. *Proceedings of the National Academy of Sciences of the USA* **83** 6682-6686

25 Montminy M R & Bilezikjian L M (1987) Binding of nuclear protein to the cyclic-AMP response element of the somatostatin gene. *Nature* **328** 175-178

26 Yamamoto K K, Gonzalez G A, Biggs III W H & Montminy M R (1988) Phosphorylation-induced binding and transcriptional efficacy of nuclear factor CREB. *Nature* **334** 494-498

27 Andrisani O M & Asem E K (1992) Luteinizing hormone regulates cyclic AMP responsive-element binding protein (CREB) in granulosa cells. Abstract 1482. *74th Annual Meeting of the Endocrine Society, San Antonio USA*

28 Cohen P & Klee C B, Eds (1988) *Calmodulin.* Elsevier: New York

29 Hertelendy F, Zakár T, Fischer V W & Raab B (1982) Steroidogenesis in avian granulosa cells: effects of ammonium on oLH and dibutyryl cyclic AMP-promoted progesterone production. *Aspects of Avian Endocrinology: Practical and Theoretical Implications* pp 83-92. Eds C G Scanes, M A Ottinger, A D Kenny, J Balthazart, J Cronshaw & I Chester Jones. Lubbock: Graduate Studies Texas Technical University

30 Asem E K, Zakár T, Biellier H V & Hertelendy F (1984) Progesterone production in granulosa cells of the domestic fowl: effects of incubation media, pH, cell density and some other factors. *Domestic Animal Endocrinology* **1** 235-249

31 Asem E K & Hertelendy F (1986) Role of calcium in luteinizing hormone-induced progesterone and cyclic AMP production in granulosa cells of the hen (*Gallus domesticus*). *General and Comparative Endocrinology* **62** 120-128

32 Putney Jr J W (1986) A model for receptor-regulated calcium entry. *Cell Calcium* **7** 1-12

33 Asem E K, Molnár M & Hertelendy F (1987) Luteinizing hormone-induced intracellular calcium mobilization in granulosa cells: comparison with forskolin and 8-bromo-cAMP. *Endocrinology* **120** 853-859

34 Asem E K & Hertelendy F (1986) Trifluoperazine inhibits progesterone and cyclic AMP production in granulosa cells of the hen (*Gallus domesticus*). *General and Comparative Endocrinology* **64** 107-111

35 Jamaluddin M & Hertelendy F (1992) Biphasic effect of calcium on luteinizing hormone-stimulated cyclic AMP production in granulosa cells of the fowl (*Gallus domesticus*). *Biology of Reproduction* **46** 698-704

36 Zobell R, Hertelendy F & Fischer V W (1987) The effects of ionophores on steroidogenesis and morphology of avian granulosa cells. *Cell and Tissue Research* **248** 551-558
37 Tilly J L & Johnson A L (1988) Attenuation of hen granulosa cell steroidogenesis by a phorbol ester and 1-oleoyl-2-acetylglycerol. *Biology of Reproduction* **39** 1-8
38 Hokin L E (1985) Receptors and phosphoinositide-generated second messengers. *Annual Review of Biochemistry* **54** 205-235
39 Nishizuka Y (1986) Studies and perspectives on protein kinase C. *Science* **233** 305-312
40 Kikkawa U, Kishimoto A & Nishizuka Y (1989) The protein kinase C family: heterogeneity and its implications. *Annual Review of Biochemistry* **58** 31-44
41 Hertelendy F, Nemecz G & Molnár M (1989) Influence of follicular maturation on luteinizing hormone and guanosine 5′-0-thiotriphosphate-promoted breakdown of phosphoinositides and calcium mobilization in avian granulosa cells. *Biology of Reproduction* **40** 1144-1151
42 Asem E K & Hertelendy F (1987) Influence of follicular maturation on 3-hydroxy-methylglutaryl coenzyme A reductase activity in hen granulosa cells. *Journal of Reproduction and Fertility* **79** 215-221
43 Hertelendy F & Molnár M (1989) Protein kinase C activity in chicken granulosa cells during follicular maturation. Abstract 3. *Biology of Reproduction* **40** Supplement 1
44 Morley P, Whitfield J F, Vanderhyden B C, Tsang B K & Schwartz J-L (1992) A new, nongenomic estrogen action: the rapid release of intracellular calcium. *Endocrinology* **131** 1305-1312
45 Hertelendy F, Todd H, Molnár M & Magier S (1988) Phorbol esters inhibit granulosa cell functions. *FASEB Journal* **2** A1685
46 Asem E K & Tsang B K (1989) Phorbol ester inhibits luteinizing hormone-, forskolin- and dibutyryl cyclic AMP-induced progesterone production in chicken granulosa cells. *Canadian Journal of Physiology and Pharmacology* **67** 122-125
47 Jamaluddin M, Molnár M, Marrone B L & Hertelendy F (1990) Diverse effects of staurosporine (ST) a protein kinase C (PKC) inhibitor, on LH-stimulated cAMP and progesterone (P) production in chicken granulosa cells (GC). Abstract 377. *23rd Annual Meeting of Society for the Study of Reproduction*
48 Kikuchi M & Ishii S (1987) Granulosa cell LH receptors of the Japanese quail. *Proceedings of 1st Congress Asia and Oceania Society for Comparative Endocrinology* pp 82-83. Nagoya University Corporation
49 Hertelendy F, Molnár M & Jamaluddin M (1991) Luteinizing hormone desensitizes both adenylate cyclase and phospholipase C-mediated signal transmission in hen granulosa cells. Abstract 626. *73rd Endocrine Society Meeting*
50 Sibley D R, Nambi P, Peters J R & Lefkowitz R J (1984) Phorbol diesters promote ß-adrenergic receptor phosphorylation and adenylate cyclase desensitization in duck erythrocytes. *Biochemical and Biophysical Research Communications* **121** 973-979
51 Naor Z (1991) Is arachidonic acid a second messenger in signal transduction? *Molecular and Cellular Endocrinology* **80** C181-C186
52 Needleman P, Turk J, Jakschik B A, Morrison A R & Lefkowith J B (1986) Arachidonic acid metabolism. *Annual Revue of Biochemistry* **55** 69-102
53 Burgoyne R D & Morgan A (1990) The control of free arachidonic acid levels. *Trends in Biochemical Sciences* **15** 365-366

54 Axelrod J, Burch R M & Jelsema C L (1988) Receptor-mediated activation of phospholipase $A_2$ via GTP-binding proteins: arachidonic acid and its metabolites as second messengers. *Trends in Neurosciences* **11** 117-123

55 Brown A M & Birnbaumer L (1988) Direct G protein gating of ion channels. *American Journal of Physiology* **254** H401-H410

56 Hirata F (1983) Lipomodulin: a possible mediator of the action of glucocorticoids. *Advances in Prostaglandin Thromboxane and Leukotriene Research* **11** 73-80

57 Exton J H (1990) Signalling through phosphatidylcholine breakdown. *Journal of Biological Chemistry* **265** 1-4

58 Liscovitch M & Amsterdam A (1989) Gonadotropin-releasing hormone activates phospholipase D in ovarian granulosa cells. Possible role in signal transduction. *Journal of Biological Chemistry* **264** 11762-11767

59 Hertelendy F, Molnár M & Jamaluddin M (1991) Regulation of arachidonic acid (AA) release in hen granulosa cells during the ovulatory cycle: effect of arginine vasotocin (AVT). Abstract 299. *Biology of Reproduction* **44** Supplement 1

60 Habenicht A J R, Salbach P, Goerig M, Zeh W, Janssen-Timmen U, Blattner C, King W C & Glomset J A (1990) The LDL receptor pathway delivers arachidonic acid for eicosanoid formation in cells stimulated by platelet-derived growth factor. *Nature* **345** 634-636

61 Hertelendy F, Todd H & Molnár M (1992) Influence of chicken and human lipoproteins on steroidogenesis in granulosa cells of the domestic fowl (*Gallus domesticus*). *General and Comparative Endocrinology* **85** 335-340

62 Krumins S A & Roth T F (1981) High-affinity binding of lower-density lipoproteins to chicken oocyte membranes. *Biochemical Journal* **196** 481-488

63 Wolf B A, Colca J R, Turk J, Florholmen J & McDaniel M L (1988) Regulation of $Ca^{2+}$ homeostasis by islet endoplasmic reticulum and its role in insulin secretion. *American Journal of Physiology* **254** E121-E136

64 Alila H W, Corradino R A & Hansel W (1990) Arachidonic acid and its metabolites increase cytosolic free calcium in bovine luteal cells. *Prostaglandins* **39** 481-496

65 Hertelendy F, Molnár M & Jamaluddin M (1992) Dual action of arachidonic acid on calcium mobilization in avian granulosa cells. *Molecular and Cellular Endocrinology* **83** 173-181

66 Wang J & Leung P C K (1988) Role of arachidonic acid in luteinizing hormone-releasing hormone action: stimulation of progesterone production in rat granulosa cells. *Endocrinology* **122** 906-911

67 Johnson A L & Tilly J L (1990) Arachidonic acid inhibits luteinizing hormone-stimulated progesterone production in hen granulosa cells. *Biology of Reproduction* **42** 458-464

68 Johnson A L & Tilly J L (1990) Evidence that arachidonic acid influences hen granulosa cell steroidogenesis and plasminogen activator activity by a protein kinase C-independent mechanism. *Biology of Reproduction* **43** 922-928

69 Johnson A L, Tilly J L & Levorse J M (1991) Possible role for arachidonic acid in the control of steroidogenesis in hen theca. *Biology of Reproduction* **44** 338-344

70 Van Der Kraak G & Chang J P (1990) Arachidonic acid stimulates steroidogenesis in goldfish preovulatory ovarian follicles. *General and Comparative Endocrinology* **77** 221-228

71 Majercik M H & Puett D (1991) Epidermal growth factor modulates intracellular arachidonic acid levels in MA-10 cultured Leydig tumour cells. *Molecular and Cellular Endocrinology* **75** 247-256

Avian Endocrinology
Ed. P J Sharp, pp 345-358
Journal of Endocrinology Ltd, Bristol (1993)

# Regulation of plasma 1,25$(OH)_2D_3$, of osteocalcin and of intestinal and uterine calbindin in hens

Y Nys
INRA, Station de Recherches Avicoles, 37380 Nouzilly, France

## Introduction

The laying hen requires large amounts of calcium because each egg contains about 10% of total body calcium. The demand for calcium for shell formation is met by large calcium fluxes within the intestine and the uterus, and from increased bone turnover. Shell deposition occurs 10 to 20 h after the preceding oviposition and severely challenges the maintenance of normal calcium concentrations of calcium in body fluids. Maintenance of plasma $Ca^{++}$ homeostasis during shell formation is achieved through the actions of vitamin D, parathyroid hormone (PTH) and secondarily calcitonin, which coordinate intestinal $Ca^{++}$ absorption, bone $Ca^{++}$ resorption and $Ca^{++}$ reabsorption by the kidney (1). Parathyroid hormone regulates $Ca^{++}$ flux by acting on bone and exerts long-term effects by stimulating the production of 1,25 dihydroxycholecalciferol (1,25$(OH)_2D_3$), an active metabolite of vitamin D which controls intestinal absorption of calcium (2-4). Moreover, hens have systems for sequestering calcium in anticipation of changes in calcium fluxes. Consumption of calcium increases daily as a consequence of a calcium appetite during the early stages of shell calcification and, during the prelaying period, hens form a medullary bone which acts as a labile calcium reservoir during the laying cycle (1, 3). The appearance of medullary bone is dependent on the combined influence of oestrogen and testosterone and coincides with increased production of 1,25$(OH)_2D_3$ in the kidney and a consequent enhancement in calcium retention. The mechanisms by which sex steroids regulate 1,25$(OH)_2D_3$ after sexual maturity remain uncertain.

Vitamin D, through its hormonally active metabolite 1,25$(OH)_2D_3$, is therefore of utmost importance in facilitating the supply of calcium for shell formation. In birds, 1,25$(OH_2)D_3$ regulates intestinal absorption of calcium mainly by the induction of RNA transcription and synthesis of proteins such as calbindin $D_{28k}$ (4). The physiological function of calbindin $D_{28k}$ is not established but its concentration in the intestine reflects the ability of this tissue to absorb calcium (5). A similar protein is present at a high concentration in the uterus (6) which transfers a large mass of calcium through its wall. The uterus of the laying hen contains receptors for 1,25$(OH)_2D_3$ (7), but there is substantial evidence to show that uterine calbindin is partly independent of 1,25$(OH)_2$-$D_3$ status. This suggests that uterine calbindin is also regulated by a tissue specific mechanism.

1,25 dihyroxycholecalciferol may also facilitate bone formation by inducing the biosynthesis of osteocalcin. The biological function of this non-collagenous protein in bone mineralization is not fully understood but osteocalcin appears to be a specific product of osteoblasts during bone formation (8).

Advances in understanding the vitamin D endocrine system have been achieved using the vitamin D- deficient chick model which is highly sensitive to vitamin $D_3$ or its metabolites. The laying hen provides a valuable model to establish the physiological significance of changes induced by vitamin D in transepithelial calcium transport, in bone calcium mobilization, and in the associated fluctuations in calcium-binding protein synthesis. This review covers recent advances in understanding, in the hen, of the regulation of $1,25(OH)_2D_3$ production and of vitamin D calcium binding proteins in the intestine, uterus (calbindin) and bone (osteocalcin).

**Regulation of plasma 1,25 dihydroxycholecalciferol in laying hens**

Changes in intestinal calcium absorption are regulated by $1,25(OH)_2D_3$ (1-4). An understanding of the control of $1,25(OH)_2D_3$ production is therefore of importance in understanding the way this tissue adapts to fluctuations in the hen's requirement for calcium. In growing pullets, renal production of $24,25(OH)_2D_3$ is 7- to 9-fold higher than that of $1,25(OH)_2D_3$ (9) but kidney 25-hydroxycholecalciferol 1$\alpha$ hydroxylase activity, plasma concentrations of $1,25(OH)_2D_3$ (10, 11), and concentrations of intestinal $1,25(OH)_2D_3$ (12) increase markedly with the onset of laying in maturing hens, in contrast to $25(OH)D_3$ 24-hydroxylase. Numerous factors have been shown to stimulate the production of $1,25(OH)_2D_3$ including parathyroid hormone, $1,25(OH)_2D_3$ itself, sex steroids and prolactin (13). At sexual maturity, increases in gonadal steroids and in the calcium demand for shell formation may be responsible for an increase in circulating $1,25(OH)_2D_3$. Hypocalcemia increases the production of $1,25(OH)_2D_3$ indirectly through PTH which substantially increases the production of $1,25(OH)_2D_3$ *in vivo* (14) or in *vitro* (15). However, hypocalcemia may also directly affect $1,25(OH)_2D_3$ production. Oestrogens stimulate renal $25(OH)D_3$ hydroxylase *in vivo* (16, 17) but are less potent in this respect when added to cultures of kidney cells (18). Circulating $1,25(OH)_2D_3$ increases in laying hens during the prelaying period and again at onset of egg production. The first increment coincides with an increase in circulating oestrogens which induce vitellogenesis and medullary bone formation. The skeleton increases in weight by about 25% during the prelaying period and Bar and Hurwitz (19) have suggested that increased $1,25(OH)_2D_3$ production results from calcium deficiency induced by oestrogens. The observation that the stimulatory effect of exogenous oestrogen on $25(OH)D_3$ 1$\alpha$ hydroxylase is eliminated in birds raised on a low calcium diet (19) supports this hypothesis.

Hens laying shell-less eggs are a useful model in which to dissociate the respective roles of reproductive hormones and of calcium demand for shell formation in the control of $1,25(OH)_2D_3$ production. Plasma profiles of gonadal steroids are similar in control hens and in hens induced to lay eggs without shells (20). These hens are hypercalcaemic due to a high calcium intake and so PTH secretion is likely to be suppressed under these conditions. The circulating concentrations of $1,25(OH)_2D_3$

increase five-fold in the period following ovulation in hens laying shell-less or hard-shelled eggs as compared with those in immature pullets (21). This observation suggests that increased production of $1,25(OH)_2D_3$ associated with the onset of sexual maturation involves a pathway that is not induced by calcium deficiency, but depends on endocrine changes associated with ovulation. During the ovulatory cycle there are preovulatory increases in plasma progesterone and testosterone (22) and inconsistent evidence for a similar surge of plasma oestradiol (20). Since exogenous oestradiol activates $1,25(OH)_2D_3$ production *in vivo* but progesterone or testosterone do not, increased plasma oestradiol may be responsible for high concentrations of $1,25(OH)_2D_3$ observed just after ovulation. However, this hypothesis is not supported by the observations that oestradiol does not stimulate kidney $25(OH)D_3$ $1\alpha$ hydroxylase activity *in vitro* and that the decrease in the activity of this enzyme in hens fed nicarbazin (23) to inhibit ovulation, is not associated with a decrease in plasma oestradiol.

Prolactin is another putative modulator of $25(OH)D_3$ $1\alpha$ hydroxylase since prolactin stimulates $1,25(OH)_2D_3$ production *in vivo* (10) and *in vitro* (24). The concentration of plasma prolactin increases in immature pullets treated with oestrogen and in laying hens (25), but the existence of a preovulatory surge during the ovulatory cycle has not been established. A large increase in plasma prolactin occurs in incubating turkey hens but circulating $1,25(OH)_2D_3$ is low in these birds (26) suggesting that prolactin is not involved in regulation of $25(OH)D_3$ $1\alpha$ hydroxylase. The activity of this enzyme decreases when ovulation is inhibited and it is possible that there are other modulators of $1,25(OH)_2D_3$ production which may change during the ovulatory cycle. One possible modulator may be parathyroid hormone related-peptide which activates $25(OH)D_3$ $1\alpha$ hydroxylase *in vitro* and may be secreted in the hypercalcemic condition (27).

The balance of evidence supports the view that PTH mediates the further stimulation of $1,25(OH)_2D_3$ production associated with the calcium demand for shell formation (21). Plasma $1,25(OH)_2D_3$ substantially decreases, hour to hour following suppression of shell formation and increases during the subsequent resumption of calcification, while concentrations of plasma ionized calcium show reciprocal changes (21). The increase in concentrations of plasma $1,25(OH)_2D_3$ does not occur if hens are parathyroidectomized before shell deposition resumes. The daily increase in plasma $1,25(OH)_2D_3$, which occurs during the period of shell calcification (21, 28) and coincides with a decrease in plasma ionized calcium and increase in plasma PTH (29, 30), is consistent with the view that the production of $1,25(OH)_2D_3$ is regulated by PTH. The upregulation of PTH receptors by oestrogen in the kidney (31) may potentiate this effect. On the other hand, PTH may reduce the number of intestinal vitamin D receptors by inhibiting gene transcription, as observed in rats (32), and consequently counteract the up-regulation of $1,25(OH)_2D_3$ receptors by $1,25(OH)_2D_3$ (33). This has been shown in rats fed a calcium-deficient diet (34). The down-regulation of vitamin D receptor mRNA (35) and increased percentage of occupied receptors (36) in chickens fed a low calcium diet support this hypothesis. In hens, total intestinal vitamin D receptor measured by exchange assay (36) is not modified by sexual maturity but vitamin D receptor occupancy increases with an increase in serum $1,25(OH)_2D_3$ (12).

**Regulation of intestinal calbindin in hens**

Avian intestinal calbindin $D_{28k}$ is a single polypeptide chain consisting of 261 amino-acid residues with a molecular mass of 28 kDa (37). This cytosolic protein is present in a high concentration and is strictly dependent on vitamin D in the intestine. It belongs to a family of proteins that binds calcium with high affinity due to the presence of repeats of an octahedral calcium-binding domain termed an EF hand (38). Calbindin $D_{28k}$ contains 6 EF hands but two sites have lost their ability to bind calcium (39). Similarly, in mammals a vitamin D-dependent calbindin $D_{9k}$ has been identified in the intestine. It contains 2 EF hands but shows no amino-acid sequence homology with calbindin $D_{28k}$. Calbindin $D_{28k}$ has been detected in mammalian and avian species in many other tissues not involved in the transcellular transport of calcium where it may be not influenced by vitamin D (38).

In intestinal cells, the precise function of calbindin, aside from its ability to bind calcium has not been established. Its localization in the cytosol and the strong correlation between the rate of intestinal calcium transport and the calbindin concentration in many physiological situations support the hypothesis that calbindin acts as an aqueous intracellular transporter (40). In addition to facilitating transport, intestinal calbindin may also act as an intracellular buffer to prevent the build-up of toxic concentrations of calcium during its translocation through the cell interior. Whatever its biochemical function(s), this protein is a key marker of $1,25(OH)_2D_3$ mediated gene expression (4).

The chicken calbindin $D_{28k}$ gene is 19 kb long and contains 11 coding exons and 10 introns (41). A putative vitamin D-responsive element (VDRE) has been located by DNase I footprinting analysis in chicks (42) and has been recently identified in rat (43). However, the presence of a VDRE in the calbindin-D gene remains to be firmly established.

Three species of calbindin $D_{28k}$ mRNA (approximately of 2.1, 2.8, 3.1 kb) (44, 45) have been identified in the chicken intestine with the smallest being the most abundant. They are transcripts of a single gene and differ in the length of their 3′ non-coding regions (39). In chicks deficient in vitamin D, calbindin $D_{28k}$ mRNA is barely detectable in the intestine (44, 46) and increases approximately 10-fold after injection of $1,25(OH)_2D_3$. Increased calbindin $D_{28k}$ gene transcription has been demonstrated using nuclear run-on assay (46) and by demonstrating the presence of calbindin pre-mRNA shortly after injection of $1,25(OH)_2D_3$ (47).

In hens, sexual maturity and the onset of egg production are associated with increased intestinal absorption of calcium (23) and elevated concentrations of intestinal calbindin $D_{28k}$ and its mRNA (3, 48). These changes occur in two stages: before the onset of laying after becoming somatically mature, and at the time of formation of the first egg shell (48). They coincide with increased concentrations in plasma $1,25(OH)_2$-$D_3$ (11, 21). Similarly, concentrations of plasma $1,25(OH)_2D_3$, intestinal calbindin and its mRNA increase in response to oestrogen and testosterone and increase further at the onset of egg laying (49). In the absence of vitamin D in the diet, these changes do not occur in immature hens treated with sex steroids (49) or in egg-laying quail (48), demonstrating that the production of intestinal CaBP mRNA is dependent on $1,25(OH)_2D_3$. These observations are consistent with the view that $1,25(OH)_2D_3$ is an important factor regulating intestinal calbindin synthesis in hens. This conclusion is supported by the decrease in concentration of $1,25(OH)_2D_3$ and of duodenal calbindin following suppression of shell formation (49) and by the direct relationship between duodenal calbindin and dietary $1,25(OH)_2D_3$ (50).

The large increase in intestinal calbindin mRNA at onset of egg laying may be regulated primarily at the transcriptional level as shown in vitamin D-deficient chicks treated with $1,25(OH)_2D_3$ (44, 46, 51) but numerous observations suggest the involvement of post-transcriptional control mechanisms. Suppression or resumption of shell formation alters plasma $1,25(OH)_2D_3$, and the concentration of calbindin in the intestine, but has little effect on concentration of calbindin mRNA level. This suggests that shell formation may be associated with more effective translation of the calbindin mRNA or altered mRNA stability. Moreover, the concentration of intestinal calbindin mRNA does not decline rapidly after suppression of shell formation (49) or after chemical arrest of egg production (45). Consequently, the high concentration of intestinal calbindin mRNA may result from improved mRNA stability as demonstrated in the intestine of the vitamin D-deficient chick (44, 51) and in kidney cells (52) treated with $1,25(OH)_2D_3$. An unidentified protein may be necessary for prolonging the half-life of the calbindin mRNA since $1,25(OH)_2D_3$ affects mRNA stability in a cycloheximide-sensitive manner *in vivo* (51, 53) and *in vitro* (52). The absence of a decrease in calbindin mRNA associated with the decrease in plasma $1,25(OH)_2D_3$ induced by suppressing shell formation or by parathyroidectomy before resumption of shell formation (49), suggests that a factor(s) in addition to $1,25(OH)_2D_3$ may be involved in stabilizing the calbindin mRNA. Calcium may be implicated since it affects the regulation of intestinal calbindin $9_k$ in the rat (54) and post-transcriptional expression of calbindin $D_{28k}$ in kidney cells (52). This hypothesis is consistent with the maintenance of calbindin mRNA in the intestine where calcium is continuously transported, unlike the uterus, where daily or artificial arrest of shell deposition totally suppresses calcium transport and calbindin mRNA accumulation (48, 49, 55).

**Regulation of uterine calbindin $D_{28k}$**

The large transfer of calcium through the wall of the uterus is associated with high concentrations of uterine calbindin $D_{28k}$ (2, 3). The identity of uterine and intestinal calbindin $D_{28k}$ has been demonstrated by their identical molecular weights, mobility in electrophoresis, amino-acid composition and their immunological identity (6). Moreover, the same species of calbindin $D_{28k}$ mRNA is observed in both tissues (45, 55). The function of uterine calbindin $D_{28k}$ is not precisely established but no physiological situation is known where calcium is translocated across the uterine wall in the absence of calbindin $D_{28k}$ (2, 3). However, the regulation of uterine calbindin is largely independent of $1,25(OH)_2D_3$ suggesting alternative control mechanisms in this tissue.

The amount of uterine calbindin and its mRNA increases in immature pullets treated with oestrogen (55, 57), and is higher in hens laying shell-less eggs (55, 56). It is also higher in mature non-laying hens and quail than in immature birds (48, 50). At sexual maturity, the development of the oviduct and increased uterine calbindin (23) coincide with the increase in the concentrations of oestrogen observed two weeks before the onset of egg production (58). The production of $1,25(OH)_2D_3$ is also stimulated by oestrogen (16, 17, 55) but the induction of calbindin synthesis in the absence of vitamin D by oestrogen (50, 56, 57) or at sexual maturity (45, 48) demonstrates that vitamin D is not essential for the constitutive concentration of uterine calbindin in the uterus. Oestrogen affects the expression of several genes in the uterus (59) and an oestrogen-responsive element (ERE), recognized by the oestrogen receptor, has been reported in the ovalbumin (60) and vitellogenin (61) gene promoters. Oestrogen receptors (62) and calbindin (63) are localized in cells of the tubular glands but an ERE is apparently not present in the 800 bases of the 5′ end of the calbindin sequence reported by Wilson *et al* (41). However, the distal sequence TGACC of the ERE is found in the calbindin gene. This base sequence binds the oestrogen receptor *in vitro* but is not functional *in vivo* (61). There is little evidence, therefore, for a direct effect of oestrogen on the calbindin gene and the stimulation of calbindin synthesis at sexual maturity appears to be a function of the oviduct development. Testosterone facilitates these oestrogen effects but, acting alone, does not stimulate the production of calbindin (56, 57).

Concentrations of uterine calbindin and its mRNA increase most markedly during formation of the first shell (48, 45) or at resumption of shell formation in hens previously laying shell-less eggs (55). Concentrations of uterine calbindin $D_{28k}$ increase further during the period of the formation of the second shell (64) and remain high in laying hens or quail (23, 64). The concentration of uterine calbindin mRNA changes during the ovulatory cycle, being maximal during the period of shell deposition (45, 48, 55, 56). The absence of related daily changes in uterine calbindin concentration may result from the much lower degradation rate of the protein compared with that of its mRNA (45). Suppression of shell formation by premature egg expulsion (45, 55, 56) and resumption of shell formation (56) elicit major changes in uterine calbindin $D_{28k}$ mRNA but do not alter circulating concentrations of gonadal steroids (20). This observation shows that gonadal sex steroids are not responsible for stimulating calbindin mRNA during shell calcification. There is convincing evidence that the stimulator is not $1,25(OH)_2D_3$, in spite of the presence of the specific receptor for $1,25(OH)_2D_3$ (7). A rise in calbindin $D_{28k}$ mRNA at the onset of egg production occurs in the vitamin D-deficient quail (48) and, reciprocally, increases in exogenous or endogenous $1,25(OH)_2D_3$ levels do not enhance concentrations of uterine calbindin (2, 3) or of its mRNA (45, 48, 55, 56), in contrast to the intestine. This establishes the tissue specificity of calbindin regulation in the uterus. After resumption of shell formation, uterine calbindin is still enhanced in parathyroidectomized hens when the plasma concentrations of $1,25(OH)_2D_3$ decrease (21), suggesting that PTH too, is not involved in stimulating calbindin synthesis and calcium transport during shell formation. In the uterus, secretion of calcium is closely related to temporal changes in calbindin mRNA concentrations both during the laying cycle and following suppression or resumption of shell formation (45, 48, 55, 56). Consequently, calcium or a calcium dependent factor may cause the increase in calbindin mRNA concentration. Such an involvement of calcium in gene regulation has been observed in chick kidney cells for calbindin $D_{28k}$ (52) and in primary cultures of pituitary prolactin cells (65). The decrease in calbindin synthesis when uterine calcium transfer is reduced by feeding hens a low calcium diet (56) or acetazolamide (45) supports this hypothesis. The magnitude and rapidity of modifications in mRNA concentrations with changes in calcium secretion suggest involvement of transcriptional effect mediated possibly by regulatory proteins and cAMP as described in other systems (66). An *in vitro* system for studying calcium transport in the shell gland is required for further progress to be made in understanding regulation of calbindin expression.

The putative dependence of calbindin expression in the uterus on calcium secretion questions the nature of regulation of calcium transfer. Both parameters vary in parallel whatever the experimental situation and, as a consequence, gonadal steroids, parathyroid hormone and $1,25(OH)_2D_3$ are unlikely to be stimulators of uterine calcium secretion. The parathyroid hormone related-protein, which is a signal for calcium translocation by the ovine placenta (67) and is expressed in the hen shell gland (68) may be of importance. However, its localization in muscle layer of arterioles suggests rather a role as modulator of vascular smooth muscle activity (68). Alternatively, the relationship between the occurrence of ovulation and the timing of uterine calcium secretion (69) and the evidence that duration of shell formation is hormonally controlled (70) suggest the existence of an additional hormonal stimulator associated with ovulation or follicular maturation.

**Regulation of osteocalcin**

The structure and properties of osteocalcin have been reviewed in detail by Price (71) and Hauschka *et al* (8). Briefly, osteocalcin (bone $\gamma$-carboxy-glutamic acid protein) is the most abundant non-collagenous protein associated with the mineralized matrix of bone. This small $Ca^{2+}$ binding protein (Mr, 5,500) contains 49 amino acid residues and three Gla residues. This protein has been purified from bone and sequenced for several species, including chicken (8). Its precise function remains unknown but osteocalcin binds calcium and shows adsorption affinity for hydroxyapatite (8, 71) suggesting its involvement in the mineral dynamics of bone. Osteocalcin is predominantly synthesized by osteoblasts and partially released into the circulation. Because of its rapid turnover (71), plasma osteocalcin provides a convenient index of bone turnover (72), which reflects new osteoblast synthesis rather than release of bone matrix protein during bone resorption (8, 72).

$1,25(OH)_2D_3$ stimulates osteocalcin synthesis in osteoblast-like cells by binding to specific vitamin D-responsive promoter elements and by enhancing osteocalcin gene transcription (73) and, in several model systems, decreased or increased concentrations of plasma of $1,25(OH)_2D_3$ are paralleled by changes in concentrations of osteocalcin (8, 71). However, osteocalcin synthesis, in contrast to intestinal calbindin, is modulated by, rather than dependent upon, vitamin D since there is substantial osteocalcin synthesis in vitamin D-deficient chicks (74).

In hens, 30-40 % of the calcium incorporated into the egg shell is mobilized from medullary bone which develops during the reproductive period under the control of ovarian hormones (1). Matrix formation of medullary bone is induced by sex steroids regardless of the vitamin D status of the chicks (75), but its mineralization is observed only when vitamin $D_3$ is administered together with the gonadal steroids (75). Changes in plasma osteocalcin recently measured in hens (76) confirm the relationship between osteocalcin synthesis, $1,25(OH)_2D_3$ production, and bone turnover. Concentrations of plasma osteocalcin increase during the period of shell formation in hens fed a low calcium diet and decrease in hens laying shell-less eggs (76). Changes in plasma oesteocalcin therefore parallel changes in blood $1,25(OH)_2D_3$ and periods of active bone resorption which coincide with increased formation of medullary bone matrix (77). In contrast, in laying hens and in immature pullets treated with gonadal steroids, plasma osteocalcin decreases at a time when medullary bone formation is induced (1) and $1,25(OH)_2D_3$ production increases (11). This paradoxical observation is in agreement with an oestrogen-mediated reduction in bone formation and osteocalcin synthesis in rats (78) and in the frequency of remodelling activation in human (79). In contrast, oestrogens stimulate osteocalcin secretion in response to $1,25(OH)_2D_3$ in osteoblast-like cells by increasing vitamin D receptor expression (80). Oestrogen effects on bone remain therefore a matter of controversy.

**Acknowledgments**

Many thanks to Dr M J Peddie (Southampton) for her help with preparation of the manuscript.

## References

1 Taylor T G & Dacke C G (1984) Calcium metabolism and its regulation. *Physiology and Biochemistry of the Domestic Fowl* pp 125-170. Eds B E Freeman. London: Academic Press

2 Hurwitz S (1989) Calcium homeostasis in birds. *Vitamins and Hormones* **45** 173-221

3 Nys Y (1990) Régulation endocrinienne du métabolisme calcique chez la poule et calcification de la coquille. *Thèse de l'Université de Paris* **6** 1-109

4 Norman A, Nemere I, Zhou L, Bishop J, Lowe K, Maiyar A, Collins E, Taoka T, Sergeev I & Farach-Carson M (1992) $1,25(OH)_2$ vitamin $D_3$, a steroid hormone that produces biologic effects via both genomic and nongenomic pathways. *Journal of Steroid Biochemistry and Molecular Biology* **41** 231-240

5 Wasserman R H & Fullmer C S (1983) Calcium transport, proteins, calcium absorption and vitamin D. *Annual Review of Physiology* **45** 375-390

6 Fullmer C S, Brindak M E, Bar A & Wasserman R H (1976) The purification of calcium-binding protein from the uterus of laying hens. *Proceedings of the Society for Experimental Biology and Medicine* **152** 237-241

7 Coty W A (1980) A specific, high affinity binding protein for 1,25-dihydroxyvitamin D in the chick oviduct shell gland. *Biochemical and Biophysical Research Communications* **93** 285-292

8 Hauschka P V, Lian J B, Cole D E C & Gundberg C M (1989) Osteocalcin and matrix Gla protein: vitamin K dependent protein in bone. *Physiological Reviews* **69** 990-1034

9 Montecuccoli G, Bar A, Cohen A & Hurwitz S (1977) The role of 25-hydroxycholecalciferol-1-hydroxylase in the response of calcium absorption to reproductive activity in birds. *Comparative Biochemistry and Physiology* **57A** 335-334

10 Spanos E, Pike J W, Haussler M R, Colston I M A, Evans A M, Goldner N M, McCain T A & McIntyre I (1976) Circulating 1-α,25dihydroxyvitamin D in the chicken: enhancement by injection of prolactin and during egg laying. *Life Science* **19** 1751-1756

11 Castillo L, Tanaka Y, Wineland M J, Jowsey J O & De Luca H F (1979) Production of 1,25 dihydroxyvitamin $D_3$ and formation of medullary bone in the egg-laying hen. *Endocrinology* **104** 1598-1601

12 Bar A & Norman A W (1981) Studies on the mode of action of calciferol. XXXIV. Relationship of the distribution of 25-hydroxyvitamin $D_3$ metabolites to gonadal activity and egg shell formation in the quail. *Endocrinology* **109** 950-955

13 Henry H L & Norman A W (1984) Vitamin D: metabolism and biological actions. *Annual Review of Nutrition* **4** 493-520

14 Garabedian M, Holick M F, De Luca H F & Boyle I T (1972) Control of 25 hydroxycholecalciferol metabolism by parathyroid glands. *Proceedings of the National Academy of Sciences of the USA* **69** 1673-1676

15 Trechsel U, Bonjour J P & Fleisch H (1979) Regulation of the metabolism of 25-hydroxyvitamin D3 in primary cultures of chick kidney cells. *Journal of Clinical Investigation* **64** 206-217

16 Baksi S N, Kenny A D, Galli-Gallardo S M & Pang P K T (1978) Vitamin D metabolism in bullfrogs and Japanese quail: effects of estradiol and prolactin. *General and Comparative Endocrinology* **35** 258-262

17 Tanaka Y, Castillo L, Wineland M J & De Luca H F (1978) Synergetic effect of progesterone, testosterone and estradiol in the stimulation of chick renal 25-hydroxyvitamin D3 *in vitro* in the Japanese quail. *Biochemical Pharmacology* **26** 2439-2443

18 Spanos E, Barrett D, Chong K T & McIntyre I (1978) Effect of oestrogen and 1,25 dihydroxycholecalciferol metabolism in primary chick kidney cell culture. *Biochemical Journal* **724** 231-236

19 Bar A & Hurwitz S (1979) The interaction between dietary calcium and gonadal hormones in their effect on plasma calcium, bone, 25-hydroxycholecalciferol-1-hydroxylase, and duodenal calcium-binding protein, measured by a radioimmunoassay in chicks. *Endocrinology* **104** 1455-1460

20 Nys Y, N'Guyen T, Williams J & Etches R J (1986) Blood levels of ionized calcium, inorganic phosphorus, $1{,}25(OH)_2D_3$ and gonadal hormones in hens laying hard-shelled or shell-less eggs. *Journal of Endocrinology* **111** 151-157

21 Nys Y, Van Baelen H & Bouillon R (1992) Plasma 1,25 dihydroxycholecalciferol and its free index are potentiated by ovulation dependent factors and shell formation induced hypocalcemia in the laying hens. *Domestic Animal Endocrinology* **9** 37-47

22 Etches R J, Croze F & Ducke C E (1981) Plasma concentration of luteinizing hormone, progesterone, testosterone and oestradiol in follicular and venous plasma during the ovulation cycle of the hen. *Recent Advances in Avian Endocrinology.*

*Advances in Physiological Sciences* pp 89-98. Eds G Pethes, P Peczely & P Rudas. Budapest: Hungarian Academy of Sciences

23 Bar A, Cohen A, Edelstein S, Shemesh M, Montecuccoli G & Hurwitz S (1978) Involvement of cholecalciferol metabolism in birds in the adaptation of calcium absorption to the needs during reproduction. *Comparative Biochemistry and Physiology* **B59** 245-249

24 Spanos E, Colston K W, Evans I M A, Galante L S, McAulay S J & McIntyre I (1976) Effect of prolactin on vitamin D metabolism. *Molecular and Cellular Endocrinology* **5** 163-169

25 El Halawani M E, Burke W H, Millam J R, Fehrer S C & Hargis B M (1984) Regulation of prolactin and its role in gallinaceous bird reproduction. *Journal of Experimental Zoology* **232** 521-529

26 Nys Y, Guémené D & Bouillon R (1992) (unpublished results)

27 Broadus A E, Burtis W J, Ikeda K, Insogna K L, Lu C, Orloff J J, Stewart A F, Vasavada R, Weir C E & Mangin M (1990) Parathyroid hormone-related peptide. *Calcium Regulation and Bone Metabolism* pp 80-88. Eds D V Cohn, F H Glorieux & J J Martin. Amsterdam: Elsevier

28 Abe E, Ranabe R, Yoshiki S, Horikawa J H, Masumura T & Sugahara M (1979) Circadian rhythm of 1,25 dihydroxyvitamin D, production in egg laying hens. *Biochemical and Biophysical Research Communications* **88** 500-507

29 Van de Velde J P, Loveridge N & Vermeiden J P W (1984) Parathyroid hormone responses to calcium stress during eggshell calcification. *Endocrinology* **115** 1901-1904

30 Singh R, Joyner C J, Peddie M J & Taylor T G (1986) Changes in the concentration of parathyroid hormone and ionic calcium in the plasma of laying hens during the egg cycle in relation to dietary difficiencies of calcium and vitamin D. *General and Comparative Endocrinology* **61** 20-28

31 Forte L R, Langeluttig S G, Bieillier H V, Poelling R E, Magliola L & Thomas M L (1983) Upregulation of kidney adenylate cyclase in the egg-laying hen: role of estrogen. *American Journal of Physiology* **245** E273-E280

32 Reinhardt T A & Horst R L (1990) Parathyroid hormone down regulates 1,25 dihydroxyvitamin D receptors (VDR) and VDR messenger ribonucleic acid *in vitro* and blocks homologous up regulation of VDR *in vivo*. *Endocrinology* **127** 942-948

33 Costa E M & Feldman D (1986) Homologous up-regulation of the 1,25$(OH)_2$ vitamin $D_3$ receptor in rats. *Biochemical and Biophysical Research Communications* **137** 742-747

34 Goff I P, Reinhardt T A, Beckman M J & Horst R L (1990) Contrasting effects of exogenous 1,25 dihydroxyvitamin D(1,25-$(OH)_2$D) versus endogenous 1,25-$(OH)_2$D, induced by dietary calcium restriction, on vitamin D receptors. *Endocrinology* **126** 1031-1035

35 Meyer J, Fullmer C S, Wasserman R H, Komm B S & Haussler M R (1992) Dietary restriction of calcium, phosphorus, and vitamin D elicits differential regulation of the mRNAs for avian intestinal calbindin-$D_{28k}$ and the 1,25-dihydroxyvitamin $D_3$ receptor. *Journal of Bone and Mineral Research* **7** 441-447

36 Hunziker W, Walters M R, Bishop J E & Norman A W (1982) Effect of vitamin D status on the equilibrium between occupied and unoccupied 1,25 dihydroxyvitamin D intestinal receptors in the chick. *Journal of Clinical Investigation* **69** 826-834

37 Fullmer C S & Wasserman R H (1987) Chicken intestinal 28-kilodalton calbindin-D: complete amino acid sequence and structural considerations. *Proceedings of the National Academy of Sciences of the USA* **84** 4772-4776

38 Christakos S, Gabroemodes C & Rhote W B (1989) Vitamin D dependent calcium binding proteins: chemistry, distribution, functional considerations, and molecular biology. *Endocrine Reviews* **10** 3-26

39 Hunziker W (1986) The 28 kDa vitamin D-dependent calcium binding protein as a six domain structure. *Proceedings of the National Academy of Sciences of the USA* **83** 7578-7582

40 Wasserman R H & Chandler J S (1985) Molecular mechanisms of intestinal calcium absorption. *Bone Mineral Research* **3** 181-201

41 Wilson P W, Rogers J, Harding M, Pohl V, Pattyn G & Lawson D E M (1988) Structure of chick chromosomal genes for calbindin and calretinin. *Journal of Molecular Biology* **200** 615-625

42 Boland R, Minghetti P P, Lowe K E & Norman A W (1991) Sequences near the CCAAT region and putative 1,25 dihydroxyvitamin $D_3$ response element and further upstream novel regulatory sequences of calbindin $D_{28K}$ promoter show DNase I footprinting protection. *Molecular and Cellular Endocrinology* **75** 57-63

43 Darwish H M & Deluca H F (1992) Identification of a 1,25-dihydroxyvitamin $D_3$-response element in the 5′-flanking region of the rat calbindin $D\text{-}_{9k}$ gene. *Proceedings of the National Academy of Sciences of the USA* **89** 603-607

44 Mayel-Afshar S, Lane S M & Lawson D E M (1988) Relationship between the levels of calbindin synthesis and calbindin mRNA in chick intestine. Quantification of calbindin mRNA. *Journal of Biological Chemistry* **263** 4355-4361

45 Bar A, Striem S, Vax E, Talpaz H & Hurwitz S (1992) Regulation of calbindin mRNA and calbindin turnover in intestine and shell gland of the chicken. *American Journal of Physiology* **262** R800-R805

46 Theofan G, Nguyen A P & Norman A W (1986) Regulation of calbindin $D_{28K}$ gene expression by 1,25 dihydroxyvitamin $D_3$ is correlated to receptor occupancy. *Journal of Biological Chemistry* **261** 16943-16947

47 Ferrari S & Battini R (1990) Identification of chicken calbindin-$D_{28k}$ pre-messenger RNA sequences by polymerase chain reaction. *Biochemical and Biophysical Research Communications* **168** 430-436

48 Striem S & Bar A (1991) Modulation of quail intestinal and egg shell gland calbindin (Mr 28,000) gene expression by vitamin $D_3$, 1,25-dihydroxyvitamin $D_3$ and egg laying. *Molecular and Cellular Endocrinology* **75** 169-177

49 Nys Y, Baker K, Bouillon R, Van Baelen H & Lawson D E M (1992) Regulation of calbindin $D_{28K}$ and its mRNA in the intestine of the domestic hen. *General and Comparative Endocrinology* **86** 460-468

50 Bar A, Striem S, Mayel-Afshar S & Lawson D E M (1990) Differential regulation of calbindin $D_{28K}$ mRNA in the intestine and eggshell gland of the laying hen. *Journal of Molecular Endocrinology* **4** 93-99

51 Clemens T L, McGlade S A, Garrett K P, Horiochi N & Hendy G N (1988) Tissue-specific regulation of avian vitamin D dependent calcium binding protein 28KDa mRNA by 1,25 dihydroxyvitamin $D_3$. *Journal of Biological Chemistry* **263** 13112-13116

52 Enomoto H, Hendy G N, Andrews & Clemens T L (1992) Regulation of avian calbindin-$D_{28k}$ gene expression in primary chick kidney cells: importance of posttranscriptional mechanism and calcium ion concentration. *Endocrinology* **130** 3467-3474

53 Theofan G & Norman A W (1986) Effects of α-amanitin and cycloheximide on 1,25 dihydroxyvitamin $D_3$ dependent calbindin $D_{28k}$ and its mRNA in vitamin $D_3$ replete chick intestine. *Journal of Biological Chemistry* **261** 7311-7315

54 Brehier A & Thomasset M (1990) Stimulation of calbindin $D_{9K}$ (CaBP 9K) gene expression by calcium and 1,25 dihydroxycholecalciferol in fetal rat duodenal organ culture. *Endocrinology* **125** 580-587

55 Nys Y, Mayel-Afshar S, Bouillon R, Van Baelen H & Lawson D E M (1989) Increases in calbindin $D_{28k}$ mRNA in the uterus of the domestic fowl induced by sexual maturity and shell formation. *General and Comparative Endocrinology* **76** 322-329

56 Nys Y, Baker K & Lawson D E M (1992) Estrogen and a calcium flux dependent factor modulate the calbindin gene expression in the uterus of laying hens. *General and Comparative Endocrinology* **87** 87-94

57 Navickis R J, Katzenellenbogen B S & Nalbandov A V (1979) Effects of sex steroid and vitamin $D_3$ on calcium-binding protein in chick shell gland. *Biology of Reproduction* **21** 1153-1162

58 Senior B E (1974) Estradiol concentration in the peripheral plasma of the domestic hen from 7 weeks of age until the time of sexual maturity. *Journal of Reproduction and Fertility* **41** 107-112

59 Rories C & Spelsberg C (1989) Ovarian steroid action on gene expression: mechanism and models. *Annual Review of Physiology* **51** 653-681

60 Tora L, Gaub M P, Mader S, Dierich A, Bellard M & Chambon P (1988) Cell-specific activity of a GGTCA half-palindromic oestrogen-responsive element in the chicken ovalbumin gene promoter. *EMBO Journal* **7** 3771-3778

61 Martinez E, Givel F and Wahli W (1987) The estrogen-responsive element as an inducible enhancer: DNA sequence requirement and conversion to a glucocorticoid-responsive element. *EMBO Journal* **6** 3719-3727

62 Isola J J (1990) Distribution of estrogen and progesterone receptors and steroid-regulated gene products in the chick oviduct. *Molecular and Cellular Endocrinology* **69** 235-243

63 Jande S S, Tolnai S & Lawson D E M (1981) Immunohistochemical localization of vitamin D-dependent calcium-binding protein in duodenum, kidney, uterus and cerebellum of chickens. *Histochemistry* **71** 99-116

64 Nys Y, Parkes C O & Thomasset M (1986) Effect of suppression and resumption of shell formation and parathyroid hormone on uterine calcium-binding protein, carbonic anhydrase activity, and intestinal calcium absorption in hens. *General and Comparative Endocrinology* **64** 293-299

65 Day R N & Maurer R A (1990) Pituitary calcium channel modulation and regulation of prolactin gene expression. *Molecular Endocrinology* **4** 736-742

66 Habener J H (1990) Cyclic AMP response element binding proteins: a cornucopia of transcription factors. *Molecular Endocrinology* **4** 1087-1093

67 Abbas S K, Picard D W, Rodda C P, Heath J A, Hammonds R D, Wood W I, Caple W, Martin T J & Care A D (1989) Stimulation of ovine placenta calcium transport by

purified natural and recombinant parathyroid hormone-related protein (PTHrP) preparations. *Journal of Experimental Physiology* **74** 549-552

68 Thiede M A, Harm S C, McKee R L, Grasser W A, Duong L E T & Leach M Jr (1991) Expression of the parathyroid hormone-related protein gene in the avian oviduct: potential role as a local modulator of vascular smooth muscle tension and shell gland motility during the egg-laying cycle. *Endocrinology* **129** 1958-1966

69 Eastin W C Jr & Spaziani E (1978) On the control of calcium secretion in the avian shell gland (uterus). *Biology of Reproduction* **19** 493-504

70 Nys Y (1987) Progesterone and testosterone elicit increases in the duration of shell formation in domestic hens. *British Poultry Science* **28** 57-68

71 Price P A (1985) Vitamin K-dependent formation of bone gla protein (osteocalcin) and its function. *Vitamins and Hormones* **42** 65-108

72 Delmas P D (1992) Clinical use of biochemical markers of bone remodeling in osteoporosis. *Bone* **13** S17-S21

73 Lian J B & Stein G S (1992) Transcriptional control of vitamin D-regulated proteins. *Journal of Cellular Biochemistry* **49** 37-45

74 Lian J B, Glimcher M J, Roufosse A H, Hauschka P V, Gallop P M, Cohen-Solal L & Reit B (1982) Alterations of the gamma-carboxyglutamic acid and osteocalcin concentrations in vitamin D deficient chick bone. *Journal of Biological Chemistry* **257** 4999-5003

75 Takahashi N, Shinki T, Abe E, Morinchi N, Yamaguchi A, Yoshiki S & Suda T (1983) The role of vitamin D in the medullary bone formation in egg-laying Japanese quails and in immature male chicks treated with sex hormone. *Calcified Tissue International* **35** 465-471

76 Allewaert K, Nys Y & Bouillon R (1992) (unpublished results)

77 Van de Velde J P, Vermeiden J P W, Touw J J A & Veldhuijzen J P (1984) Changes in activity of chicken medullary bone cell populations in relation to the egg-laying cycle. *Metabolism and Bone Disease Related Research* **5** 191-193

78 Turner R T, Backup P, Sherman P J, Hill E, Evans G L & Spelsberg T C (1992) Mechanism of action of estrogen on intramembranous bone formation: regulation of osteoblast differentiation and activity. *Endocrinology* **131** 883-889

79 Hasling C, Eriksen E F, Melkko J, Risteli L, Charles P, Mosekilde L & Risteli J (1991) Effect of a combined estrogen-gestagen regimen on serum levels of the carboxy-terminal propeptide of human type I procollagen in osteoporosis. *Journal of Bone and Mineral Research* **6** 1295-1300

80 Liel Y, Kraus S, Levy J & Shany S (1992) Evidence that estrogens modulate activity and increase the number of 1,25-dihydroxyvitamin D receptors in osteoblast-like cells. *Endocrinology* **130** 2597-2601

Avian Endocrinology
Ed. P J Sharp, pp 359-366
Journal of Endocrinology Ltd, Bristol (1993)

# Ovarian prostaglandin and arginine vasotocin in relation to oviposition

N Saito[1], K Shimada[1] and T I Koike[2]
[1]Department of Animal Physiology, Nagoya University, Chikusa, Nagoya 464-01, Japan; [2]Department of Physiology and Biophysics, University of Arkansas for Medical Sciences, 4301 West Markham, Little Rock, AR 72205, USA

## Introduction

Prostaglandins (PG) are unsaturated fatty acids derived from dihomo-g-linolenic, arachidonic or eicosapentaenoic acids. Arginine vasotocin (AVT), a nanopeptide, is one of the avian neurohypophysial hormones. In the domestic hen, both hormones play a pivotal role in the regulation of oviposition because they are potent stimulators of uterine contraction. The production and actions of ovarian PG and AVT are reviewed in relation to the control of oviposition.

## Changes in concentrations of ovarian prostaglandins during the ovulatory cycle

*Midsequence ovipositions*

Oviposition in the middle of a sequence of eggs is followed, within about 30 minutes, by ovulation. Day and Nalbandov (1) were the first to measure concentrations of immunoreactive prostaglandin F (PGF) in pre-ovulatory and post-ovulatory (ruptured) follicles during a 26-h laying cycle and they demonstrated that there is a significant, 4-fold increase in the concentration of PGF in the largest pre-ovulatory (F1) follicle 4 to 6 h prior to the expected time of ovulation. In addition, the most recent post-ovulatory (R1) follicle shows a 100-fold increase in PGF about 24 h after ovulation. It was suggested that the increase in PGF in the R1 follicle may play a physiological role in increasing contractions of the uterine muscle and causing expulsion of the egg (1). The concentration of PGF in the F1 follicle increases 20-fold to 627.0 ± 106.4 ng/g shortly after oviposition/ovulation while the concentrations of PGF in the two most recent post-ovulatory follicles fluctuates between 9.2 ± 2.1 and 131 ± 12.7 ng/g during the ovulatory cycle (2). In addition, concentrations of PGF in the two largest pre-ovulatory follicles and the two most recently ruptured follicles increase during the 12 h period before a midsequence oviposition (3). The concentration of PGF in the F1 follicle is similar to that in the R1 follicle. Thus, an increase in concentrations of PGF in the ovary occurs not only in the R1 follicle but also in the F1 follicle.

The concentrations of PGF in isolated theca and granulosa layers, the principle tissues of the ovarian wall, have been measured during the oviposition cycle (4, 5). In the theca layer, the concentrations of PGF in the F1 follicle and the R1 follicle increase at oviposition. In contrast, there are no changes in the concentrations of PGF in the granulosa layers of these follicles. The concentrations of PGF in the theca layers from the fourth largest pre-ovulatory follicle to the R1 follicle are significantly higher than the concentrations in the associated granulosa layers in the same follicles.

Unlike PGF, the concentrations of prostaglandin E (PGE) do not increase in the ovarian follicles at oviposition. Furthermore, concentrations of PGE change in a different manner to those of PGF during the oviposition cycle. This suggests that the concentrations of the two prostaglandins are regulated separately, possibly by different mechanisms.

*Terminal oviposition*

The concentration of PGF in the R1 follicle increases at the time of a terminal oviposition (3). During the 12-h period before the terminal oviposition, the concentrations of PGF in the two largest pre-ovulatory and two most recent post-ovulatory follicles remain unchanged. Similarly, the concentrations of PGF in the theca layer of the R1 follicle are higher than in the F1 pre-ovulatory follicle (5). Because a terminal oviposition is not followed by ovulation, unlike a midsequence oviposition, the pre-ovulatory surge of gonadotrophins and steroid hormones that precede ovulation does not occur. The F1 follicle present at a terminal oviposition may not be mature enough to produce PGFs or may not produce them because of the absence of pre-ovulatory surges of hormones.

*First ovulation*

Uterine contractions increase 0.5 to 1.0 h before a first ovulation; i.e. in the absence of oviposition (6). During this increase in uterine contractions, concentrations of PGF in the F1 follicle are significantly higher than those of the R1 follicle and the second largest pre-ovulatory follicle (7, 8). Since the R1 follicle is already about 2 days old at the time of the first ovulation of the sequence, the follicle has regressed more and may not be as functionally active as an R1 follicle at a midsequence oviposition. This may explain why the concentrations of PG in the R1 follicle present during the first ovulation of a sequence are lower than those in the F1 follicle (7, 8).

## Concentrations of ovarian prostaglandins and follicular growth

The chicken ovary contains various sized follicles arranged in a hierarchy with respect to maturation and concentrations of PGF in these follicles increase during follicular maturation (3). The gradual increase in concentrations of PGF during follicular maturation occurs in the theca layer (4). In the granulosa layer, the concentration of PGF does not change in relation to follicular maturation. Thecal PGF may therefore play a role in the maturation of the follicle.

## Control of ovarian prostaglandin synthesis and secretion

Concentrations of PGF in isolated theca layers are higher than those in the granulosa layers (4). However, the concentrations of PGF in the media of primary cultures of granulosa cells from the F1 follicle are significantly greater than in media containing 4 times as many theca cells (9), suggesting that the granulosa cells of the F1 follicles are the major intraovarian source of prostaglandins. This view is supported by an immunocytochemical study showing that PG synthetase is localized in the granulosa cells of small pre-ovulatory follicles of less than 5 mm diameter (10). It therefore seems that PG may be synthesized mainly in the granulosa layer and then stored in the theca layer. Additional studies are required to establish the site (i.e. theca or granulosa layer) at which PG production increases at the time of oviposition.

Several studies have been carried out to identify hormones which might stimulate chicken granulosa cells to produce prostaglandins (9, 11). Exposure of primary cultures of granulosa cells to 0.1 and 5 mg ovine luteinizing hormone/ml for up to 7 h does not stimulate the production of PGF, PGE or 6-keto-PGF (9, 11), although the Ca ionophore, A23187, causes a small increase in PGF secretion (9). When granulosa cells are collected 6 h before oviposition, treatment with LH followed by A23187 causes a 15- to 20-fold increase in PGF secretion (9). In contrast, granulosa cells collected within 5 min after oviposition are not stimulated by this treatment (9). These data suggest that PGF production in granulosa cells of the F1 follicle may be regulated by the pre-ovulatory increase in the concentrations of plasma of LH and steroid hormones.

The increase in PG production in the R1 follicle during a terminal oviposition is not directly associated with pre-ovulatory LH surge and accompanying ovulation. Thus, the regulation of PG production in the ruptured follicle may differ from that in the pre-ovulatory follicles.

**Actions of ovarian prostaglandins**

Prostaglandin administration induces oviposition and stimulates uterine contractility *in vitro* (12-14) and *in vivo* (15-17) while concentrations of plasma PG increase at the time of midsequence ovipositions (3, 4, 18). Furthermore, the concentration of PG in ovarian venous plasma at the time of oviposition is 5 to 20 times higher than that in peripheral plasma (3, 4). These observations indicate that the release of ovarian PG is involved in the stimulation of uterine contractility during a midsequence oviposition.

The concentrations of PG in peripheral and follicular venous plasma also increase markedly during a first ovulation of a sequence (7, 8) and the concentration of 13,14 dihydro-15-keto-$PGF_{2\alpha}$ (PGFM) in the peripheral circulation increases during a terminal oviposition of a sequence (19). Unlike midsequence ovipositions, however, the concentrations of PGF in the venous plasma draining large pre-ovulatory follicles do not increase before a terminal oviposition (3, 4). Although the concentrations of PG in venous plasma from post-ovulatory follicles have not been measured, it is possible that PG from these follicles plays a role in initiating terminal ovipositions.

Neither PGF nor PGE stimulate the secretion of progesterone and oestradiol from granulosa cells *in vitro* and *in vivo* (11) and large doses of indomethacin, an inhibitor of PG synthesis, do not inhibit spontaneous ovulation (1). However, evidence for a role for PG in ovulation has been obtained from studies on the perfused chicken ovary *in vitro* in which the perfusate is recycled (20). In this system, doses of 0.1 and 1 mg

PGF/litre induced 25% and 30% respectively of perfused ovaries to ovulate. The same doses of PGE induced ovulation in 60% and 63% of perfused ovaries respectively, while indomethacin partially blocked gonadotrophin-induced ovulation (20). Prostaglandin stimulates both cell-associated and secreted plasminogen activator, a substance that converts inactive plasminogen to plasmin (active form) (21). Since plasminogen activator and plasmin have been implicated in the process of follicular rupture and ovulation, prostaglandin may play a supportive role in causing rupture of the follicle.

**Characterization of ovarian arginine vasotocin**

The avian neurohypophysial peptides are arginine vasotocin (AVT) and mesotocin (MT). An analysis of extracts of chicken ovary and ruptured follicles by high performance liquid chromatography and radioimmunoassays of eluates demonstrated that they contain arginine vasotocin, mesotocin and other neuropeptides (22). Extracts of theca and granulosa layers from chicken pre-ovulatory follicles dilute out parallel with the hormone standards in specific radioimmunoassays for arginine vasotocin and mesotocin (23).

An ovarian oviposition-inducing factor (OOIF) is present in pre-ovulatory and ruptured follicles (24, 25). The oxytocic activity of the OOIF is not inactivated by tryptic digestion but is completely destroyed by thioglycollate treatment. It is possible that OOIF is a hormone or hormone-like substance that is chemically similar to neurohypophysial peptides (26).

**Changes in concentrations of ovarian arginine vasotocin during the ovulatory cycle**

The theca layers of the three largest pre-ovulatory follicles and of the three ruptured follicles contain similar amounts of AVT (0.8-2.5 ng) (23). The arginine vasotocin content of theca layer changes significantly during the oviposition cycle and is lowest 2 h before expected oviposition while the AVT content in the granulosa layer (about 1 ng/tissue) does not change during the oviposition cycle (23). The amounts of MT in the follicular layers are an order of magnitude lower than the amounts of AVT (23). The changes in the amounts of AVT in the theca layer during the oviposition cycle suggest that this peptide may play a role in the control of oviposition or ovulation.

**Biosynthesis of ovarian arginine vasotocin**

There is no direct evidence that AVT is synthesized in the ovarian follicle in chickens. The highest amounts of AVT in the theca and granulosa layers are $2.5 \pm 0.6$ and $1.3 \pm 0.2$ ng, respectively (23). If it is assumed that the peptides are distributed in the extracellular compartment in these tissues (20 % of the weight) then the concentration of AVT in theca and granulosa layers would be about 46 and 96 ng/ml respectively. These concentrations are over a thousand times greater than basal, circulating concentrations of AVT in laying hens (27 to 31 pg/ml). This suggests that the AVT is synthesized in the ovarian follicle rather than being sequestered from the peripheral circulation. This possibility should be considered because in the cow, oxytocin mRNA has been identified in the corpus luteum using a bovine hypothalamic oxytocin cDNA as a probe (27). It was also observed that luteal and hypothalamic oxytocin mRNAs, when translated in cell-free, rabbit reticulocyte lysate system, give rise to a similar protein that is immunoprecipitable with anti-oxytocin neurophysin (27). The nucleotide sequence of luteal and hypothalamic oxytocin cDNAs are identical. Similar studies, using avian hypothalamic arginine vasotocin cDNA as a probe, have not been done in birds.

**Actions of ovarian arginine vasotocin**

The concentrations of AVT in the peripheral plasma increase at the time of oviposition, supporting the view that that this peptide participates in the regulation of oviposition (28-30). However, the concentration of AVT in the ovarian follicular venous plasma dose not increase at oviposition and is not higher than that in peripheral plasma (31). This observation suggests that ovarian AVT does not stimulate directly uterine contractility at the time of oviposition. In mammals, there are some reports that ovarian oxytocin may play an role in controlling ovarian steroidogenesis (32) and uterine PG production (33, 34). The possibility that AVT in the chicken ovary may have similar functions remains to be determined.

**References**


1 Day S L & Nalbandov A V (1977) Presence of prostaglandin F (PGF) in hen follicles and its physiological role in ovulation and oviposition. *Biology of Reproduction* **16** 486-494

2 Hammond R W, Olson D M, Frankel R B, Biellier H V & Hertelendy F (1980) Prostaglandins and steroid hormones in plasma and ovarian follicles during the ovulation cycle of the domestic hen (*Gallus domesticus*). *General and Comparative Endocrinology* **42** 195-202

3 Olson D M, Shimada K & Etches R J (1986) Prostaglandin concentrations in peripheral plasma, and ovarian and uterine plasma and tissue in relation to oviposition in hens. *Biology of Reproduction* **35** 1140-1146

4 Saito N, Sato K & Shimada K (1987) Prostaglandin levels in peripheral and follicular plasma, the isolated theca and granulosa layers of pre- and postovulatory follicles, and the myometrium and mucosa of the shell gland (uterus) during a midsequence-oviposition cycle of the hen (*Gallus domesticus*). *Biology of Reproduction* **36** 89-96

5 Saito N & Shimada K (1988) Ovarian prostaglandin levels at the midsequence oviposition and the terminal oviposition and first ovulation of a sequence of the hen (*Gallus domesticus*). *Japanese Poultry Science* **25** 296-304

6 Shimada K & Asai I (1978) Uterine contraction during the ovulatory cycle of the hen. *Biology of Reproduction* **19** 1057-1062

7 Shimada K, Olson D M & Etches R J (1984) Follicular and uterine prostaglandin levels in relation to uterine contraction and the first ovulation of a sequence in the hen. *Biology of Reproduction* **31** 76-82

8 Shimada K, Olson D M & Etches R J (1986) The effect of indomethacin on ovarian prostaglandin release in hens. *Biology of Reproduction* **35** 1147-1153

9 Etches R J, Kelly J D, Anderson-Langmuir C E & Olson D M (1990) Prostaglandin production by the largest preovulatory follicles in the domestic hen (*Gallus domesticus*). *Biology of Reproduction* **43** 378-384
10 Gilbert A B & Davidson M F (1984) Location of prostaglandin synthetase in the ovary of the domestic hen (*Gallus domesticus*). *IRCS Medical Science* **12** 1107-1108
11 Hertelendy F & Hammond R W (1980) Prostaglandins do not affect steroidogenesis and are not being produced in response to oLH in chicken granulosa cells. *Biology of Reproduction* **23** 918-923
12 Verma O P, Prasad B K & Slaughter J (1976) Avian oviduct motility induced by prostaglandin $E_1$. *Prostaglandins* **12** 217-227
13 Wechsung E & Houvenaghel A (1976) A possible role of prostaglandins in the regulation of ovum transport and oviposition in the domestic hen. *Prostaglandins* **12** 599-608
14 Olson D M, Biellier H V & Hertelendy F (1978) Shell gland responsiveness to prostaglandin $F_{2\alpha}$ in relation to oviposition in the domestic hen (*Gallus domesticus*). *Biology of Reproduction* **24** 496-504
15 Hertelendy F (1972) Prostaglandin-induced premature oviposition in the coturnix quail. *Prostaglandins* **2** 269-279
16 Hertelendy F, Yeh M & Biellier H V (1974) Induction of oviposition in the domestic hen by prostaglandins. *General and Comparative Endocrinology* **22** 529-531
17 Hertelendy F, Biellier H V & Todd H (1975) Effects of the egg cycle and route of administration on prostaglandin-induced oviposition of hens and Japanese quail. *Journal of Reproduction and Fertility* **44** 579-582
18 Wechsung E, Korteweg M, Verdonk G & Houvenaghel A (1978) Plasma levels of prostaglandin F related to oviposition in the domestic hen. *Archives Internationales de Pharmacodynamie et de Therapeutic* **236** 331-333
19 Olson D M & Hertelendy F (1981) Plasma levels of 13, 14-dihydro-15-keto prostaglandin $F_{2\alpha}$ in relation to oviposition and ovulation in the domestic hen (*Gallus domesticus*). *Biology of Reproduction* **24** 496-504
20 Tanaka K, Higuchi T, Tawarazumita M & Hertelendy F (1992) The effects of prostaglandins and indomethacin on ovulation in the perfused fowl ovary. *Asia-Australias Journal of Animal Science* **5** 29-32
21 Tilly J L & Johnson A L (1987) Presence and hormonal control of plasminogen activator in granulosa cells of the domestic hen. *Biology of Reproduction* **37** 1156-1164
22 Watkin W B & Choy V J (1988) Identification of neurohypophysial peptides in the ovaries of several mammalian and nonmammalian species. *Peptides* **9** 927-932
23 Saito N, Kinzler S & Koike T I (1990) Arginine vasotocin and mesotocin levels in theca and granulosa layers of the ovary during the oviposition cycle in hens (*Gallus domesticus*). *General and Comparative Endocrinology* **79** 51-63
24 Tanaka K & Nakada T (1974) Participation of the ovarian follicle in the control of the time of oviposition in the domestic fowl. *Poultry Science* **53** 2120-2125
25 Tanaka K (1976) Oviposition-inducing activity in the ovarian follicles of different sizes in the laying hen. *Poultry Science* **55** 714-716
26 Tanaka K & Goto K (1976) Partial purification of the oviposition-inducing factor and estimation of its chemical nature. *Poultry Science* **55** 1774-1778

27 Ivell R & Richter D (1984) The gene for the hypothalamic peptide hormone oxytocin is highly expressed in the bovine corpus luteum: biosynthesis, structure and sequence analysis. *The EMBO Journal* **3** 2351-2354

28 Arad Z & Skadhauge E (1984) Plasma hormone (arginine vasotocin, prolactin, aldosterone and corticosterone) in relation to hydration state, NaCl intake, and egg laying. *Journal of Experimental Zoology* **232** 707-714

29 Nouwen E J, Decuypere E, Kühn E R, Michels H, Hall T & Chadwick A (1984) Effect of dehydration, haemorrhage and oviposition on serum concentrations of vasotocin, mesotocin and prolactin in the chicken. *Journal of Endocrinology* **102** 345-351

30 Tanaka K, Goto K, Yoshioka T, Terao T & Koga O (1984) Changes in the plasma concentration of immunoreactive arginine vasotocin during oviposition in the domestic fowl. *British Poultry Science* **25** 589-595

31 Shimada K, Saito N, Itogawa K & Koike T I (1987) Changes in plasma concentrations of arginine vasotocin and prostaglandin F after intrauterine injections of prostaglandin $F_{2\alpha}$ and acetyl choline at various times during oviposition cycle of the domestic fowl (*Gallus domesticus*). *Journal of Reproduction and Fertility* **80** 143-150

32 Tan G J S, Tweedale R & Biggs J S G (1982) Effects of oxytocin on the bovine corpus luteum of early pregnancy. *Journal of Reproduction and Fertility* **66** 75-78

33 McCracken J A, Schamm W & Okulicz W C (1984) Hormone receptor control of pulsatile secretion of $PGF_{2\alpha}$ from the ovine uterus during luteolysis and its abrogation in early pregnancy. *Animal Reproduction Science* **7** 31-55

34 Flint A P F & Sheldrick E L (1986) Ovarian oxytocin and the maternal recognition of pregnancy. *Journal of Reproduction and Fertility* **76** 831-839

# Part Seven
# Techniques for Transgenesis

Avian Endocrinology
Ed. P J Sharp, pp 367-380
Journal of Endocrinology Ltd, Bristol (1993)

# The production of transgenic birds from primordial germ cell manipulated embryos

K Simkiss, Y Li, G Luke, N Page and L Vick
Department of Pure and Applied Zoology, School of Animal and Microbial Sciences, University of Reading, Reading RG6 2AJ, UK

## Introduction

In 1980 Gordon *et al.* found that foreign DNA could be injected into the fertilized mouse egg and subsequently recovered from all the cells of the newborn animal (1). This discovery established a new experimental approach to both genetics and embryology, encapsulated in the concept of producing transgenic offspring. The background to this work lay in the studies of viral and DNA-mediated gene transfer into cells, and it was soon built upon with the use of aggregation chimeras in embryology (2). These approaches have continued to make fundamental contributions to the subject and three cellular systems have been commonly employed utilizing two types of vector. The cellular methods are sperm coating, microinjection into pronuclei and stem cell chimeras. The vectors are plasmid DNA and retroviruses. The use of primordial germ cells in avian transgenesis is a particular aspect of these systems. It is, therefore, important to realize the advantages and constraints of these different approaches and in particular how they relate to the rather unusual features of avian reproduction.

The simplest way to introduce foreign DNA into an egg would be to coat the surface of the fertilizing sperm. There was, therefore, enormous interest in the claim of Lavitrano *et al.* that they had succeeded in producing transgenic mice by this method (3). Attempts by other workers to repeat the experiment have so far failed (4) and even when liposomes were used to introduce the DNA into the sperm cell it was not possible to produce transgenic mice (5). The use of this approach in birds has so far failed to transfer foreign DNA into the egg in a way that produces transformed progeny (6). At the present time it is, therefore, an appealing but unproven approach.

The second and most popular way of inserting DNA into the genome of an animal is by direct injection, usually into the male pronucleus of a freshly fertilized ovum (1, 7). This method is highly successful in mice but more difficult in larger species where the pronuclei are difficult to see unless the eggs are first stratified by centrifugation (7). Attempts to use this approach in birds have so far failed to produce nuclear integration (8, 9).

The final approach to transgenesis is based on the use of chimeric embryos and has received a major boost from recent advances in the understanding of stem cells. The most significant of these arose from the demonstration that cells from the inner cell mass of the mammalian embryo could be transferred to other embryos, where they participated in the formation of chimeras and occasionally colonized the germ line. Even more importantly, these embryonic stem cells could be maintained for prolonged periods in tissue culture. This enabled them to be used for gene insertion and gene selection before introducing them into chimeras where they could express this foreign DNA (10) and produce germ cells (11). Similar techniques are being developed in avian species where blastoderm cells have been used to produce somatic and germ line chimeras (12). An alternative approach, which is the subject of this review, is to use a cell line from the embryo that is committed, developmentally, to producing the gametes. These cells are the primordial germ cells of the embryo (13, 14).

Two types of vector are commonly used in these genetic manipulation studies. Plasmid DNA, which is usually linearized before use, can be introduced directly into avian cells either by using electroporation (15) or by using cationic liposomes such as Lipofectin (16). Alternatively, retroviruses can be used since these organisms have efficient methods for gaining entry into cells and integrating their nucleic acids into the host's chromosomes as part of their normal life cycle (17). Suitable subgroups of replication-competent retroviruses such as reticuloendotheliovirus (REV) and Rous associated virus (RAV) readily infect avian cells (18). These retroviruses have three protein-coding genes, i.e. gag (viral encapsidation), pol (reverse transcription) and env (envelope proteins) that can be deleted so as to make the viruses replication-defective. By using helper cell lines it is possible to construct such replication-defective retroviruses that will insert genes into avian cells without the possibility of them subsequently infecting other cells (19). Each of the two types of vector have their own advantages and disadvantages. Plasmids can be used for gene targeting but they have low rates of transfection and integration. Retroviruses are efficient at integrating and expressing DNA but have size limitations and can be unstable (49). The particular value of each of these vectors depends upon how gene insertion is attempted in the bird.

## Four unusual aspects of avian reproduction

### *Fertilization*

About 2 h before ovulation the avian oocyte undergoes its first reduction division to form a secondary oocyte and the first polar body. Fertilization occurs almost immediately after ovulation in the infundibulum region of the oviduct and it has been suggested that this is essential before the second maturation division can occur to form the female pronucleus (20). Birds differ, however, from mammals in that polyspermy is common and in the fowl an average of around 10 sperm will enter the oocyte, lose their cytoplasm and swell to about 50 times their volume to form the male pronuclei (21). At this stage these additional supernumerary nuclei are indistinguishable from the one that is destined to form the zygote. Thus, the large opaque oocyte and the presence of supernumerary pronuclei make it extremely difficult to introduce foreign DNA into the bird by microinjection of the egg (Fig. 1a).

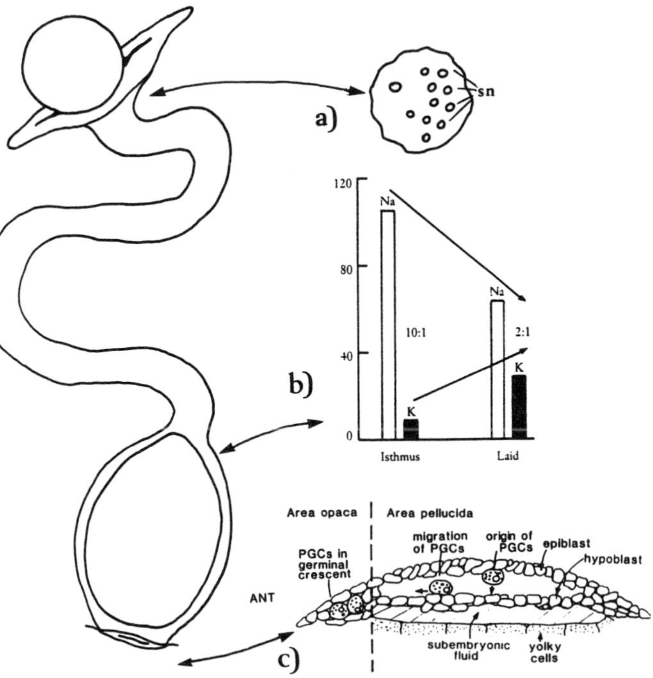


Fig. 1 Avian reproduction involves 4 unusual aspects of development: a) the fertilized ovum contains many supernumerary sperm (Sn) that form male pronuclei prior to fertilization, making microinjection inefficient; b) the albumen changes in composition (mmol/l) during its formation making it difficult to culture early stages; c) the blastoderm contains large numbers of cells by the time of oviposition. Some of these already appear to be differentiated as far as primordial germ cells (PGCs), making the search for suitable stem cells require earlier stages of development. Finally the migration of PGCs to the germinal crescent takes them outside the embryonic tissues.

*Oviducal changes*

In the domestic fowl the fertilized egg passes down the magnum region of the oviduct, where the albumen is secreted, in about 2.5 h and reaches the isthmus or membrane-secreting region by the time of the first cell division. The egg then enters the shell gland where it is 'plumped' and increases in size while the eggshell is being secreted (22). This process takes roughly 18 h during which time cell division continues so that when the egg is laid it contains about 50 000 cells, half of which are probably extra embryonic (23).

Many of the features of cleavage and blastoderm formation that are of interest to molecular biologists occur in birds before the egg is laid. For this reason a number of attempts have been made to culture fertilized eggs that are removed prematurely from the oviduct (24, 25). The results of this approach have been frustrating and complex, probably for the reason shown in Figure 1b. The albumen that is initially secreted around the zygote has a high Na to low K ratio of about 10:1. During the shell-gland plumping phase this is modified to a ratio of about 2:1 (26). Thus, the initial albumen that is secreted by the magnum appears to be unable to sustain normal embryology, so that it needs to be replaced or dialysed against a high K saline before it achieves a composition in which development is possible (27). This unusual feature probably explains why existing culture methods are so complex (24, 25).

*Blastodermal stem cells*

For the reasons explained above, microinjection and the manipulation of early cell lines are difficult procedures to use on the avian system of development. Emphasis has, therefore, shifted to the possibility of producing transgenics from stem cells introduced into chimeric embryos. At the time of laying, the cells of the avian blastoderm are described as being pluripotent (28). It is possible to dissociate this tissue and inject up to 500 cells from such a donor into the blastoderm of a recipient embryo (12). In experiments of this type roughly 11% of such treatments produced chimeric birds of which 4 individuals hatched. One male hatchling was used for breeding and produced 719 chicks of which 2 showed germ line transmission of the donor cell phenotype. This demonstrates that blastoderm cells can be used to form germ line chimeras. At the present time this method looks as if it could be used to produce transgenics if foreign genes were inserted into the donor blastoderm cells. The attraction of this approach is that such stem cells might be capable of being maintained in tissue culture in an undifferentiated state for extended periods of time. This would enable screening to be undertaken to ensure that the foreign genes had integrated into the cells before they were injected into recipient embryos. The disadvantage of the current procedure is that considerably more than 500 cells may have to be injected if the number of germ line chimeras is to exceed more than a few percent.

*Primordial germ cells (PGCs)*

Primordial germ cells are the progenitor cells that will give rise to the gametes, so it is clearly into these cells that any foreign DNA needs to be inserted if it is to enter the avian germ line. In all vertebrates PGCs arise at extragonadal sites but birds are particularly interesting in that these cells can easily be detected in the germinal crescent which is an extraembryonic site to the anterior of the early embryo (29). This makes them easy to collect and thus easy to manipulate. It is interesting to note that the formation of PGCs is apparently already determined in the preincubation (stage X) blastoderm (30, 31). The PGCs arise in the epiblast layer at the centre of the blastoderm of the developing embryo (32) and migrate to the hypoblast at stage XII. They continue to migrate anteriorly and laterally during gastrulation until they arrive at the germinal crescent (33) where they accumulate in large numbers after about 2 days of incubation (Fig. 1c).

## Manipulating the primordial germ cell (PGC) system

*Primordial germ cell migration*

Primordial germ cells are identifiable in the blastoderm of the new-laid egg. During incubation they are carried anteriorly by morphogenetic movements until they come to rest in the germinal crescent region (34). Estimates vary as to the number of cells that are involved at this stage but values of 150-300 are typical for the fowl at stage 11 of development (35). As blood islands develop in the germinal crescent region the PGCs settle in these spaces and, with the establishment of the vascular system, they are carried into the circulation where they reach a peak number of about 2 $\mu l^{-1}$ at stage 15-16 (36). The PGCs leave the blood and pass through the wall of the blood vessel as they circulate past the local thickening of the coelomic wall that is known as the germinal ridge. Here they gather in increasing numbers from stage 16 onwards (37), apparently as a result of a chemoattractant. The coelomic wall, with possible contributions from the mesonephros (38), forms the stroma of the gonad in which the primordial germ cells divide and eventually differentiate into the gametes. After their settlement in the germinal ridge the primordial germ cells proliferate rapidly and differentiate into germ cells. In the female chick the number exceeds half a million by day 17 of incubation but the population then crashes so that the 4d old hatchling contains only a few thousand that will persist into adult life (39).

*Primordial germ cell signalling*

In a strict sense a stem cell is one that divides asymmetrically; one daughter cell remaining as a self-renewing cell line while the other differentiates into a specialized cell type (64). PGCs are not typical stem cells in that they seem to form by the progressive differentiation of a population of cells. These differences are illustrated in Figure 2. The relevant features of PGCs are:

a) they are formed as a 'one off' event during development and if they are destroyed they cannot be replaced (29);
b) they are produced rapidly between stages X & XIV of development (54) when the numbers rise from about 27 to 162 (33) over a period of about 6 h.

This would suggest that they form by a differentiation process rather than by a series of sequential stem cell divisions. It is difficult to relate these features to 'embryonic stem cells' which are defined as pluripotent cell lines isolated from the delayed blastocysts of mammals (40, 41). Clearly such cells may exist in the avian blastocyst but their properties and relationship to PGC production are not clear. It has, for example, recently been shown in *Drosophila* that the number of primordial germ cells that are formed in the embryo is dictated by a specific gene product (*oskar*) and that the production of *oskar* RNA at other sites induces ectopic germ-cell formation (62). This provides a basis for the general concept that the germ line is identified and segregated very early in development. In amphibians, where similar evidence exists for the localization of the germ cell line, there is information on two additional properties. By marking and transferring these cells it has been shown 1), that primordial germ cells are not irreversibly determined (i.e. PGCs remain pluripotent and can contribute to a variety of somatic tissues) but 2), migratory PGCs are clearly showing some transient gene expression because they will not repeat the migration to the germinal ridge if they are transplanted at a later stage of development (63).

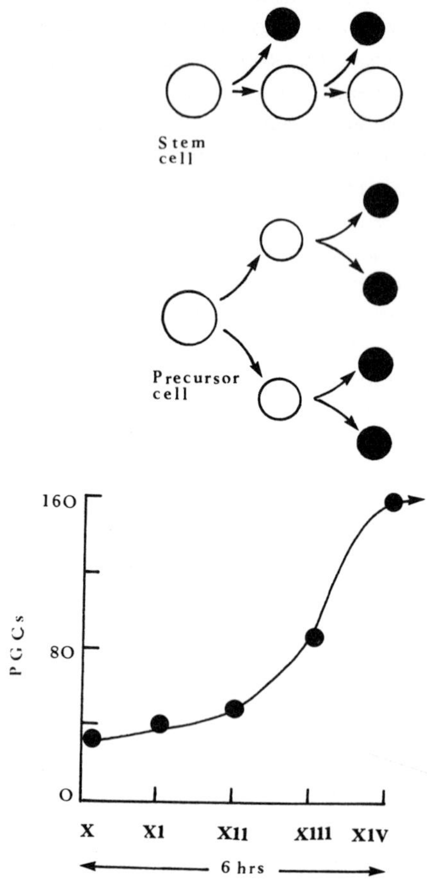


Fig. 2 Normal stem cells (top) show asymmetrical division producing a self-renewing cell and a specialized type (solid circle). Differentiation (centre) can be regarded as the development of specialized cell types in daughter populations. Primordial germ cells occur rapidly in large numbers in the blastoderm within 6 h of the egg being laid and incubated (33) suggesting that they form by differentiation from precursor cells that are already committed (bottom).

The approach to the genetic manipulation of birds by forming chimeras of various sorts is based upon the mammalian model. It should be realized however that the new-laid egg of 50,000 cells is considerably greater in cell number than an equivalent mammal at the inner-cell mass stage and that at this stage the avian PGCs appear to be determined. The crucial question, of course, is the relation of any stem cells to the germ line and whether germ-line cells can be maintained in culture indefinitely without destroying their ability subsequently to populate the germinal ridge and form gametes. There is increasing evidence that by the time the egg is laid the blastoderm already contains primordial germ cells or their immediate precursors (31, 33, 35). There is clearly a major experimental problem here and as Watanabe *et al.* (42) write, it is not possible *"to answer the question of whether somatic chimerism and germ line chimerism ... are due to the random incorporation and differentiation of pluripotent cells or due to the selective incorporation of committed cells"*. The question is not just of academic interest for it dictates this whole approach to producing avian transgenics by using chimeras. If the use of blastoderm chimeras and primordial germ cell chimeras are effectively the same experiments, the question becomes which are the easiest experiments to perform? If they are not the same experiments, how far back in development does the experimenter have to go to find a true embryonic stem cell and can it be maintained in either a clonal culture or a mixed cell culture?

*The pessimistic view*

Several attempts have recently been made to determine the effectiveness of transferring primordial germ cells from one avian embryo to another. Successful transfers were achieved when the cell lineages were traced with phenotypic markers (43) or DNA probes (44) but in analysing the offspring of 59 germ line chimeras Petitte *et al.* (45) found no offspring of the transferred PGCs from among 3117 chicks produced. This negative result was attributed either to technical problems or to germ cell-stroma interactions. It is known, for instance, that male PGCs may degenerate if introduced into the embryonic ovary and if this is true, at least 25% of the transfers may have been moribund in these experiments since the sex of donor and recipient embryos was not determined prior to the transfer. It has also been suggested from experiments in mammals that endogenous and transferred PGCs may compete with each other so that the progeny are always produced from only one component (46). Both these experimental problems raise doubts about the efficiency of transferring PGCs into recipient embryos, but there are further problems in transforming these cells. To date the only successful experiments in producing transgenic offspring in birds have used either competent or defective retroviruses as vectors (47). Plasmid DNA has so far not been injected or chemically transferred into extra gonadal PGCs (48) although there are reports of DNA insertion into gonad-derived cells (61). Since there are size constraints on the amount of nucleic acid that can be introduced by retroviruses and public resistance to the use of these organisms, the use of non-viral DNA has many attractions. If these difficulties can be overcome, the final problem is that the production of offspring that are chimeras means that these animals have to be reared to sexual maturity before they can be screened to see if they have genetically modified gametes and this is a serious constraint to large scale use.

*The optimistic view*

The infection of cells by retroviruses depends upon there being suitable receptors on the cell surface to which the envelope glycoproteins of the virus can attach (49). These are clearly available on avian primordial germ cells which are therefore easy to transfect (50).

Primordial germ cells can be isolated from either the germinal crescent or the blood of suitably staged embryos, transfected with replication-defective retroviruses and then injected into the circulation of stage 15 recipient embryos. Hatchlings are produced from about 25% of these operations and foreign DNA is present in the sperm of most of these birds. Transgenic chicks are found in about 2-3% of the offspring (51). This is roughly in keeping with expectations since a typical injection introduces 5-10 PGCs into an embryo that would have 2-300 endogenous PGCs settling in its germinal ridge. This rough calculation is encouraging since it indicates that cell selection between donor and host PGCs does not occur (46). Two aspects of the technique that need to be improved are: 1) the ability to screen PGCs for integrated foreign DNA prior to transferring them to chimeras, and 2) the low rate of transformed germ-cell representation in the recipient gonad.

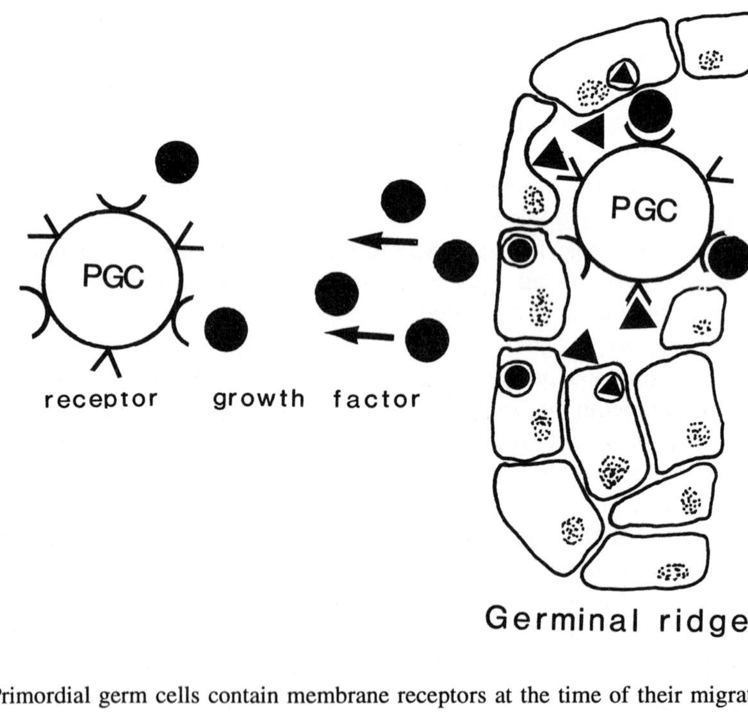


Fig. 3 Primordial germ cells contain membrane receptors at the time of their migration to the germinal crescent which produces various growth factors. Some of these appear to be involved in directing the PGCs to the ridge while others induce mitogenic activity once they arrive.

Primordial germ cells are extremely difficult to culture *in vitro*. Both mammalian and avian PGCs can be maintained in culture for 1-2 weeks but they rarely divide and eventually die (52, 53). This makes it difficult to use reporter genes for selecting those cells that have been successfully transformed. Several decades of work on the dominant-white spotting (W) and steel (Sl) loci of mice have shown however that these genes control migratory stem cell development (54). A major advance in the understanding of these effects occurred with the identification of the W locus with the proto oncogene c-*kit* (55) and the Sl locus with the c-*kit* ligand (56). The c-*kit* ligand is also known as stem cell factor and it initiates a number of responses in cells that express its receptor. It appears that stem cell factor is probably one of a number of synergistic growth factors that are expressed along migratory cell routes and stimulate cells, such as PGCs, that possess a receptor to it. Thus, both the direction of migration of PGCs to the germinal ridge and their subsequent proliferation are probably directed by such growth factors. Identification of these sets of interactions should, therefore, facilitate the future cloning and screening of PGCs (Fig. 3).

The second aspect of PGC biology relevant to the production of transgenics concerns the efficiency of integration of donor PGCs relative to recipient cells in the gonad. The problem can be expressed in the form:

$$\%\ \text{transformed gametes} = \frac{\text{dPGC}}{\text{rPGC}} \times 100te$$

| | | | |
|---|---|---|---|
| where | dPGC | = | number of donor PGCs |
| | rPGC | = | number of recipient embryo's PGCs |
| | t | = | transfection rate |
| | e | = | rate of expression of transfected cells |

The ratio dPGC/rPGC can obviously be increased if the recipient embryos are sterile. Unfortunately no mutations are known in birds which influence this but a number of approaches have been made to sterilize embryos either by irradiation (57), by producing sterile hybrids (43) or by using chemosterilants. This latter approach is particularly promising (58, 59) and has concentrated on the use of busulphan which destroys PGCs during their migration to the germinal ridges. The transfection rate (t) of cells exposed to retroviruses approaches unity with only 1 copy per cell normally being integrated (49). There is also every reason to believe that the rate of expression of transfected cells (e) may be quite high although this will obviously depend on the vector construct that is used. The factor e is obviously a complex feature including any competition effects between donor and endogenous PGCs but from the arguments already given this does not seem to be an important constraint.

The screening of potential transgenics from among the offspring of chimeric birds can be facilitated by using donor PGCs with a different phenotype from the recipient embryo. If PGCs from a White Leghorn embryo are transferred to a Rhode Island Red recipient the resulting chimera will produce offspring of both phenotypes. If the donor PGCs have been transformed with foreign DNA these cells will be apparent as White

Leghorns. Such birds will be transgenics if t = 1 (51) so that preliminary screening can be done without the need to use molecular techniques.

A final aspect of PGC biology offers particularly valuable opportunities for this technology. The germinal crescent, where the PGCs accumulate prior to their migration to the germinal ridge, is not only extragonadal but actually extraembryonic. It is therefore possible to use the ballistic approach (60) to fire nucleic acids directly into this region. Any material that penetrates through the germinal crescent ends up either in the yolk or in the lateral extensions of the mesoderm, where it causes little effect, while the embryo itself can be easily covered to protect it from damage. This technique is extremely simple, requiring no surgical intervention into the embryo and introducing foreign DNA directly into the germ line (65). The method is equally applicable to defective retroviruses or to plasmid DNA on microprojectiles and it is potentially capable of being used by an inexperienced operator on a production line. It is quite possible, therefore, that instead of being one of the more difficult animals to engineer genetically, the unusual features of avian development will actually result in it being the organism of choice for such studies.

## References


1 Gordon J W, Scangos G A, Plotkin D J, Barbosa J A & Ruddle F H (1980) Genetic transformation of mouse embryos by microinjection of purified DNA. *Proceedings of the National Academy of Sciences of the USA* **77** 7380-7384

2 Gordon J W (1989) Transgenic animals. *International Review of Cytology* **115** 171-229

3 Lavitrano M, Camaioni A, Fazio V M, Dolci S, Farace M G & Spadafora C (1989) Sperm cells as vectors for introducing foreign DNA into eggs: genetic transformations of mice. *Cell* **57** 717-723

4 Brinster R L, Sandgren E P, Behringer R R & Palmiter R D (1989) No simple solution for making transgenic mice. *Cell* **59** 239-241

5 Bachiller D, Schellander K, Peli J & Ruther U (1991) Liposome-mediated DNA uptake by sperm cells. *Molecular Reproduction and Development* **30** 194-200

6 Gavora J S, Benkel B, Sasada H, Cantwell W J, Fiser P, Teather R M, Nagai J & Sabour M P (1991) An attempt at sperm-mediated gene transfer in mice and chickens. *Canadian Journal of Animal Science* **71** 287-291

7 Brinster R L & Palmiter R D (1986) Introduction of genes into the germ line of animals. *Harvey Lectures* **80** 1-38

8 Sang H & Perry M M (1989) Episomal replication of cloned DNA injected into the fertilized ovum of the hen, *Gallus domesticus*. *Molecular Reproduction and Development* **1** 98-106

9 Perry M M, Morrice D, Hettle S & Sang H (1991) Expression of exogenous DNA during the early development of the chick embryo. *Roux's Archives of Developmental Biology* **200** 312-319

10 Lovell-Badge R H, Bygrave A E, Bradley A, Robertson E, Evans M J & Cheah K S E (1985) Transformation of embryonic stem cells with the human type II collagen gene and its expression in chimeric mice. *Cold Harbor Symposium on Quantitative Biology* **50** 707-711

11 Robertson E, Bradley A, Kuehn M & Evans M (1986) Germ line transmission of genes introduced into cultured pluripotential cells by retroviral vectors. *Nature* **323** 445-448

12 Petitte J N, Clark M E, Liu G, Verrinder Gibbins A M & Etches R J (1990) Production of somatic and germ line chimeras in the chicken by transfer of early blastodermal cells. *Development* **108** 185-189

13 Simkiss K (1991) Primordial germ cells and the scope for genetic manipulation using embryos. *Avian Incubation* pp 125-136. Ed. S Tullett. London: Butterworth-Heinemann

14 Wentworth B C, Tsai H, Hallett J H, Gonzales D S & Rajcic-Spasojevic G (1989) Manipulation of avian primordial germ cells and gonadal differentiation. *Poultry Science* **68** 999-1010

15 Buerstedde J-M & Takeda S (1991) Increased ratio of targeted to random integration after transfection of chicken B cell lines. *Cell* **67** 179-188

16 Brazolot C L, Petitte J N, Etches R J & Verrinder Gibbins A M (1991) Efficient transfection of chicken cells by Lipofectin, and introduction of transfected blastodermal cells into the embryo. *Molecular Reproduction and Development* **30** 304-312

17 Varmus H E (1988) Retroviruses. *Science* **240** 1427-1435

18 Salter D W, Smith E J, Hughes S H, Wright S E, Witter R L & Crittenden L B (1986) Gene insertion into the chicken germ line by retroviruses. *Poultry Science* **65** 1445-1458

19 Bosselman R A, Hsu R-Y, Briskin M J, Boggs T, Hu S, Nicolson M, Souza L M, Schultz J A, Rishell W & Stewart R G (1990) Transmission of exogenous genes into the chicken. *Journal of Reproduction and Fertility* **41** 183-195

20 Gilbert A B (1971) The ovary. *Physiology and Biochemistry of the Domestic Fowl* vol 3 pp 1163-1208. Eds D J Bell & B M Freeman. London: Academic Press

21 Perry M M (1987) Nuclear events from fertilization to the early cleavage stages in the domestic fowl (*Gallus domesticus*). *Journal of Anatomy* **150** 99-109

22 Simkiss K (1967) *Calcium in Reproductive Physiology*. London: Chapman & Hall

23 Spratt N T & Haas H (1960) Morphogenetic movements in the lower surface of the incubated and early chick blastoderm. *Journal of Experimental Zoology* **144** 139-157

24 Perry M M (1988) A complete culture system for the chick embryo. *Nature* **331** 70-72

25 Naito M, Nirasawa K & Oishi T (1990) Development in culture of the chick embryo from fertilized ovum to hatching. *Journal of Experimental Zoology* **254** 322-326

26 Simkiss K (1991) Fluxes during embryogenesis. *Egg Incubation: Its Effects on Embryonic Development in Birds and Reptiles* pp 47-52. Eds C D Deemlng & M W J Ferguson. Cambridge: Cambridge University Press

27 Simkiss K (1993) Surrogate eggs, chimeric embryos & transgenic birds. *Comparative Biochemistry & Physiology* **104A** 411-417

28 Eyal-Giladi H (1984) The gradual establishment of cell commitments during the early stages of chick development. *Cell Differentiation* **14** 245-255

29 Nieuwkoop P D & Sutasurya L A (1979) *Primordial germ cells in the chordates.* Cambridge: Cambridge University Press

30 Ginsburg M & Eyal-Giladi H (1987) Primordial germ cells of the young chick blastoderm originate from the central zone of the area pellucida irrespective of the embryo forming process. *Development* **101** 209-219

31 Pardanaud L, Buck C & Dieterlen-Lièvre F (1987) Early germ cell segregation and distribution in the quail blastodisc. *Cell Differentiation* **22** 47-60

32 Urven L E, Erickson C A, Abbot U K & McCarrey J (1988) Analyses of germ line development in the chick embryo using an antimouse EC cell antibody. *Development* **103** 299-304

33 Muniesa P & Dominguez L (1990) A morphological study of primordial germ cells at pregastrular stages in the chick embryo. *Cell Differentiation & Development* **31** 105-117

34 Eyal-Giladi H, Ginsburg M & Farbarov A (1981) Avian primordial germ cells are of epiblastic origin. *Journal of Embryology & Experimental Morphology* **65** 139-147

35 Ginsburg M & Eyal-Giladi H (1987) Primordial gem cells of the young chick blastoderm originate from the central zone of the area pellucida irrespective of the embryo-forming process. *Development* **101** 209-219

36 Al-Thani R & Simkiss K (1991) Effects of an acute *in vivo* application of concanavalin A on the migration of avian primordial germ cells. *Protoplasm* **161** 52-57

37 Meyer D B (1964) The migration of primordial germ cells in the chick embryo. *Developmental Biology* **10** 154-190

38 Rodemez-Lenz E (1989) On cell contribution to gonadal stroma formation in quail-chick chimeras during the indifferent stage of gonadal development. *Anatomy & Embryology* **179** 237-242

39 Hughes G G (1963) The population of germ cells in the developing female chick. *Journal of Embryology & Experimental Morphology* **11** 513-536

40 Evans M J & Kaufman M (1981) Establishment in culture of pluripotent cells from mouse embryos. *Nature* **292** 154-156

41 Nichols J, Evans E P & Smith A G (1990) Establishment of germ-line-competent embryonic stem (ES) cells using differentiation inhibiting activity. *Development* **110** 1341-1348

42 Watanabe M, Kinutani M, Naito M, Ochi O & Takashuma Y (1992) Distribution analysis of transferred donor cells in avian blastodermal chimeras. *Development* **114** 331-338

43 Wentworth B C, Tsai H, Hallett J H, Gonzales D S & Rajcic-Spasojevic G (1989) Manipulation of avian primordial germ cells and gonadal differentiation. *Poultry Science* **68** 999-1010

44 Simkiss K, Rowlett K, Bumstead N & Freeman B M (1989) Transfer of primordial germ cell DNA between embryos. *Protoplasma* **151** 164-166

45 Petitte J N, Clark M E & Etches R J (1991) Assessment of functional gametes in chickens after transfer of primordial germ cells. *Journal of Reproduction and Fertility* **92** 225-229

46 McLaren A (1984) Germ cell lineages. *Chimeras in Developmental Biology* pp 111-129. Eds N Le Douarin & A McLaren. London: Academic Press

47 Shuman R (1991) Production of transgenic birds. *Experientia* **47** 897-905

48 Shuman R (1990) Genetic Engineering. *Poultry Breeding & Genetics* pp 585-598. Ed. R D Crawford. Amsterdam: Elsevier

49 Temin H M (1986) Retrovirus vectors for gene transfer: efficient integration into and expression of exogenous DNA in vertebrate cell genomes. *Gene Transfer* pp 149-187. Ed. R Kucherlapati. New York: Plenum Press

50 Savva D, Page N, Vick L & Simkiss K (1991) Detection of foreign DNA in transgenic embryos using the polymerase chain reaction. *Research in Veterinary Science* **50** 131-133

51 Vick L, Li Y & Simkiss K (1993) Transgenic birds from transformed primordial germ cells. *Proceedings of the Royal Society (London) B* **251** 179-182

52 Tsai H & Wentworth B C (1992) Long-term culture of chick primordial germ cells isolated from the blastodisc and germinal crescent. Abstract 72. *Poultry Science* **71**

53 Donovan P J, Stott D, Cairns L A, Heasman J & Wylie C C (1986) Migratory and post migratory mouse primordial germ cells behave differently in culture. *Cell* **44** 831-838

54 Eyal-Giladi H & Kochav S (1976) From cleavage to primitive streak formation. A complimentary normal table and a new look at the first stages of the development of the chick 1. General morphology. *Developmental Biology* **49** 321-337

55 Chabot B, Stephenson D A, Chapman V M, Besmer P & Bernstein A (1988) The photo-oncogene c kit encoding a transmembrane tyrosine receptor maps to the mouse W locus. *Nature* **335** 88-89

56 Witte O N (1990) Steel locus defines a new multipotent growth factor. *Cell* **63** 5-6

57 Aige-Gil V & Simkiss K (1991) Sterilizing embryos for transgenic chimeras. *British Poultry Science* **32** 427-438

58 Aige-Gil V & Simkiss K (1991) Sterilization of avian embryo with busulphan. *Research in Veterinary Science* **50** 139-144

59 Hallett J S & Wentworth B C (1991) The effect of busulphan on gonadal differentiation in Japanese quail (*Coturnix coturnix japonica*). *Poultry Science* **70** 1619-1623

60 Wright R S, Johnson S A, Riedy M, DeVit M J, McElligott S G & Sanford J C (1991) Introduction of foreign genes into tissues of living mice by DNA coated microprojectiles. *Proceedings of the National Academy of Sciences of the USA* **88** 2726-2730

61 Wentworth A L, Frank A & Wentworth B C (1992) Transfection of Japanese quail primordial germ cells with DNA expressing gene product Lac Z. Abstract 73. *Poultry Science* **71**

62 Ephrussi A & Lehmann R (1992) Induction of germ cell formation by oskar. *Nature* **358** 387-392

63 Wylie C C, Heasman J, Snape A, O'Driscoll M & Holwell S (1985) Primordial germ cells of *Xenopus laevis* are not irrevocably determined early in development. *Developmental Biology* **112** 66-72

64 Wolpert L (1988) Stem cells: a problem in asymmetry. *Journal of Cell Science* **Supplement 10** 1-9

65 Li Y, Behnam J & Simkiss K (1994) Ballistic infection of avian primordial cells *in vivo* (submitted)

Avian Endocrinology
Ed. P J Sharp, pp 381-396
Journal of Endocrinology Ltd, Bristol (1993)

# Novel approaches to studies of avian endocrinology using transgenic chickens produced via chimeric intermediates

R J Etches, R S Carsience, M E Clark, R A Fraser, A Toner, D L Shaw and A M Verrinder Gibbins
Department of Animal and Poultry Science, University of Guelph, Guelph, Ontario N1G 2W1, Canada

## Introduction

Spontaneously arising genetic variation among chickens has been collected for centuries by poultry breeders whose goals have been the creation of practically useful and aesthetically pleasing chickens. During the twentieth century, this collection of genetically diverse birds has also been used to elucidate the cellular and molecular mechanisms that are involved in many physiological processes, including those regulating reproduction, intermediary metabolism, and the immune system. The recessive, sex-linked dwarfing gene described by Hutt (1), for example, has provided a very useful model to study the endocrine interactions that limit growth [see review by Decuypere *et al.* (2)]. The DNA sequence for this genetic variant is probably equivalent to the region of the genome coding for the growth hormone (GH) receptor (3). Confirmation of the suggestion that a small deletion in the Z chromosome results in a disruption of the structure of the hepatic GH receptor and, consequently, impairment of the binding and uptake of GH by hepatocytes in sex-linked dwarfs would provide the information that is required to make specific modifications to the genome that would confer sex-linked dwarfism. The ability to make precise modifications to the genome would provide endocrinologists with the tools to investigate the effects of alterations to the intracellular, transmembrane and extra-cellular domains of the GH receptor molecule and its ability to regulate hepatic function.

The feasibility of using precisely and intentionally modified genotypes to study physiological systems that are regulated by endocrine processes has been limited by a lack of knowledge of the DNA sequences coding for proteins that are believed to control these systems and by the rudimentary techniques that are available to generate the required changes in the genome. During the past few years, however, DNA sequences have become available for many of the proteins that play a regulatory role in endocrine systems including vasoactive intestinal peptide (4), growth hormone, insulin-like growth factor I (IGF-I), and the IGF-I receptor [see review by Goddard *et al.* (5)], the GH receptor (3), prolactin [see review by Shimada (6)], the gonadotrophins [see review by Ishii (7)], and gonadotrophin-releasing hormone (8). In addition, the

DNA sequences coding for the egg yolk and egg white proteins have become available as information regarding the influence of the steroid hormones on the cellular and molecular interactions regulating the expression of these genes has been acquired [see review by Deeley *et al.* (9)]. At the same time, there has been significant progress in the development of systems that could be used to make modifications to the genome (10). It is possible, therefore, that transgenic chickens with altered endocrine function may be available for the acquisition of information about the endocrine control of reproduction and growth in the near future. Transgenic models are particularly attractive in endocrine studies which, by definition, investigate hormonally-mediated interactions between tissues. The availability of transgenic animals will allow unprecedented opportunity to investigate these endocrine interactions in intact animals using experimental paradigms that are not available using existing *in vitro* or *in vivo* systems.

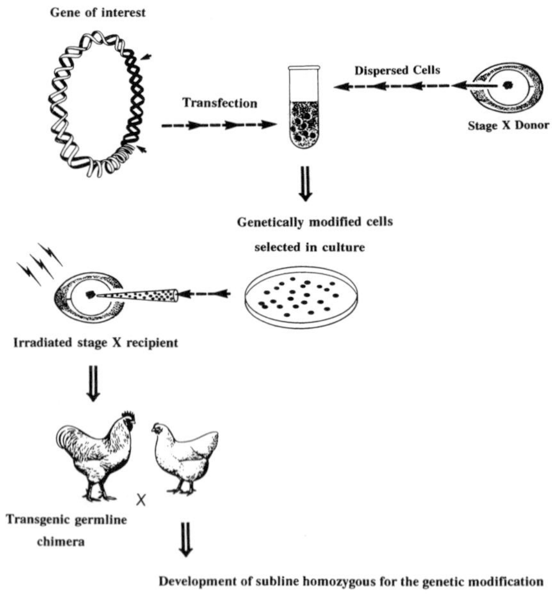


Fig. 1 A strategy to introduce genetic material into the genome of chickens. Cells from a stage X (31) embryo are removed and dispersed. The dispersed cells are transfected with a gene of interest and genetically modified cells are selected in culture. The genetically modified cells are injected into a recipient embryo that is compromised by exposure to irradiation at the same stage of development and a chimera is produced. If genetically modified cells are incorporated into the gonad of the recipient, the genetic modification will be passed to the next generation in a classical Mendelian pattern and a new line of chickens bearing the trait can be produced (from 32).

**The development of a strategy for producing transgenic chickens**

To introduce heritable changes into the genome of domestic birds, simple techniques amenable to routine use are required to access cells within the developmental pathway leading to the formation of the ova and spermatozoa. The precise timing has not been clearly established in the sequence of developmental events that separate primordial germ cells, which are the first morphologically identifiable cells in the germline lineage, from the radially symmetric dome of morphologically uniform cells that make up the embryo in a newly-laid egg. Nevertheless, cells capable of entering the germline are still present in the stage X embryo and these cells can be used to gain access to the germline of chickens (11-13). Manipulation of the genome in cells isolated from a stage X embryo can be accomplished using lipofection and these genetically altered cells will express the modification after they have been injected into a host embryo (14). These attributes of stage X blastodermal cells have been exploited in the development of the strategy to produce transgenic chickens that is illustrated in Figure 1.

The strategy outlined in Figure 1 has several advantages over other methods that have been proposed for introducing DNA into the genome of chickens and meets most of the theoretical requirements of an ideal system for gene transfer. For example, this strategy does not require access to the newly fertilized egg to inject DNA into either the pronuclei or the newly fertilized ovum (15, 16). Injection of DNA into either of these structures is expensive because a hen must be killed to obtain each ovum or zygote, and it is technically difficult to identify the male and female pronuclei among the supernumary spermatozoa that enter the ovum at fertilization (17). Furthermore, exogenous DNA does not integrate into the genome at a high rate when injected into the newly fertilized zygote (15, 16, 18) and it is technically difficult to return the manipulated ovum to the oviduct of a fistulated hen (19) or to maintain it in a surrogate egg (20). Taken together, these pragmatic considerations make manipulation of blastodermal cells contained in the newly laid egg appealing because fertile eggs are readily available and the embryo contained within is more accessible than the pronuclei or zygote in a newly fertilized ovum.

Transfer of DNA into the germline of chickens has been accomplished using retroviral vectors (21-23) and to date, only this method for introducing DNA into the germline of chickens has proven to be successful. Widespread use of this technology is limited, however, by the potential acquisition of pathogenic viruses from the vector in the transgenic line of birds (22), the inability to direct the insertion of DNA sequences into a specific location within the genome, and the practical restriction in the size of the exogenous sequence to about 1000-2000 base pairs. By contrast, much larger DNA sequences, including those designed to promote specific integration within the genome via homologous recombination, can be accommodated by liposome-mediated gene transfer. In combination with cell culture systems designed to isolate blastodermal cells in which specific modifications have been made, liposome-mediated gene transfer should facilitate the production of germline transgenic chickens that transmit specific genetic modifications to their offspring.

## The fate of dispersed, cultured, and transfected blastodermal cells in chimeric embryos

Execution of the strategy for the production of transgenic chickens outlined in Figure 1 depends on the ability to isolate blastodermal cells from stage X embryos, introduce genetic modifications to them, select the genetically modified cells in culture, and incorporate them into the germline of a chimera. Evidence to support the feasibility of each of the component parts of this strategy has been tested independently in a series of experiments summarized in Figures 2-4.

The principle that germline chimeras could be made by combining blastodermal cells from stage X donor embryos and recipients at the same stage of development was first established by Petitte *et al.* (11) using the dominant and recessive alleles at the

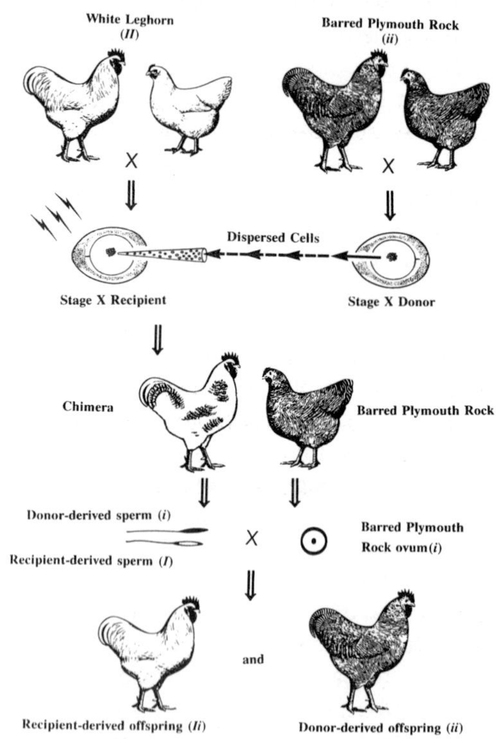


Fig. 2 The paradigm used to assess the colonization of recipient embryos by donor cells. Embryos were removed from eggs laid by Barred Plymouth Rock hens, dispersed, and injected into White Leghorn recipients according to the methods described by (11) and modified by Carsience *et al.* (13). Recipient embryos were exposed to 490-680 rads from a $^{60}$Co source within one hour before the donor cells were injected as described by Carsience *et al.* (13). The resulting chimeras were mated to Barred Plymouth Rocks to determine the proportion of donor-derived (black) and recipient-derived (yellow) offspring (from 32).

*I* locus to mark the genome of White Leghorn recipient embryos and Barred Plymouth Rock donor embryos, respectively (Fig. 2). This pair of alleles yields an indication of the contribution of the donor embryo to the melanocyte population in the down of the newly hatched chick from the proportion of black pigmentation. At hatch, chicks were classified as somatic chimeras if black pigmentation was evident and putative chimeras in the absence of donor-derived melanocytes. After rearing 85 putative and 21 somatic chimeras to sexual maturity one somatic and one putative chimera were identified that produced donor-derived offspring (11, 12). The ratio of donor-derived to recipient-derived offspring was approximately 1:250 in the putative chimera and 1:350 in the somatic chimera. While these data demonstrated that germline chimeric chickens could be produced, they also revealed that donor cells would contribute to the germline only rarely under these experimental conditions and that the rate of transmission was low when chimeras were constructed using these techniques.

The ratio of donor-derived to recipient-derived offspring was dramatically increased by exposing the recipient embryos to approximately 500-700 rads from a $^{60}$Co source immediately before injection of the donor cells (13). When 200-400 donor cells were injected into recipient embryos that were compromised by exposure to irradiation, 8 of the 14 chicks that hatched were germline chimeras (Table 1). In addition to yielding a much higher proportion of germline chimeras, the ratio of donor-derived and recipient-derived offspring from chimeras made with compromised recipients was also significantly increased. The proportion of donor-derived offspring from male chimeras made by injecting 200-400 cells into compromised recipients, for example, ranges from a minimum of 3% to a maximum of 100% (Table 1). Using a slightly modified protocol, a second male chimera has been produced in which the germline is derived entirely from donor cells (data not shown). These two birds, which produced only donor-derived offspring, provide the first demonstration that the germline in a chimera can be derived entirely from the donor cell lineage and establish the possibility that genetic modifications could be transmitted to all of the offspring of a chimera made with genetically modified blastodermal cells.

Transfection of blastodermal cells using liposomes introduces the genetic modification into 5-40% of the cells and it is estimated that the genetic modification will become stably incorporated into the genome of about 10% of the genetically modified cells (14, 24). If such a population of donor cells were injected into a chimera, the expected proportion of the donor-derived offspring carrying the desired genetic alteration would range from 0.5 to 4%. It is evident, therefore, that *in vitro* culture techniques such as those that are utilized to select transfected cells from non-transfected embryonic stem cells in mice (25) would enhance the likelihood of obtaining transgenic offspring from chimeras made with genetically modified chicken blastodermal cells. The utility of blastodermal cells would also be enhanced if single cells could be cloned using *in vitro* systems to provide a homogeneous population of genetically modified donor cells. The availability of these *in vitro* capabilities could provide a homogeneous population of donor cells which could colonize the entire germline of a chimera and, therefore, ensure transmission of the genetic modification

Table 1 The number of donor- and recipient-derived offspring from matings to Barred Plymouth Rocks of somatic and putative chimeras made using compromised White Leghorn recipients. Data from (13).

| Description of the chimera | Approx. # donor cells injected | Percentage of black pigmentation | Number of donor-derived offspring | Number of recipient-derived offspring | Percentage of donor-derived offspring |
|---|---|---|---|---|---|
| **Somatic ♂** | | | | | |
| 1585-1586 | 100 | 50 | 0 | 984 | 0 |
| 1583-1584 | 100 | 45 | 0 | 946 | 0 |
| 119-1592 | 100 | 30 | 0 | 830 | 0 |
| 101-102 | 100 | 30 | 0 | 833 | 0 |
| 103-104 | 100 | 85 | 0 | 616 | 0 |
| 114-116 | 100 | 20 | 0 | 212 | 0 |
| 142-143 | 100 | 35 | 0 | 752 | 0 |
| 1587-120 | 100 | 5 | 41 | 2147 | 1.9 |
| 105-106 | 100 | 45 | 104 | 1929 | 5.1 |
| **Somatic ♀** | | | | | |
| 1579-1580 | 100 | 45 | 0 | 199 | 0 |
| 110-111 | 100 | 25 | 0 | 75 | 0 |
| 125-126 | 100 | 5 | 0 | 68 | 0 |
| 134-139 | 100 | 85 | 1 | 73 | 1.3 |
| 163-165 | 100 | 20 | 0 | 90 | 0 |
| **Putative ♂** | | | | | |
| 1595-1596 | 100 | 0 | 0 | 809 | 0 |
| 1597-1598 | 100 | 0 | 0 | 688 | 0 |
| 121-122 | 100 | 0 | 0 | 607 | 0 |
| 167-168 | 100 | 0 | 0 | 612 | 0 |
| **Putative ♀** | | | | | |
| 1589-1590 | 100 | 0 | 0 | 207 | 0 |
| 107-108 | 100 | 0 | 0 | 171 | 0 |
| 076-077 | 100 | 0 | 0 | 172 | 0 |
| 112-173 | 100 | 0 | 0 | 90 | 0 |
| 203-204 | 100 | 0 | 0 | 90 | 0 |
| 201-174 | 100 | 0 | 0 | 131 | 0 |
| **Somatic ♂** | | | | | |
| 065-066 | 275 | 20 | 398 | 539 | 42.5 |
| 075-078 | 400 | 30 | 377 | 708 | 34.7 |
| 048-049 | 275 | 99 | 9 | 271 | 3.2 |
| 042-195 | 300 | 99 | 869 | 0 | 100.0 |
| 068-198 | 275 | 45 | 62 | 1001 | 5.8 |
| 082-083 | 250 | 20 | 0 | 449 | 0 |
| **Somatic ♀** | | | | | |
| 050-055 | 275 | 99 | 1 | 43 | 2.3 |
| 021-022 | 350 | 85 | 3 | 54 | 5.3 |
| 028-029 | 300 | 3 | 15 | 45 | 25.0 |
| 046-047 | 300 | 60 | 0 | 96 | 0 |
| 032-033 | 300 | 95 | 0 | 113 | 0 |
| 034-035 | 300 | 35 | 0 | 103 | 0 |
| **Putative ♀** | | | | | |
| 063-064 | 275 | 0 | 0 | 111 | 0 |
| 175-176 | 400 | 0 | 0 | 77 | 0 |

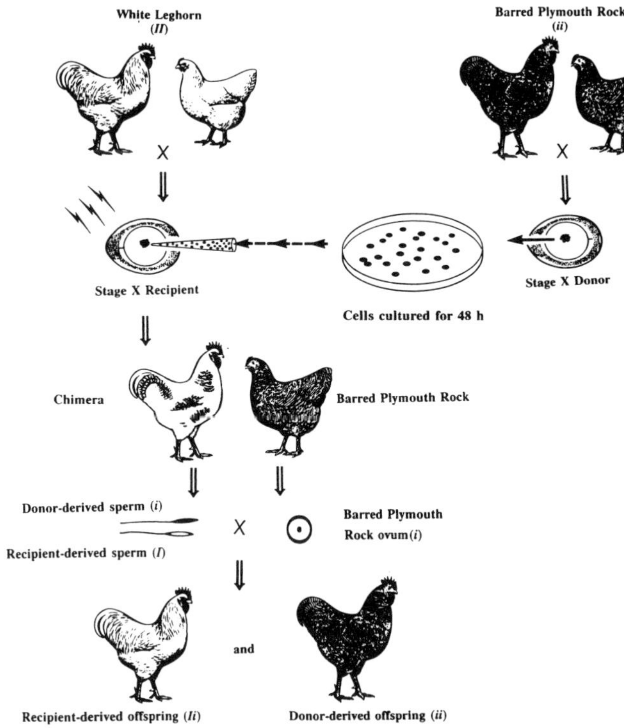


Fig. 3 The paradigm used to assess the ability of cells that had been cultured for 48 h to contribute to chimeras. Embryos were removed from eggs laid by Barred Plymouth Rock hens, and cultured as explants or as a dispersed monolayer of cells for 48 h. After 48 h in culture, the cells were dispersed and injected into irradiated White Leghorn recipients as described by Carsience *et al.* (13). The resulting chimeras were mated to Barred Plymouth Rocks to determine the proportion of donor-derived (black) and recipient-derived (yellow) offspring (from 32).

to all of its offspring. As a first step in the development of these techniques, experiments were conducted to establish the *in vitro* conditions in which blastodermal cells would retain their ability to colonize the germline. Barred Plymouth Rock embryos were cultured as explants and as monolayers of dissociated cells for 48 h. The cells were then dispersed and 50-100 cells were injected into irradiated White Leghorn recipient embryos. The proportion of recipient-derived and donor-derived offspring was estimated from the proportion of yellow and black offspring, respectively, from both putative and somatic chimeras that hatched and were reared to sexual maturity (Fig. 3). Two somatically chimeric males and one hen, which showed no evidence of incorporation of donor cells at hatch, have produced donor-derived offspring. These chimeras transmitted the donor-derived genotype to 30 of 1011 (3.0 %), 31 of 651 (6.1%) and 1 of 47 (2.2%) of their offspring, respectively, providing proof that cells capable of entering the germline can be maintained *in vitro* for at least 48 h.

The identity of cells capable of entering somatic and germline tissues after culture of stage X embryos has not yet been determined although it is evident that they are present after 48 h of culture in the explants and in the population of cells in the monolayer derived from dispersed cells from stage X embryos. It is not yet apparent if pluripotential cells contribute to somatic and germline tissues or if the explants and the monolayers contain several cell lineages, each of which contributes to a specific tissue in the chimera. Nevertheless, it is evident that cells derived from explants of stage X embryos will provide a means of accessing the genome of chickens during at least a 48 h period and further investigation is warranted to determine the characteristics of these cells. In particular, information is required that will assist in the isolation and culture of these cells to provide pure populations of genetically modified cells that can be injected into the recipient embryo and colonize the germline.

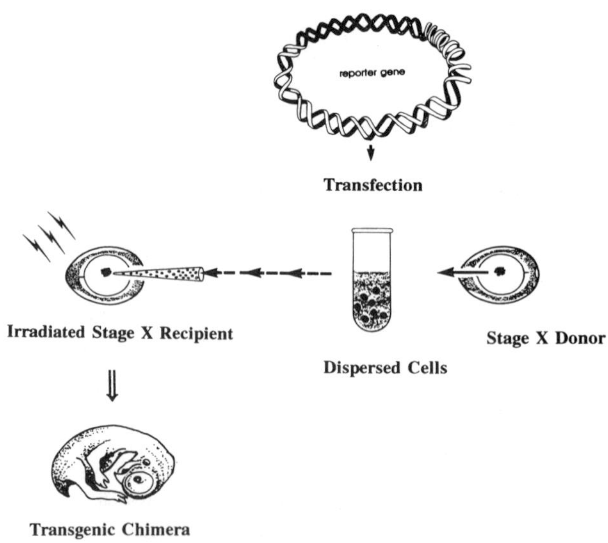


Fig. 4 The paradigm used to assess the expression of the reporter gene *lacZ* in blastodermal cells that were transfected using a modification of the methods described by Brazolot *et al.* (14). Following transfection, the cells were injected into irradiated stage X recipients as described by Carsience *et al.* (13) and the putative chimera was incubated for 96 hours. After exposure to the substrate X-gal, transfected cells were revealed in chimeric embryos (from 32).

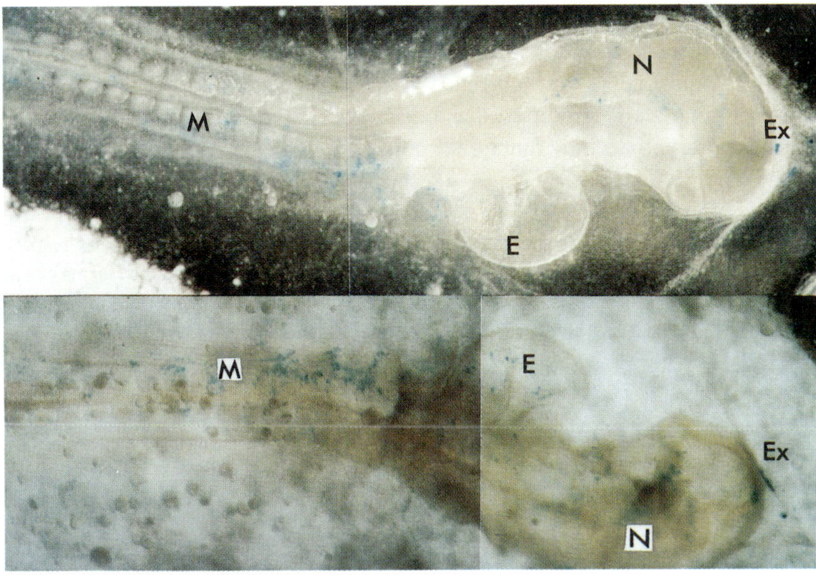


Fig. 5 The dorsal (upper panel) and ventral (lower panel) surfaces of a chimera incubated for 96 h after the injection of dispersed cells from a stage X embryo that were transfected with *lacZ*. Cells expressing the bacterial enzyme, β-galactosidase are revealed upon incubation with the substrate, X-gal, by blue staining. It is apparent from this figure that cells expressing *lacZ* have colonized ectodermal tissue in the central nervous system (N), mesodermal tissue in the somites (M) and endodermal tissue in the epithelium of the heart (E). In addition, transfected cells have colonized the extraembryonic tissues (Ex) (from 32).

The production of genetically modified cells for injection into recipient embryos requires an efficient procedure that promotes integration of the desired sequence of DNA into the genome and the potential benefits of using lipofection to accomplish this goal have been described above. Assessment of the conditions that optimize the incorporation of transfected cells into which the reporter gene *lacZ* has been introduced has been conducted in our laboratory using the paradigm illustrated in Figure 4. The first demonstration that genetic modifications could be introduced into blastodermal cells using lipofection and that transfected cells could be incorporated into and expressed in chimeric embryos was reported by Brazolot *et al.* (14). Subsequently, donor cells expressing the gene, *lacZ,* were injected into recipients that were compromised by exposure to irradiation and, as expected, the proportion of donor cells that were incorporated into the chimeras was substantially increased (24). An example of one of these embryos is shown in Figure 5 and it is evident that transfected cells were incorporated into the chimera and that the *lacZ* transgene was expressed in ectodermal, mesodermal, and endodermal tissues. It is not possible to distinguish cells within the germline in an embryo at the stage illustrated in Figure 5 and, therefore, the contribution of transfected cells to the germline cannot be assessed from these studies. There is, however, no reason to expect that transfected cells will be excluded from the germline and experiments are currently underway to obtain definitive evidence indicating that genetically modified blastodermal cells enter the germline of chimeric chickens and differentiate into functional gametes.

Table 2 Degree of chimerism at various stages of development of male chickens resulting from the injection into stage X Shaver Leghorn recipients of dispersed blastodermal cells from stage X Barred Plymouth Rock embryos, judged by the proportions of donor and recipient feather pigmentation and by the proportion of female erythrocytes[1].

| Male Chimera | Number of embryos used to make the donor cell pool | Donor feather pigmentation pattern observed (%)* at | | Age at sampling | Proportion of erythrocytes that were female (%)** |
|---|---|---|---|---|---|
| | | hatch | sampling | | |
| O.C.# | 5 | 10 | 0 | 42 months | 0 |
| BRSC2# | 2 | 25 | 0 | 20 months | 0 |
| BRSC4# | 4 | 20 | 0 | 18.5 months | 0 |
| BRSC7# | 3 | 10 | 0 | 18.25 months | 0 |
| BRSC8# | 3 | 10 | 0 | 19 months | 0 |
| 1583-1584 | 2 | 45 | 30 | 3 weeks | .067 |
| 1585-1586 | 2 | 50 | 40 | 3 weeks | .025 |
| 1587-1588 | 1 | 5 | 5 | 8 weeks | 0 |
| 1591-1592 | 1 | 25 | 25 | 7 weeks | .083 |
| 1595-1596 | 1 | 0 | 0 | 8 weeks | 0 |
| 1597-1598 | 1 | 0 | 0 | 8 weeks | .077 |
| 101-102 | 4 | 30 | 30 | 6 weeks | 0 |
| 103-104 | 4 | 85 | 85 | 13 weeks | .333 |
| 105-106 | 4 | 45 | 50 | 8 weeks | .020 |
| 114-116 | 1 | 20 | 15 | 5 weeks | .033 |
| chick that died 4 days post-hatch | 5 | 30 | 30 | 4 days | 1.000 |

[1]Stage X of embryo development as described in (31). *A subjective measure. **Estimated by examination of approximately 6000 erythrocytes following *in situ* hybridization using a digoxigenin-labeled W probe and alkaline phosphatase detection. #Non-irradiated recipient embryo. Data from (28).

## Sexual differentiation in normal and chimeric chickens

Sexual differentiation in normal male and female birds is directed by the ZZ and ZW genotypes, respectively. In the absence of oestrogens, both genetically male and female chickens revert to a male-like phenotype, indicating that oestrogen is an obligatory requirement for the development of the female phenotype (26). More recent experiments using an aromatase inhibitor to block oestrogen synthesis during embryonic development have confirmed this conclusion by demonstrating that inhibition of oestrogen production in developing embryos induces the formation of testes in genetically female chickens (27).

Table 3 Proportion of female erythrocytes in blood samples taken at 10 days of incubation from the chorio-allantois of putative chimeric embryos formed by injection into irradiated stage X Shaver Leghorn embryos of dispersed blastodermal cells pooled from 5 stage X Barred Plymouth Rock embryos.

| **Putative chimeric embryo*** | **Proportion of erythrocytes that were female (%)**** |
|---|---|
| male 1 | 0 |
| male 2 | 0 |
| male 3 | 0 |
| male 4 | 5.0 |
| male 5 | 20.0 |
| male 6 | 90.0 |
| male 7 | 100.0 |
| female 1 | 70.0 |
| intersex 1 | 100.0 |

[1] Stage X of embryo development is described in (31). * The sex of the embryo was assigned according to the appearance of the gonads; the intersex embryo had one right testis and one left ovary. ** Estimated by *in situ* hybridization as for Table 1. Data from (28).

The sexual differentiation of chimeras provides an interesting situation in which genetically male and female cells may be adjacent and, therefore, subjected to hormonal signals and a gonadal milieu that is inconsistent with their chromosomal constitution. It is interesting to note, however, that both male and female chimeras are produced and there is no clear indication at the present time of a skewed sex ratio. In addition, phenotypically intersex birds have never been observed among 47 adult chimeras examined to date. To investigate the contribution of male and female cells within some of these chimeras, the presence of female erythrocytes was determined by *in situ* hybridization (28) using a W-chromosome specific DNA probe (29). Analysis of male chimeras ranging in age from 4 days to 42 months revealed that none, or only a very small proportion, of the erythrocytes contained the W-chromosome (Table 2). By contrast, analysis of erythrocytes in male chimeras at 10 days of incubation revealed that up to 100 % of the erythrocytes could be derived from female donor embryos (Table 3). Finally, analysis of 11 female chimeras demonstrated that their erythrocytes could be derived from both male and female sources (Table 4). In one female chimera, all of the erythrocytes were male-derived; in 5 female chimeras all of the erythrocytes were female-derived; and in the remaining 5 birds, between 30 and 95% of the erythrocytes were female-derived. These data cannot be resolved into firm conclusions about the fate of donor cells in mixed-sex chimeras because the sex of the recipient was unknown and the donor cells were collected from several embryos of undetermined sex. In most cases, however, 4 or 5 embryos contributed to the pool of donor cells (Tables 2-4) and, therefore, it is likely that both male and female cells were injected into most of the recipients. The observation that juvenile and adult male chimeras are essentially devoid of female cells, whereas embryonic males contain both male and female erythrocytes suggests that either the male and female combination is lethal when the phenotypic sex is male or that female cells are excluded as development proceeds in a male embryo. Assuming that the chromosomal complement of erythrocytes is an accurate reflection of the chromosomal complement of spermatogonia, it might be surmised that male chimeras will not produce W-bearing spermatozoa because female cells will be excluded from the gonad during development.

Table 4 Degree of chimerism at various stages of development of female chickens resulting from the injection into stage X Shaver Leghorn recipients of dispersed blastodermal cells from stage X Barred Plymouth Rock embryos, judged by the proportions of donor and recipient feather pigmentation and by the proportion of female erythrocytes[1].

| Female Chimera | Number of embryos used to make the donor cell pool | Donor feather pigmentation pattern observed (%)* at | | Age at sampling (weeks) | Proportion of erythrocytes that were female (%)** |
|---|---|---|---|---|---|
| | | hatch | sampling | | |
| 611-612 | 2 | 90 | 90 | 5 | 100 |
| 601-614 | 2 | 100 | 100 | 6 | 40 |
| 748-749 | 2 | 60 | 50 | 6 | 50 |
| 021-022 | 4 | 65 | 70 | 16 | 35 |
| 009-010 | 4 | 98 | 100 | 16 | 100 |
| 050-055 | 4 | 98 | 90 | 20 | 50 |
| 034-035 | 4 | 35 | 15 | 20 | 0 |
| 032-033 | 4 | 95 | 80 | 20 | 100 |
| 046-047 | 4 | 60 | 60 | 21 | 95 |
| 163-165 | 2 | 20 | 10 | 21 | 100 |
| 134-139 | 1 | 85 | 80 | 23 | 100 |

[1] Stage X of embryo development is described in (31). * A subjective measure.
** Estimated by examination of approximately 6000 erythrocytes following *in situ* hybridization using a digoxigenin-labeled W probe and alkaline phosphatase detection.

In view of the data above, the derivation of the two male chimeras which produce only donor-derived offspring warrants speculation since the observation that their germline is colonized only by donor-derived cells is unexpected. A plausible cause for the exclusion of the recipient-derived germline in the testis might be anticipated if the recipient was female and the majority of donor cells were male. If this were the case, it is possible that female cells would be excluded from the gonad and, therefore, the chimera would produce only donor-derived gametes. A potentially analogous interaction between male embryonic stem cells and female recipient embryos exists in mice. In this species, male embryonic stem cells can masculinize the somatic components of the female recipient gonad and prevent differentiation of the female oogonia into spermatozoa (30). Experiments in which the sex of donor and recipient chicken embryos can be determined are currently underway to resolve the enigmas of male chimeras that lack female erythrocytes and have only donor-derived spermatozoa. If female recipients can be masculinized by injection of genetically modified cells from male chickens, and ZW oogonia cannot differentiate into spermatozoa, male chimeras might be produced from female recipients that transmit only donor-derived spermatozoa. The current investigations of mixed-sex chimeras, therefore, have the potential to expedite the routine production of transgenic chickens from chimeric intermediates bearing genetically modifed blastodermal cells.

## Acknowlegements

This work was supported by the Natural Sciences and Engineering Research Council of Canada, the Alberta Heritage Foundation for Medical Research, the Ontario Ministry of Agriculture and Food, the Ontario Egg Producers' Marketing Board, the University of Guelph Research Enhancement Fund, Agriculture Canada, and the University Research Incentive Fund.

## References

1 Hutt F B (1949) *Genetics of the Fowl.* New York: McGraw-Hill Book Co

2 Decuypere E, Huybrechts L M, Kuhn E R, Tixier-Boichard M & Merat P (1991) Physiological alteration associated with the chicken sex-linked dwarfing gene. *Critical Reviews in Poultry Biology* **3** 191-221

3 Burnside J, Liou S S & Cogburn L (1991) Molecular cloning of the chicken growth hormone receptor complementary deoxyribonucleic acid: mutation of the gene in sex-linked dwarfs. *Endocrinology* **6** 3183-3192

4 Talbot R, Dunn I C, Sang H & Sharp P J (1992) Cloning of chicken VIP cDNA. Abstract P7. *Proceedings of the Fifth International Symposium on Avian Endocrinology.* September 1992. Edinburgh: AFRC Roslin Institute

5 Goddard C, Gray A, Gilhooley H & O'Neill I E (1993) The molecular biology and genetic control of growth in poultry. *Manipulation of the Avian Genome* pp 165-184. Eds R J Etches & A M Verrinder Gibbins. Boca Raton: CRC Press

6 Shimada K, Ohkubo T, Saito N, Talbot R T & Sharp P J (1993) Molecular biology of prolactin. *Avian Endocrinology* pp 135-148. Ed. P J Sharp. Bristol UK: Journal of Endocrinology Ltd

7 Ishii S (1993) The molecular biology of avian gonadotropins. *Poultry Science* **72** 856-866

8 Dunn I C, Chen Y, Sharp P J & Sang H (1993) The molecular cloning of the chicken pre-pro-gonadotrophin releasing hormone-I gene. *Avian Endocrinology* pp 1-10. Ed. P J Sharp. Bristol UK: Journal of Endocrinology Ltd

9 Deeley R G, Burtch Wright R, Grant C E, Hoodless P A, Ryan A K & Schader T J (1993) Synthesis and deposition of egg proteins. *Manipulation of the Avian Genome* pp 205-223. Eds. R J Etches & A M Verrinder Gibbins. Boca Raton: CRC Press

10 Etches R J & Verrinder Gibbins A M (Eds) (1993) *Manipulation of the Avian Genome.* Boca Raton: CRC Press

11 Petitte J N, Clark J N, Liu G, Verrinder-Gibbins A M & Etches R J (1990) Production of somatic and germline chimeras in the chicken by transfer of early blastodermal cells. *Development* **108** 185-189

12 Petitte J N, Brazolot C L, Clark M E, Liu G, Verrinder Gibbins A M & Etches R J (1993) Accessing the genome of the chicken using germline chimeras. *Manipulation of the Avian Genome* pp 81-102. Eds. R J Etches & A M Verrinder Gibbins. Boca Raton: CRC Press

13 Carsience R S, Clark M E, Verrinder Gibbins A M & Etches R J (1993) Germline chimeric chickens from dispersed donor blastodermal cells and compromised recipient embryos. *Development* **117** 669-675

14 Brazolot C L, Petitte J N, Etches R J & Verrinder Gibbins A M (1991) Efficient transfection of chicken cells by lipofection and introduction of transfected blastodermal cells into the embryo. *Molecular Reproduction and Development* **30** 304-312

15 Sang H M & Perry M M (1989) Episomal replication of cloned DNA injected into the fertilised ovum of the hen, *Gallus domesticus. Molecular Reproduction and Development* **1** 98-106

16 Naito M, Agata K, Otsuka K, Kino K, Ohta M, Hirose K & Eguchi G (1991) Expression of actin-lacZ hybrid gene injected into the fertilized ovum of the domestic fowl. *International Journal of Developmental Biology* **35** 69-75

17 Perry M M (1987) Nuclear events from fertilization to the early cleavage stages in the domestic fowl (*Gallus domesticus*). *Journal of Anatomy* **150** 99-109

18 Perry M M, Morrice D, Hettle S & Sang H M (1991) Expression of exogenous DNA during the early development of the chick embryo. *Roux's Archives of Developmental Biology* **200** 312-319

19 Pancer Z, Snapir N, Robinzon B, Friedman A & Eyal-Giladi H (1989) Recovery of ova and their re-insertion into the hen's oviduct through a fistula. *British Poultry Science* **30** 953-957

20 Perry M M (1988) A complete culture system for the chick embryo. *Nature* **331** 70-72

21 Shuman R M (1991) Production of transgenic birds. *Experientia* **47** 897-904

22 Petropoulos C J, Payne W, Salter D W & Hughes S H (1992) Using avian retroviral vectors for gene transfer. *Journal of Virology* **66** 3391-3397

23 Salter D, Payne W, Crittenden L, Federspiel M, Petropoulos C, Bradac J & Hughes S (1993) Avian leukosis retroviruses and gene transfer into the avian genome. *Manipulation of the Avian Genome* pp 135-150. Eds. R J Etches & A M Verrinder Gibbins. Boca Raton: CRC Press

24 Fraser R A, Carsience R S, Clark M E, Etches R J & Verrinder Gibbins A M (1993) Efficient incorporation of transfected blastodermal cells into chimeric chicken embryos. *International Journal of Developmental Biology* (in press)

25 Mansour S L, Thomas K R & Capecchi M R (1988) Disruption of the proto-oncogene *int-2* in mouse embryo-derived stem cells: a general strategy for targeting mutations to non-selectable genes. *Nature* **366** 248-252

26 Finlay G F (1925) Studies of sex differentiation of fowls. *British Journal of Experimental Biology* **2** 439-468

27 Elbrecht A & Smith G G (1992) Aromatase enzyme activity and sex determination in chickens. *Science* **255** 467-470

28 Shaw D L, Carsience R S, Etches R J & Verrinder Gibbins A M (1992) The fate of female donor blastodermal cells in male chimeric chickens. *Biochemistry and Cell Biology* **70** 1218-1229

29 Tone M, Nakano N, Takao E, Narisawa S & Mizuno S (1982) Demonstration of W chromosome specific repetitive DNA sequences in the domestic fowl *Gallus gallus domesticus*. *Chromosoma* **86** 551-569

30 Bradley A, Evans M, Kaufman M H & Roberston E (1984) Formation of germ-line chimeras from embryo-derived teratocarcinoma cell lines. *Nature* **309** 255-256

31 Eyal-Giladi H & Kochav S (1976) A complementary normal table and a new look at the first stages of the development of the chick. 1. General morphology. *Developmental Biology* **49** 321-337

32 Etches R J, Carsience R S, Clark M E, Fraser R A, Toner A & Verrinder Gibbins A M (1993) Chimeric chickens and their use in manipulation of the chicken genome. *Poultry Science* **72** 882-889

## Author Index

**Author Index**

## Subject index

Page numbers in *italics* and **bold** indicate figures and tables respectively appearing away from their text.

## Subject index